Y0-AAA-432

MATH/STAT
LIBRARY

Group Theory

Group Theory

Proceedings of the Singapore Group Theory Conference held at the National University of Singapore, June 8-19, 1987

Edited by

Kai Nah CHENG and Yu Kiang LEONG

Walter de Gruyter
Berlin · New York 1989

o 324-4994
MATH.-STAT.

Editors

Kai Nah Cheng
Yu Kiang Leong
Department of Mathematics
National University of Singapore
Kent Ridge, Republic of Singapore

Library of Congress Cataloging-in-Publication Data

Singapore Group Theory Conference (1987 : National University of Singapore)
Group theory : proceedings of the Singapore Group Theory Conference held at the National University of Singapore, June 8–19, 1987 / edited by Kai Nah Cheng and Yu Kiang Leong.
p. cm.
-- Pref.
ISBN 0-89925-406-3 (U.S. : alk. paper)
1. Groups, Theory of--Congress. I. Cheng, Kai Nah, 1947–
II. Leong, Yu Kiang, 1946– . III. Title.
QA 171 . S 619 1987
512'. 22--dc 19

Deutsche Bibliothek Cataloging-in-Publication Data

Group theory : proceedings of the Singapore Group Theory Conference held at the National University of Singapore, June 8 - 19, 1987 / ed. by Kai Nah Cheng and Yu Kiang Leong. -
Berlin ; New York : de Gruyter, 1989
ISBN 3-11-011366-X
NE: Cheng, Kai Nah [Hrsg.]; Group Theory Conference <1987, Singapore> ; National University <Singapore>

(∞) Printed on acid free paper.

© Copyright 1989 by Walter de Gruyter & Co.. Berlin. All rights reserved, including those of translation into foreign languages. No part of this book may be reproduced in any form – by photoprint, microfilm or any other means nor transmitted nor translated into a machine language without written permission from the publisher. Printing: Ratzlow-Druck, Berlin, Binding: Dieter Mikolai, Berlin. Cover design: Thomas Bonnie. Printed in Germany.

Preface

QA171
S6191
1987
MATH

This volume brings together in written form most of the talks delivered at the Singapore Group Theory Conference in the National University of Singapore from 8 to 19 June 1987. As was our tradition since 1981, a workshop was held during the first three days of the conference. Although it was originally intended to provide instructional material for the non-specialists, it soon became clear that the workshop lectures would become more of an introduction to the conference proper. The wide range of topics covered by the contributed talks made it necessary to hold two sessions in parallel, and we hope that this volume would in some sense make up for talks missed. For participants with computational inclinations there were hands-on sessions for learning the computer language CAYLEY.

The conference was a joint effort by the Department of Mathematics of the National University of Singapore and the Singapore Mathematical Society, and was sponsored by the Southeast Asian Mathematical Society. We are grateful to the workshop and invited speakers for setting the direction of the conference, to all the speakers of contributed talks for maintaining the stimulating atmosphere and, above all, to all our participants for making the event so unforgettable.

On the material side, we would like to acknowledge with appreciation the generous financial support from the Lee Kong Chian Centre for Mathematical Research and the Singapore Turf Club. In addition, UNESCO provided us with a grant which was used for the travel of regional participants, and the French Academic Exchange Programme provided support for one of our invited speakers. A donation from Shell Eastern Petroleum (Pte) Ltd was received, and two Sun-3/200 machines were on loan from Sun Microsystems, Inc. during the period of the conference.

We are thankful to the National University of Singapore for the use of their facilities, many of which were extended to our participants. Mention should be made of the excellent administrative support given by the non-academic staff of the Department of Mathematics; in particular, by Madam Yvonne Chow, Miss Michelle Goh, Miss Rubiah bte Tukimin, Miss Stephanie Sng, Inche Ahmad bin Haji Yasin and Mr Ramasamy Thevarajan.

All said and done, only these proceedings will bear final witness to an eventful conference and they represent the collective effort of our tireless contributors together with the selfless service of our many referees. We owe to the patience and skill of Professor Chen Chuan Chong, Madam Khoo Siew Ee and Miss Angeline Yee the high quality of the camera-ready manuscripts typed in TEX. And, of course, we are grateful to the staff of Walter de Gruyter & Co. for their understanding, cooperation and encouragement in the course of our editing. We have tried to make these proceedings as complete as possible and any shortcomings remain ours.

CHENG Kai Nah
LEONG Yu Kiang

Organising Committee

Chairman	PENG Tsu Ann
Vice-Chairman	Malcolm WICKS
Secretaries	CHENG Kai Nah LEONG Yu Kiang
Treasurer	TAN Sie Keng
Committee Members	Thomas BIER CHAN Kai Meng CHEN Chuan Chong CHEW Tuan Seng Daniel FLATH KOH Khee Meng Roger K.S. POH SHEE Sze Chin

Contents

Workshop lectures

Invited lectures

Appendices

J.–P. Serre

C. Cassidy D.J.S. Robinson G. Higman

SINGAPORE GROUP THEORY CONFERENCE
8-19 JUNE 1987

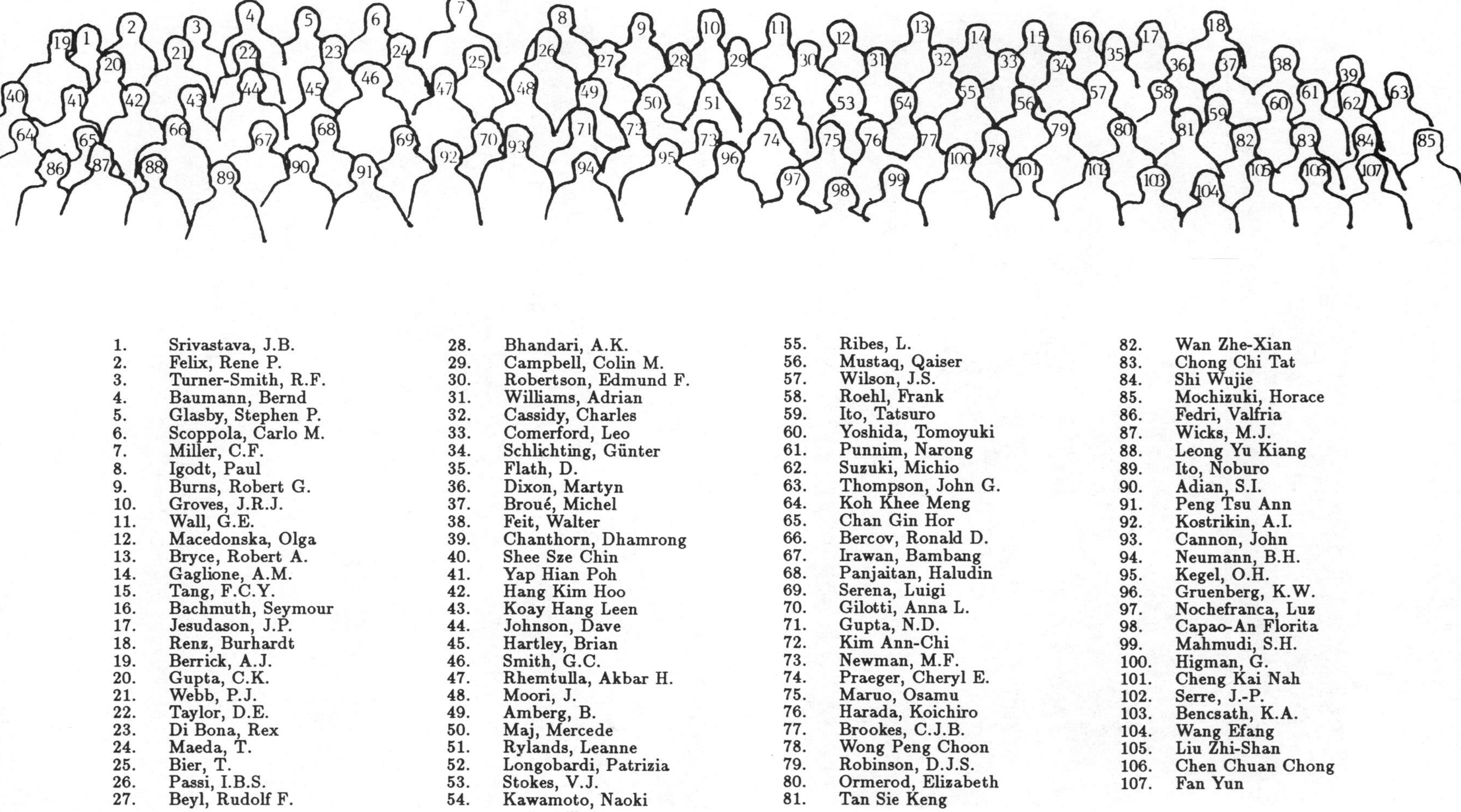

1. Srivastava, J.B.
2. Felix, Rene P.
3. Turner-Smith, R.F.
4. Baumann, Bernd
5. Glasby, Stephen P.
6. Scoppola, Carlo M.
7. Miller, C.F.
8. Igodt, Paul
9. Burns, Robert G.
10. Groves, J.R.J.
11. Wall, G.E.
12. Macedonska, Olga
13. Bryce, Robert A.
14. Gaglione, A.M.
15. Tang, F.C.Y.
16. Bachmuth, Seymour
17. Jesudason, J.P.
18. Renz, Burhardt
19. Berrick, A.J.
20. Gupta, C.K.
21. Webb, P.J.
22. Taylor, D.E.
23. Di Bona, Rex
24. Maeda, T.
25. Bier, T.
26. Passi, I.B.S.
27. Beyl, Rudolf F.
28. Bhandari, A.K.
29. Campbell, Colin M.
30. Robertson, Edmund F.
31. Williams, Adrian
32. Cassidy, Charles
33. Comerford, Leo
34. Schlichting, Günter
35. Flath, D.
36. Dixon, Martyn
37. Broué, Michel
38. Feit, Walter
39. Chanthorn, Dhamrong
40. Shee Sze Chin
41. Yap Hian Poh
42. Hang Kim Hoo
43. Koay Hang Leen
44. Johnson, Dave
45. Hartley, Brian
46. Smith, G.C.
47. Rhemtulla, Akbar H.
48. Moori, J.
49. Amberg, B.
50. Maj, Mercede
51. Rylands, Leanne
52. Longobardi, Patrizia
53. Stokes, V.J.
54. Kawamoto, Naoki
55. Ribes, L.
56. Mustaq, Qaiser
57. Wilson, J.S.
58. Roehl, Frank
59. Ito, Tatsuro
60. Yoshida, Tomoyuki
61. Punnim, Narong
62. Suzuki, Michio
63. Thompson, John G.
64. Koh Khee Meng
65. Chan Gin Hor
66. Bercov, Ronald D.
67. Irawan, Bambang
68. Panjaitan, Haludin
69. Serena, Luigi
70. Gilotti, Anna L.
71. Gupta, N.D.
72. Kim Ann-Chi
73. Newman, M.F.
74. Praeger, Cheryl E.
75. Maruo, Osamu
76. Harada, Koichiro
77. Brookes, C.J.B.
78. Wong Peng Choon
79. Robinson, D.J.S.
80. Ormerod, Elizabeth
81. Tan Sie Keng
82. Wan Zhe-Xian
83. Chong Chi Tat
84. Shi Wujie
85. Mochizuki, Horace
86. Fedri, Valfria
87. Wicks, M.J.
88. Leong Yu Kiang
89. Ito, Noburo
90. Adian, S.I.
91. Peng Tsu Ann
92. Kostrikin, A.I.
93. Cannon, John
94. Neumann, B.H.
95. Kegel, O.H.
96. Gruenberg, K.W.
97. Nochefranca, Luz
98. Capao-An Florita
99. Mahmudi, S.H.
100. Higman, G.
101. Cheng Kai Nah
102. Serre, J.-P.
103. Bencsath, K.A.
104. Wang Efang
105. Liu Zhi-Shan
106. Chen Chuan Chong
107. Fan Yun

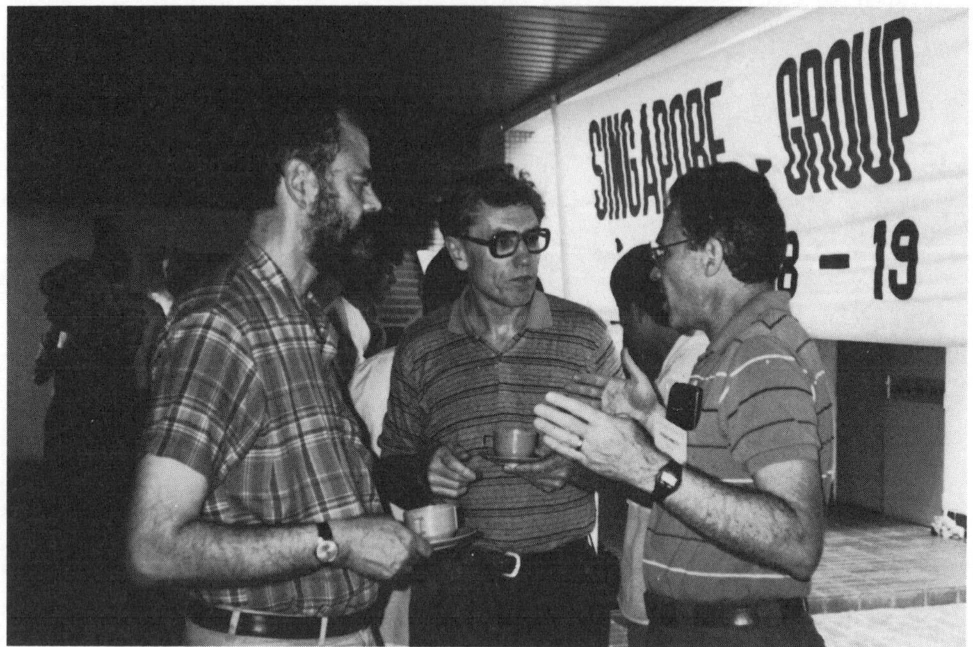

J.G. Thompson   A.I. Kostrikin   S. Bachmuth

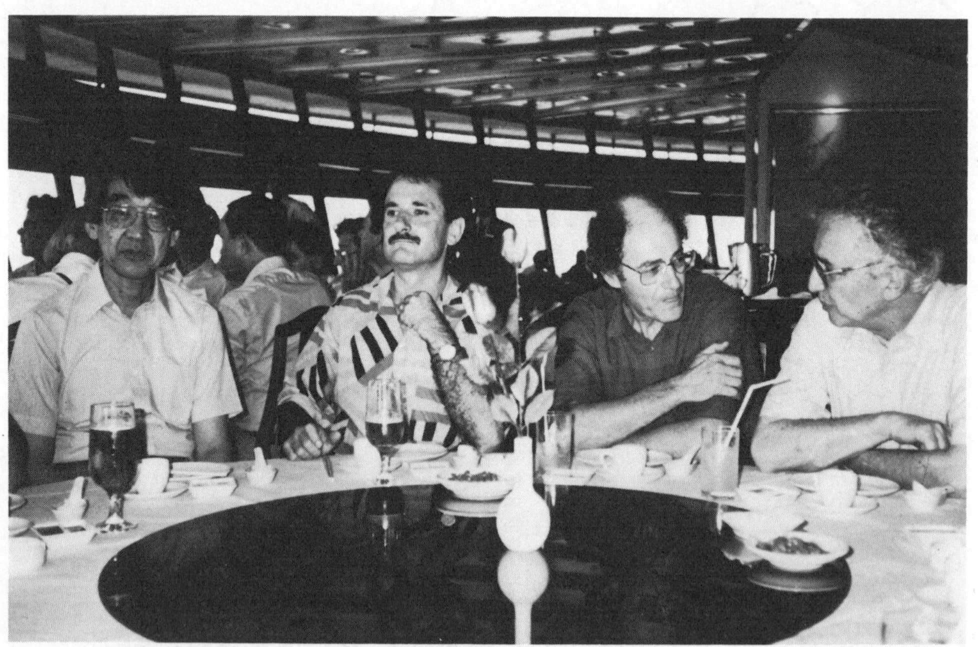

M. Suzuki   M. Broué   K.W. Gruenberg   W. Feit

J.S. Wilson O.H. Kegel

M.F. Newman E.F. Robertson C.M. Campbell

Workshop lectures

Four lectures on Sylow theory in locally finite groups

O.H. Kegel

Introduction

In their wording some theorems about finite groups also make sense for infinite groups; however, since there are counter examples, most of them cannot be proved. Narrowing down the class of groups considered more theorems of finite origin will become provable. In general, assuming the validity of some theorem about finite groups for some infinite group is a strong finiteness condition, and one might wonder what other finiteness conditions may be deduced from it. A good class in which to study such phenomena is the class of locally finite groups, i.e. the class of those groups in which every finite set of elements generates a finite subgroup. Such groups abound. Although locally finite groups are just direct limits of finite groups – and thus locally all the theory of finite groups seems available, only some results about finite groups carry over to all locally finite groups. Such a result is the theorem of Frobenius and Thompson :

Let H be a non-trivial subgroup of the locally finite group G such that $H \cap H^g = \langle 1 \rangle$ for all elements $g \in G \setminus H$, then the set K of all elements of G not conjugate to an element of H forms, together with the identity element, a nilpotent normal subgroup of G.

That this statement cannot be true for infinite groups in general is seen taking as G any non-abelian free group and as H any maximal abelian subgroup of G contained in the commutator subgroup.

The basic result on finite groups is the *Sylow Theorem* stating that for a fixed prime p the maximal p-subgroups of a finite group G are conjugate in G. This statement makes sense in arbitrary groups, but it is false in general, as we shall see even in locally finite groups. We shall refer to the discussion of the validity of the Sylow Theorem for the prime p or some variation or generalisation of it in some class of infinite groups as *Sylow Theory.* More than anyone else, Brian Hartley has contributed to Sylow Theory in locally finite groups. If not only in the locally finite group G itself, but also in every subgroup of G, the maximal p-subgroups are conjugate we shall say that G satisfies the *strong Sylow Theorem* for the prime p. For locally finite, locally p-soluble groups satisfying the strong Sylow Theorem for the prime p Hartley exhibited in [10] a very strong finiteness property, if $p \neq 2$. An extension of a weak form of Hartley's finiteness result is

If for the prime $p \geq 5$ the locally finite group G satisfies the strong Sylow Theorem then there is a finite series of normal subgroups N_i of G with

$$G = N_0 \supseteq \ldots \supseteq N_i \supseteq N_{i+1} \supseteq \ldots \supseteq N_k = \langle 1 \rangle$$

such that the factors N_i/N_{i+1} of this series are either a direct product of finitely many linear simple groups or locally p-soluble.

Hartley's result allows one to refine the series so that the locally p-soluble factors N_i/N_{i+1} are either p-groups or p'-groups.

It is this result that I want to explain in these lectures. I hope that on the way the audience will get a few glimpses of the landscape of locally finite groups in general and of the importance that definite results in the theory of finite groups may have for the theory of locally finite groups. Occasionally, questions on finite groups have been motivated by possible applications to locally finite groups.

Lecture I : Variations on Sylow's Theorem

Fix a prime p. We shall consider the question whether and when the maximal p-subgroups of the locally finite group G are all conjugate in G. If this is so, we shall say that G satisfies the *Sylow Theorem for the prime p*. Thus we suppress the arithmetic part of the classical Sylow Theorem for finite groups. We shall say that G satisfies the ***strong Sylow Theorem for the prime p*** if every subgroup of G satisfies the Sylow Theorem for the prime p. In general, the validity of the Sylow Theorem for the prime p does not imply the validity of the strong Sylow Theorem for p, as we shall see. But clearly a necessary condition for the set $\mathcal{M}_p(G)$ of all maximal p-subgroups of G to be one orbit under conjugation is the cardinal inequality

$$|\mathcal{M}_p(G)| \leq |G|.$$

We shall see that this inequality entails the validity of the strong Sylow Theorem for p for groups G with $|G| < 2^{\aleph_0}$. If G is a countably infinite locally finite group then there exists an ascending sequence $\{G_i\}_{i\in \mathbf{N}}$ of finite subgroups G_i of G such that $G = \bigcup_{i\in\mathbf{N}} G_i$. For such an approximating sequence choose Sylow p-subgroups P_i of the finite groups G_i inductively such that for $i < j$ one has $P_i \subseteq P_j$. Put $P = \bigcup_{i\in\mathbf{N}} P_i$.

(1.1) Lemma. *The subgroup P is a maximal p-subgroup of G.*

Proof. Arguing by contradiction, let $g \neq 1$ be an element of G such that $\langle P, g\rangle$ is a p-subgroup of G properly containing P. There is an index i such that $g \in G_i$. But as $g \notin P_i \subseteq P$, the subgroup $\langle P_i, g\rangle$ of $\langle P, g\rangle$ cannot be a p-group. This contradiction shows that there cannot be such an element g and hence that P is maximal among the p-subgroups of G. □

It may be worthwhile to point out that a countably infinite locally finite group G may have maximal p-subgroups which cannot be obtained as the union of Sylow p-subgroups P_i of such an approximating sequence $\{G_i\}_{i\in\mathbf{N}}$ of finite subgroups of G. – If in the countably infinite locally finite group G with the approximating sequence $\{G_i\}_{i\in\mathbf{N}}$ of finite subgroups for infinitely many indices i the Sylow p-subgroup P_i of G_i is contained in at least two Sylow p-subgroups of G_{i+1}, then there are $2^{\aleph_0}$ such maximal p-subgroups in G, and the maximal p-subgroups of G cannot all be conjugate. – This remark also applies if for infinitely many indices i there is an index $j > i$ such that the Sylow p-subgroup P_i of G_i is contained in at least two Sylow p-subgroups of G_j. – If the group G satisfies Sylow's Theorem for p then for the approximating sequence $\{G_i\}_{i \in \mathbf{N}}$ there is an index i_0 such that for all pairs of indices $j \geq i \geq i_0$ every Sylow p-subgroup P_i is contained in a unique Sylow p-subgroup of G_j.

(1.2) Theorem. *For the countable locally finite group G and the prime p the following properties are equivalent :*

(a) *There exists an approximating sequence of finite groups G_i for G and an index $i_0 \in \mathbf{N}$ such that for every pair $j \geq i \geq i_0$ of indices every Sylow p-subgroup P_i of G_i is contained in a unique Sylow p-subgroup of G_j;*

(b) *There exists a finite p-subgroup P_0 of G such that for every finite subgroup F of G containing P_0 there is a unique Sylow p-subgroup of F containing P_0;*

(c) *Any two maximal p-subgroups of G that are obtained as unions of Sylow p-subgroup of approximating sequences of finite subgroups of G are conjugate in G.*

Proof. Assume (a) and put $P_0 = P_{i_0}$ a Sylow p-subgroup of G_{i_0}. Let F be any finite subgroup of G containing P_0. For every index j such that $F \subseteq G_j$ the unique Sylow p-subgroup of G_j containing P_0 must contain a Sylow p-subgroup of F, and no other Sylow p-subgroup of F can contain P_0. Thus (b) is proved. – Clearly (b) implies (a). – As we saw in the discussion preceding this theorem, the conjugacy of the maximal p-subgroups of G obtained from one approximating sequence $\{G_i\}$ of finite subgroups of G or of any subsequence yields property (a). Thus (c) entails (a), and hence (b). – If now (b) holds and P and Q are maximal p-subgroups obtained as unions of Sylow p-subgroups of approximating sequences $\{G_i\}$ and $\{H_i\}$ of finite subgroups of G. Without loss of generality one may assume that $P_0 \subseteq G_1 \cap H_1$. Let P_i and Q_i be the unique Sylow p-subgroups of G_i, respectively H_i, containing P_0. For given $i \in \mathbf{N}$ there exists an index i' such that $H_i \subseteq G_{i'}$. But then $Q_i \subseteq P_{i'}$; consequently $\bigcup_{i \in \mathbf{N}} Q_i = P$, by (1.1). On the other hand Q and $\bigcup_{i\in\mathbf{N}} Q_i$ are conjugate already in one of the subgroups H_i. $\square$

That the equivalent properties of (1.2) are not equivalent to the validity of Sylow's Theorem for the prime p in evey countably infinite locally finite group will be shown in one of the examples to be considered now.

Examples. (1) There are $2^{\aleph_0}$ maximal p-subgroups in the countable locally finite group $G = FS(\mathbf{N})$ of all finitary permuations of the set $\mathbf{N}$ of natural numbers. To see this in the vein of the discussion preceding (1.2) put $G_i = S_{2ip}$, the symmetric group of degree $2ip$ having as support the first $2ip$ numbers and acting as the identity on the rest of $\mathbf{N}$. Obviously, one has $G = \bigcup_{i\in\mathbf{N}} G_i$. In the group G_{i+1} there is the subgroup $G_i \times S_{2p}$ where the symmetric group S_{2p} has as support the numbers $2ip+1,\dots,2(i+1)p$ and acts trivially on the rest of $\mathbf{N}$. Since the symmetric group S_{2p} has at least three Sylow p-subgroups every Sylow p-subgroup of G_i is contained in at least three Sylow p-subgroups of $G_i \times S_{2p}$, and hence of G_{i+1}. (Suprunenko [24] showed that the group $FS(\mathbf{N})$ actually has $2^{\aleph_0}$ isomorphism types of maximal p-subgroups.)

(2) Let F be a finite group with at least two Sylow p-subgroups. Take countably many copies F_i, $i \in \mathbf{N}$, of F and consider the Cartesian product $C = \prod_{i\in\mathbf{N}} F_i$ and its normal subgroup the restricted direct product $D = \prod^{\circ}_{i\in\mathbf{N}} F_i$. Both groups are locally finite, and their maximal p-subgroups are Cartesian products and restricted direct products of the Sylow p-subgroups of the F_i, respectively. The maximal p-subgroups of C are conjugate; however the $2^{\aleph_0}$ many maximal p-subgroups of the countable group D cannot be conjugate. (They form one orbit, however, under the group of locally inner automorphims of D induced by C.) Thus conjugacy of maximal p-subgroups is not inherited by (normal) subgroups, and the validity of the Sylow Theorem for the prime p is genuinely weaker than the validity of the strong Sylow Theorem for p.

(3) Consider the group C of example (2). Let P_0 be a Sylow p-subgroup of the diagonal subgroup ($\cong F$) in C. Then P_0 is contained in a unique maximal p-subgroup of the subgroup $G = DP_0$ of C, namely the split extension of the restricted direct product of the projections of P_0 into the F_i with P_0.

It is easy to verify the property (a) of (1.2) for the subgroup P_0 of G. – This example shows that the properties (a), (b), and (c) of (1.2) do not entail the validity of the Sylow Theorem for p even in countable locally finite groups.

The next result is due, independently, to Asar [1] and Hartley [10]; it shows that the validity of the strong Sylow Theorem for the prime p is a countably recognisable property of locally finite groups.

(1.3) Theorem. *If the maximal p-subgroups P and Q of the locally finite group G are not conjugate, then there is a countably infinite subgroup H of G with* $|\mathcal{M}_p(H)| = 2^{\aleph_0}$.

Proof. We first show that at least one of the groups P and Q has the embedding property that

(*) each of its finite subgroups is contained in at least two maximal p-subgroups of G.

Suppose that X is a finite subgroup of P and that Y is a finite subgroup of Q such that Q is the only maximal p-subgroup of G containing Y. Sylow's Theorem applied to the finite group $\langle X, Y\rangle$ shows that for some element $g \in G$ the subgroup $\langle X, Y^g\rangle$ is a p-group. By the assumption on Y it follows that $\langle X, Y^g\rangle \subseteq Q^g \neq P$. Thus, if Q does not satisfy (*) then P does. – We shall assume that P satisfies (*). If X is any finite subgroup of P there exists a maximal p-subgroup $R \neq P$ of G containing X. Let $x \in R \setminus P$, and put $T = P \cap \langle X, x\rangle$. Since $\langle X, x\rangle$ is a finite p-subgroup of R there is an element t of $\langle X, x\rangle \setminus T$ normalising T. This element t cannot normalise P, since – if that were the case – the maximality of P would force $t \in P$, which is not so. Thus the subgroup $\langle P, P^t\rangle$ is not a p-subgroup, and so there exists a finite subgroup X_0 of P containing X so that with $X_1 = X_0^t$ the subgroup $\langle X_0, X_1\rangle$ is not a p-group. (Note that $X \neq X_0$ since $\langle X, X^t\rangle \subseteq T$, a p-group.) We may repeat the above construction with X replaced by X_0 (and hence also by its conjugate X_1) and thus construct $X_{00}, X_{01}, X_{10}, X_{11}$. We then repeat the construction with each of the X_{ij}. Doing this infinitely often, we construct $2^{\aleph_0}$ sequences

$$X \subseteq X_{i_1} \subset X_{i_1 i_2} \subset \ldots \subset X_{i_1 i_2 \ldots i_n} \subset \ldots$$

of finite p-subgroups of G where each i_k is either 0 or 1, and for every $n \in \mathbf{N}$

the finite group

$$\langle X_{i_1 \ldots i_n i}, X_{i_1 \ldots i_n j} \rangle$$

is not a p-subgroup whenever $i \neq j$. Hence the unions of two distinct such chains are countable p-subgroups that are not both contained in any single p-subgroup of G. It follows that the countable subgroup H of G generated by the countably many finite subgroups $X_{i_1 \ldots i_n}$ contains $2^{\aleph_0}$ maximal p-subgroups. □

From this one immediately gets

(1.3′) *The locally finite group G satisfies the strong Sylow Theorem for the prime p if and only if in every countable subgroup H of G there are fewer than $2^{\aleph_0}$ maximal p-subgroups.*

Thanks to this property of "countable control" we can prove an important inheritance property for the class of all locally finite groups satisfying the strong Sylow Theorem for the prime p.

(1.4) Theorem. *If the locally finite group G satisfies the strong Sylow Theorem for the prime p, then so does every section A/B, $B \trianglelefteq A \leq G$.*

Proof. Because of (1.3′) we only have to check an arbitrary countable subgroup C of A/B. Choose representatives, one for every element of C, in A and denote by D the countable subgroup generated by these representatives. Put $N = B \cap D$. Let Q/N be a maximal p-subgroup $D/N \cong C$. We want to show that there is a maximal p-subgroup $\bar{P}$ of D with $Q/N = \bar{P}N/N$. In the countable group Q choose an approximating sequence $\{Q_i\}_{i \in \mathbf{N}}$ of finite subgroups Q_i and in each Q_i choose a Sylow p-subgroup P_i inductively such that for $i < j$ one has $P_i \subseteq P_j$; put $P = \bigcup_{i \in \mathbf{N}} P_i$. Then one has $Q = PN$. Let now $\bar{P}$ be a maximal p-subgroup of D containing P, then $\bar{P}N \supseteq PN = Q$. Since Q/N is a maximal p-subgroup of C one also has $\bar{P}N = Q$: *Every maximal p-subgroup of C is the image of a maximal p-subgroup of D.* – By assumption, any two maximal p-subgroups of D are conjugate in D, thus their images in C must be conjugate in C. □

Definition. The finite p-subgroup P of the locally finite group G is *singular in* G if for every finite subgroup F of G containing P there is a unique Sylow p-subgroup of F containing G.

If the group G is finite then its Sylow p-subgroups are singular in G. For countable locally finite groups G the connection of the existence of a singular p-subgroup P with the validity of a weak form of the Sylow Theorem for the prime p has been established in (1.2). Although the existence of a finite singular p-subgroup does not imply the Sylow Theorem for the prime p, let alone the strong Sylow Theorem, in G, we find the notion of a singular p-subgroup useful, at least for our exposition, and interesting even for finite groups.

(1.5) Theorem. *If the locally finite group G satisfies the strong Sylow Theorem for the prime p there exists a finite p-subgroup P which is singular in G.*

Proof. Assume, by contradiction, that every finite p-subgroup of G is contained in at least two Sylow p-subgroups of some finite subgroup of G. Let G_1 be any countable subgroup of G. Assume that we have already chosen countable subgroups G_i of G for $1 \leq i \leq n$, with $G_1 \subseteq \ldots \subseteq G_i \subseteq G_{i+1} \subseteq \ldots \subseteq G_n$. The subgroup G_n has countably many finite p-subgroups P; for each such P choose a finite subgroup F of G such that there are two Sylow p-subgroups of F containing P. Put G_{n+1} the subgroup of G generated by G_n and these countably many finite subgroups F; then none of the finite p-subgroups of G_n can be singular in G_{n+1} .... Put $G_\infty = \bigcup_{n \in \mathbb{N}} G_n$. As every finite p-subgroup of G_∞ is already contained in one of the subgroups G_n, it is – by the choice of G_{n+1} – contained in at least two Sylow p-subgroups of some finite subgroup of G_∞. But then the countable subgroup G_∞ has $2^{\aleph_0}$ maximal p-subgroups, which – by (1.3′) – contradicts the validity of the strong Sylow Theorem for the prime p in G. □

A question we could not settle is : *If every subgroup S of the locally finite group G contains a finite p-subgroup which is singular in S, does G then satisfy the strong Sylow Theorem for the prime p?*

If the finite p-subgroup of the locally finite group G is singular in G

it is also singular in any subgroup H of G containing it. Similarly to (1.4) there is also invariance under homomorphisms.

(1.6). *If the finite p-subgroup P is singular in the locally finite group G then for every normal subgroup N of G the finite p-subgroup PN/N of G/N is singular in G/N.*

Proof. Let F be any finite subgroup G/N containing PN/N, and choose as $\tilde{F}$ any finite subgroup of G containing P such that $\tilde{F}/(N \cap \tilde{F}) \cong \tilde{F}N/N = F$. Since a Sylow p-subgroup of F is the image of a Sylow p-subgroup of $\tilde{F}$ under the map $\tilde{F} \to \tilde{F}/(N \cap \tilde{F})$, one only has to consider the situation $(N \cap \tilde{F})Q \supseteq P$, where Q is a Sylow p-subgroup of $\tilde{F}$. By Sylow's Theorem one may assume that $P \subseteq Q$, hence that Q is the only Sylow p-subgroup of $\tilde{F}$ containing P. Thus, the image of Q in F is the only Sylow p-subgroup of F containing PN/N. $\square$

An important class of locally finite groups satisfying the strong Sylow Theorem for the prime p was exhibited independently by Platonov [19] and Wehrfritz [27] :

(1.7). *If the locally finite group G is linear (i.e. there exist a natural number n and a field F such that G is isomorphic to a subgroup of $GL(n, F)$) then it satisfies the strong Sylow Theorem for every prime p.*

Since p-subgroups of locally finite linear groups have very special structure, this result of Platonov and Wehrfritz served as starting point for investigations into the Sylow theory for the prime p for locally finite groups satisfying the minimal condition for p-subgroups (cf. [15] Chapter III, and [29]). – A generalisation of linear groups are the CZ-groups, which satisfy the minimum condition on centralisers. For locally finite CZ-groups Bryant and Ti Yen [4] established the Sylow Theorem for every prime p.

Lecture II : Singular p-subgroups and simple locally finite groups

In the last lecture we saw that the validity of some form of the Sylow Theorem for the prime p in the locally finite group G is closely linked to the existence of a finite p-subgroup that is singular in G. If the following question on finite groups admits a quantitative answer this might be very useful in the study of locally finite groups, in particular in the context of Sylow Theory.

Question. *Let the p-subgroup P be singular in the finite group G. Is the structure of G somehow restricted in terms of P?*

Here one should view the fixed subgroup P as "small", while G should be viewed as variable and "big". The importance of this question for locally finite groups has been pointed out by J.G. Thompson and by B. Hartley [10].

To my knowledge there is only one paper in the literature that directly addresses this question :

(2.1) Theorem. (A. Rae [20]) : *If for the odd prime p the finite p-subgroup P of order p^k is singular in the finite p-soluble group G then the p-length of G is bounded in terms of P by $2^{k+2} - 1$.*

A result of this type might well be true also for the prime 2, perhaps with a much larger bound. Rae's arguments break down since they depend on the ideas of P. Hall and G. Higman [8] as expanded by J. Shamash and E. Shult [22], and above all by E. Dade [6]. There is a partial result if P is a cyclic 2-group [20].

By standard methods (2.1) yields a result for locally finite groups.

(2.2) Theorem. *If for the odd prime p the finite p-group P of order p is singular in the locally finite, locally p-soluble group G then the group G is p-separable of p-length $\ell_p(G) \leq 2^{k+2} - 1$.*

Since every finite subgroup F of G (containing P) satisfies $\ell_p(F) \leq 2^{k+2} - 1$, (2.2) follows immediately from

(2.2′) *If every finite subgroup F of the locally finite, locally p-soluble group G satisfies $\ell_p(F) \leq \ell$ then G has a finite series of normal subgroups G_i, $0 < i \leq 2\ell + 1$, with*

$$\langle 1 \rangle = G_0 \subseteq G_1 \subseteq \ldots \subseteq G_i \subseteq G_{i+1} \subseteq \ldots \subseteq G_{2\ell+1} = G$$

such that G_{i+1}/G_i is a p'-group if i is odd, and a p-group if i is even.

Proof. For every finite subgroup S of G denote by $\mathbf{r}_i S$ the i-th term of the upper p-series of the p-soluble group S. The assignment $\mathbf{r}_i$ is a radical rule on the finite subgroups of G in the sense of [15], pp. 12 - 17. Put $S_i := \cap \mathbf{r}_i T$ for all finite subgroups T of G containing S. The subgroups S_i form a local system of the subgroup G_i which they generate. These characteristic subgroups G_i of G have the desired property. □

Actually B. Hartley [10] proves without direct reference to a finite p-subgroup singular in G the remarkable

(2.3) Theorem. *If for the odd prime p the locally finite and locally p-soluble group G satisfies the strong Sylow Theorem then $G/\mathbf{O}_{p,p',p}$ is finite and $G/\mathbf{O}_p(G)$ satisfies the minimum condition for p-subgroups.*

Thus, in the situation of (2.3), there is a subgroup of finite index in G which has p-length two.

From the classification of finite simple groups one knows that there are exactly seven (two parameter) families of finite simple groups in which the rank parameter is unbounded. The result on finite groups that we shall need comes from the inspection of these seven unbounded families.

(2.4) Theorem. *Let the p-subgroup P be singular in the finite simple group S which belongs to one of the seven rank-unbounded families. Then the rank of S is bounded in terms of P.* □

In order to get at the structure of infinite locally finite simple groups "knowing" the finite ones we need two reductions, the first to the countable case, the second showing that a countable simple locally finite group is a limit – of sorts – of finite simple groups.

(2.5) Theorem. *The infinite group G is simple if and only if it has a local system $\sum$ consisting of countably infinite simple subgroups of G.*

This result, 4.4 of [15], is an observation of P. Hall (cf. [16]).

Proof. The existence of a local system $\sum$ of G consisting of simple subgroups automatically makes the group G simple. Thus, we need only show that every infinite simple group has such a local system. – Let G be an infinite simple group. For every pair c, d of elements of G with $c \neq 1 \neq d$ choose a finite set $S(c,d)$ of elements of G such that

$$c \in \langle d^x;\ x \in S(c,d)\rangle;$$

since the simple group G is generated by the set of elements conjugate to the element d, such a choice is always possible. With respect to this choice we define for every countably infinite subgroup C of G another countably infinite subgroup C^* by

$$C^* := \langle C, S(c,d);\ c,d \in C,\ c \neq 1 \neq d\rangle.$$

If M is any normal subgroup of C^* with $M \cap C \neq \langle 1\rangle$, pick an element $d \neq 1$ in $M \cap C$. Then one has $c \in \langle d^x;\ x \in S(c,d)\rangle \subseteq M$ for every element $c \in C$. Thus $C \subseteq M$. Now put $C = C_1$ and choose, inductively, $C_{i+1} := (C_i)^*$ for every natural number i. The group $\bar{C} := \bigcup_{i \in \mathrm{N}} C_i$ is countable, being the union of (countably many) countable subgroups. The set $\{C_i\}$ of subgroups of the group $\bar{C}$ is a local system of $\bar{C}$. As the normal subgroup $M \neq \langle 1\rangle$ of $\bar{C}$ must contain every one of the subgroups C_i, the group $\bar{C}$ is simple. □

(2.6) Theorem. *If G is a countably infinite, locally finite simple group then there exists a strictly ascending sequence $\{R_i\}$ of finite subgroups of G with $G = \bigcup_{i \in \mathrm{N}} R_i$, such that for each natural number i there exists a maximal normal subgroup M_{i+1} of R_{i+1} satisfying $M_{i+1} \cap R_i = \langle 1\rangle$.*

Proof. For this let R be any non-trivial finite subgroup of G. By the simplicity of G there exists a finite subgroup S of G containing R such that for every element $r \neq 1$ of R one has $R \subseteq \langle r^S \rangle$. Thus every normal subgroup of S intersects R trivially. Choose a finite subgroup T of G containing S such that $S \subseteq \langle R^T \rangle$. Let X be any maximal normal subgroup of $\langle R^T \rangle$. The intersection $Y = \bigcap_{t \in T} X^t$ is normal in T and does not contain R; thus $R \cap Y = \langle 1 \rangle$. The factor group $\langle R^T \rangle / Y$ is a direct product of isomorphic simple groups. Let M be a normal subgroup of $\langle R^T \rangle$ containing Y and maximal with respect to $R \cap M = \langle 1 \rangle$. If then N/M is a minimal normal subgroup of $\langle R^T \rangle / Y$ then N/M is simple and $R \subseteq N$. – So if – inductively – we put $R = R_i$, we could choose this N as R_{i+1}. □

According to their classification, the finite simple groups are distributed into finitely many (two parameter) families. Thus – by the pigeon hole principle – one may assume that the simple sections $S_i = R_i/M_i$ in (2.6) all belong to the same family. – Assuming further that the rank parameter of the groups S_i is bounded, which implies that the finite simple groups S_i have faithful linear representations of bounded degree, we obtain that the group G is locally linear of bounded degree. By a theorem of Mal'cev [17] G then admits a faithful linear representation of finite degree. Thus, thanks to (2.4), the existence of a singular p-subgroup will have strong structural consequences in a simple locally finite group.

(2.7) Theorem. *For the locally finite simple group G the following are equivalent :*

(i) *Every countable simple subgroup of G contains a singular p-subgroup;*

(ii) *G satisfies the strong Sylow theorem for the prime p;*

(iii) *G is linear.*

Proof. If G is linear it satisfies the strong Sylow Theorem for every prime p by (1.7). By (1.2) the strong Sylow Theorem for the prime p implies the existence of a singular p-subgroup in every countable subgroup of G. – To show that (i) implies the linearity of the simple group G we consider the local system $\sum$ of all countable simple subgroups of G. For $S \in \sum$, let P be a singular p-subgroup. By the argument outlined before stating this theorem, S is linear of minimal degree d_S. Assuming that the minimal degrees of the groups belonging to $\sum$ are unbounded, one can find in $\sum$ an ascending sequence of countable simple groups $S = S_0 \subseteq S_1 \subseteq \ldots \subseteq S_i \subseteq S_{i+1}$, $i \in \mathbf{N}$, with $d_{S_i} < d_{S_{i+1}}$. The group $T = \bigcup_{i \in \mathbf{N}} S_i$ is a countable simple subgroup of G which cannot have a non-trivial finite-dimensional representation. This contradicts the assumption that T contains a singular p-subgroup. Consequently the minimal degree of a faithful linear representation of a simple countable subgroup of G is bounded, and G is locally linear of bounded degree. By Mal'cev's theorem [17], the simple group G is itself linear. □

One should add several remarks on the structure of locally finite simple linear groups :

(2.8). *Locally finite simple linear groups are countable.*

This was first pointed out by Zalesskii [31] and then, independently, by Winter [30].

(2.9). *Locally finite simple linear groups have a local system consisting of finite simple subgroups.*

This strengthening of (2.6) for linear groups may be found as 4.6 in [15]. That this cannot be achieved for simple locally finite groups in general was first shown by Zalesskii and Serejkin [32], see also Hickin [13] and Phillips [18].

By (2.9) and the classiciation of the finite simple groups, an infinite linear locally finite simple group has a local system consisting of finite Chevalley groups (possibly of twisted type) all of the same type. The structural information for this direct limit was obtained by S. Thomas [25], [26] and by Hartley and Shute [12], compare also Borovik [3] and Belayev [2].

(2.10). *If the infinite locally finite simple group G has a non-trivial linear representation of finite degree, then G is a Chevalley group (possibly of twisted type) over some infinite locally finite field.* □

Lecture III : The study of crucial configurations

In this lecture we shall study certain sequences of finite subgroups of a locally finite group and try to decide whether the union of this sequence satisfies the strong Sylow Theorem for the prime p. This study will be crucial for the main finiteness result vaguely announced in the introduction and to be presented in the next lecture.

The sequence $\{F_i\}_{i\in\mathbf{N}}$ of subgroups $F_i \neq \langle 1\rangle$ of the group G will be called a *split sequence with associated ascending sequence* $\{H_i\}_{i\in\mathbf{N}}$ if for every natural number i the subgroup $H_i := \langle F_j;\ 1 \leq j \leq i\rangle$ normalises the subgroup F_{i+1} and satisfies $H_i \cap F_{i+1} = \langle 1\rangle$. – In the context of Sylow Theory for the prime p in a locally finite group G we are interested in split sequences consisting of non-trivial perfect subgroups generated by their p-elements; we shall call such finite subgroups *p-perfect.* The split sequence $\{F_i\}_{i\in\mathbf{N}}$ of finite p-perfect subgroups will be called *straight* with associated ascending sequence $\{H_i\}_{i\in\mathbf{N}}$ if for every number i the group F_{i+1} does not contain any proper H_i-invariant normal p-perfect subgroup. – In a straight split sequence of finite p-perfect groups the subgroups F_{i+1} have a unique maximal H_i-invariant normal subgroup $M_{i+1} \neq F_{i+1}$, this is p-soluble.

Observe that no structural assumption was made for F_1; thus, starting with a straight split sequence $\{F_i\}_{i\in\mathbf{N}}$ of finite p-perfect subgroups for every natural number n the sequence $\{F_i^n\}_{i\in\mathbf{N}}$ is also a straight split sequence, if we put $F_1^n = H_{n+1}$ and $F_i^n = F_{i+n}$ for $i > 1$.

(3.1). *From any split sequence $\{F_i\}_{i\in\mathbf{N}}$ of finite p-perfect subgroups of the group G one can obtain a straight split sequence $\{F_i'\}_{i\in\mathbf{N}}$ of finite p-perfect subgroups with $F_i' \subseteq F_i$ for all $i \in \mathbf{N}$.*

For this put $F_1 = F_1'$ and, having chosen F_i' and $H_i' = \langle F_j';\ 1 \leq j \leq i\rangle$ already, select F_{i+1}' minimal among the H_i'-invariant p-perfect subgroups $(\neq \langle 1\rangle)$. □

The straight split sequence $\{F_i\}_{i\in\mathbf{N}}$ of finite p-perfect subgroups of the group G is called *simple* if all the factor groups F_{i+1}/M_{i+1} are simple.

(3.2). *If in the straight split sequence $\{F_i\}_{i\in\mathbf{N}}$ of finite p-perfect sub-*

groups of the group G for infinitely many indices i the factor groups F_i/M_i are simple, then there is a simple straight split sequence of finite p-perfect groups in G.

For this let J be the set of all those $i \in \mathbf{N}$ with $i > 1$ and F_i/M_i simple. Choose $F_1 = F'_1$ and suppose F'_k already chosen. Let i_{k+1} be the smallest index in J not yet considered, put $H'_k = \langle F'_i;\ 1 \leq i \leq k\rangle$ and choose F'_{k+1} minimal among the H'_k-invariant p-perfect subgroups of $F_{i_{k+1}}$. □

For the simple straight split sequence $\{F_i\}_{i\in\mathbf{N}}$ of finite p-perfect subgroups of the group G with the associated ascending sequence $\{H_i\}_{i\in\mathbf{N}}$ we shall consider the centralisers $C_{i,j} := \mathbf{C}_{H_i}(F_j/M_j)$ for $j > i$. If for all triples i, j, k of natural numbers with $i < j < k$ one has $C_{i,j} = C_{i,k}$, we shall call the sequence *smooth* and denote the $C_{i,j}$ – which now do not depend on j – simply by C_i.

(3.3). *If in the group G there exists a simple straight split sequence of finite p-perfect subgroups, then there exists also a smooth simple straight split sequence of finite p-perfect subgroups.*

From the sequence $\{F_i\}_{i\in\mathbf{N}}$ assumed to exist we select and adapt by diagonal induction the desired sequence $\{F'_i\}_{i\in\mathbf{N}}$. For this we choose $F_1 = F'_1 = H'_1$. Put $N_0 = \mathbf{N}$. By the finiteness of H'_1 there is a normal subgroup C_1 of H'_1 which is of the form $C_1 = \mathbf{C}_{H_i}(F_j/M_j)$ for infinitely many $j > 1$. The normal subgroup C_1 chosen will determine the infinite subset N_1 of N_0 of those indices $j > 1$ such that C_1 is precisely the centraliser in H'_1 of the simple group F_j/M_j. – Now assume by induction that the subgroups $F'_1, \ldots, F'_k$ with $H'_k = \langle F'_1, \ldots, F'_k\rangle$ and the infinite subset N_{k-1} of the natural numbers are already chosen. By the finiteness of H'_k there is a normal subgroup C_k of H'_k which is the centraliser in H'_k of infinitely many of the simple groups F_j/M_j, these j's forming an infinite subset N_k of N_{k-1}. Let j_0 be the smallest number in N_k; choose F'_{k+1} minimal among the H'_k-invariant p-perfect subgroups of F_{j_0}. – The sequence $\{F'_i\}_{i\in\mathbf{N}}$ so selected will be a smooth straight split sequence of p-perfect finite subgroups of G. □

For the simple straight sequence $\{F_i\}$ of finite p-perfect subgroups of the group G one has by the validity of Schreier's conjecture (which is a consequence of the Classification Theorem) that the group H_i of the associated sequence induces a p-perfect group of inner automorphisms in the simple group F_{i+1}/M_{i+1}. By induction on i one shows that the p-perfect group H_i/S_i is a direct product of simple p-perfect groups; here S_i denotes the largest normal p-soluble subgroup of H_i.

(3.4) Theorem. *If $\{F_i\}_{i\in \mathbf{N}}$ is a smooth simple straight split sequence of finite p-perfect subgroups of the group G, then the countably infinite group $U = \langle F_i;\ i \in \mathbf{N}\rangle$ has $2^{\aleph_0}$ maximal p-subgroups.*

Proof. If the group U has fewer than $2^{\aleph_0}$ maximal p-subgroups, then by (1.3′) it would satisfy the (strong) Sylow Theorem for the prime p. We shall assume this in order to derive a contradiction. Without loss of generality we may assume that the first term $F_1 = H_1$ of our smooth simple straight split sequence $\{F_i\}_{i\in \mathbf{N}}$ has its Sylow p-subgroups singular in U. Because of (1.6) every Sylow p-subgroup of F_1 will leave invariant exactly one Sylow p-subgroup of F_j/M_j for every $j > 1$. As the group of outer automorphisms of the simple group F_j/M_j is soluble, the perfect group F_1 must induce a group of inner automorphisms of F_j/M_j. Thus the factor group F_1/C_1 may be identified with a subgroup of F_j/M_j, and $C_1 \neq F_1$ since F_j/M_j has more than one Sylow p-subgroup and a Sylow p-subgroup of F_1/C_1 leaves exactly one Sylow p-subgroup of F_j/M_j invariant. By (2.4) the Chevalley rank of all the simple groups F_j/M_j, $j > 1$ is bounded in terms of the Sylow p-subgroup of F_1/C_1. But the finite simple groups of Chevalley rank $\leq r$ have a faithful linear representation of degree at most $d(r)$. By a result of Kargapolov (cf. [28], 9.29) the subgroup H_i/C_i of the simple groups F_j/M_j, $j > i$, can have only a bounded number of non-abelian composition factors, bounded in terms of $d(r)$. Thus the number of p-perfect composition factors of the groups C_i must grow with i. As a normal subgroup of H_i the group C_i has the same structure as H_i, that is, modulo its largest normal p-soluble normal subgroup C_i is a direct product of simple p-perfect groups. In fact, the subgroups C_iS_j are normal subgroups of H_j for $j \geq i$; here S_j denotes largest p-soluble normal subgroup of H_j. Thus, if the number of p-perfect composition factors of C_j, $j > i$, is larger than that of C_i, a Sylow p-subgroup of C_i is by (1.6) contained in (at least) two Sylow p-subgroups of C_j. Thus one obtains that the countable subgroup $C = \cup C_i$ contains $2^{\aleph_0}$ maximal p-subgroups. A contradiction. □

A useful corollary for us of the result (3.4) is the fact that in a locally finite group satisfying the strong Sylow Theorem for the prime p there cannot exist any simple straight split sequence of p-perfect subgroups. Thus, if in the locally finite group G satisfying the strong Sylow Theorem for the prime p there exists a straight split sequence $\{F_i\}_{i\in \mathbf{N}}$ of finite p-perfect subgroups, then all but finitely many of the factor groups F_j/M_j must be proper direct products. Our aim is to show that this cannot occur either.

For this we shall study the permutation representation which the group F_i induces on the simple director factors of F_j/M_j for $j > i$. Let $P_{i,j}$ be the kernel of this permutation representation. Starting with a straight split sequence $\{F_i\}_{i\in \mathbf{N}}$ of finite p-perfect subgroups of G we may find – by an inductive diagonal selection and adaption process similar to the one used to obtain a smooth sequence from a simple one – a straight split sequence $\{F_i'\}_{i\in \mathbf{N}}$ which satisfies the identity $P'_{i,j} = P_{i,k}$ for all indices i, j, k with $i < j < k$.

(3.5) Theorem. *Let the straight split sequence $\{F_i\}_{i\in \mathbf{N}}$ of finite p-perfect subgroups of G satisfy $P_{i,j} = P_{i,k} =: P_i$ for every triple i, j, k with $i < j < k$. Then the countable subgroup $U = \{F_i;\ i \in \mathbf{N}\}$ contains $2^{\aleph_0}$ many maximal p-subgroups, if $p \geq 5$.*

Proof. If there were infinitely many indices i such that $P_i = F_i$, then from this subsequence one would construct a simple straight sequence of finite p-perfect subgroups; by (3.4) this would yield a subgroup of U containing $2^{\aleph_0}$ maximal p-subgroups. Thus, enlarging F_1 if necessary, we may assume that for all $i > 1$ we have the inequality $P_i \subseteq M_i \subset F_i$.

Assuming now, for contradiction, that U contains fewer than $2^{\aleph_0}$ maximal p-subgroups, we may assume that a Sylow p-subgroup S_1 of F_1 is singular in U. Thus it will select in each of the subgroups F_j a unique S_1-invariant Sylow p-subgroup S_j and its normaliser $N_j := \mathbf{N}_{F_j}(S_j)$. Clearly the normalisers $\{N_i\}_{i\in \mathbf{N}}$ form a split sequence of finite p-soluble subgroups of U. We shall consider the ascending sequence $\{Q_i\}_{i\in \mathbf{N}}$ of finite p-soluble subgroups $Q_i := \{N_j;\ 1 \leq j \leq i\}$ of U. By A. Rae's result (2.1) the p-

length $\ell_p(Q_i)$ of these groups will be bounded in terms of the order of S_1, if p is odd. On the other hand, the p-length of the p-soluble group Q_i is i, at least for $p \geq 5$. This follows by induction from the fact that the normaliser $\mathbf{N}_S(P)$ of the non-trivial Sylow p-subgroup P of a finite simple group S is not of the form $P \cdot \mathbf{C}_S(P)$. Assume by induction that $\ell_p(Q_k) = k$ and that $P_k \subseteq \mathbf{O}_{p',p}(Q_k)$. Since the group P_k permutes the simple direct factors of the factor group F_{k+1}/M_{k+1} non-trivially, it acts in the same way on $A = N_{k+1}M_{k+1}/M_{k+1}$ and on $A/\mathbf{O}_{p'}(A)$. Since $p \geq 5$, $A/\mathbf{O}_{p'}(A)$ is not a p-group, so the split extension $A \leftthreetimes P_k$ has p-length 2. Now it is obvious that $\ell_p(Q_{k+1}) = k+1$, and not smaller. Choosing i large enough, one obtains a contradiction against the singularity of P_1 in Q_i. This contradiction proves (3.5).

Summing up the results of this lecture we have

(3.6). *If for the prime $p \geq 5$ there is a straight split sequence $\{F_i\}_{i \in \mathbf{N}}$ of finite p-perfect subgroups in the group G, then the countable subgroup $U = \langle F_i;\ i \in \mathbf{N} \rangle$ contains $2^{\aleph_0}$ many maximal p-subgroups. Hence U does not satisfy the (strong) Sylow Theorem for the prime p.*

Lecture IV : The strong finiteness results

For any locally finite group denote by $\mathbf{S}_p(G)$ the largest normal locally p-soluble subgroup of G. This subgroup is not only globally defined, it may be described by a radical rule on the finite subgroups of G (cf. [15], pp. 12 - 15) : For the finite subgroup F of G denote by $\mathbf{r}(F) = F \cap (\cap \mathbf{S}_p(X))$ for all the finite subgroups X of G containing F. The subgroups of the form $\mathbf{r}(F)$ with F ranging over all the finite subgroups of G, form a local system for the group $\mathbf{S}_p(G)$. Hartley's result [10] stated as (2.3) will be restated here :

(4.1) Theorem. *If the locally finite and locally p-soluble group G satisfies the strong Sylow Theorem for the odd prime p then the factor group $G/\mathbf{O}_{p,p',p}(G)$ is finite, and the factor group $G/\mathbf{O}_p(G)$ satisfies the minimal condition for p-subgroups.*

Thus, for some subgroup H of finite index in the locally finite locally p-soluble group G satisfying the strong Sylow Theorem for the odd prime p the p-length $\ell_p(H)$ is at most two. It might be of interest to search for a representation theoretic condition on the action of a Černikov p-group P on a locally finite group not containing any elements of order p that ensures that the split extension $G = N\lambda P$ of N by P satisfies the strong Sylow Theorem for the (odd) prime p.

If the locally finite group considered is not locally p-soluble, it will contain finite p-perfect subgroups and also normal p-perfect subgroups; these are generated by finite p-perfect subgroups. The main step in proving a stronger form of the finiteness result stated in the Introduction is to show that if the locally finite group G satisfies the strong Sylow Theorem for the prime $p \geq 5$, then G satisfies the minimal condition for normal p-perfect subgroups. We shall prove this rather for the factor group $X = G/\mathbf{S}_p(G)$.

Assume this factor group X has an infinite descending sequence $\{N_n\}_{n\in\mathbf{N}}$ of normal p-perfect subgroups N_n. Denote by I the intersection $I = \bigcap_{n\in\mathbf{N}} N_n$ and consider the factor group $Y = X/I$ with its infinite descending sequence $\{\bar{N}_n\}_{n\in\mathbf{N}}$, $\bar{N}_n = N_n/I$, of normal p-perfect subgroups satisfying $\bigcap_{n\in\mathbf{N}} \bar{N}_n = \langle 1 \rangle$. For every finite subgroup F of Y there is an $\bar{N}_n$ in this sequence such that $F \cap \bar{N}_n = \langle 1 \rangle$. It is this property that we shall exploit now : Choose in the normal subgroup $\bar{N}_1$ a finite p-perfect subgroup F_1; assume that we have inductively chosen subgroups $F_1, \ldots, F_k$ already and put $H_k = \langle F_1, \ldots, F_k \rangle$. Then there is a normal p-perfect subgroup $\bar{N}_n$ in our sequence with $H_k \cap \bar{N}_n = \langle 1 \rangle$. Choose F_{k+1} minimal among the finite H_k-invariant p-perfect subgroups of $\bar{N}_n$. – Thus we obtain a straight split sequence $\{F_i\}_{i \in \mathbf{N}}$ of finite p-perfect subgroups of Y with associated ascending sequence $\{H_i\}_{i \in \mathbf{N}}$. According to (3.6) the countable subgroup $U = \langle F_i;\ i \in \mathbf{N} \rangle = \bigcup_{i \in \mathbf{N}} H_i$ of Y contains $2^{\aleph_0}$ maximal p-subgroups. Thus we have proved

(4.2). *If the locally finite group G contains an infinitely descending sequence of normal p-perfect subgroups it cannot satisfy the strong Sylow Theorem for the prime $p \geq 5$.*

Now assume that the locally finite group G is not locally p-soluble and satisfies the strong Sylow Theorem for the prime $p \geq 5$, then so does the factor group $X = G/\mathbf{S}_p(G)$ and there is a minimal normal p-perfect subgroup M of X. Since $\mathbf{S}_p(M)$ is a characteristic subgroup of M, it is normal in X. But this means $\mathbf{S}_p(M) = \langle 1 \rangle$. Thus M is a minimal normal subgroup of X. Applying the same argument to M instead of X, one finds that M has a minimal normal p-perfect subgroup N which in fact is a minimal normal subgroup of M. But then the minimal normal subgroup M of X is a (restricted) direct product of simple subgroups conjugate to N. By the discussion of direct products in the first lecture we know that – in order that the strong Sylow Theorem for the prime p may hold – we can only have finitely many such direct factors. Taken together with (2.7) this yields

(4.3) Theorem. *If the locally finite group G is not p-soluble and satisfies the strong Sylow Theorem for the prime $p \geq 5$, then the socle* $\mathrm{soc}(X)$ *of the factor group $X = G/\mathbf{S}_p(G)$ is the direct product of finitely many linear simple p-perfect subgroups. Also the centraliser* $\mathbf{C}_X(\mathrm{soc}(X))$ *is trivial.*

This is our main contribution to this discussion. – As the socle of X has only finitely many simple direct factors, the kernel P of the permutation representation of G on these has finite index in G; in fact P is a characteristic subgroup of finite index in G. For the infinite simple direct factors of the socle of X we have the structural result (2.10) : they are all Chevalley groups (possibly of twisted type) over suitable locally finite fields. For the finite simple direct factors the classification of the finite simple groups tells us that beside the Chevalley groups (possibly of twisted type) and the alternating groups further 26 sporadic groups have to be considered. For all these simple groups there is rather strong information on their groups of outer automorphisms (cf. Steinberg [23], §§10 and 11) : they are soluble with a pro-cyclic normal subgroup of finite index bounded by a function of their Chevalley ranks. (This pro-cyclic normal subgroup represents essentially the automorphism group of the underlying locally finite field. We shall only be interested in its torsion subgroup which is locally and residually cyclic.) – Taking all this together, one obtains, essentially as a corollary to (4.3) and (4.1) :

(4.4) Theorem. *If the locally finite group G satisfies the strong Sylow Theorem for the prime $p \geq 5$, then there are characteristic subgroups*

$$\langle 1\rangle \subseteq \mathbf{O}_p(G) \subseteq \mathbf{O}_{p,p'}(G) \subseteq \mathbf{O}_{p,p',p}(G) \subseteq \mathbf{S}_p(G) \subseteq S \subseteq A \subseteq P \subseteq G$$

such that $S/\mathbf{S}_p(G) = \mathbf{soc}(G/\mathbf{S}_p(G)$ is a direct product of finitely many locally finite simple linear groups, A/S is an abelian group of rank bounded by the number of simple direct factors of $S/\mathbf{S}_p(G)$, the factor group P/A is a finite soluble group of order bounded by the number and a function of the types of the simple direct factors of $S/\mathbf{S}_p(G)$, and the factor group G/P permutes these direct factors faithfully. If none of the characteristics of the underlying locally finite fields of the infinite simple direct factors of $S/\mathbf{S}_p(G)$ is p, then the factor group $G/\mathbf{O}_p(G)$ satisfies the minimum condition for p-subgroups. In any case, the factor group $G/\mathbf{O}_{p,p',p}(G)$ is countable.

Making a suitable definition for the p-length of the group $S/\mathbf{S}_p(G)$ – possibly simply the number of simple direct p-perfect factors – one gets that with this extended notion of p-length the locally finite group G satisfying the strong Sylow Theorem for the prime $p \geq 5$ will have finite p-length.

It seems desirable to remove the restrictions on the prime p in the above result. This would mean finding an argument in the finite soluble case to prove a qualitative result like (2.1) and finding a rather different argument to prove (3.5) which might use different properties of finite simple groups.

Acknowledgements. For the presentation of these ideas I am indebted to the efforts of my former student Felix Flemisch. – B. Hartley pointed out an embarrassing error in Lecture III.

References

[1] A.O. Asar, A conjugacy theorem for locally finite groups, *J. London Math. Soc.* (2) **6** (1973), 358-360.

[2] V.V. Belayev, Locally finite Chevalley groups, *Publication on group theory*, Akad. Nauk SSSR Ural Scientific Centre, (1984), 39-50.

[3] A.V. Borovik, Classification of the periodic linear groups over fields of odd characteristic, *Sibirskii Mat. Z* **25** (1984), 67-83 = Engl. Transl., *Siberian Math. J.* **25** (1984), 221-235.

[4] R.M. Bryant & Ti Yen, Sylow subgroups of locally finite CZ-groups. *Arch. Math.* **31** (1978), 117-119.

[5] M. Curzio, Some problems of Sylow type in locally finite groups, *Institutiones Mathematicae V*, Academic Press, (1979).

[6] E. Dade, Carter subgroups and Fitting heights of finite soluble groups, *Illinois J. Math.* **13** (1969), 449-514.

[7] G. Glauberman, Factorisation in local subgroups of finite groups, *Regional Conference Series in Mathematics* No. 33, A.M.S., (1977).

[8] P. Hall & G. Higman, On the p-length of p-soluble groups and reduction theorems for Burnside's problem, *Proc. London Math. Soc.* (3) **6** (1956), 1-42.

[9] B. Hartley, Sylow subgroups of locally finite groups, *Proc. London Math. Soc.* (3) **23** (1971), 159-192.

[10] B. Hartley, Sylow p-subgroups and local p-solubility, *J. Algebra* **23** (1972), 347-365.

[11] B. Hartley, Sylow theory in locally finite groups, *Com. Math.* **25** (1972), 263-280.

[12] B. Hartley & G. Shute, Monomorphisms and direct limits of finite groups of Lie type, *Quarterly J. Math.* (Oxford) **35** (1984), 49-71.

[13] K. Hickin, Universal locally finite central extension of groups, *Proc. London Math. Soc.* (3) **52** (1986), 53-72.

[14] O.H. Kegel, Chain conditions and Sylow's Theorem in locally finite groups, *Symposia Mathematica*, vol XVII, Academic Press (1976), 251-259.

[15] O.H. Kegel & B.A.F. Wehrfritz, *Locally finite groups*, North-Holland (1973).

[16] R.D. Kopperman & A.R.D. Mathias, Some problems in group theory, 131-138, in *The syntax and semantics of infinitary languages* (ed. J. Barwise), Lecture Notes in Mathematics No. 72, Springer (1968).

[17] A.I. Mal'cev, On isomorphic matrix representations of infinite groups, *Mat. Sbornik* **8** (1940), 405-422 = Engl. Transl., *A.M.S. Transl.* (2) **45** (1965), 1-18.

[18] R.E. Phillips, Existentially closed locally finite central extensions, multipliers and local systems, *Math. Z.* **187** (1984), 383-392.

[19] V.P. Platonov, The theory of algebraic groups and periodic groups, *Isv. Akad. Nauk SSSR ser. Mat.* **30** (1966), 573-620 = Engl. Transl., *A.M.S. Transl.* (2) **69** (1968), 61-110.

[20] A. Rae, Sylow p-subgroups of finite p-soluble groups, *J. London Math. Soc.* (2) **7** (1973), 117-123; **11** (1975), 11.

[21] A. Rae, Groups of type (p, p) acting on p-soluble groups, *Proc. London Math. Soc.* (3) **31** (1975), 331-363.

[22] J. Shamash & E. Shult, On groups with cyclic Carter subgroups, *J. Algebra* **11** (1969), 564-597.

[23] R. Steinberg, Lectures on Chevalley groups, Yale University, (1967).

[24] D.A. Suprunenko, On locally nilpotent subgroups of infinite symmetric groups, *Dokl. Akad. Nauk SSSR* 167 (1966), 302-304 = Engl. Transl. *Soviet Math. Doklady* **7** (1966), 392-394.

[25] S. Thomas, An identification theorem for the locally finite non-twisted Chevalley groups, *Arch. Math.* **40** (1983), 21-31.

[26] S. Thomas, The classification of the simple periodic linear groups, *Arch. Math.* **41** (1983), 103-116.

[27] B.A.F. Wehrfritz, Sylow theorems for periodic lienar groups, *Proc. London. Math. Soc.* (3), **18** (1968), 125-140.

[28] B.A.F. Wehrfritz, *Infinite linear groups*, Springer (1973).

[29] J.S. Wilson, On groups satisfying Min-p, *Proc. London Math. Soc.* (3) **26** (1973), 226-248.

[30] D.J. Winter, Representations of locally finite groups, *Bull. A. M. S.* **74** (1968), 145-148.

[31] A.E. Zalesskii, Maximal periodic subgroups of the full linear group over fields of positive characteristic, *Vesc : Akad. Nauk BSSR* **2** (1966), 121-123.

[32] A.E. Zalesskii & V.N. Serejkin, Linear groups generated by transvections, *Izv. Akad. Nauk SSSR Ser. Mat.* **40** (1976), 26-49 = Engl. Transl. *Math. USSR Izv.* **10** (1976), 25-46.

Mathematisches Institut der
Albert-Ludwigs-Universität
Albertstraße 23b
D-7800 Freiburg i. Br.
Federal Republic of Germany

Cohomology in infinite group theory

Derek J. S. Robinson

§0. Introduction

It is now forty five years since cohomology first made its appearance in group theory via the theory of group extensions. In a series of fundamental articles [7], [8], [21] Eilenberg and Mac Lane (see also Eckmann [6]) showed how the cohomology groups $H^n(G, M)$, $n = 0, 1, 2, 3$, occur naturally in group theory. The higher dimensional cohomology groups had to wait until the 1970's for convincing group theoretic interpretations. Cohomology began to be used to prove purely group theoretic results in the 1950's and 1960's, and increasingly so after the publication of the influential lecture notes of Gruenberg [12] and Stammbach [33]. Today no group theorist, finite or infinite, can afford to dispense with cohomological tools.

Our intention here is first to give an account of the various group theoretic interpretations of the cohomology groups, following Holt [15] and Huebschmann [17] in dimensions $n \geq 3$. Thereafter we shall consider how cohomology can be applied to prove group theoretic results; topics discussed include conjugacy of maximal torsion subgroups, existence of outer automorphisms, complete groups, cohomological vanishing theorems and Kropholler's theorem on soluble minimax groups. We make no claim to completeness and the choice of topics reflects the author's preference. Usually some indications of proof are given, but often in abbreviated form. As general references we cite the standard books on homological algebra [5], [14], [22], [31]. Further applications to group theory can be found in [12], [33] and [28].

§1. Group theoretic interpretations of the cohomology groups

Let G be a group and M a (right) G-module. If $\mathcal{P} \twoheadrightarrow \mathbf{Z}$ is a projective G-resolution of the trivial G-module $\mathbf{Z}$, the cohomology of G with coefficients in M is defined to be the cohomology of the complex $\mathrm{Hom}_G(\mathcal{P}, M)$; thus

$$H^n(G,M) = H^n(\mathrm{Hom}_G(\mathcal{P}, M)).$$

As is well-known the groups $H^n(G,M)$ are independent of the choice of resolution. If $\mathcal{P}$ is taken to be the bar resolution, we obtain the familiar description of $H^n(G,M)$ as n-cocycles modulo n-coboundaries,

$$H^n(G,M) = Z^n(G,M)/B^n(G,M).$$

We shall begin with a brief review of the classical interpretations of H^0, H^1, H^2.

(0) H^0 and fixed points

There is little mystery about the identity of $H^0(G,M)$; this is the set of G-fixed points $M^G = \{a \in M \mid ag = a, \forall g \in G\}$, as can be seen from a resolution of the form $\ldots \longrightarrow P_1 \longrightarrow \mathbf{Z}G \longrightarrow\!\!\!\!\rightarrow \mathbf{Z}$.

(1) H^1: derivations, complements and automorphisms

There are two well-known and useful interpretations of H^1.

(a) Consider the natural semidirect product $E = G \ltimes M$, and let K be a complement of M in E, i.e. $E = KM$ and $K \cap M = 1$. Then K determines a *derivation* (=1-cocycle) $\delta_k : G \to M$ by means of the rule $gg^\delta \in K$, $(g \in G)$. It is easy to see that the assignment $K \mapsto \delta_K$ is a bijection from the set of all complements of M in E to the group of all derivations $\mathrm{Der}(G,M) = \mathbf{Z}^1(G,M)$ in which G corresponds to 0. Moreover two complements are conjugate precisely when their derivations differ by an inner derivation (= 1-coboundary).

Proposition 1. *There is a bijection between the set of conjugacy classes of complements of M in $G \ltimes M$ and the group $H^1(G,M)$ in which the conjugacy class of G maps to 0.*

Thus all the complements of M are conjugate if and only if $H^1(G,M) = 0$. A related concept that has proved useful in infinite group theory is that of near conjugacy. The complements of M in E are said to be *nearly conjugate* if, modulo some finite G-invariant subgroup of M, the complements fall into finitely many conjugacy classes. There is a good cohomological expression of this property.

Proposition 2. ([28], (1.3)). *Assume that for each $\ell > 0$ the endomorphism $a \to \ell a$, $(a \in M)$, has finite kernel and cokernel. Then the complements of M in $G \ltimes M$ are nearly conjugate if and only if $H^1(G,M)$ is bounded as an abelian group.*

(b) The second group theoretic interpretation of H^1 involves automorphisms. Let E be a group with a normal subgroup N, and write M for $\mathbf{Z}(N)$, the centre of N. Then M is a module over $G := E/N$ via conjugation. If α is an automorphism of G which operates trivially on both N and G, it is readily seen that the map $xN \to x^{-1}x^{\alpha}$ is a derivation δ_α from G to M. Furthermore $\alpha \mapsto \delta_\alpha$ is an isomorphism from $\mathbf{C}_{\mathrm{Aut}\,E}(N) \cap \mathbf{C}_{\mathrm{Aut}\,E}(G)$ to $\mathrm{Der}(G,M)$. In case N contains its centralizer in E, i.e. $\mathbf{C}_E(N) = M$, it is clear that α is inner if and only if δ_α is an inner derivation.

Proposition 3. *There is an isomorphism between $A := \mathbf{C}_{\mathrm{Aut}\,E}(N) \cap \mathbf{C}_{\mathrm{Aut}\,E}(G)$ and $\mathrm{Der}(G,M)$. If $\mathbf{C}_E(N) = M$, this induces an isomorphism $A/A \cap \mathrm{Inn}(E) \simeq H^1(G,M)$.*

This simple fact is frequently used in the study of automorphisms of composite groups.

(2) H^2 and group extensions

(a) We recall the familiar link between H^2 and group extensions. Let M be a G-module and let

$$S^2(G,M)$$

denote the class of all extensions of M by G, i.e. short exact sequences

$$\mathcal{E} : 0 \longrightarrow M \xrightarrow{\tau_1} E \xrightarrow{\tau_2} G \longrightarrow 1$$

of groups and homomorphisms in which conjugation in E is compatible with the G-module structure of M.

Extensions $\mathcal{E}$ and $\mathcal{D}$ are said to be *equivalent* if there is an isomorphism $\theta : E \longrightarrow D$ making the following diagram commute:

$$\begin{array}{ccccc} & & E & & \\ & {}^{\tau_1}\nearrow & & \searrow^{\tau_2} & \\ 0 \longrightarrow M & & \downarrow \theta & & G \longrightarrow 1 \\ & \searrow^{\sigma_1} & & {}^{\sigma_2}\nearrow & \\ & & D & & \end{array}$$

Obviously this is an equivalence relation; let

$$H^2(G, M)^*$$

denote the *set* of all equivalence classes $[\mathcal{E}]$.

There is a binary operation on $S^2(G, M)$ known as the *Baer sum*. Here $\mathcal{E} + \mathcal{D}$ is defined to be $\mathcal{L}$ where $\mathcal{L}$ is constructed by forming the pull-back $E\overset{G}{\wedge}D$

$$\begin{array}{ccc} E\overset{G}{\wedge}D & \longrightarrow & D \\ \downarrow & & \downarrow \\ E & \longrightarrow & G \end{array}$$

and defining $L := E\overset{G}{\wedge}D/B$ where $B = \{(m\tau_1, (-m)\sigma_1) \mid m \in M\}$. The maps $M \longrightarrow L$ and $L \longrightarrow G$ are the obvious ones.

It can be verified that the Baer sum induces an abelian group structure in $H^2(G, M)^*$, the zero element being the equivalence class of the split extension $0 \longrightarrow M \longrightarrow G \ltimes M \longrightarrow G \longrightarrow 1$.

To identify $H^2(G, M)^*$ with $H^2(G, M)$ choose a transversal function $\lambda : G \longrightarrow E$ for the extension $0 \longrightarrow M \xrightarrow{\tau} E \longrightarrow G \longrightarrow 1$, and define the *factor set* (=2-cocycle) $\varphi : G \times G \longrightarrow M$ in the usual way, $x^\lambda y^\lambda = (xy)^\lambda((x, y)\varphi)\tau$. Then the assignment $[\mathcal{E}] \mapsto \varphi + B^2(G, M)$ yields an isomorphism $H^2(G, M)^* \simeq H^2(G, M)$. Hence every extension of M by G splits if and only if $H^2(G, M) = 0$.

There is also the useful notion of a nearly split extension. We say that an extension $0 \longrightarrow M \xrightarrow{\tau} E \longrightarrow G \longrightarrow 1$ *nearly splits* if there is a subgroup K such that $|E : K(\operatorname{Im}\tau)|$ and $|K \cap \operatorname{Im}\tau|$ are both finite. The cohomological meaning is supplied by

Proposition 4 ([28], (2.5)). *Assume that for each $\ell > 0$ the endomorphism $a \mapsto \ell a$, $(a \in M)$, has finite kernel and cokernel. Then all extensions of M by G nearly split if and only if $H^2(G, M)$ is a torsion group.*

This has proved to be a useful tool in infinite soluble group theory, where near-splitting occurs much more frequently than splitting; for some applications see [28], §3.

(b) Non-abelian extensions

Let $0 \longrightarrow N \longrightarrow E \longrightarrow G \longrightarrow 1$ be a group extension in which N need not be abelian. This extension is equipped with a *coupling*, i.e. a homomorphism $\chi : G \longrightarrow \mathrm{Out}\, N$ arising from conjugation in E. Notice that $M := Z(N)$ is a G-module.

Now suppose we are given a homomorphism $\chi : G \longrightarrow \mathrm{Out}\ N$ (sometimes called an *abstract kernel*). There need not be an extension of N by G with coupling χ. Indeed it is at this point that the Eilenberg-Mac Lane interpretation of $H^3(G,M)$ becomes relevant; for χ determines an element $\hat{\chi}$ - the obstruction - of $H^3(G,M)$, and χ is realizable as the coupling of an extension precisely when $\hat{\chi} = 0$. Moreover when this occurs, the equivalence classes of extensions with coupling χ correspond bijectively - but non-naturally - to elements of $H^2(G,M)$.

(c) Automorphisms of group extensions

Consider a (non-abelian) group extension $\mathcal{E} : O \longrightarrow N \overset{\tau}{\longrightarrow} E \longrightarrow G \longrightarrow 1$. The *automorphism group of the extension* is defined by

$$\mathrm{Aut}\,\mathcal{E} = \mathbf{N}_{\mathrm{Aut}E}(\mathrm{Im}\ \tau).$$

By inducing in N and G we obtain a natural homomorphism $\theta : \mathrm{Aut}\ \mathcal{E} \longrightarrow \mathrm{Aut}\ N \times \mathrm{Aut}\ G$ whose kernel consists of automorphisms acting trivially on N and G and is therefore isomorphic with $\mathrm{Der}(G,M)$ where $M = Z(N)$. It is more difficult to describe the image of the homomorphism θ. One necessary condition for (ν,κ) to belong to $\mathrm{Im}\,\theta$ is the *compatibility condition*

$$\chi\tilde{\nu} = \kappa\chi,$$

where $\chi : G \longrightarrow \mathrm{Out}\, N$ is the coupling of the extension and $\tilde{\nu}$ is conjugation by ν in $\mathrm{Out}\, N$. The set of all compatible pairs (ν,κ) is a subgroup

$$\mathrm{Comp}\,(\chi)$$

of $\operatorname{Aut} N \times \operatorname{Aut} G$.

However not every compatible pair belongs to $\operatorname{Im}\theta$. It can be shown that each compatible pair determines an element of $H^2(G,M)$ which must vanish if the pair is to belong to $\operatorname{Im}\theta$. These observations are the basis for the following result of Wells.

Proposition 5 ([36], [28]). *Given an extension $\mathcal{E} : 0 \longrightarrow N \longrightarrow E \longrightarrow G \longrightarrow 1$ with coupling χ there is an exact sequence*

$$0 \longrightarrow \operatorname{Der}(G,M) \longrightarrow \operatorname{Aut}\mathcal{E} \longrightarrow \operatorname{Comp}(\chi) \longrightarrow H^2(G,M)$$

in which the right hand map is a set map.

A useful special case occurs when χ is injective, i.e. $\mathbf{C}_E(N) = M$.

Corollary. *If χ is injective, there is an exact sequence*

$$0 \longrightarrow H^1(G,M) \longrightarrow \operatorname{Out}\mathcal{E} \longrightarrow \mathbf{N}_{\operatorname{Out}N}(G^\chi)/G^\chi \longrightarrow H^2(G,M)$$

where $\operatorname{Out}\mathcal{E} = \operatorname{Aut}\mathcal{E}/\operatorname{Inn}E$.

(3) H^n and crossed $(n-1)$-fold extensions

Our aim is to describe $H^n(G,M)$ in terms of crossed $(n-1)$-fold extensions of M by G following the accounts of Holt [15] and Huebschmann [17]. But first we return briefly to Eilenberg and Mac Lane's concept of an abstract kernel, i.e. a homomorphism $\chi : G \longrightarrow \operatorname{Out} N$ where N, G are arbitrary groups. If $\nu : \operatorname{Aut} N \longrightarrow \operatorname{Out} N$ is the canonical homomorphism, let E be the pull-back

$$\begin{array}{ccc} E & \longrightarrow & G \\ \big\downarrow & & \big\downarrow \chi \\ \operatorname{Aut} N & \xrightarrow{\nu} & \operatorname{Out} N \end{array}$$

This gives rise to an exact sequence

$$O \longrightarrow M \longrightarrow N \xrightarrow{\tau} E \longrightarrow G \longrightarrow 1$$

in which $M = \mathbf{Z}(N)$ and the maps are the obvious ones. Note that an action of E on N is implied. Here N, E and τ constitute what is called a crossed module and the above sequence is a crossed 2-fold extension.

Let us now proceed to the definitions. A *crossed module* is a triple (N, E, τ) where N, E are groups with a right action of E on N, and $\tau : N \longrightarrow E$ is an E-operator homomorphism such that

$$x^{-1}yx = (y)x^{\tau}, \quad (x, y \in N).$$

This generalizes the notion of an E-module (let $\tau = 0$) and of a normal subgroup of E (let τ be inclusion). Observe that $\mathrm{Im}\,\tau \triangleleft E$, $\mathrm{Ker}\,\tau \leq \mathbf{Z}(N)$ and $\mathbf{Z}(N)$ is an $E/\mathrm{Im}\,\tau$-module.

Let M be a module over a group G. If $n \geq 2$, a *crossed* $(n-1)$*-fold extension of* M *by* G is an exact sequence of groups and homomorphisms

$$0 \longrightarrow M \xrightarrow{\tau_{n-1}} M_{n-2} \xrightarrow{\tau_{n-2}} M_{n-3} \longrightarrow \cdots \longrightarrow M_2 \xrightarrow{\tau_2} M_1 \xrightarrow{\tau_1} E \xrightarrow{\tau_0} G \longrightarrow 1$$

such that

(i) $M_2 \ldots M_{n-2}$ are G-modules,
(ii) (M_1, E, τ_1) is a crossed module,
(iii) $\tau_2, \tau_3, \ldots, \tau_{n-1}$ are G-operator homomorphisms.

Note that $\mathrm{Im}\,\tau_2 \leq \mathbf{Z}(M_1)$, so the former is a module over $E/\mathrm{Im}\,\tau_1 \simeq G$, and (iii) is meaningful for τ_2.

Thus the concept of an crossed 2-fold extension is a natural generalization of an abstract kernel. (It can be shown that any 2-fold crossed extension in which M maps onto $\mathbf{Z}(M_1)$ is equivalent–in the sense described below–to one which arises from an abstract kernel).

Let $S^n(G, M)$ denote the class of all crossed $(n-1)$-fold crossed extensions of M by G. A relation $\longrightarrow$ on $S^n(G, M)$ is defined as follows: $\mathcal{E} \longrightarrow \mathcal{E}^*$ if there are homomorphisms $M_i \longrightarrow M_i^*$ making the diagram

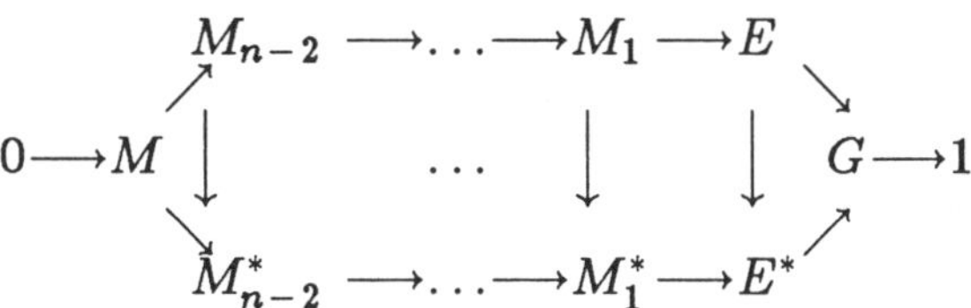

commute and preserving the various structures. Since $\longrightarrow$ is not symmetric, we pass to the equivalence closure $\sim$; thus $\mathcal{E} \sim \mathcal{E}^*$ if and only if there is a finite sequence of crossed $(n-1)$-fold extensions $\mathcal{E} = \mathcal{E}(0), \mathcal{E}(1), \ldots, \mathcal{E}(m) =$

$\mathcal{E}^*$ such that $\mathcal{E}(i) \longrightarrow \mathcal{E}(i+1)$ or $\mathcal{E}(i+1) \longrightarrow \mathcal{E}(i)$. Denote the equivalence class of $\mathcal{E}$ by $[\mathcal{E}]$ and write

$$H^n(G,M)^*$$

for the class (actually a set) of all equivalence classes.

For $n \geq 3$ the *Baer sum* of two extensions $\mathcal{E}$, $\mathcal{E}^*$ is defined to be $\mathcal{E}^+$ where M^+_{n-2} is the push-out $M_{n-2} \underset{M}{\vee} M^*_{n-2}$, E^+ is the pull-back $E \overset{G}{\wedge} E^*$, and $M_i^+ = M_i \oplus M_i^*$ if $1 \leq i \leq n-3$; the maps are the natural ones.

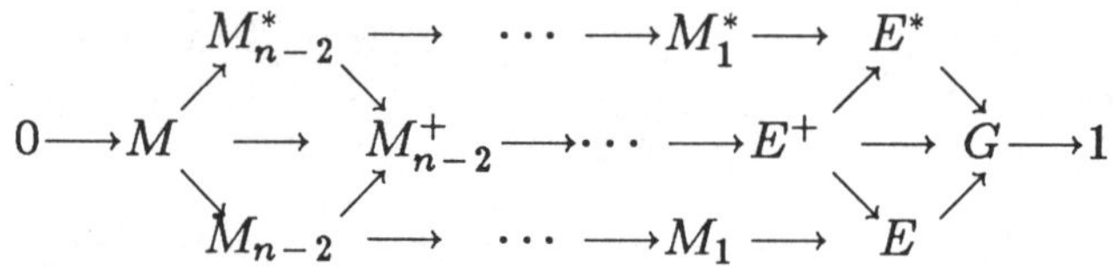

$$\begin{array}{ccccccccc} & & M^*_{n-2} & \longrightarrow & \cdots \longrightarrow M^*_1 & \longrightarrow & E^* & & \\ & \nearrow & & \searrow & & \nearrow & & \searrow & \\ 0 \longrightarrow M & \longrightarrow & & & M^+_{n-2} \longrightarrow \cdots \longrightarrow E^+ & \longrightarrow & & & G \longrightarrow 1 \\ & \searrow & & \nearrow & & \searrow & & \nearrow & \\ & & M_{n-2} & \longrightarrow & \cdots \longrightarrow M_1 & \longrightarrow & E & & \end{array}$$

The Baer sum induces a binary operation on $H^n(G,M)^*$. To show that there is a resulting abelian subgroup structure it is best to construct a bijection between $H^n(G,M)^*$ and $H^n(G,M)$. This is very much in the spirit of the identification of the Yoneda version of Ext^n with the usual Ext^n (see [22]). The ensuing discussion is abbreviated: for details see [17].

Let G be a group. A *free (projective) crossed resolution* of G

$$\mathcal{C} \longrightarrow\!\!\!\!\!\rightarrow G$$

is an exact sequence of groups and homomorphisms

$$\cdots \longrightarrow C_n \xrightarrow{\partial_n} C_{n-1} \longrightarrow \cdots \longrightarrow C_2 \xrightarrow{\partial_2} C_1 \xrightarrow{\partial_1} F \xrightarrow{\partial_0} G \longrightarrow 1$$

in which F is a free group, (C_1, F, ∂_1) is a *free (projective) crossed F-module* (for this see [4]), $C_n, n \geq 2$, is a free (projective) G-module and the ∂_n, $n \geq 2$, are G-operator homomorphisms. Such crossed resolutions behave in many respects like ordinary resolutions. Thus every group has a free crossed resolution. Also if M is a G-module and $\mathcal{C}$ is a free (projective) crossed resolution of G, the cohomology of the complex $\mathrm{Hom}(\mathcal{C}, M)$, i.e. of

$$\mathrm{Der}(F,M) \longrightarrow \mathrm{Hom}_F(C_1,M) \longrightarrow \mathrm{Hom}_G(C_2,M) \longrightarrow \cdots,$$

is given by

$$H^0(\mathrm{Hom}(\mathcal{C},M)) = \mathrm{Der}(G,M) \text{ and } H^{n-1}(\mathrm{Hom}(\mathcal{C},M)) = H^n(G,M), n \geq 2.$$

The connection between $H^n(G,M)^*$ and $H^n(G,M)$ can now be demonstrated. Given an $(n-1)$-fold crossed extension $\mathcal{E}$ of M by G and a free crossed resolution $\mathcal{C} \twoheadrightarrow G$, we can lift the identity map on G to obtain a commutative diagram

$$\begin{array}{ccccccccccccccc}
\cdots \longrightarrow & C_{n+1} & \longrightarrow & C_n & \xrightarrow{\partial_n} & C_{n-1} & \longrightarrow & C_{n-2} & \longrightarrow \cdots \xrightarrow{\partial_2} & C_1 & \xrightarrow{\partial_1} & F & \xrightarrow{\partial_0} & G & \longrightarrow 1 \\
& \downarrow & & \downarrow & & \downarrow \varphi_n & & \downarrow & & \downarrow & & \downarrow & & \| & \\
\cdots \longrightarrow & 0 & \longrightarrow & 0 & \longrightarrow & M & \longrightarrow & M_{n-2} & \longrightarrow \cdots \longrightarrow & M_1 & \longrightarrow & E & \longrightarrow & G & \longrightarrow 1
\end{array}$$

where $\varphi_n \in \mathrm{Hom}_G(C_{n-1}, M)$ or $\varphi_2 \in \mathrm{Hom}_F(C_1, M)$ if $n = 2$. Since φ_n vanishes on $\mathrm{Im}\ \partial_n$, its class $[\varphi_n]$ belongs to $H^n(G,M)$. Then $[\mathcal{E}] \mapsto [\varphi_n]$ is a map θ from $H^n(G,M)^*$ to $H^n(G,M)$. It is not hard to construct an inverse of θ and show that $([\mathcal{E}]+[\mathcal{E}^*])\theta = ([\mathcal{E}])\theta + ([\mathcal{E}^*])\theta$. Thus $H^n(G,M)^*$ is an abelian group with respect to the Baer sum; the zero element is the class of the crossed $(n-1)$-fold extension

$$0 \longrightarrow M \longrightarrow M \longrightarrow 0 \longrightarrow \cdots \longrightarrow 0 \longrightarrow G \longrightarrow G \longrightarrow 1, \quad (\text{if } n \geq 2).$$

Theorem 1 ([15], [17]). *Let M be a module over a group G and let $n \geq 2$. Then the equivalence classes of crossed $(n-1)$-fold extensions of M by G form a set which is a group $H^n(G,M)^*$ with respect to the Baer sum. Furthermore $H^n(G,M)^* \simeq H^n(G,M)$.*

Remarks. 1. In case $n = 3$ it can be shown that every crossed 2-fold extension of M by G is equivalent to one in which M maps *onto* $\mathbf{Z}(M_1)$ (see [15]). Thus the link between crossed 2-fold extensions and abstract kernels is complete.

2. Despite the above interpretations no real applications of H^n, $n \geq 4$, to group theory seem to have been made.

3. In §3 we shall discuss vanishing theorems for cohomology groups; typically such theorem assert that $H^n(G,M) = 0$ for large enough n, provided that M satisfies a finiteness condition, M is relatively non-trivial as a G-module and G is close to being nilpotent or supersoluble.

§2. Applications

We begin with a straightforward application of H^1 to prove a near-conjugacy theorem due originally to Mal'cev.

(1) Near-conjugacy of maximal torsion subgroups

Theorem 2 ([23], [3], [28]). *Let G be a group with a series $1 = G_0 \lhd G_1 \lhd \ldots G_m = G$ in which each G_{i+1}/G_i is either cyclic or locally finite. Then the maximal torsion subgroups of G fall into finitely many conjugacy classes.*

Sketch of Proof. In the first place it may be assumed that G has no non-trivial normal torsion subgroups. Then it is easy to show that there is a *normal* series in G whose infinite factors are torsion-free abelian of finite rank. Assuming the result to be false for G and using induction on the length of the normal series, one reduces quickly to the situation $G = H_1A = H_2A = \ldots$ where A is a torsion-free abelian normal subgroup of finite rank and the H_i are infinitely many non-conjugate finite subgroups. If $|G : A| = |H_i| = m$, then $m.H^1(G/A, A) = 0$ by a standard theorem. Also A/A^ℓ is finite for all $\ell > 0$. Proposition 2 now shows that the H_i fall into finitely many conjugacy classes.

(2) Existence of outer automorphisms

Recall that a group G is said to be *complete* if the natural map $G \longrightarrow \operatorname{Aut} G$ is an isomorphism.

Theorem 3. *An infinite supersoluble group G cannot be complete .*

Proof. Suppose that G is complete and let A be a maximal abelian normal subgroup. Then $A = \mathbf{C}_G(A)$. Since $\mathbf{Z}(G) = 1$ and G is supersoluble, it is easy to see that A cannot have elements of order 2. Hence there is an exact sequence $1 \longrightarrow A \xrightarrow{\alpha} A \longrightarrow A/A^2 \longrightarrow 1$ in which the mapping α is $a \mapsto a^2$. Now $H^1(G/A, A) = 0$ by Proposition 3. Hence, on applying the cohomology functor to the exact sequence, one obtains $(A/A^2)^G = 0$. The supersolubility of G implies that $A = A^2$ and A is finite. Hence G is finite.

The following result is probably the most famous application of cohomology to group theory. The original form is due to Gaschütz.

Theorem 4 ([11], [32], [35]). *If G is finite non-abelian p-group, then $\mathbf{C}_{\mathrm{Out}G}(\mathbf{Z}(G))$ contains elements of order p. Hence p divides $|\mathrm{Out}\, G|$.*

Rather than repeat the entire proof we shall sketch the idea behind it. After a few elementary reductions one arrives at the following situation: there is a non-abelian subgroup M of index p such that $\mathbf{C}_G(M) = \mathbf{Z}(M) > \mathbf{Z}(G)$. Let $Q = G/M$. Now we can assume $H^1(Q, \mathbf{Z}(M)) = 0$ by Proposition 3. The formulae for cohomology of cyclic groups allow us to deduce that $H^2(Q, \mathbf{Z}(M)) = 0$. At this point the Wells sequence comes to our aid. By the Corollary to Proposition 5 we see that each element of $N_{\mathrm{Aut}M}(Q)$ lifts to an automorphism of G. By induction on the group order $L := N_{\mathrm{Out}M}(\mathbf{Z}(M))$ has elements of order p. A routine argument using Sylow's theorem shows that Q fixes some element $\nu(\mathrm{Inn}\, M)$ of L with order p, so that ν extends to an automorphism of G acting trivially on $\mathbf{Z}(M)$ and so on $\mathbf{Z}(G)$. Finally ν is outer since $\mathbf{Z}(M) \neq \mathbf{Z}(G)$.

Remarks. (a) Recently, Theorem 4 has been extended to infinite p-groups by Menogazzo and Stonehewer [24], who showed that an infinite non-abelian nilpotent p-group always has p-automorphisms that are outer; their proof is not homological. (b) There are a number of unsolved problems in the area–does every finitely generated infinite nilpotent group have an outer automorphism? (see [35]) The same question for torsion-free supersoluble groups.

(3) Finite complete groups

The Wells sequence can be used effectively to investigate complete groups that are not simple.

Proposition 6. *Let $\mathcal{E}$ be a group extension $1 \longrightarrow N \longrightarrow E \longrightarrow G \longrightarrow 1$ where E is finite, $M := \mathbf{C}_E(N) = \mathbf{Z}(N)$ and G is nilpotent. Then* $\mathrm{Out}\, \mathcal{E} = 1 = \mathbf{Z}(E)$ *if and only if $M^G = 1$ and G is self-normalizing as a subgroup of* $\mathrm{Out}\, N$ *via the natural embedding* $G \hookrightarrow \mathrm{Out}\, N$.

Proof. Let Out $\mathcal{E} = 1 = \mathbf{Z}(E)$; then certainly $M^G = 1$. Since G is nilpotent, it follows that $H^2(G, M) = 0$; this is a consequence of the Vanishing Theorem 5 below, but it can also be deduced from the well-known Gashütz-Schenkman splitting theorem. Applying the Corollary to Proposition 5 we obtained $G = N_{\text{Out } N}(G)$. Conversely if these conditions hold, then $H^1(G, M) = 0$ by Theorem 5, so Out $\mathcal{E} = 1$ by the same Corollary, while it is clear that $\mathbf{Z}(E) = 1$.

If E is a finite soluble group, an appropriate choice for N would be $LC_E(L)$ where L is the final term of the lower central series of E. Then E is complete if and only if $G := E/N$ is a Carter subgroup of Out N and $\mathbf{Z}(N)^G = 1$. Further progress depends upon knowledge of such Carter subgroups. Cases where complete groups have been completely classified include finite metabelian groups ([10]) and finite abelian-by-nilpotent groups ([9]). For a detailed account of recent work on complete groups see [27].

§3. Vanishing theorems

We shall describe a series of cohomological vanishing theorems for locally supersoluble groups. Common features of these theorems are that the module is subjected to a finiteness condition, namely a chain condition or finiteness of rank, and is, in some sense, highly non-trivial.

The theorem of Gaschütz and Schenkman mentioned at the end of §2 can be regarded as the most primitive instance of the theorems. Group theoretic applications arise in dimensions ≤ 3 and include splitting and conjugacy theorems, and the construction of automorphisms.

All the theorems are also valid for homology, but we shall not discuss this here.

Theorem 5 ([29]). *Let G be a locally nilpotent group, R a ring with identity, and M an RG-module such that $G/\mathrm{C}_G(M)$ is a hypercentral group. If either*

(i) *M is RG-noetherian and $M_G = 0$*

or

(ii) *M is RG-artinian and $M^G = 0$, then $H^n(G, M) = 0$ for all $n \geq 0$.*

Theorem 6 ([29]). *Let G be a locally nilpotent group, R a Dedekind domain and M an RG-module. If either*

(i) *M is torsion-free with finite rank as an R-module and $M_G = 0$*

or

(ii) *M is torsion with finite total rank as an R-module and $M^G = 0$, then $H^n(G, M) = 0$ for all $n \geq 0$.*

Some explanation of the term "rank" is in order here. Let R be a Dedekind domain with quotient field F and let M be an R-module. The *torson-free rank* of M is simply $r_0(M) = \dim_F(M \underset{R}{\otimes} F)$. If P is a non-zero prime ideal in the associated set of prime ideals $\mathrm{Ass}_R(M)$, then $A[P] = \{a \in M | aP = 0\}$ is a vector space over the field R/P, so we may define the *P-rank* of M to be $r_P(M) = \dim_{R/P}(M[P])$ and the total rank to be $r_{tot}(M) = \sum_P r_P(M) + r_0(M)$. It can be shown that an R-torsion module has finite total rank if and only if it is artinian ([29]).

A more general condition than finite total rank is that the ranks $r_0(M)$ and $r_P(M)$ be finite for every P in $\mathrm{Ass}_R(M)$. This is referred to as *finite R-ranks*; here infinitely many primary components may be non-zero. There is a vanishing theorem for modules of this type with a weaker conclusion.

Theorem 7 ([29]). *Let G be a locally nilpotent torsion group, R a Dedekind domain and M an RG-module. Assume that $|G| = \aleph_m$, with $m < \infty$, the characteristic of R does not belong to $\pi(G)$ (the set of prime divisors of orders of elements of G) and M is torsion with finite ranks as an R-module. If $M^G = 0$, then $H^n(G, M) = 0$ for all $n \geq m + 2$.*

The bound $m + 2$ is best possible when $m = 0$, and perhaps in all cases. In applications made so far the ring R has been either $\mathbf{Z}$ or the group algebra $\mathbf{F}_p\langle t \rangle$ of an infinite cyclic group. Another case of interest is where R is the ring of integers in an algebraic number field.

There are similar theorems for locally supersoluble groups. The main difference in these results is that the conditions $M_G = 0$ and $M^G = 0$ must be replaced by stronger conditions that forbid the occurrence of non-zero R-cyclic RG-quotients or RG-submodules. The situation is not quite so tidy - for example, we still do not know if the exact analogue of Theorem 5 holds.

We shall be content to quote just two results.

Theorem 8 ([30]). *Let G be a supersoluble group, R a ring with identity and M an RG-module. If either*

(i) *M is RG-noetherian and has no non-zero R-cyclic RG-quotients*

or

(ii) *M is RG-artinian and has no non-zero R-cyclic RG-submodules, then $H^n(G, M) = 0$ for all $n \geq 0$.*

Theorem 9 ([30]). *Let G be a locally supersoluble group, R a Dedekind domain and M an RG-module which has finite total rank as an R-module. If M has no non-zero R-cyclic RG-submodules or quotients, then $H^n(G, M) = 0$ for all $n \geq 0$.*

The proofs of all these results employ spectral sequences; indeed, whenever one has a composite group or a group that is expressed as a direct limit, spectral sequences can provide valuable information about the structure of the cohomology groups. We shall review two spectral sequences that are especially useful in this context.

(i) The Lyndon-Hochschild-Serre (LHS) spectral sequence

This is the spectral sequence best known to group theorists. Consider a group extension $1 \longrightarrow N \longrightarrow E \longrightarrow G \longrightarrow 1$ and an E-module M. Then E acts on $H^q(N, M)$ by a combination of conjugation in N and the module action in such a way that N operates trivially. Thus $H^q(N, M)$ is a G-module and we may form the mixed cohomology group

$$H^p(G, H^q(N, M)),$$

which for technical reasons is written E_2^{pq}; these groups constitute the so-called *E_2-page* of the LHS-spectral sequence. The spectral sequence relates these mixed cohomology groups to $H^n(G, M)$. It turns out that there is a series in $H^n(G, M)$

$$0 = H(0) \leq H(1) \leq \ldots \leq H(n+1) = H^n(G, M)$$

in which $H(i+1)/H(i)$ is isomorphic with a section of $E_2^{n-i,i}$, denoted by $E_\infty^{n-i,i}$. The LHS-spectral sequence is said to *converge* to $H^n(G, M)$, in symbols

$$E_2^{pq} \underset{p+q=n}{\Longrightarrow} H^n(G, M).$$

The actual identity of the sections E_∞^{pq} is usually hard to discover, and will not concern us here.

(ii) The inverse limit spectral sequence

Let $(S_i)_{i\in I}$ be a direct system of groups and homomorphisms and write G for the direct limit $\varinjlim(S_i)$. If M is a G-module, then M becomes an S_i-module via the natural map $S_i \longrightarrow G$. Moreover for each $n \geq 0$ the set $(H^n(S_i, M))_{i\in I}$ is an *inverse* system in which the maps are the obvious induced ones. Now the inverse limit of this system is not isomorphic with $H^n(G, M)$, essentially because the functor $\varprojlim$ is not right exact. However this failure is compensated for by a spectral sequence whose E_2-terms are

$$E_2^{pq} = \varprojlim{}^{(p)}(H^q(S_i, M))$$

and which converges to $H^n(G, M)$ in the sense described above. Here $\varprojlim^{(p)}$ is the pth right derived functor of $\varprojlim$. In practice these derived functors are hard to compute. Therefore the following results are invaluable.

Proposition 7 (Jensen). *Let $(S_i)_{i\in I}$ be an inverse system of artinian R-modules where R is any ring with identity. Then $\varprojlim^{(p)}(S_i) = 0$ for all $p > 0$.*

Proposition 8 (Goblot). *Let $(S_i)_{i\in I}$ be an inverse system of abelian groups with $|I| = \aleph_m$ and m finite. Then $\varprojlim^{(p)}(S_i) = 0$ if $p \geq m + 2$.*

For these facts and an illuminating treatment of the functors $\varprojlim^{(p)}$ see the lectures notes [18] of Jensen. As a simple application of the inverse limit spectral sequence we mention a result of Holt.

Theorem 10 ([16]). *Let G be a locally finite group with $|G| = \aleph_m$ and m finite. Let M be a G-module such that M is torsion as an abelian group and $\pi(G) \cap \pi(M) = \emptyset$. Then $H^n(G, M) = 0$ for all $n \geq m+2$.*

Proof. Let $(S_i)_{i \in I}$ be the direct system of all finite subgroups of G; then $|I| = \aleph_m$. Now $H^q(S_i, M) = 0$ for all $q \geq 0$ since multiplication by $|S_i|$ is an automomorphism of M but kills $H^q(S_i, M)$. Also $\varprojlim^{(n)}(M^{S_i}) = 0$ if $n \geq m+2$ by Proposition 8, and the result follows from the spectral sequence.

Since $\aleph_{-1}$ can be interpreted as a finite cardinal, Theorem 10 formally includes Schur's theorem. The case $|G| = \aleph_0$ is also well known. Holt shows by example that the bound $m+2$ is best possible (in *J. London Math. Soc.* (2) 24(1981),129-134.)

We shall sketch the proofs of two of the above vanishing theorems.

Proof of Theorem 7. Let $(S_i)_{i \in I}$ be the direct system of all finite subgroups of G. By the inverse limit spectral sequence it is enough to prove that $E_2^{pq} = 0$ whenever $p + q = n \geq m+2$.

Since $A = \coprod_P A_P$ and S_i is finite, we have $H^q(S_i, A) \simeq \coprod_P H^q(S_i, A_P)$. In addition for $q > 0$ almost all the $H^q(S_i, A_P) = 0$ since $|S_i| \neq 0$ in R. It is clear from the cocycle description that $H^q(S_i, A_P)$ is R-isomorphic with a section of a finite direct power of A_P. Since the latter is R-artinian, so must be $H^q(S_i, A_P)$. Hence $H^q(S_i, A)$ is R-artinian since almost all $H^q(S_i, A_P)$ are zero. It follows from Proposition 7 that $E_2^{pq} = 0$ if $p > 0$ and $q > 0$. Also $E_2^{n0} = 0$ by Proposition 8. Thus we are left with E_2^{0n}. Now $\varprojlim$ is left exact and commutes with products; hence

$$E_2^{0n} \hookrightarrow \prod_P \varprojlim H^n(S_i, A_P).$$

Now fix P. Since $A^G = 0$ and A_P is artinian, $A_P^{S_i} = 0$ for all S_i containing some S_0. Therefore $H^n(S_i, A_P) = 0$ for $S_1 \supseteq S_0$ by Theorem 5; this leads to $\varprojlim H^n(S_i, A_P) = 0$ and $E_2^{0n} = 0$, as required.

Proof of Theorem 8. In the proof we shall need the following fact.

Proposition 9. *Let $N = \langle g \rangle$ be a normal subgroup of a group G and let M be a G-module. Then the action of G on $H^n(N, M)$ (when the latter is identified with a section of M in the usual way) is given by $[a] \cdot x = [ax]s^{\left[\frac{n+1}{2}\right]}$ where $xgx^{-1} = g^s$.*

This can be proved quite straightforwardly by computing the change of groups (ξ, α) where $\xi : N \longrightarrow N$ and $\alpha : M \longrightarrow M$ are given by $g^\xi = xgx^{-1}$ and $a\alpha = ax$ (see [22]. pp. 350-1).

We shall prove part (i) of Theorem 8 by induction on the length of a shortest normal series of G with cyclic factors, $\ell = \ell(G)$ say; we can suppose $\ell > 0$. By the noetherian condition, the $\mathbf{Z}$-torsion-subgroup of M is bounded, which allows us to assume that M is either torsion-free or an elementary abelian p-group.

Find $N = \langle g \rangle \triangleleft G$ such that $\ell(G/N) = \ell - 1$. Denote by α the R-endomorphism $a \mapsto a(g-1)$; observe that α maps G-submodules to G-submodules and that Ker α^i is a G-submodule. Now Ker α^i = Ker α^{i+1} for some i, by the noetherian condition again, so Im $\alpha^i \cap$ Ker $\alpha = 0$. We can now reduce to two cases, (a) $\alpha = 0$, and (b) α is injective. If $\ell = 1$, then $G = \langle g \rangle$ and $M = M(g-1)$, in which case the result follows from the formulae for the cohomology of cyclic groups; so let $\ell > 1$.

Now $H^q(N, M) \simeq 0, M, M/\mathrm{Im}\,\alpha$ or M/Mm (if $|g| = m < \infty$). By Proposition 9 the hypotheses on M are inherited by $H^q(N, M)$. Therefore $H^p(G/N, H^q(N, M)) = 0$ for all p, q, by induction on ℓ. The LHS-spectral sequence now yields $H^n(G, M) = 0$ for all $n \geq 0$.

§4. Kropholler's criterion for a finitely generated soluble group to be minimax

Recall that a *minimax group* is a group with a series of finite length whose factors satisfy *either* the maximal condition *or* the minimal condition for subgroups. It is well-known that a soluble group G is minimax precisely when it has a series of finite length whose factors are either cyclic or quasi-cyclic (i.e. p^∞ for some prime p); the number of infinite factors in the series is an invariant of G called the *minimax length*, $m(G)$. Soluble minimax groups were introduced by Baer [1]; for an account of their properties see

[25].

Finitely generated soluble minimax groups constitute a significant subclass of the class of finitely generated soluble groups; in fact it was shown in [26] that they are precisely the groups with finite abelian section rank, a theorem whose proof requires cohomological techniques.

It is clear that a standard wreath product $C_p \operatorname{wr} C_\infty$ cannot be a minimax group. A remarkable theorem of Kropholler asserts that sections of this type are the only obstacle to a finitely generated soluble group being minimax.

Theorem 11 ([19]). *A finitely generated soluble group E is minimax if and only if it has no sections of type $C_p \operatorname{wr} C_\infty$ for any prime p.*

The proof of this theorem makes essential use of cohomological results concerning $H^n(G, M)$ when the module is expressed as a direct limit. (Compare this with the proof of Theorem 7 above where we encountered the same situation but with G a direct limit).

We shall give an account of the proof of Theorem 11. In this the following elementary – but non-obvious – result plays an essential part.

Proposition 10. *A finitely generated soluble minimax group can be expressed as a product of finitely many cyclic subgroups.*

Suppose that we are trying to prove Theorem 11 (in the non-trivial direction) by induction on the derived length. We quickly arrive at the following situation; E is a finitely generated soluble group with a normal abelian subgroup M which is either torsion or torsion-free, and E/M is minimax. Let $a \in M$ and $x \in E$; if M is torsion, then $a^{\langle x \rangle}$ must be finite, otherwise a wreath product of forbidden type will occur as a subgroup of E; if M is torsion-free, then $a^{\langle x \rangle}$ must have finite rank for the same reason. Now $G := E/M = \langle g_1 \rangle \dots \langle g_k \rangle$ for some g_i by Proposition 10. Hence

$$a^G = (\dots (a^{\langle g_1 \rangle}) \dots)^{\langle g_k \rangle}$$

will be finite or have finite rank. Consequently either M is a locally finite G-module or $M \otimes \mathbf{Q}$ is a locally finite dimensional $\mathbf{Q}G$-module.

The next step now is to obtain information about the extension $1 \longrightarrow M \longrightarrow E \longrightarrow G \longrightarrow 1$. For example, if this were to split or even nearly split, one sees immediately that E would have finite abelian section rank and would thus be minimax. While this need not be true, it does suggest a line of attack. Of course M is the direct limit of its finite (respectively finite rank) submodules in the two cases. Thus two basic situations face us.

The cohomological results required

To deal with the two situations just referred to, two results are needed.

Proposition 11. *Let G be a soluble minimax group and $(M_i)_{i\in I}$ a direct system of finite dimensional $\mathbf{Q}G$-modules. Then the natural map*

$$\varinjlim(H^n(G,M_i)) \longrightarrow H^n(G,\varinjlim M_i)$$

is an isomorphism for all $n \geq 0$.

This is relatively straightforward to prove by induction on the length of a cyclic-quasicyclic series of G, with the aid of the LHS-spectral sequence, and we shall say no more about it.

The second auxiliary result is more complex to state and to prove. Let G be a soluble minimax group with an abelian normal subgroup A, and let M be a locally finite G-module. Define $M_0 = 0$ and $M_{i+1}/M_i = (M/M_i)^A$, and write $U = \bigcup_i M_i$. Let

$$\nu_n : \varinjlim(H^n(G,M_i)) \longrightarrow H^n(G,M)$$

be the composite of the natural maps

$$\varinjlim(H^n(G,M_i)) \longrightarrow H^n(G,U) \text{ and } H^n(G,U) \longrightarrow H^n(G,M).$$

Proposition 12. (i) *ν_0 and ν_1 are isomorphisms and ν_2 is a monomorphism.*

(ii) *If G is finitely generated, then ν_2 is isomorphism and ν_3 a monomorphism.*

Sketch of Proof. The basic idea is to introduce a finitely generated subgroup A_0 of A such that A/A_0 is a divisible torsion group; this exists because G is minimax.

The first observation is that $H^n(G,U) \longrightarrow H^n(G,M)$ is an isomorphism for all $n \geq 0$, which amounts to showing that $H^n(G,\bar{M}) = 0$ where $\bar{M} = M/U$. Applying the LHS-spectral sequence first for $1 \longrightarrow A \longrightarrow G \longrightarrow G/A \longrightarrow 1$ and then for $1 \longrightarrow A_0 \longrightarrow A \longrightarrow A/A_0 \longrightarrow 1$, we see that it is sufficient to prove that $H^n(A_0,\bar{M}) = 0$. Since M is locally finite and A/A_0 is divisible, $\bar{M}^{A_0} = \bar{M}^A = 0$. Also it follows from Theorem 5 (or by a direct argument) that $H^n(A_0,F) = 0$ where F is a finite A_0-submodule of $\bar{M}$. Now $H^n(A_0,-)$ commutes with direct limits since A_0 is a finitely generated abelian group ([2]). Hence $H^n(A_0,\bar{M}) = 0$. By this observation we can assume that $U = M$.

Apply the functor $H^*(G,-)$ to $0 \longrightarrow M_j \longrightarrow M \longrightarrow M/M_j \longrightarrow 0$ and take $\varinjlim$. One then recognizes that it is enough to establish the following statements:

(i) $\varinjlim (M/M_j)^G = 0$,

(ii) $\varinjlim H^1(G,M/M_j) = 0$

and, if G is finitely generated,

(iii) $\varinjlim H^2(G,M/M_j) = 0$.

Of these (i) is quite clear. We shall indicate how (ii) can be proved.

Apply the LHS-spectral sequence for $1 \longrightarrow A \longrightarrow G \longrightarrow G/A \longrightarrow 1$ in conjunction with $\varinjlim$. The terms that are relevant for $H^1(G,M/M_j)$ are

$$E_2^{10} = \varinjlim H^1(G/A,(M/M_j)^A)$$

and

$$E_2^{01} = \varinjlim (H^1(A,M/M_j))^G.$$

Now $(M/M_j)^A = (M/M_j)^{A_0} = M_{j+1}/M_j$, and in the direct system (M_{j+1}/M_j) all the maps are zero; therefore $E_2^{10} = 0$.

To analyze $E_2^{01} = 0$ apply the LHS-spectral sequence in the direct limit form for $1 \longrightarrow A_0 \longrightarrow A \longrightarrow A/A_0 \longrightarrow 1$. The relevant terms are

$$\varinjlim H^1(A/A_0, (M/M_j)^{A_0}) = \varinjlim H^1(A/A_0, M_{j+1}/M_j) = 0$$

and

$$\varinjlim (H^1(A_0, M/M_j))^A \hookrightarrow H^1(A_0, \varinjlim(M/M_j)),$$

since $\varinjlim$ is exact and A_0 finitely generated. Finally $\varinjlim(M/M_j) = 0$, so the result follows.

To prove (iii) one proceeds in a similar manner, the essential point to establish being that $\varinjlim(H^2(A, M/M_j))^G = 0$. One argues using the spectral sequence for $1 \longrightarrow A_0 \longrightarrow A \longrightarrow A/A_0 \longrightarrow 1$ that this is isomorphic with $(\varinjlim \text{Hom}(A, H^1(A, M/M_j)))^G$ and it is not hard to see that the latter vanishes.

Proof of Theorem 11. (i) *Case: E has finite torsion-free rank.* By induction on the derived length of E reduce to the situation where there is a normal abelian torsion subgroup M such that $G := E/M$ is minimax. Let E be a counter example chosen so that $m(E)$ is minimal and G has an abelian normal subgroup $A := B/M$ with $m(A)$ maximal. Define M_i and A_0 as in Proposition 12. Then the natural map $\varinjlim(H^2(G, M_j)) \longrightarrow H^2(G, M)$ is an isomorphism.

From this it follows that there is an integer $i \geq 0$ such that E/M_i splits over M/M_i. Since E is finitely generated, M/M_i is finite. Writing B_0 for $\mathbf{C}_B(M/M_i)$, we observe that B/B_0 is finite and B_0 is nilpotent. We can now pass to the group E/B_0'; in short we can replace B by B_0/B_0' and assume that B is abelian; also M can be assumed to be the torsion-subgroup of B. A further simple reduction allows us to suppose that A is rationally irreducible with respect to G, and also that B is finitely generated as a G-module.

Write $Q = E/B$ and observe that the semidirect product $Q \ltimes B$ is also a counterexample and hence can be substituted for E. Let $C = \mathbf{C}_E(A)$. If C/B were finite, then, since E/C is abelian by-finite, being an irreducible $\mathbf{Q}$-linear group. E would be metabelian-by-finite; however E would then satisfy the maximal condition on normal subgroups by a well-known theorem of P. Hall [13], and so M would be finite.

Therefore C/B is infinite and thus contains an infinite abelian normal subgroup C_0/B. Clearly C_0/M is abelian and yet $m(C_0/M) = m(A) + m(C_0/B) > m(A)$, a contradiction.

(ii) *The general case.* It suffices to prove that E has finite torsion-free rank. Arguing by induction on the derived length, one reaches the situation where M is a torsion-free abelian normal subgroup of E and $G := E/A$ is minimax. Put $\bar{M} := \bar{M} \otimes \mathbf{Q}$; the obvious map $M \to \bar{M}$ leads to a commutative diagram

$$\begin{array}{ccccccccc} 1 & \longrightarrow & M & \longrightarrow & E & \longrightarrow & G & \longrightarrow & Z \\ & & \downarrow & & \downarrow & & \| & & \\ 1 & \longrightarrow & \bar{M} & \longrightarrow & \bar{E} & \longrightarrow & G & \longrightarrow & Z. \end{array}$$

Here $E \longrightarrow \bar{E}$ is a monomorphism. By Proposition 11, $\bar{E}/F$ splits over $\bar{M}/F$ for some finite dimensional $\mathbf{Q}G$-submodule F of $\bar{M}$. This implies that $\bar{M}$ has finite dimension and M finite rank.

Concluding remarks

In a recent paper [20] Kropholler has obtained the following improved version of Proposition 12. Recall that a group is of *type* $(FP)_n$ if there is a G-projective resolution $P \twoheadrightarrow \mathbf{Z}$ with P_i finitely generated for $i \leq n$.

Theorem 12. *Let G be a soluble minimax group which is of type $(FP)_n$ and let M be a locally finite G-module. If (M_i) is the direct system of all G-submodules of finite total rank, then the natural map*

$$\varinjlim(H^{n+1}(G, M_i)) \longrightarrow H^{n+1}(G, M)$$

is an isomorphism.

This allows a speedier proof of Theorem 11 by eliminating the need for the group theoretic arguments in part (i) of the proof given above. For since G is finitely generated, it is of type $(FP)_1$, and so Theorem 12 may be applied with $n = 1$. The conclusion is that for some i the module M/M_i is finite and thus M has finite total rank. Hence E has finite abelian section rank and the theorem follows.

There is however a price to be paid for this handier proof; for Theorem 12 lies deeper and requires a careful analysis of the homology groups $H_n(G, \overline{\mathbf{Z}_p G})$ where G is a nilpotent minimax group and $\overline{\mathbf{Z}_p G}$ is the completion of $\mathbf{Z}_p G$ at the unique maximal ideal containing the augmentation ideal.

References

[1] R. Baer, Polyminimaxgruppen, *Math. Ann.* **175** (1968), 1-43.

[2] R. Bieri, *Homological Dimension in Discrete Groups*, Mathematics Notes, Queen Mary College, London (1976).

[3] J. F. Bowers and S. Stonehewer, A theorem of Mal'cev on periodic subgroups of soluble groups, *Bull. London Math. Soc.* **5** (1973), 323-324.

[4] R. Brown and P. J. Higgins, On the connection between the second relative homotopy groups of some related spaces, *Proc. London Math. Soc.* (3) **36** (1978), 193-212.

[5] H. Cartan and S. Eilenberg, *Homological Algebra*, Princeton (1956).

[6] B. Eckmann, Der Cohomologie-Ring einer beliebigen Gruppe, *Comment. Math. Helv.* **19** (1945-6), 232-282.

[7] S. Eilenberg and S. MacLane, Group extensions and homology, *Ann. Math.* **43** (1942), 757-831.

[8] ——————————, Cohomology theory in abstract groups I, II, *Ann. Math.* **48** (1947), 51-78, 326-341.

[9] T. M. Gagen, Some finite soluble groups with no outer automorphisms, *J. Algebra* **65** (1980), 84-94.

[10] T. M. Gagen and D. J. S. Robinson, Finite metabelian groups with no outer automorphisms, *Arch. Math.* (Basel) **32** (1979), 417-423.

[11] W. Gaschütz, Nichtabelsche p-Gruppen besitzen äussere p-Automorphismen, *J. Algebra* **4** (1966), 1-2.

[12] K. W. Gruenberg, *Cohomological topics in Group Theory*, Lecture Notes in Mathematics, Vol. 143, Springer, Berlin (1970).

[13] P. Hall, Finiteness conditions for soluble groups, *Proc. London Math. Soc.* (3) **4** (1954), 419-436.

[14] P. J. Hilton and U. Stammbach, *A Course in Homological Algebra*, Springer, New York (1970).

[15] D. F. Holt, An interpretation of the cohomology groups $H^n(G,M)$, *J. Algebra* **60** (1979), 307-320.

[16] ________, On the cohomology of locally finite groups, *Quart. J. Math.* (2) **32** (1981), 165-172.

[17] J. Huebschmann, Crossed n-fold extensions of groups and cohomology, *Comment. Math. Helv.* **55** (1980), 302-314.

[18] C. U. Jensen, *Les Foncteurs Dérivé de* $\varprojlim$ *et Leur Applications en Theorie des Modules*, Lecture Notes in Mathematics, Vol 254, Springer, Berlin, (1972).

[19] P. H. Kropholler, On finitely generated soluble groups with no large wreath product sections, *Proc. London Math. Soc.* (3) **49** (1984), 155-169.

[20] ____________, The cohomology of soluble groups of finite rank, *Proc. London Math. Soc.* (3) **53** (1986), 453-473.

[21] S. Mac Lane, Cohomology theory in abstract groups III, *Ann. Math.* (2) **50** (1949), 736-761.

[22] _________, *Homology*, Springer, Berlin (1967).

[23] A. I. Mal'cev, On certain classes of infinite soluble groups, *Mat. Sb.* **28** (1951), 567-588. *Amer. Math. Soc. Translation* (2) **2** (1956), 1-21.

[24] F. Menogazzo and S. E. Stonehewer, On the automorphism group of a nilpotent p-group, *J. London Math. Soc.* (2) **31** (1985), 272-276.

[25] D. J. S. Robinson, *Finiteness Conditions and Generalized Soluble Groups*, 2 vols., Springer, Berlin (1972).

[26] ______________, On the cohomology of soluble groups of finite rank, *J. Pure Appl. Algebra* **6** (1975), 155-164.

[27] ______________, Recent results on complete groups, in *Algebra, Carbondale 1980*, Lecture Notes in Mathematics, Vol. 848, Springer, Berlin (1981), 178-185.

[28] ______________, Applications of cohomology to the theory of groups, in *Groups St. Andrews 1981*, London Math. Soc. Lecture Notes, Vol. 71 (1982), 46-80.

[29] ______________, Cohomology of locally nilpotent groups, *J. Pure Appl. Algebra*, **48** (1987), 281-300.

[30] ______________, Homology and cohomology of locally supersoluble groups, *Math. Proc. Cambridge Philos. Soc.*, **102** (1987), 233-250.

[31] J. Rotman, *An Introduction to Homological Algebra*, Academic Press, New York (1979).

[32] P. Schmid, Normal p-subgroups in the group of outer automorphisms of a finite p-group, *Math. Z.* **147** (1976), 271-277.

[33] U. Stammbach, *Homology in Group Theory*, Lecture Notes in Mathematics, Vol. 359, Springer, Berlin (1973).

[34] U. H. M. Webb, Outer automorphisms of some finitely generated nilpotent groups I, *J. London Math. Soc.* (2) **21** (1980), 216-224.

[35] ______________, An elementary proof of Gaschütz's theorem, *Arch. Math.* (Basel) **35** (1980), 23-26.

[36] C. Wells, Automorphisms of group extensions, *Trans. Amer. Math.* Soc. **155** (1971), 189-194.

Department of Mathematics
University of Illinois at Urbana-Champaign
Urbana, IL 61801
U. S. A.

Invited lectures

Upper and lower bounds for nilpotency classes of Lie algebras with Engel conditions

S.I. Adian, A.A. Razborov and *N.N. Repin*

In 1902 W. Burnside [1] formulated the following problem:

"Is the group $B(r,n)$ defined by r generators and an identical relation $x^n = 1$ one of finite order, and if so what is its order?"

A negative answer to this question has been obtained in [2], in which it is proved that for any $r > 1$ and odd $n \geq 4381$ the group $B(r,n)$ is infinite. In [3] this result has been improved for odd $n \geq 665$. Obviously, from $|B(r,n)| = \infty$ it follows that $|B(r,kn)| = \infty$ for any $k \geq 1$. The finiteness of $B(r,n)$ has been proved earlier for $n \leq 3$ [1], $n = 4$ [4], and $n = 6$ [5]. An important question of existence of a natural number k such that the group $B(r,2^k)$ is infinite for some $r > 1$ is still open. The finiteness of $B(r,5^k)$ for $r > 1$ is another interesting open question.

The reader can find more detailed information on the history of investigations on the Burnside problem in the book [6] and in the survey articles [7] and [8]. The great interest of algebraists in the Burnside problem and the difficulties that arise from the investigation of this problem lead to several Burnside-type problems for groups and for algebraic systems (see, for instance, [9] and [10]. One of the most important problems of the Burnside type is the following question:

R_n: *For any natural number m find an integer $\beta_{m,n}$ such that any finite group of exponent n generated by m elements has order less than or equal to* $\beta_{m,n}$ (see [11]).

This is equivalent to a question of the existence of a maximal finite m-generator group $R(r,n)$ of exponent n, such that all finite m-generator groups of exponent n are quotient groups of $R(r,n)$. This problem is known as *the restricted Burnside problem* (a term introduced by Magnus in 1950 [12]). Without rejecting this term we propose to use for the same problem the term "*the Burnside-Magnus problem*" which expresses correctly the history of the problem.

Apparently W. Magnus was the first to have mentioned a close connection between groups and Lie algebras. In [13] he showed that the lower central series of any group G defines a Lie algebra $L(G)$ that is connected with G. The terms of the lower central series of a given group G are defined by the equations:

$$G_1 = G \quad \text{and} \quad G_{i+1} = [G_i, G] \quad \text{for all} \quad i \geq 1.$$

A Lie ring $L(G)$ corresponding to the group G is a direct sum

$$L = \bigoplus_{i=1}^{\infty} L_i(G)$$

of additively written quotients $L_i(G) = G_i/G_{i+1}$ where the operation of addition is the direct sum; a Lie product $[x, y]$ of given elements $x \in L_i(G)$ and $y \in L_j(G)$ is defined as a commutator $[x, y]$ in G modulo G_{i+j+1} and this product is extended to all sums by distributivity. If the identity $x^n = 1$ holds in the group G then the identity $nx = 0$ holds in the additive group $L(G)$. For the case when $n = p$ is prime Zassenhaus [14] showed that the identity $[y, x^{p-1}] = 1$ holds in G, where $[y, x^{p-1}]$ denotes the following left sided normalized commutator

$$[\dots[[y, x], x], \dots, x].$$

In the Lie algebra $L(G)$ we have the corresponding identity

$$[y, x^{p-1}] = [\dots[[y, x], x], \dots, x] = 0$$

where $[y, x]$ is the product in $L(G)$. The last identity is usually called *the Engel condition* and is denoted by E_{p-1}. Therefore to any periodic group G of prime exponent p there corresponds a Lie algebra over the field $\mathbf{Z}_p$ that satisfies the Engel condition E_{p-1}. If one has proved that any Lie algebra with r generators and the Engel condition E_{p-1} over the field $\mathbf{Z}_p$ is nilpotent, then from the nilpotency of the Lie algebra $L(R(r,p))$, the nilpotency of the group $R(r,p)$ follows.

The first result on the Burnside-Magnus problem has been obtained by O. Grün [15]. Grün claimed that the lower central series of $B(2,5)$

is stablised at the seventh term. But later Sanov [16] showed that this statement is not correct. In 1956 A. I. Kostrikin [17] and G.Higman [18] independently proved that for any number r of generators there exists a maximal finite group $R(r, 5)$.

In 1958 Kostrikin [19] published a result that for any prime p any finitely generated Lie algebra over the field $\mathbf{Z}_p$ with the Engel condition E_{p-1} is nilpotent. In 1959 he published a proof of this result [20] but as it was ascertained later this proof contained an essential gap. For this reason Kostrikin has published in 1979 an appendix [21] to the article [20]. One can find a complete proof of Kostrikin's result in the book [22] that has been published recently in "Nauka". While Higman's proof in [8] for $p = 5$ gives a linear upper bound for the nilpotency classes of the groups $R(r, 5)$, Kostrikin's proof is *a proof by contradiction* and gives no information about the nilpotency classes of the Lie algebras under consideration and, therefore, of the corresponding groups. The absence of a direct or an effective proof of Kostrikin's theorem in the book [22] (see p.17) was mentioned in the following way: "... *an approach nominated below, connected with the consideration of a locally nilpotent radical in L, still remains unique. It is absolutely noneffective in order to obtain upper bounds for nilpotency classes but its introduction is quite natural.*"

An effective proof of Kostrikin's theorem has been given for the first time in the article [23] of S. I. Adian and A. A. Razborov. This proof allows one to find primitive recursive upper bounds for the nilpotency classes of the Lie algebras $L(r, p)$ for any integer r and prime p. From this result one can find easily effective bounds for the orders of the maximal finite groups $R(r, p)$. This result constitutes the contents of the first part of the present lecture. It shows that a consideration of a locally nilpotent radical is not necessary for the solution of the given problem.

It is interesting to compare this result with the results of the article [28] for the groups $B(r, n)$ for $r \geq 66$ and odd $n \geq 665$. We formulate them below shortly.

For an arbitrary word A in the alphabet

$$a_1, a_2, \cdots, a_r, a_1^{-1}, a_2^{-1}, \cdots, a_r^{-1}, \tag{1}$$

we denote by $\tilde{A}$ the word obtained from A as a result of replacing each letter a^σ $(\sigma = \pm 1)$ by the letter $a^\sigma_{i+_r 1}$, where $+_r$ is addition modulo r.

Example: for $r = 3$ and $A = a_2 a_1^4 a_3^{-1} a_2$ we have $\tilde{A} = a_3 a_2^4 a_1^{-1} a_3$. We call an uncancellable word E in alphabet (1) a *metaperiodic word of exponent* r if it has a letter-for-letter representation of the form

$$E = A_1 A_2 \dots A_r,$$

where $A_{i+_r 1} = \tilde{A}_i$ for all $i \leq r$. We call the words A_i $(i \leq i \leq r)$ *metaperiods* of E.

Let $X_i = a_1 a_2 a_1^{-1}$ and $X_{i+1} = X_i a_2 X_i^{-1}$ for $i \geq 1$ and

$$\mathcal{M} = \{X_i \mid i = 1, 2, 3, \dots, \}.$$

We denote by $B_{\mathcal{M}}(r, n)$ the group defined by the generators (1) for $r \geq 66$, the identical relation $X^n = 1$ for odd $n \geq 665$, and an infinite system of complementary defining relations of the form

$$A_1 A_2 \dots A_{r-1} A_r = 1, \tag{2}$$

whose left sides run through all ascending metaperiodic words of exponent k with metaperiods $A_i \in \mathcal{M}$. The following theorem has been proved in [28].

For any $r \geq 66$ *and odd* $n \geq 665$ *the defining relations* (2) *for the group* $B_{\mathcal{M}}(r, n)$ *are independent; i.e., none of them follows from the remaining relations* (2) *and the identity* $X^n = 1$.

The proof of this theorem given in [28] is based on some generalisation of the method of a classification of periodic words developed in [2] and [3]. This method enables us to construct groups with various given properties (see [7]).

The following corollaries can be easily deduced from the theorem.

1) *For a free periodic group* $B(r, n)$ *of any odd exponent* $n \geq 665$ *with* $r \geq 66$ *generators one can find infinite ascending and descending chains of normal subgroups.*

2) *For each odd* $n \geq 665$ *and* $r \geq 66$ *there exist a continuum of distinct periodic groups of exponent* n *with* r *generators.*

3) *For any odd $n \geq 665$ and $r \geq 66$ one can find a factor group G of the periodic group $B(r,n)$ defined by a recursive set of complementary defining relations and having an unsolvable word problem.*

The second part of this lecture is devoted to lower bounds of nilpotency classes of relatively free Lie algebras with the Engel condition E_n as functions of n. The first result on this problem was obtained by Chang in 1961 [24]. He proved that a nilpotency class $\gamma_2(p-1)$ of the relatively free 2-generator Lie algebras with the Engel condition E_{p-1} over the field $\mathbf{Z}_p$ grows faster than any linear function. Later F.M. Malyshev [25] extended this result to the case of any field. It has been considered as a refutation of Kostrikin's conjecture in [26] (p.41). However one must mention that Malyshev's proof is based on a statement incorrect for $n \geq p$ (see (6) on p.56 in [25]).

In a manuscript of the book [22] Kostrikin had proposed a new conjecture that $\gamma_2(n)$ grows as a quadratic function of n (the first author of the present article was a reviewer of the book [22]). An exponential lower estimate for $\gamma_2(n)$ has been published in [27]. In the second part of the present lecture we give a somewhat simpler proof of the result from [27]. This result also takes care of the problem of correction of the mistake from Malyshev's article [25] that has been transferred into the Russian edition of the book [22] (see p.144, formula (6.1)).

§1. Upper bounds

The following theorem is proved in [23].

Theorem 1. *Any r generator Lie algebra with the Engel condition E_n over a field F of characteristic $p > n$ or 0 is nilpotent and its class of nilpotency is less than or equal to*

$$F(n,r,N(n,10)\cdot 6^{n+12}),$$

where $F(n,r,\alpha)$ and $N(n,r)$ are primitive recursive functions defined by the following equations:

$$F(n,r,0)=1, \quad F(n,r,\alpha+1)=n\cdot r^{3\cdot F(n,r,\alpha)}; \tag{3}$$
$$N(n,4)=6, \quad N(n,r+1)=F(n,r+1,N^2(n,r)\cdot 3^{(n+6)/2}).$$

Let

$$L(r,n) = \langle x_1, x_2, \ldots, x_r;\ \forall u, v\ ([v, u^n] = 0)\rangle$$

be a relatively free r-generator Lie algebra with the Engel condition E_n over a field F of characteristic $p > n$ or 0. As is known, the algebra $L = L(r,n)$ is a homogeneous algebra with r generators $x_1, x_2, \ldots, x_r$. It decomposes into a direct sum of vector spaces L_i of rank i

$$L = \bigoplus_{i=1}^{\infty} L^i,$$

where L^1 is a space over F with basis $\{x_1, x_2, \ldots, x_r\}$ and $L^{i+1} = [L^i, L^1]$. By Witt's formula the dimension of the homogeneous component L^d of the algebra L has an upper bound

$$\dim L^d = \frac{1}{d} \sum_{i|d} \mu(i) r^{d/i} \le \frac{1}{d} r^d. \tag{4}$$

This component is generated by the basic commutators of weight d.

Any nonzero element $u \in L$ has a unique representation of the form

$$u = \sum_{i=i_0}^{k} u_i,$$

where $u_i \in L^i$ and $u_{i_0} \neq 0$. For such $u \neq 0$ we denote $|u| = i_0$ and $\delta(u) = u_{i_0} = 0$. Let

$$\delta_j(u) = \begin{cases} \delta(u), & \text{if } j = |u| \\ 0, & \text{otherwise.} \end{cases} \tag{5}$$

A mapping $\delta_j : L \longrightarrow L^j$ has the following useful property:

$$\delta_j(u) \neq 0 \quad \& \quad \delta_j(v) \neq 0 \Rightarrow \delta_j(u+v) = \delta_j(u) + \delta_j(v).$$

For any j and any linear subspace $V \subseteq L$ one can naturally define its image $\delta_j(V)$ that is a linear subspace of L^j. The following main lemma plays an essential role in the proof of Theorem 1.

Lemma. *If the relatively free Lie algebra $L = L(r,n)$ has a chain of ideals*

$$L = T_0 \supseteq T_1 \supseteq T_2 \supseteq \dots T_{\alpha-1} \supseteq T_\alpha = 0, \tag{6}$$

where for $0 \le i \le \alpha$, T_i/T_{i+1} is the sum of abelian ideals of the algebra L/T_{i+1}, then any r-generator Lie algebra with Engel condition E_n over a field F of characteristic $p > n$ or 0 has a nilpotency class $\le F(n,r,\alpha)$.

A similar statement is true also for the Lie algebra

$$\bar{L}(r,n) = \langle x_1, \cdots, x_r; \forall u, v([v,u^n] = 0 \mathop{\&}_{i=1}^{r} [v, x_i^2] = 0)\rangle,$$

generated by the elements x_1, $\cdots$, x_r of nil-index 2.

Here we shall give a sketch of the proof of the lemma for the algebra $L(r,n)$. The proof for $\bar{L}(r,n)$ is similar to this one.

By induction on t, we prove for $t \le \alpha$ the equality

$$L^{F(n,r,t)} = \delta_{F(n,r,t)}(T_t). \tag{7}$$

From (7) and $T_\alpha = 0$ follows $L^{F(n,r,\alpha)} = 0$, that is, the algebra $L(r,n)$ has a nilpotency class $\le F(n,r,\alpha)$.

For $t = 0$ the equality (7) is clear. Let $0 < t \le \alpha$ and the equality

$$L^{F(n,r,t-1)} = \delta_{F(n,r,t-1)}(T_{t-1}) \tag{8}$$

be true. Denote by M the set of all commutators of the form

$$[\cdots[g_1, g_2], \cdots, g_s],$$

where g_i $(1 \le i \le s)$ are commutators of weights $|g_i| \le 2 \cdot F(n,r,t-1)$ on generators x_1, x_2, $\cdots$, x_r and

$$\sum_{i=1}^{s} |g_i| = F(n,r,t).$$

Clearly, M generates a homogeneous component $L^{F(n,r,t)}$ of the space L. Therefore to prove (7) it is sufficient to prove the inclusion

$$M \subseteq \delta_{F(n,r,t)}(T_t). \tag{9}$$

Suppose that (7) is not true. Choose an element

$$[g_1, g_2, \cdots, g_s] \in M \backslash \delta_{(n,r,t)}(T_t) \tag{10}$$

with the minimal possible value of s. Denote

$$I = \{i | 1 \leq i \leq s \quad \& \quad |g_i| \geq F(n,r,t-1)\}.$$

Let i_1, i_2 be two neighbouring indices from I, i.e. i_1, $i_2 \in I$, $i_1 < i_2$ and

$$\forall i(i_1 < i < i_2 \Rightarrow |g_i| < F(n,r,t-1)). \tag{11}$$

At first we shall prove the inequality

$$\left.\begin{array}{l} \displaystyle\sum_{i=i_1+1}^{i_2-1} |g_i| \leq (n-1) \cdot \sum_{d=1}^{F(n,r,t-1)-1} d \cdot \left(\frac{1}{d} r^d\right) \\ \qquad\qquad\qquad < (n-1) r^{F(n,r,t-1)}. \end{array}\right\} \tag{12}$$

Firstly, one can assume that in (10) all g_i with the condition (11) are the basic commutators. Secondly, by definition (5), the condition of minimality of s and by the equality

$$g_i g_{i+1} = [g_i, g_{i+1}] + g_{i+1} g_i,$$

the result of a permutation of two neighbouring multipliers between g_{i_1} and g_{i_2} in a commutator (10) satisfies (10) too. Therefore any basic commutator can occur in the interval between g_{i_1} and g_{i_2} less than n times as in $L(r,n)$ the identity $[u, v^n] = 0$ holds. Since the number of such basic commutators is bounded by (4) and (11) we easily obtain inequality (12).

For an estimation of card (I) we consider representation of the elements g_i for $i \in I$ by the generators x_1, x_2, $\cdots$, x_r:

$$g_i = \sum_j \lambda_{i,j} [x_{\alpha_i(j,i)}, x_{\alpha_i(j,2)}, \cdots, x_{\alpha_i(j,|g_i|)}], \tag{13}$$

where $\lambda_{i,j} \in F$ and $1 \leq \alpha_i(j,k) \leq r$. We replace in $[g_1, g_2, \cdots g_s]$ these g_i by suitable summands g_i' from (13) preserving the condition (10). Then we consider in such components having a form

$$g_i' = [x_{\alpha_i(j_i,1)} x_{\alpha_i(j_i,2)} \cdots x_{\alpha_i(j_i,|g_i|)}] \quad (i \in I)$$

the initial commutators of weight $\rho = F(n,r,t-1)$:

$$h_i = [x_{\alpha_i(j_i,1)}, x_{\alpha_i(j_i,2)}, \cdots, x_{\alpha_i(j_i,\rho)}] \quad (i \in I).$$

Using (8) for any $i \in I$ we can find an element $\hat{h}_i \in T_{t-1}$ such that $\delta(\hat{h}_i) = h_i$. By definition of the chain (6) we can represent all $\hat{h}_i$ in the form

$$\hat{h}_i = \sum_{j=1}^{m_i} u_{i,j}, \quad (i \in I) \tag{14}$$

where each $u_{i,j}$ generates an abelian ideal in L/T_t.

For any $u \in L$ denote by $\bar{u}$ the projection of u on the first $F(n,r,t-1)$ summands in the decomposition $L = \oplus_{d=1}^{\infty} L^d$. The set

$$V = \{\bar{u}_{i,j} | i \in I,\ i \leq j \leq m_i\}$$

generates in the space

$$\bigotimes_{d=1}^{F(n,r,t-1)} L^d$$

a linear subspace $\mathcal{M}$ that by (6) has dimension

$$\gamma = \dim |\mathcal{M}| \leq \sum_{d=1}^{F(n,r,t-1)} \frac{1}{d} r^d < r^{F(n,r,t-1)+1}. \tag{15}$$

Let

$$\{\bar{u}_1, \bar{u}_2, \cdots, \bar{u}_\gamma\} \subseteq V \tag{16}$$

be a generating subset for $\mathcal{M}$. Expressing all $\bar{u}_{i,j} \in V$ over this basis: $\bar{u}_{i,j} = \sum_{k=1}^{\gamma} \lambda_{i,j,k} \bar{u}_k$ we denote

$$\tilde{h}_i = \sum_{k=1}^{\gamma} \left(\sum_{j=1}^{m_i} \lambda_{i,j,k} \right) u_k, \quad (i \in I) \tag{17}$$

where u_k $(k = 1, 2, \cdots, \lambda)$ are the summands $u_{i,j}$ from (14) whose projections belong to the basis (16). Using (17), (14) we get $\bar{\tilde{h}}_i = \bar{\hat{h}}_i$ and consequently $\delta(\tilde{h}_i) = \delta(\hat{h}_i) = h_i$.

Now we shall replace in $[g_1, g_2, \ldots, g_s]$ each g_i for $i \in I$ by g_i' and the initial commutator h_i in it by $\tilde{h}_i$ using decomposition (17). For the result we shall obtain a sum of some commutators where each g_i $(i \in I)$ is replaced by some u_k $(k = 1, 2, \ldots, \lambda)$.

Suppose that $\gamma <$ card(I). Then in each of the summands obtained for $[g_1, g_2, \ldots, g_s]$ at least one of multipliers u_k must occur twice. Since u_k generates an abelian ideal in L/T all commutators obtained belong to T_t. Then from $\delta(\tilde{h}_i) = h_i$ and

$$[\delta(u), \delta(v)] \neq 0 \Rightarrow \delta([u, v]) = [\delta(u), \delta(v)]$$

follows

$$\delta([\tilde{g}_1, \tilde{g}_2, \ldots, \tilde{g}_s]) = [g_1', g_2', \ldots, g_s'] \in \delta_{F(n,r,t)}(T_t),$$

where for $i \notin I$, $\tilde{g}_i = g_i' = g_i$, and for $i \in I$, $\tilde{g}_i$ is obtained by replacing the initial component h_i by $\tilde{h}_i$ in γ_i'. This contradicts the condition (10). Consequently we have card $I \leq \gamma =$ card $\mathcal{M}$. Then using (12), (15) and $r \geq 2$ we obtain

$$\begin{aligned} F(n,r,t) = \sum_{i=1}^{s} |g_i| &< (\text{card}(I) + 1)(n-1)r^{F(n,r,t-1)} + \\ &+ 2F(n,r,t-1)\text{card}(I) < r^{F(n,r,t-1)+1}((n-1)r^{F(n,r,t-1)} + \\ &+ 2F(n,r,t-1)) \leq r^{F(n,r,t-1)+1} n r^{F(n,r,t-1)+1} \leq n r^{3F(n,r,t-1)}. \end{aligned}$$

That contradicts (3). Hence (9) is true and the lemma is proved.

After the proof of this lemma we have to prove the existence of a chain of ideals (6) and to find an upper bound for the corresponding parameter α. In [23] it was achieved by proving a large series of so-called "quasiidentities" in the algebras L and $L_i = L/T_i(L)$ $(i = 1, 2, \ldots)$, where the ideals $T_i(L)$ are defined by the conditions: $T_0(L) = 0$ and $T_{i+1}(L)/T_i(L)$, for $i \geq 0$, is the sum of all abelian ideals of $L/T_i(L)$.

A quasiidentity in a given Lie algebra L is an implication of the form

$$\left.\begin{aligned} &\forall x, u, \ldots, w(\mathop{\&}_{i=1}^{k} H_i(x, u, \ldots, w, a, \ldots, h) = 0) \Rightarrow \\ &\quad\Rightarrow \forall x, u, \ldots, w(\mathop{\&}_{j=1}^{s} G_j(x, u, \ldots, w, a, \ldots, h) = 0). \end{aligned}\right\} \quad (18)$$

where H_i, G_j are associative polynomials in the algebra of differentiations $A(L)$ of L; some of the variables $x, u, \ldots, w$ can be absent. We say that a quasiidentity (18) is true in L if it is true in $A(L)$ for any $a, b, \ldots, h$ from L when all quantifiers are running over L. In some of the quasi-identities the quantifiers can be restricted by a given ideal L, or the equality $G_j(x, u, \ldots, w, a, \ldots, h) = 0$ can be replaced by $G_j(x, u, \ldots, w, a, \ldots, h) \overset{i}{=} 0$, where $\overset{i}{=}$ means equality in the algebra $A(L_i)$.

Most of the quasiidentities that are proved in [23] are actually contained in [22]. Some of them are more or less obvious. But many of them need highly nontrivial calculations. In [23] all necessary calculations are given for the convenience of the reader; therefore he does not need to appeal to the book [22]. This has been done for the following reasons. At first, in his proof by contradiction Kostrikin considers a hypothetical object - a nontrivial Lie algebra over a field $\mathbf{Z}_p$ with the Engel condition E_{p-1} that does not contain abelian ideals. It follows from the final result that such algebras do not exist and a reader could object that one can prove anything for algebras which do not exist. Secondly, some of these conditional identities has not been formulated evidently in [22]. Thirdly, in [22] corresponding proof includes various inequalities that are connected with so-called "sandwiches". On can see from our proof that these inequalities are unnecessary.

As the result of the sequence of quasiidentities in [23] we prove that in the relatively free Lie algebra $L = L(r, n)$ with the Engel condition E_n $(n \geq 5)$ over any field of characteristic $p \geq n$ or $p = 0$, in the chain of ideals T_α described above, for $\alpha = N(n, 10) \cdot 6^{n+12}$ the term T_α coincides with L.

Together with the lemma this gives a proof of Theorem 1. From Theorem 1 we obtain the following upper bound for the order of any finite r-generator group G of prime exponent p:

$$|G| < p^{rF(p-1, r, N(p-1, 10) \cdot 6^{n+12})}.$$

§2 Lower Bounds

Let $\gamma_r(n)$ be the nilpotency class of a relatively free r-generator Lie algebra $L(r, n)$ over an arbitrary field F with the Engel identity $[X, Y^n] = 0$.

If $L(r,n)$ is not nilpotent we put $\gamma_r(n) = \infty$. We shall find an exponential lower bound for $\gamma_2(n)$. A similar result can be proved in the same way for any $r \geq 2$.

Theorem 2. *There exists a natural number N such that for all $n > N$*

$$\gamma_2(N) > 2^{cn}, \qquad (c = 0.074).$$

Proof. In [27] Theorem 2 has been proved for $c = \frac{1}{15}$. Here we shall give a simpler proof for $c = 0.074$.

Represent the algebra $L = L(2,n)$ as a direct sum of homogeneous components:

$$L = L^1 \oplus L^2 \oplus \dots .$$

L^k is a linear space over F on the set of all basic elements u of weight $|u| = k$. From Witt's theorem it follows that

$$\frac{1}{2k}2^k \leq \dim L^k \leq \frac{1}{k}2^k. \tag{19}$$

Denote by I_n^k the linear space over F generated by elements of the form

$$\left\{u_0, \overset{u_1}{\underset{j_1}{\cdot}}, \dots, \overset{u_r}{\underset{j_r}{\cdot}}\right\} = \sum_{(\alpha_1,\dots,\alpha_n)} [u_0, u_{\alpha_1}, \dots, u_{\alpha_n}], \tag{20}$$

where $u_0, u_1, \dots, u_r$ are some distinct basic commutators, $j_1, \dots, j_r$ are positive integers,

$$j_1 + \dots + j_r = n, \quad |u_0| + j_1|u_1| + \dots + j_r|u_r| = k, \tag{21}$$

and the summing is over all n-tuples of natural numbers $(\alpha_1, \dots, \alpha_n)$ such that $1 \leq \alpha_i \leq r$ and each number $s \leq r$ occurs in $(\alpha_1, \dots, \alpha_n)$ precisely j_s times. A direct sum

$$I_n = I_n^{n+1} \oplus I_n^{n+2} \oplus \dots$$

is an ideal of L. This follows from the formulas:

$$\left[\left\{u_0, \begin{matrix} u_1 \\ \cdot \\ j_1 \end{matrix}, \ldots, \begin{matrix} u_r \\ \cdot \\ j_r \end{matrix}\right\}, x\right] = \left\{[u_0, x], \begin{matrix} u_1 \\ \cdot \\ j_1 \end{matrix}, \ldots, \begin{matrix} u_r \\ \cdot \\ j_r \end{matrix}\right\}$$

$$+ \sum_{i=1}^{r} \left\{u_0, \begin{matrix} u_1 \\ \cdot \\ j_1 \end{matrix}, \ldots, \begin{matrix} u_i \\ \cdot \\ j_i - 1 \end{matrix}, \begin{matrix} [u_i, x] \\ 1 \end{matrix}, \ldots, \begin{matrix} u_r \\ \cdot \\ j_r \end{matrix}\right\},$$

$$u_s = u_t \to \left\{u_0, \begin{matrix} u_1 \\ \cdot \\ j_1 \end{matrix}, \ldots, \begin{matrix} u_r \\ \cdot \\ j_r \end{matrix}\right\} =$$

$$\binom{j_s + j_t}{j_s} \left\{u_0, \begin{matrix} u_1 \\ \cdot \\ j_1 \end{matrix}, \ldots, \begin{matrix} u_s \\ \cdot \\ j_s + j_t \end{matrix}, \ldots, \begin{matrix} u_{t-1} \\ \cdot \\ j_{t-1} \end{matrix}, \begin{matrix} u_{t+1} \\ \cdot \\ j_{t+1} \end{matrix}, \ldots, \begin{matrix} u_r \\ \cdot \\ j_r \end{matrix}\right\}.$$

All elements of the form $[X, Y^n]$ are contained in I_n. Indeed, by linearity of $[X, Y^n]$ for X we can assume that $X = u_0$ is a basic element. Further, if $Y = a_1 u_1 + \ldots + a_r u_r$, where $a_i \in F$ and the u_i are distinct basic elements, then we have

$$[X, Y^n] = [u_0, (a_1 u_1 + \ldots, + a_r u_r)^n] =$$

$$= \sum_{j_1 + \ldots + j_r = n} a_1^{j_1} \ldots a_r^{j_r} \left\{u_0, \begin{matrix} u_1 \\ \cdot \\ j_1 \end{matrix}, \ldots, \begin{matrix} u_r \\ \cdot \\ j_r \end{matrix}\right\}.$$

Consequently, the Engel ideal E_n generated by all elements of the form $[X, Y^n]$ is contained in I_n. Then

$$E_n^k = E_n \cap L^k \subseteq I_n \cap L^k = I_n^k.$$

Let us estimate the dimension of I_n^k. It is enough to estimate the number of elements of the form (20) with the above-mentioned conditions (21). Each sum (20)

$$\left\{u_0, \begin{matrix} u_1 \\ \cdot \\ j_1 \end{matrix}, \ldots, \begin{matrix} u_r \\ \cdot \\ j_r \end{matrix}\right\}$$

corresponds to a formal symbol

$$\left(u_0, \begin{matrix} u_1 \\ \cdot \\ j_1 \end{matrix}, \ldots, \begin{matrix} u_r \\ \cdot \\ j_r \end{matrix}\right). \tag{22}$$

For given $u_0, u_1, \ldots, u_r$, $j_1, \ldots, j_r$ there are $r!$ distinct symbols of the form

$$\left(u_0, \begin{matrix} u_{\sigma(1)} \\ \cdot \\ j_{\sigma(1)} \end{matrix}, \ldots, \begin{matrix} u_{\sigma(r)} \\ \cdot \\ j_{\sigma(r)} \end{matrix}\right) \quad (\sigma \in S_r)$$

that correspond to the same sum of the form (20). By (19), for a given r the number of all symbols of the form (22), where $u_0, u_1, \ldots, u_r$ are basic elements, $j_1, \ldots, j_r$ - positive integers, satisfying (21), is less than or equal to

$$\sum_{\substack{j_1+\ldots+j_r=n\\ j_1>0,\ldots,j_r>0}} \quad \sum_{k_0+j_1k_1+\ldots+j_rk_r=k} \frac{1}{k_0}2^{k_0}\frac{1}{k_1}2^{k_1}\ldots\frac{1}{k_r}2^{k_r}.$$

By alternating r, we obtain the following estimation:

$$\dim I_n^k \le \sum_{r=1}^{n}\frac{1}{r!}\cdot \sum_{\substack{j_1+\ldots+j_r=n\\ j_1>0,\ldots,j_r>0}} \quad \sum_{k_0+j_1k_1+\ldots+j_rk_r=k} \frac{2^{k_0+k_1+\ldots+k_r}}{k_0k_1\ldots k_r}$$

From $k_0 + j_1k_1 + \ldots + j_rk_r = k$ and $j_1 + \ldots + j_r = n$, we obtain

$$\begin{aligned} k_0 + k_1 + \ldots + k_r &= k - (j_1-1)k_1 - \ldots - (j_r-1)k_r \\ &\le k - (j_1-1) - \ldots - (j_r-1) = k - n + r. \end{aligned}$$

Using that, we deduce the following inequalities:

$$\begin{aligned} \dim I_n^k &\le \sum_{r=1}^{n}\frac{1}{r!}\cdot \sum_{\substack{j_1+\ldots+j_r=n\\ j_1>0,\ldots,j_r>0}} \quad \sum_{k_0+j_1k_1+\ldots j_rk_r=k} \frac{2^{k-n+r}}{k_0k_1\ldots k_r} \\ &\le 2^{k-n}\sum_{r=1}^{n}\frac{2^r}{r!}\cdot \sum_{\substack{j_1+\ldots+j_r=n\\ j_1>0,\ldots,j_r>0}} \quad \sum_{\substack{1\le k_i\le k\\ i=1,\ldots,r}} \frac{1}{k_0k_1\ldots k_r} \\ &= 2^{k-n}\sum_{r=1}^{n}\frac{2^r}{r!}\cdot\binom{n-1}{r-1}\cdot\left(1+\frac{1}{2}+\ldots+\frac{1}{k}\right)^{r+1} \\ &\le 2^{k-n}\sum_{r=1}^{n}\frac{2^r(\ln(k)+1)^{r+1}}{r!}\cdot\binom{n-1}{r-1} \\ &= 2^{k-n}\left(2(\ln(k)+1) + \sum_{r=2}^{n}\frac{2^r(\ln(k)+1)^{r+1}}{r!}\binom{n-1}{r-1}\right) \\ &\le 2^{k-n+1}(\ln(k)+1)^2\cdot\sum_{r=1}^{n-1}\frac{2^r(\ln(k)+1)^r}{r!}\binom{n}{r}. \end{aligned}$$

Since $\dim E_n^k \leq \dim I_n^k$ we obtain

$$\dim E_n^k \leq 2^{k-n+1}(\ln(k)+1)^2 \sum_{r=1}^{n-1} \frac{2^r(\ln(k)+1)^r}{r!}\binom{n}{r}. \tag{23}$$

By Stirling's formula we have

$$\sqrt{2\pi m}\left(\frac{m}{e}\right)^m \leq m! \leq e\sqrt{2\pi m}\left(\frac{m}{e}\right)^m.$$

By substituting these inequalities into (23) we get:

$$\begin{aligned}\dim E_n^k &\leq 2^{k-n+1}(\ln(k)+1)^2 \sum_{r=1}^{n-1} \frac{2e(\ln(k)+1)^r e\sqrt{2\pi n}n^n}{\sqrt{2\pi r}r^r 2\pi\sqrt{r(n-r)}r^r(n-r)^{n-r}} \\ &< 2^{k-n+1}(\ln(k)+1)^2 \sum_{r=1}^{n-1} \frac{2en(\ln(k)+1)^r}{r^{2r}}\left(\frac{n}{n-r}\right)^{n-r}.\end{aligned}$$

Since $(\frac{n}{n-r})^{n-r} = (1+\frac{r}{n-r})^{n-r} \leq e^r$, we obtain:

$$\dim E_n^k < 2^{k-n+1}(\ln(k)+1)^2 \sum_{r=1}^{n-1} \frac{2e^2 n(\ln(k)+1)^r}{r^{2r}}. \tag{24}$$

Denote $\gamma_2(n)+1$ by $\gamma(n)$. Clearly $\gamma(n)$ is a minimal h such that $L^h \subseteq E_n^h$. Therefore

$$\dim L^{\gamma(n)} \leq \dim E_n^{\gamma(n)} \tag{25}$$

For Theorem 2 it is enough to prove that there exists N such that for all $n > N$,

$$\gamma(n) \geq e^{cn\ln 2}, \qquad c = 0.0741. \tag{26}$$

Suppose that it is not true. Then there exists an infinite increasing sequence

$$n_1 < n_2 < \ldots < n_i < \ldots,$$

such that for $\alpha = 0.0741\ln 2$,

$$\gamma(n_i) < e^{\alpha n_i} \qquad (i = 1, 2, \ldots). \tag{27}$$

Using (24) for $n = n_1$ and $k = \gamma(n_1)$, we obtain

$$\dim E_n^{\gamma(n_1)} < 2^{\gamma(n_1)-n_1+1}(\alpha n_i + 1)^2 \sum_{r=1}^{n-1} \frac{2e^2 n_i(\alpha n_i + 1)^r}{r^{2r}}$$

$$\leq 2^{\gamma(n_i)-n_i+1} n_i(\alpha n_i + 1)^2 \max_{1 \leq x \leq n_i} \frac{2e^2 n_i(\alpha n_i + 1)^x}{x^{2x}}.$$

It is easy to check that a derivation of the function occurring under symbol "max" has a value 0 at the point

$$x_i^0 = \sqrt{2n_i(\alpha n_i + 1)}.$$

Clearly, this point lies in the interval $[1, n_{i-1}]$ for all $n_i > 5$ and the above-mentioned function has a maximum at $x = x_i^0$. Therefore for $n_i > 5$ we have

$$\dim E_n^{\gamma(n_i)} < 2^{\gamma(n_i)-n_i+1} n_i^3 e^{2\sqrt{2n_1(\alpha n_i+1)}}. \tag{28}$$

Using (19), (25) and (28), we obtain

$$\frac{1}{2\gamma(n)} 2^{\gamma(n_i)} < 2^{\gamma(n_i)-n_i+1} n_i^3 e^{2\sqrt{2n_i(\alpha n_i+1)}}$$

and then

$$\gamma(n_i) \geq e^{n_i \ln 2 - 2\sqrt{2n_i(\alpha n_i+1)} - \ln(4n_i^9)}.$$

Further, by (27) we obtain

$$\alpha n_i > n_i \ln 2 - 2\sqrt{2n_i(\alpha n_i + 1)} - \ln(4n_i^9).$$

Dividing by n_i and taking the limit as $i \to \infty$, we get

$$\alpha \geq \ln 2 - 2\sqrt{2\alpha}.$$

It is easy to check that $\alpha = 0.0741 \ln 2$ does not satisfy last inequality, i.e. we have obtained a contradiction. Therefore inequality (26) must be true for suitable N and all $n > N$. Theorem 2 is proved.

References

[1] W. Burnside, On an unsettled question in the theory of discontinuous *groups, Quart. J. Math.* **33** (1902), 230 - 238.

[2] P.S. Novikov and S.I. Adian, Infinite periodic groups I. II, III, *Izv. Akad. Nauk SSSR Ser. Mat.* **32** (1968), 212 - 244, 251 - 524, 709 - 731; English transl. in Math. USSR Izv. **2** (1968).

[3] S.I. Adian, *The Burnside problem and identities in groups*, "Nauka", Moscow, 1975; English transl. Springer, 1978.

[4] I.N. Sanov, Solution of Burnside's problem for exponent 4, Leningrad. *Gos. Univ. Uchen. Zap. Ser. Mat.* **10** (1940), 166 - 170.

[5] M. Hall Jr., Solution of the Burnside problem for exponent six. *Illinois J. Math.* **2** (1958), 764 - 786.

[6] B. Chandler and W. Magnus, *The history of combinatorial group theory: A case study in the history of ideas*, Springer, 1982.

[7] S.I. Adian, Investigations on the Burnside problem and questions connected with it, *Trudy Mat. Inst. Steklov Akad. Nauk SSSR* **168** (1984), 171 -196; English transl. in *Proc. Steklov Inst. Math.* 1986, Issue 3 (**168**).

[8] M.R. Vaughan-Lee, The restricted Burnside problem, *Bull. London Math. Soc.* **17** (1985), 113 - 133,

[9] A.G. Kurosh, Problems in ring theory connected with the Burnside problem on periodic groups, *Izv. Akad. Nauk SSSR Ser. Mat.* **5** (1941), 233 - 240 (Russian).

[10] E.S. Golod, Some problems of Burnside type, Proceedings Internat. Congr. *Math. (Moscow, 1966)*, "Mir", Moscow, 1968, pp.284 - 289; English transl. in Amer. Math. Soc. Transl. (2) **84** (1969).

[11] P. Hall and G. Higman, On the p-length of p-soluble groups and reduction theorems for Burnside's problem, *Proc. London Math. Soc.* **6** (1956), 1 - 42.

[12] W. Magnus, A connection between the Baker-Hausdorff formula and a problem of Burnside, *Ann. Math.* **5** (1950), 111 - 126.

[13] W. Magnus, Über Gruppen und zugeordnete Liesche Ringe, *J. reine angew. Math.* **182** (1940), 142 - 149.

[14] H. Zassenhaus, Über Liesche Ringe mit Primzahlcharakteristik, *Abh. Math. Sem. Hans. Univ. Hamburg* 13 1939, 1 - 100.

[15] O. Grün, Zusammenhang zwischen Potenzbildung und Kommutatorbildung, *J. reine angew. Math.* **182** (1940), 158 - 177.

[16] I.N. Sanov, Establishment of a connection between periodic groups with period a prime number and Lie rings, *Izv. Akad Nauk SSSR Ser. Mat.* **16** (1952), 23 - 58 (Russian).

[17] A.I. Kostrikin, Lie rings satisfying the Engel condition, *Dokl. Akad. Nauk SSSR* **108** (1956), 580 - 582; English transl. in *Amer. Math. Soc. Transl.* (2) **45** (1965).

[18] G. Higman, On finite groups of exponent five, *Proc. Cambridge Philos. Soc.* **52** (1956), 381 - 390.

[19] A.I. Kostrikin, The Burnside problem, *Dokl. Akad. Nauk SSSR* **119** (1958), 1081 - 1084 (Russian).

[20] A.I. Kostrikin, The Burnside problem, *Izv. Akad. Nauk SSSR Ser. Mat.* **23** (1959), 3 - 34; English transl. in *Amer. Math. Soc. Transl.* (2) **36** (1964).

[21] A.I. Kostrikin, Sandwiches in Lie algebras, *Mat. Sbornik* **110** (1979), 3 - 12; English transl. in *Math. USSR Sb.* **38** (1981).

[22] A.I. Kostrikin, *Around Burnside*, "Nauka", Moscow, 1986 (Russian).

[23] S.I. Adian and A.A. Razborov, Periodic groups and Lie algebras, *Uspekhi Mat. Nauk* **42:2** (1987), 3 - 68; English transl. in *Russian Math. Surveys* **42:2** (1987).

[24] B. Chang, On Engel rings of exponent p-1 over GF(p), *Proc. London Math. Soc.* (Ser. 3) **11** (1961), 203 - 212.

[25] F.M. Malyshev, The class of nilpotence of Engel Lie algebras, *Moscow Gos. Univ. Vestnik Ser. Mat. Mekh.* **2** (1980), 55 - 58.

[26] A.I. Kostrikin, Some related questions in the theory of groups and Lie

algebras, *Proc. Second Internat. Conf. Theory of Groups (Canberra, 1973)*, Springer, 1974, pp.40 - 46.

[27] S.I. Adian and N.N. Repin, Exponential lower estimate of the degree of nilpotency of Engel Lie algebras, *Mat. Zametki* **39** (1986), 444 - 452; English transl. in *Math. Notes Acad. Sciences USSR* **39** (1986).

[28] S.I. Adian, Normal subgroups of free periodic groups, *Izv. Akad. Nauk SSSR Ser. Mat.* **45** (1981), 931 - 947.

Steklov Mathematical Institute
Moscow
U.S.S.R.

Recent progress on rewritability in groups*

Russell D. Blyth and *Derek J.S. Robinson*

1. The rewriting properties $\mathcal{P}_n, \mathcal{Q}_n$

Let n be an integer > 1. Then a group G is said to have the property $\mathcal{Q}_n$, (or to be *n-rewritable*), if, for each n-element subset $\{x_1, \ldots, x_n\}$ of G, there are distinct permutations $\pi, \sigma \in S_n$ such that

$$x_{\pi(1)} x_{\pi(2)} \cdots x_{\pi(n)} = x_{\sigma(1)} x_{\sigma(2)} \cdots x_{\sigma(n)}.$$

If one of π, σ can be always chosen to be the identity, then G has the property $\mathcal{P}_n$, (or is *totally n-rewritable*). We shall also use $\mathcal{P}_n$ and $\mathcal{Q}_n$ to denote the classes of groups with these properties.

Obviously $\mathcal{P}_2 = \mathcal{Q}_2 =$ the class of abelian groups; equally clear are the inclusions $\mathcal{P}_n \subseteq \mathcal{P}_{n+1}$, $\mathcal{Q}_n \subseteq \mathcal{Q}_{n+1}$, $\mathcal{P}_n \subseteq \mathcal{Q}_n$. In addition, we define $\mathcal{P} = \bigcup_{n=2,3,\ldots} \mathcal{P}_n$ and $\mathcal{Q} = \bigcup_{n=2,3,\ldots} \mathcal{Q}_n$, so that $\mathcal{P} \subseteq \mathcal{Q}$.

2. Background from automata theory

Rewriting properties first arose in the context of semigroups - see Restivo and Reutenauer [11]. The motivation seems to have come from automata theory. We give a brief explanation of the connection and refer the reader to [1] for a detailed account of the role of semigroups in automata theory.

Let A be a state-output automaton with input set I, state set S and next state function $\mu : S \times I \to S$ (it is understood that the output is just the next state). Then there is a corresponding submonoid of the monoid $\mathrm{Fun}(S)$ of all functions on S, namely the submonoid generated by all $\varphi_i \in \mathrm{Fun}(S)$, where $(s)\varphi_i = \mu(s, i)$, $s \in S$, $i \in I$. Conversely, if M is a submonoid of $\mathrm{Fun}(S)$, where S is any non-empty set, there is a corresponding state-output

*This is an extended version of the conference lecture given by the second author.

automaton with input set and state set both equal to S and next state function μ given by $\mu(s,i) = si$, $(i, s \in S)$. Thus there is a correspondence between state-output automata with state set S and submonoids of Fun(S).

Suppose, for example, that a monoid M has a rewriting property $\mathcal{P}_n$. Identify M with a submonoid of Fun(M), by right multiplication, and form the corresponding automaton A_M. Then A_M has the property that, for any sequence of n input strings $u_1, u_2, \ldots, u_n$, the same next state is produced by some permutation of these strings, $u_{\pi(1)}, u_{\pi(2)}, \ldots, u_{\pi(n)}$, $1 \neq \pi \in S_n$. Conversely, this property of input strings for A_M implies that M has $\mathcal{P}_n$.

3. Basic results

We list some fundamental - and for the most part elementary - results on rewritability. Unless otherwise indicated they can be found in [3] and [4].

(i) *If a group G does not have $\mathcal{Q}_n$, then $|G| > n!$. Hence the symmetric group S_n has $\mathcal{Q}_n$.*

That S_n does not have $\mathcal{Q}_{n-1}$ is shown by the subset $\{(1,2), (1,3), \ldots, (1,n)\}$. Similarly it can be shown that the alternating group A_n does not have $\mathcal{Q}_{n-2}$, $n \geq 4$, by considering $\{(1,2,3),\ (1,2,4), \ldots, (1,2,n)\}$. It seems likely that A_n does not even have $\mathcal{Q}_{n-1}$ if $n \geq 4$.

Rewriting numbers

If G is a group with $\mathcal{P}$, define the *$\mathcal{P}$-rewriting number* of G,

$$p(G),$$

to be the smallest m such that G has $\mathcal{P}_m$.

Also let

$$p(n) = sup\{p(G) \mid |G| = n\}.$$

(There are, of course, corresponding numbers $q(G)$, $q(n)$ for the property $\mathcal{Q}$).

The following results provide useful, if crude, estimates for rewriting numbers.

(ii) *Let H be a subgroup with finite index m in a group G. If H has $\mathcal{P}_r$ (respectively, $\mathcal{Q}_r$), then G has $\mathcal{P}_{mr}$ (respectively, $\mathcal{Q}_{mr}$).*

(iii) *Let G be a group such that $m = |G'|$ is finite. Then G has $\mathcal{P}_{m+1}$. Also, if $m < n!$, then G has $\mathcal{Q}_n$.*

(iv) That $p(G)$ is usually much smaller than $|G|$ is shown by

Theorem 1 (Brandl [12], Piochi and Pirillo, written communcation). *Let $\epsilon > 0$. Then there is a positive integer $N(\epsilon)$ such that $\frac{p(G)}{|G|} < \epsilon$ for all finite groups G of order $\geq N(\epsilon)$. Thus $\lim_{n\to\infty} \left(\frac{p(n)}{n}\right) = 0$.*

Proof. First note the easily proved result : if every abelian subgroup of a finite group H has order $\leq m$, then $|H| \leq f(m)$ for some function f.

For a given $\epsilon > 0$ define $N(\epsilon)$ to be $1 + [f(2/\epsilon)]$. Let G be a finite group with $|G| \geq N(\epsilon)$; then G has an abelian subgroup A of order $> 2/\epsilon$. By (ii) $p(G) \leq p(A) \,.\, |G : A| = 2|G : A|$, so that $p(G)/|G| \leq 2/|A| < \epsilon$, as required.

(v) The method of the preceding proof provides upper bounds for $p(G)/|G|$ when G is finite. For example, $p(G)/|G| \leq 2$ for all G, and $p(G)/|G| \leq \frac{1}{2}$ if $|G| > 6$. The point here is that if G has an abelian subgroup A of order ≥ 4, then $p(G)/|G| \leq 2/|A| \leq \frac{1}{2}$; otherwise $|G| \leq 6$.

An immediate consequence of (ii) and (iii) is

(vi) *Every finite-by-abelian-by-finite group is in $\mathcal{P}$, i.e. $\mathcal{FAF} \subseteq \mathcal{P}$.*

In view of the structure theorems described later, it is important to confirm that

(vii) $\mathcal{P}_n \neq \mathcal{Q}_n$ *if* $n > 2$.

The relevant example is a metacyclic group $G = \langle x \rangle \ltimes \langle a \rangle$, (the semidirect product), where $|x| = j_n$, $a^x = a^2$, $|a| = 2^{j_n} - 1$ and $j_n = [\log_2(n!)]$. Since $|G'| = |a| < n!$, the group G has $\mathcal{Q}_n$. However, G does not have $\mathcal{P}_n$. Indeed for $n \geq 6$, $(xa^{n-1}, xa^{n-2}, \ldots, xa, x)$ cannot be rewritten. For smaller n other non-rewritable sequences were found by computer search ([3], [4]).

(viii) *Let the group G be a product $G_1 G_2 \ldots G_m$ of non-abelian subgroups G_i such that $G' = \underset{i=1,2,\ldots,m}{Dr} G'_i$ and $[G_i, G_j] = 1$ if $i \neq j$. If G has $\mathcal{Q}_n$, then $n(n-1) > 2m$.*

In particular, this may be applied to the direct product $\underset{i=1,2,\ldots,m}{Dr} G_i$. This result plays an essential part in the proofs of Theorems 3 and 4 below.

(ix) *If G is a finite group with $\mathcal{Q}_n$ and the Sylow p-subgroups of G are non-abelian, then $p \leq f(n)$ for some function $f(n)$ of n.*

For $n = 3, 4$ the smallest values of $f(n)$ are 5 and 23 respectively (these values arise by considering a nonabelian group of order p^3). For $n > 4$ only upper bounds are known.

4. Structure theorems

The structure of $\mathcal{P}$-groups was determined by Curzio, Longobardi, Maj and Robinson [6].

Theorem 2. *A group has the property $\mathcal{P}$ if and only if it is finite-by-abelian-by-finite; thus $\mathcal{P} = \mathcal{FAF}$.*

Subsequently Blyth [3], [4] generalized this, obtaining the structure of $\mathcal{Q}$-groups; remarkably, $\mathcal{Q}$ turned out to be the same property as $\mathcal{P}$, despite (vii) above.

Theorem 3. *A group has the property $\mathcal{Q}$ if and only if it is finite-by-abelian-by-finite; thus $\mathcal{P} = \mathcal{Q} = \mathcal{FAF}$.*

It is an interesting comment on the difference between groups and semigroups that $\mathcal{P}$ and $\mathcal{Q}$ are different properties for semigroups. Indeed Pirillo [10] has shown that even $\mathcal{Q}_3$ does not imply $\mathcal{P}$ for semigroups.

The proofs of the structure theorems, especially the second, are too complex to analyze here. However, as one might expect, the basic argument is a pigeonhole principle.

5. Semisimple groups

Since all abelian groups have $\mathcal{P}_2$, there is no bound for the order of a finite group with $\mathcal{Q}_n$ or $\mathcal{P}_n$. However, the situation is quite different for semisimple groups, i.e. groups without non-trivial abelian normal subgroups.

Theorem 4 (Blyth [3], [4]). *If G is a semisimple group with $\mathcal{Q}_n$, then G is finite and $|G| \leq f(n)$ for some function f of n.*

A similar result for $\mathcal{P}_n$ was obtained in [2]. To prove Theorem 4 one uses Theorem 3 to reduce to the case where G is finite. A simple argument using (viii) above and the completely reducible radical of G further reduces to the case where G is simple. Blyth used the classification of finite simple groups at this point. However, this can be avoided by employing the following result, found independently by R. Brandl and D. Robinson.

Proposition. *Let G be a group and let n be an integer* > 2.

(i) *If $G \in \mathcal{P}_n \setminus \mathcal{P}_{n-1}$, there is an element $c \neq 1$ in G such that* $|G : C_G(c)| \leq (n-1)(n-1)!$.

(ii) *If $G \in \mathcal{Q}_n \setminus \mathcal{Q}_{n-1}$, there is an element $c \neq 1$ in G such that*

$$|G : C_G(c)| \leq \frac{1}{2}(n-1)!((n^2-1)(n-1)! - n + 1).$$

6. The identification of $\mathcal{P}_n$ and $\mathcal{Q}_n$ for small n

In a few cases detailed information is available about $\mathcal{P}_n$ and $\mathcal{Q}_n$.

(a) *A group G has $\mathcal{P}_3$ if and only if* $|G'| \leq 2$ (Curzio, Longobardi, Maj [5]).

(b) **Theorem 5** (Longobardi and Maj [7]). *$\mathcal{P}_4$-groups are metabelian.*

Recently Longobardi and Maj, in collaboration with S. E. Stonehewer, have given a complete classification of $\mathcal{P}_4$-groups. This is too complicated to reproduce, but, briefly put, $\mathcal{P}_4$-groups either have derived subgroup of order ≤ 8 or else they have an abelian subgroup of index 2.

Current knowledge about $\mathcal{Q}_n$-groups for specific n is less precise. Longobardi and Maj have shown

(c) *$\mathcal{Q}_3$-groups are metabelian.*

A complete classification of these groups is not available at present. Concerning $\mathcal{Q}_4$ there is the following result.

(d) **Theorem 6** (Blyth [3], [4]). *$\mathcal{Q}_4$-groups are soluble.*

It seems likely that a complete classification of $\mathcal{Q}_4$-groups will be difficult.

7. Solubility and $\mathcal{P}_n$

Probably the most important open problem in rewritability theory is :

What is the largest integer n such that $\mathcal{P}_n$ implies solubility?

Thus $n+1$ is the smallest integer such that $\mathcal{P}_{n+1}$ contains a non-abelian simple group. Note that the corresponding number for $\mathcal{Q}$ is definitely 4, by Theorem 6 and the fact that A_5 has $\mathcal{Q}_5$.

An easy upper bound for n is 19. Indeed A_5 certainly has $\mathcal{P}_{20}$ because A_4 has $\mathcal{P}_4$ and (ii) can be applied. In fact, by Theorem 7 below, $n \geq 6$.

Any successful assault on the solubility problem must involve a two-pronged attack.

(I) In order to prove that $\mathcal{P}_m$ implies solubility for some m, one observes that it suffices to show that none of the minimal simple groups have $\mathcal{P}_m$.

The minimal simple groups are :

$PSL(2,2^\iota)$, ι a prime,
$PSL(2,3^\iota)$, ι an odd prime,
$PSL(2,p)$, $p=5$ or p a prime >5 with $p\equiv\pm2 \pmod 5$,
$PSL(3,3)$,
$Sz(2^{2\iota+1})$, $2\iota+1$ a prime.

The Proposition (in §5) excludes all but a finite number of these groups. The remaining groups must be dealt with on a case-by-case basis with the aid of machine computation.

Using such techniques, the authors have proved

Theorem 7. *Every $\mathcal{P}_6$-group is soluble.*

The computations needed to establish this were performed using CAYLEY on the SUN 3/50 in the Symbolic Computation Laboratory at the University of Illinois in Urbana-Champaign. (R. Brandl has informed the authors that he had shown that $\mathcal{P}_5$ implies solubility.)

The authors are presently working on the solubility problem for $\mathcal{P}_7$. In this case it is possible to eliminate all but 24 of the minimal simple groups, thereby leaving a still formidable computational problem.

(II) One tries to discover $k(A_5)$ in the hope that this will turn out to equal $n+1$. It should be emphasized that showing that A_5 belongs to a particular class $\mathcal{P}_m$ can involve a massive amount of computation. It has been shown that A_5 *does not have* $\mathcal{P}_7$; in fact the following 7-tuple does not rewrite :

$$(1,2)(3,4),\ (2,5)(3,4),\ (1,4)(2,3),\ (1,2,3,5,4),\ (1,5,4,3,2),\ (2,4)(3,5),\ (1,3)(2,4).$$

So far the computational evidence suggests a small value for $k(A_5)$, possibly 8 or 9.

8. Bounds for derived length

Quite recently, Longobardi and Maj [8] have given interesting upper bounds for the derived lengths of soluble $\mathcal{P}_n$- and $\mathcal{Q}_n$-groups.

Theorem 8. *Let G be a soluble group with derived length d. Let $n \geq 4$.*

(i) *If $G \in \mathcal{P}_n$, then $d \leq 2n - 6$.*

(ii) *If $G \in \mathcal{Q}_n$, then $d \leq \frac{1}{2}(n-2)(n^2 - n - 2)$.*

For example, Theorems 7 and 8 show that $\mathcal{P}_6$-groups are soluble with derived length ≤ 6.

The first part of Theorem 8 follows from Longobardi and Maj's result that $\mathcal{P}_4$-groups are metabelian and the following lemma, which is obviously well suited to induction arguments.

Lemma. *If a group G has $\mathcal{P}_n$, $n > 2$, and N is a non-abelian normal subgroup, then G/N has $\mathcal{P}_{n-1}$.*

Finally, we remark that Theorem 8 shows that an arbitrary $\mathcal{Q}_n$-group has a unique maximal soluble normal subgroup S. Furthermore, G/S is semisimple and so, by Theorem 4, it is finite of bounded order.

Corollary. *Let $G \in \mathcal{Q}_n$. Then there is a maximal soluble normal subgroup S of G. Moreover $|G : S| \leq f(n)$ and S has derived length $\leq g(n)$ for certain functions f, g.*

References

[1] M.F. Arbib, *Theories of abstract automata*, Prentice-Hall : Englewood Cliffs, N.J., (1969).

[2] M. Bianchi, R. Brandl, and A.G.M. Mauri, On the 4-permutational property for groups, *Arch. Math.* (Basel) **48** (1987), 281-285.

[3] R.D. Blyth, *Rewriting products of group elements*, Ph.D. thesis, University of Illinois at Urbana-Champaign (1987).

[4] ——, Rewriting products of group elements I, II, *J. Algebra*, to appear.

[5] M. Curzio, P. Longobardi, and M. Maj, Su di un problema combinatorio in teoria dei gruppi, *Atti. Accad. Naz. Lincei Rend. Cl. Sci. Fis. Mat. Natur.* (8) **74** (1983), 136-142.

[6] M. Curzio, P. Longobardi, M. Maj, and D.J.S. Robinson, A permutational property of groups, *Arch. Math.* (Basel) **44** (1985), 385-389.

[7] P. Longobardi and M. Maj, On groups in which every product of four elements can be reordered, *Arch. Math.* (Basel) **49** (1987), 273-276.

[8] ——, On the derived length of groups with some permutational properties, preprint.

[9] B. Piochi and G. Pirillo, A proof of a theorem by G. Higman, preprint.

[10] G. Pirillo, On permutational properties for semigroups, in *Group Theory*, Proceedings of a conference held at Bressanone, Lecture Notes in Math., vol. 1281, 118-119, Springer, Berlin (1987).

[11] A. Restivo and C. Reutenauer, On the Burnside problem for semigroups, *J. Algebra* **89** (1984), 102-104.

Additional reference

[12] R. Brandl, General bounds for permutability in finite groups, preprint.

Department of Mathematics and Computer Science
Saint Louis University
St. Louis, Missouri
U. S. A.

Department of Mathematics
University of Illinois
Urbana, Illinois
U. S. A.

Some finite groups with nontrivial centers which are Galois groups

Walter Feit[1]

§1. Introduction

Let G be a finite simple group with a cyclic Schur multiplier of order m. If $d|m$ then dG will denote a central extension of G with center of order d and $(dG)' = dG$. Thus dG is unique up to isomorphism. The following two theorems are the major results of this paper.

Theorem A. *Let L be a finite extension of* $\mathbf{Q}$ *such that*

$$h(z) = z^3 - 18z - 12$$

has a root in L. Then $6A_6$ is a Galois group over L.

Theorem B. *Let L be a finite extension of* $\mathbf{Q}$ *such that*

$$h(z) = z^3 - 63z + 147$$

has a root in L. Then $6A_7$ is a Galois group over L.

These theorems are consequences of stronger results, Theorems 7.4 and 8.6 below. The latter of these implies Theorem 8.7 which asserts that $6A_7$ is a Galois group over $\mathbf{Q}$. The methods used here do not seem to imply that $6A_6$ is a Galois group over $\mathbf{Q}$. However J.-F. Mestre [14] has constructed an example of a polynomial whose splitting field over $\mathbf{Q}$ can be imbedded in an extension M_0 with $\mathbf{Gal}(M_0/\mathbf{Q}) \simeq 6A_6$.

The basic approach in proving Theorems A and B is to find polynomials whose splitting fields M over L have $\mathbf{Gal}(M/L) \simeq A_n$, and extensions M_1, M_2 of M with $\mathbf{Gal}(M_1/L) \simeq 2A_n$ and $\mathbf{Gal}(M_2/L) \simeq 3A_n$. Thus $\mathbf{Gal}(M_1M_2/L) \simeq 6A_n$.

It is considerably simpler to construct extensions of number fields with Galois group $3A_n$. In fact the following result is proved in Section 2 by

[1] The work on this paper was partially supported by NSF grant DMS-8512904.

using rigidity.

Theorem C. *Let G be one of the following groups.*

$$A_6, A_7, M_{22}, J_3, McL, Suz, ON, Fi'_{24}.$$

If $G = J_3$ let L be the real subfield of the field of 9th roots of unity over $\mathbf{Q}$. *If $G \simeq McL$ let $L = \mathbf{Q}(\sqrt{-11})$. In all other cases let $L = \mathbf{Q}$.*

Then $3G$ is the Galois group of a regular extension of $L(T)$, where T is an indeterminate. Thus by the Hilbert irreducibility theorem $3G$ is a Galois group over every number field which contains L.

If M is a field constructed in Theorem C with $\mathbf{Gal}(M/\mathbf{Q}) \simeq A_n$ for $n = 6$ or 7 then M cannot be totally real, [17] Section 6. Thus this field cannot be imbedded in a field M_1 with $\mathbf{Gal}(M_1/\mathbf{Q}) \simeq 2A_n$, since $2A_n$ contains a unique involution. Hence M cannot be imbedded in a field M_0 with $\mathbf{Gal}(M_0/\mathbf{Q}) \simeq 6A_n$.

Observe that Fi_{22} is the only sporadic group whose Schur multiplier has order divisible by 3 that is missing from the list in Theorem C. If it can be shown that $\mathrm{Aut}(Fi_{22})$ is rigid with $k = 3$ conjugacy classes then the argument used in Section 2 will show that $3Fi_{22}$ is a Galois group of a regular extension $L(T)$ for a suitable number field L. This has apparently not yet been done.

Added in Proof. After seeing a preprint of this paper H. Pahlings informed me that he has shown that Aut (G) is rationally rigid for every sporadic simple group whose Schur multiplier has order divisible by 3. Hence Theorem C can be strengthened to assert the following:

If $G \simeq A_6$, A_7 or a sporadic simple group whose Schur multiplier has order divisible by 3 then $3G$ is the Galois group of a regular extension of $\mathbf{Q}(T)$.

Section 2 also contains the proof of the next result.

Theorem D. *Let $\tilde{G}$ be the universal central extension of $Sz(8)$. Thus the center of $\tilde{G}$ is noncyclic of order 4. Then $\tilde{G}$ is the Galois group of a* *regular extension of* $\mathbb{Q}(\sqrt{-3})(T)$*, where* T *is an indeterminate. Thus* $\tilde{G}$ *is a Galois group over any number field* L *which contains* $\sqrt{-3}$.

This paper owes much to correspondence and conversations with J.-P. Serre. In particular, he informed me of the results in Section 6 and pointed out the relevance of the duality theorem of Tate.

The notation used is standard. ν_p denotes the exponential p-adic valuation normalized so that $\nu_p(p) = 1$.

§2. Some consequences of rigidity

Rigidity and related concepts and their connection with the construction of Galois groups are due to Belyi [1], Fried [7], Matzat [11], and Thompson [20]. For convenience I will summarize the relevant results here. See [3] or [13].

Let C_1, C_2, C_3 be conjugacy classes of the finite group G. Define

$$A = A_G(C_1, C_2, C_3) = \{(x_1, x_2, x_3) \mid x_i \in C_i,\ x_1x_2x_3 = 1\}.$$
$$\mathbb{Q}(C_1, C_2, C_3) = \{\mathbb{Q}(\chi_u(x_i)) \mid i = 1, 2, 3;\ u = 1, 2, \ldots\},$$

where $\{\chi_u\}$ is the set of all irreducible characters of G.

If $y \in G$ and $(x_1, x_2, x_3) \in A$ then $(x_1^y, x_2^y, x_3^y) \in A$. Thus G acts as a permutation group on A.

Definition. $A = A_G(C_1, C_2, C_3)$ is *rigid* if

(i) $A \neq \emptyset$.

(ii) G acts transitively on A.

(iii) If $(x_1, x_2, x_3) \in A$ then $G = \langle x_1, x_2, x_3 \rangle$.

Lemma 2.1. *Suppose that the center of* G *is* $\langle 1 \rangle$. *Then*
$$A = A_G(C_1, C_2, C_3)$$
is rigid if and only if the following conditions are satisfied.

(i) $|G| = |A|$.

(ii) *If* $(x_1, x_2, x_3) \in A$ *then* $G = \langle x_1, x_2, x_3 \rangle$.

The following formula is well known.

$$(2.2)\qquad |A_G(C_1,C_2,C_3)| = \frac{|C_1|\,|C_2|\,|C_3|}{|G|} \sum_{\chi} \frac{\chi(x_1)\chi(x_2)\chi(x_3)}{\chi(1)}.$$

Here $x_i \in C_i$ and χ ranges over all the irreducible characters of G.

If L is an algebraic number field and T is an indeterminate, then an extension K of $L(T)$ is *regular* if L is algebraically closed in K.

Theorem 2.3. *let G be a finite group with center of order 1. Let C_1, C_2, C_3 be conjugacy classes of G such that $A_G(C_1,C_2,C_3)$ is rigid. Let L be a number field with $\mathbf{Q}(C_1,C_2,C_3) \subseteq L$. Then there exists a regular Galois extension K of $L(T)$ with $\mathrm{Gal}(K/L(T)) \simeq G$.*

Hilbert's irreducibility theorem now implies

Corollary 2.4. *Suppose that the assumptions of Theorem 2.3 are satisfied. Then G is a Galois group over L.*

Let F_0 be a function field of genus g_0 over the complex numbers and let F be a finite extension of F_0 with $[F : F_0] = n$. Let g be the genus of F. The following fundamental formula is due to Hurwitz. See e.g. [10] p. 26,

$$(2.5)\qquad 2g - 2 = n(2g_0 - 2) + \sum_{P_i}(e_i - 1),$$

where P_i ranges over all ramified places in F and e_i is the corresponding index of ramification.

For the next result see [12], p.372.

Theorem 2.6. *Let G be a finite group with center of order 1. Let $H \lhd G$ with $|G : H| = 2$ or 3. Let $C_1 \neq \{1\}$, C_2, C_3 be conjugacy classes of G with C_2, $C_3 \not\subseteq H$ such that $A_G(C_1,C_2,C_3)$ is rigid. Let K and L be as in Theorem 2.3 and let $L(T) \subseteq M \subseteq K$ where M corresponds to H. Then $M \approx L(T)$ and H is a Galois group over M. In particular H is a Galois group over L.*

An outline of the proof goes as follows. There are exactly 2 points of $\bar{L}(T)$ which ramify in $K\bar{L}$, where $\bar{L}$ is the algebraic closure of L. By the Hurwitz formula (2.5), K has genus 0. If $|G : H| = 2$, then at least one of the ramified points is rational in $L(T)$. If $|G : H| = 3$, then at least one of the ramified points has odd degree. Thus $M \simeq L(T)$.

H. Matzat has pointed out that [3] Theorem B, which is a stronger version of Theorem 2.6, has not been proved. Thus the corollary in [4] remains open.

Lemma 2.7. *Let G be a finite group with center of order 1. Let $\Phi(G)$ be the Frattini subgroup of G. If $S \subseteq G$, let $\bar{S}$ denote the image of S in $\bar{G} = G/\Phi(G)$. Let C_1, C_2, C_3 be conjugacy classes of G and let $x_i \in C_i$ for $i = 1, 2, 3$. Assume that the following are satisfied.*

(i) $\Phi(G) \cap C_i = \emptyset$ *for* $i = 1, 2, 3$.

(ii) $\Phi(G) \subseteq \mathbf{Z}(\mathbf{C}_G(x_1))$.

(iii) $\Phi(G) \cap \mathbf{C}_G(x_i) = \langle 1 \rangle$ *for* $i = 2, 3$.

(iv) $A_{\bar{G}}(\bar{C}_1, \bar{C}_2, \bar{C}_3)$ *is rigid.*

Then there exists a conjugacy class C_0 of G with $\bar{C}_0 = \bar{C}_1$ such that $A_G(C_0, C_2, C_3)$ is rigid.

Proof. Let $(\bar{y}_1, \bar{y}_2, \bar{y}_3) \in \bar{A} = A_{\bar{G}}(\bar{C}_1, \bar{C}_2, \bar{C}_3)$. Thus $y_1 y_2 y_3 = z$ is contained in $\Phi(G)$. Let $y_0 = z^{-1} y_1$ and let C_0 be the conjugacy class of G which contains y_0.

If $(u_0, u_2, u_3) \in A = A_G(C_0, C_2, C_3)$ then $(\bar{u}_0, \bar{u}_2, \bar{u}_3) \in \bar{A}$. Hence $\bar{G} = \langle \bar{u}_0, \bar{u}_2, \bar{u}_3 \rangle$ and so by a basic property of the Frattini subgroup, $G = \langle u_0, u_2, u_3 \rangle$. Hence it suffices to show that $|G| = |A|$ by Lemma 2.1.

Let X be the set of all irreducible characters of G and let Y be the set of all irreducible characters of G which have $\Phi(G)$ in their kernel. Let $x_i \in C_i$.

If $i = 2$ or 3 then (iii) implies that

$$\frac{|G|}{|C_i|} = |\mathbf{C}_G(x_i)| = |\mathbf{C}_{\bar{G}}(\bar{x}_i)| = \frac{|\bar{G}|}{|\bar{C}_i|}$$

and

$$\sum_{\chi \in X} |\chi(x_i)|^2 = |\mathbf{C}_G(x_i)| = |\mathbf{C}_{\bar{G}}(\bar{x}_i)| = \sum_{\chi \in Y} |\chi(\bar{x}_i)|^2.$$

Thus if $\chi \in X \backslash Y$ then $\chi(x_i) = 0$. Hence (2.2) yields

$$|A| = \frac{|C_0||C_2||C_3|}{|G|} \sum_{\chi \in X} \frac{\chi(x_0)\chi(x_2)\chi(x_3)}{\chi(1)}$$

$$= \frac{|C_0||\bar{C}_2||\bar{C}_3|}{|\bar{G}||\Phi(G)|} \sum_{\chi \in Y} \frac{\chi(x_1)\chi(x_2)\chi(x_3)}{\chi(1)} = \frac{|C_0||\bar{A}|}{|\bar{C}_1||\Phi(G)|} = \frac{|C_0|}{|\bar{C}_1|}|G|.$$

It remains to show that $|C_0| = |\bar{C}_1|$.

Since $\bar{C}_0 = \bar{C}_1$ it follows that $|C_0| \geq |\bar{C}_1|$. By (ii) $|C_1| = |\bar{C}_1|$. If $y_0 = z^{-1}y_1 \in C_0$ with $y_1 \in C_1$ and $z \in \Phi(G)$ then (ii) implies that $\mathbf{C}_G(y_1) \subseteq \mathbf{C}_G(y_0)$ and so $|C_0| \leq |C_1|$. □

Proof of Theorem C. Let H be one of the simple groups listed in Theorem C. In each case there exists a nonsplit central extension (unique up to isomorphism) $\tilde{H}$ of H with a center Z of order 3.

Furthermore there exists a group G with $|G : \tilde{H}| = 2$, $\mathbf{Z}(G) = \langle 1 \rangle$ and $\Phi(G) = Z$. If $H \not\simeq A_6$ then $G/Z \simeq \mathrm{Aut}(H)$. If $H \simeq A_6$ then G can be chosen so that $G/Z \simeq \Sigma_6$. See e.g. [2].

Let C_1, C_2, C_3 be the conjugacy class of Σ_n for $n \geq 5$, consisting of transpositions, $(n-1)$-cycles, n-cycles respectively. Then $A_{\Sigma_n}(C_1, C_2, C_3)$ is rigid.

In case $H \simeq M_{22}$, Suz, ON, Fi'_{24}, $A_{\bar{G}}(C_1, C_2, C_3)$ is rigid with conjugacy classes C_1, C_2, C_3 such that $\mathbf{Q}(C_1, C_2, C_3) = \mathbf{Q}$ and the hypotheses of Lemma 2.7 are satisfied. See [9], Theorem B.

In the remaining cases $A_{\bar{G}}(C_1, C_2, C_3)$ is rigid with conjugacy classes C_1, C_2, C_3 such that $\mathbf{Q}(C_1, C_2, C_3) = L$ and the hypotheses of Lemma 2.7 are satisfied. See [8] p. 280.

The result now follows from Lemma 2.7 and Theorem 2.3. □

Proof of Theorem D. Let $H = Suz(8)$. Let $\tilde{H}$ be the universal central extension of H. Then the center Z of $\tilde{H}$ is noncyclic of order 4. There exists an outer automorphism of H of order 3. By [19] this yields the existence of an outer automorphism σ of $\tilde{H}$ of order 3. [I am indebted to J. Alperin who drew my attention to this reference. He also provided an independent proof of the existence of σ.] Since $3 \not| |\bar{H}|$ this implies the existence of a group G such that

$$|G : \tilde{H}| = 3, \ \mathbf{Z}(G) = \langle 1 \rangle, \ Z = \Phi(G), \ \bar{G} = G/Z \simeq \mathrm{Aut}(H).$$

It is easily verified that $A_{\bar{G}}(\bar{C}_1, \bar{C}_2, \bar{C}_3)$ is rigid, where $\bar{C}_1, \bar{C}_2, \bar{C}_3$ consist of elements of order 2, 3, 15 respectively, and if $x_i \in C_i$ for $i = 1, 2$ then $x_2 x_3 \in \tilde{H}$. Furthermore $\mathbf{Q}(\bar{C}_1, \bar{C}_2, \bar{C}_3) = \mathbf{Q}(\sqrt{-3})$.

The result follows from Lemma 2.7 and Theorem 2.3. □

§3. An irreducibility criterion

The argument used in this section was suggested independently and at different times by M. Fried and J. -P. Serre. It depends on Faltings' theorem which asserts that a curve of genus $g > 1$ over an algebraic number field K has only finitely many points that are rational over K.

Theorem 3.1. *Let T be an indeterminate over the algebraic number field K. Let $G = G'$ be a finite group. Let $M = K(T, X(T))$ where $X(T)$ is an algebraic function of T. Let $L = K(T, f(X(T))) \subseteq M$ where f is a polynomial with coefficients in $K(T)$. Assume that M is a regular Galois extension of L with $\mathbf{Gal}(M/L) \simeq G$ and L has genus $g_0 \geq 1$.*

Then there are only finitely many elements $t \in K$ with
$$\mathbf{Gal}(K(X(t))) \,/K(f(X(t))) \not\simeq G.$$

Proof. Suppose that the result is false. Then there exist infinitely many values $t \in K$ so that $[K(X(t)) : K(f(X(t)))] < [M : L]$. Since the lattice of subgroups of G is finite this implies that there exist fields $E_i(T) = K(T, g_i(X(T))) \subseteq M$ for $i \neq 1,2$ such that $E_1 \neq E_2$ but $E_1(t) = E_2(t)$ for infinitely many values of T. Hence also $E_1(t)E_2(t) = E_2(t)$. Thus replacing $E_2(T)$ by $E_1(T)E_2(T)$ it may be assumed that $E_1(T) \subseteq E_2(T)$. Let g_i be the genus of $E_i(T)$.

By Faltings' theorem $g_2 \leq 1$. Thus $1 \leq g_0 \leq g_1 \leq g_2 \leq 1$. By Hurwitz's formula (2.5), this implies that E_2 is an unramified extension of L. Let F be the maximal unramified extension of L in M. Then F is a Galois extension of L. Thus $\mathbf{Gal}(F/L)$ is a homomorphic image of the algebraic fundamental group of the algebraic closure of L. See e.g. [13], Chapter I, Section 4. Since L has genus 1, this fundamental group is abelian. Hence $\mathbf{Gal}(F/L)$ is an abelian homomorphic image of $G = G'$. Thus $F = L$. Hence $E_1 = E_2 = L$ contrary to the choice of E_i. □

§4. Generalized Laguerre polynomials

Let λ be an indeterminate over $\mathbf{Q}$. For any integer j let $c_j = \lambda + j$. Define

$$F_n(z,\lambda) = z^n - nc_n z^{n-1} + \binom{n}{2} c_n c_{n-1} z^{n-2} + \cdots + (-1)^n c_n c_{n-1} \cdots c_1 = \sum_{j=0}^{n} (-1)^{n-j} \binom{n}{j} \Big(\prod_{i=j+1}^{n} c_i \Big) z^j. \tag{4.1}$$

Pólya and Szegö defined these polynomials and called them generalized Laguerre polynomials. Schur studied them and showed that if $\Delta_n(\lambda)$ is the discriminant of $F_n(z,\lambda)$ then

$$\Delta_n(\lambda) = n! \prod_{i=1}^{n} (ic_i)^{i-1}. \tag{4.2}$$

See [16], p. 229.

Theorem 4.3. *Let M be a splitting field of $F_n(z,\lambda)$ over $\mathbf{Q}(\lambda)$. Then M is a regular Galois extension of $\mathbf{Q}(\lambda)$ and $\mathbf{Gal}(M/\mathbf{Q}(\lambda)) \simeq \Sigma_n$.*

Proof. It suffices to show that if L is any finite extension of $\mathbf{Q}$ then $\mathbf{Gal}(ML/L) \simeq \Sigma_n$. We will show that as a permutation group on the roots of $F_n(z,\lambda)$, G contains a k cycle for all integers $k \le n$. It suffices to show that there exists $a \in L$ so that the Galois group of the splitting field of $F_n(z,a)$ over L contains a k-cycle.

Fix k. Choose a prime $p > n$ which does not ramify in L. Let $a = p - k$. Let L_0 be the completion of L at some prime divisor of p and let π be a prime in L_0. Let ν denote the extension of ν_p to L_0. Let N be a splitting field of $F_n(z,a)$ over L_0. Let H be the inertia group of N/L_0. Since N is tamely ramified, H is cyclic.

By (4.1) $F_n(z,a) \equiv z^k g(z) \pmod{\pi}$ where $g(0) \not\equiv 0 \pmod{\pi}$. Thus by Hensel's lemma $F_n(z,a) = g(z)h(z)$, where $h(z) \equiv z^k \pmod{\pi}$.

Let $\alpha_1, \cdots, \alpha_k$ be the roots of $h(z)$. Then $\nu(\alpha_i) \geq 1/k$ for each i. Thus by (4.2)

$$\nu\Big(\prod_{1 \leq i \leq j \leq k} (\alpha_i - \alpha_j)\Big) \geq \frac{k(k-1)}{2k} = \frac{k-1}{2} = \frac{1}{2}\nu(\Delta_n(a)).$$

Hence if Δ is the discriminant of $g(z)$, it follows that $\nu(\Delta) = 0$. Thus $g(z)$ defines an unramified extension of L_0, which is therefore in the fixed field of H. It remains to show that $k \mid |H|$.

Since $\nu(c_i) = 0$ for $i \neq k$ and $\nu(c_k) = 1$, (4.1) implies that the Newton polygon of $F_n(z,a)$ is the lower convex envelope of

$$(0,1), \;\ldots\; (k,1), \;(k+1,0), \;\ldots, \;(n,0).$$

This has a segment of slope $-1/k$. Hence $k \mid |H|$. □

Theorem 4.4. *Suppose that $n \geq 6$. Let L be an algebraic number field. Let $M(\lambda)$ be a splitting field of $F_n(z,\lambda)$ over $L(\lambda)$. Then* $\mathbf{Gal}(M(a)/L) \simeq$ *Σ_n or A_n for all but finitely many values of $\lambda = a$ in L.*

Proof. If $n \geq 6$ then (4.2) implies that at least 3 places ramify in $L(\lambda, \sqrt{\Delta_n}/L(\lambda))$. Thus $L(\lambda, \sqrt{\Delta_n})$ has genus $g_0 \geq 1$ by Hurwitz's formula (2.5). By Theorem 4.3 $M(\lambda)$ is a regular extension of $L(\lambda)$ with $\mathbf{Gal}(M(\lambda)/L(\lambda, \sqrt{\Delta_n})) \simeq A_n$. The result follows from Theorem 3.1. □

Theorem 4.4 implies [6] Theorem 10.5. After that paper appeared, I realized that the proof of Theorem 10.5 was incomplete. This now fills the gap.

For future reference we state here the following formulas which follow by direct computation from (4.1).

Let $w = z - c_n$ and let

$$H_n(w,\lambda) = F_n(z - c_n, \lambda). \tag{4.5}$$

Then

$$H_6(w,\lambda) = w^6 - 15c_6w^4 - 40c_6w^3 - 45c_6(\lambda+4)w^2 + 24c_6(5\lambda+24)w - 5c_6(3\lambda^2+10\lambda-24). \tag{4.6}$$

$$H_7(w,\lambda) = w^7 - 21c_7w^5 - 70c_7w^4 + 105c_5c_7w^3 + 84c_7(5\lambda+29)w^2 - 35c_7(3\lambda^2+16\lambda-11)w - 6c_7(35\lambda^2+336\lambda+757). \tag{4.7}$$

Of course $H_n(w,\lambda)$ has the same splitting field over $\mathbf{Q}(\lambda)$ as $F_n(z,\lambda)$.

§5. A class of generalized Laguerre polynomials

Consider the following two elliptic curves

$$E_6 : y^2 = 15x(x^2-4). \tag{5.1}$$

$$E_7 : y^2 = 105x(x^2-4). \tag{5.2}$$

Define

$$P_6 = (3,15)$$
$$P_7 = (5,105).$$

Then P_n is a point on E_n for $n = 6$ or 7.

For any integer m let mP_n denote the point on E_n given by the group law on the elliptic curve.

Lemma 5.3. *Let $mP_n = (x,y)$ with $y \neq 0$. Then the following hold.*

(I) *If $n = 6$ then $(x-2, x, x+2)$ is (x_1^2, x_2^2, x_3^2) if m is even and $(x_1^2, 3x_2^2, 5x_3^2)$ if m is odd for $x_i \in \mathbf{Q}^\times$.*

(II) *If $n = 7$ then $(x-2, x, x+2)$ is (x_1^2, x_2^2, x_3^2) if m is even and $(3x_1^2, 5x_2^2, 7x_3^2)$ if m is odd for $x_i \in \mathbf{Q}^\times$*

Proof. See e.g [6] Theorem 3.1. □

If $P = mP_n$ for some integer n define

$$H_n(w,P) = H_n(w, x-4) \in \mathbf{Q}[w],$$

where $H_n(w,\lambda)$ is defined by (4.1), (4.6) and (4.7).

Let $K(P)$ denote the splitting field of $H_n(w,P)$ over $\mathbf{Q}$. Then $K(P)$ is also the splitting field of $F_n(z,x-4)$ over $\mathbf{Q}$. Let $\Delta(P)$ denote the discriminant of $K(P)$.

Lemma 5.4. *Let $P=(x,y)=mP_n$. Let $x-4=a/b$, where a,b are relatively prime integers. Let $p>n$ be a prime such that $\nu_p(b)>0$. Then p does not ramify in $K(P)$.*

Proof. By Lemma 5.3 $\nu_p(b)=2s$ is even.

Let α be a root of $H_n(w,P)$ in a finite extension of $\mathbf{Q}_p$. Let ν also denote the extension of ν_p to $\mathbf{Q}_p(\alpha)$. Let $\nu(\alpha)=-t\in\mathbf{Q}$. We will show that

$$t\le s. \tag{5.5}$$

Suppose first that $n=6$. Then the values of the terms in $H_6(\alpha,P)$ are

$$-6t,\ -4t-2s,\ -3t-2s,\ -2t-4s,\ -t-4s,\ -6s. \tag{5.6}$$

If $t>s$, then $-6t$ is strictly less than any other term in (5.6) contrary to the fact that $H_6(\alpha,P)=0$.

Suppose that $n=7$. The values of the terms in $H_7(\alpha,P)$ are

$$-7t,\ -2s-5t,\ -2s-4t,\ -4s-3t,\ -4s-2t,\ -6s-t,\ -6s. \tag{5.7}$$

If $t>s$, then $-7t$ is strictly less than any other term in (5.7) contrary to the fact that $H_7(\alpha,P)=0$

Thus (5.5) is proved in all cases. Hence $p^s\alpha$ is a local integer in some finite extension of $\mathbf{Q}_p$. Hence

$$\nu(\Delta(P))\le\nu(\Delta), \tag{5.8}$$

where Δ is the discriminant of the polynomial whose roots are all of the form $p^s\alpha$ as α ranges over the roots of $H_n(w,P)$. By (4.2)

$$\Delta=p^{sn(n-1)}n!\prod_{i=1}^{n}(ic_i)^{i-1}. \tag{5.9}$$

Since $p > n$, it follows that $\nu(c_i) = -2s$ for all i. Thus (5.9) yields that

$$\nu(\Delta) = sn(n-1) - s\sum_{i=1}^{n-1} 2i = 0.$$

The result follows from (5.8). □

For future reference we state here the following result which is easily verified.

Lemma 5.10. *Let H be a subgroup of A_7 which is metacyclic such that $9 \mid |H|$ and $3 \mid |H : H'|$. Then $|H| = 9$.*

§6. Reduction to the local case

Let $L \subseteq K$ be number fields, where K is a Galois extension of L. For any valuation v of L, let L_v denote the completion at v. Similarly, let $K_{\bar{v}}$ denote the completion of K at any extension $\bar{v}$ of v.

Let $G = \mathbf{Gal}(K/L)$ and let $G_{\bar{v}} \approx \mathbf{Gal}(K_{\bar{v}}/L_v)$ denote the decomposition group of $\bar{v}$. If $\bar{v}_1, \bar{v}_2$ are extensions of v then $G_{\bar{v}_1}$ is conjugate to $G_{\bar{v}_2}$ in G.

Suppose that $\tilde{G} = \tilde{G}'$ is a group with center Z such that $\tilde{G}/Z \simeq G$. Let $\tilde{G}_{\bar{v}}$ denote the inverse image of $G_{\bar{v}}$ in $\tilde{G}$.

If $9 \not| |G|$ then $\tilde{G}_v \simeq Z \times G_v$. In view of this the next result follows for instance from [18] Section 1.

Theorem 6.1. *There exists a Galois extension M of L with $L \subseteq K \subseteq M$ and $\mathbf{Gal}(M/L) \simeq \tilde{G}$ if and only if for each valuation $\bar{v}$ of L with $9 \mid |G_v|$ there exists a Galois extension $M_{\bar{v}}$ of L_v with $L_v \subseteq K_{\bar{v}} \subseteq M_{\bar{v}}$ and an isomorphism $e_{\bar{v}} : \tilde{G}_{\bar{v}} \to \mathbf{Gal}(M_{\bar{v}}/L_v)$.*

I am indebted to J.-P. Serre who pointed out the next result:

Theorem 6.2. *Let L be a finite extension of $\mathbf{Q}$ which does not contain a primitive cube root of 1. Let $n = 6$ or 7 and let K be a Galois extension* *of L with $\mathbf{Gal}(K/L) = G \simeq A_n$. Let $\tilde{G} = 3A_n$. Let v be a valuation of L such that $9 \mid |G_v|$. If L_v does not contain a primitive cube root of 1 then there exists an extension $M_{\bar{v}}$ of L_v with $K_{\bar{v}} \subseteq M_{\bar{v}}$ and an isomorphism $e_{\bar{v}} : \tilde{G}_{\bar{v}} \to \mathrm{Gal}(M_{\bar{v}}/L)$.*

Proof. Since L_v does not contain a primitive cube root of unity, $\mathrm{Hom}(Z_3, L_v^{\times}) = \langle 1 \rangle$. The result follows from a duality theorem of Tate. See e.g. [15] p.32 Theorem 2.1. □

It should be observed that the full force of the Tate duality theorem is only needed to handle the case that v corresponds to a divisor of 3 in L.

Suppose that v does not correspond to a divisor of 3. If $9 \mid |G_v|$, $K_{\bar{v}}$ contains an unramified extension of L_v of degree 3. Thus G_p has a normal subgroup of index 3. As $9 \mid |G_p|$ and $G_p \subseteq A_7$ it follows that G_p has a normal 3-complement and so there is a Galois extension F of L_v with $\mathbf{Gal}(F/L_v)$ noncyclic of order 9. This is necessarily tamely ramified.

The Galois group of the maximal tamely ramified extension of L_v is given be $H = \langle x, y \mid x^{-1}yx = y^q \rangle$, where q is the cardinality of the residue class field of L_v. By assumption $q \not\equiv 1 \pmod 3$. Hence H has no homomorphic image which is noncyclic of order 9.

§7. The case $n = 6$

Lemma 7.1. *Let $N = \mathbf{Q}(\alpha)$, where α is a root of $h(z) = z^3 - 18z - 12$. If v is a valuation of N corresponding to a prime divisor of 2, 3, or 5 then N_v does not contain a primitive cube root of 1.*

Proof. Let ω be a primitive cube root of 1. Then $[\mathbf{Q}_p(\omega) : \mathbf{Q}_p] = 2$ for $p =$ 2,3, or 5. Thus if N_v contains ω then $h(z)$ is reducible over $\mathbf{Q}_p$ and $\mathbf{Q}_p(\omega) = N_v = \mathbf{Q}_p(\sqrt{d})$, where d is the discriminant of $h(z)$. By direct computation $d = 2^4 \cdot 3^5 \cdot 5$. Hence $\mathbf{Q}_p(\omega) = \mathbf{Q}_p(\sqrt{15})$.

If $p = 2$ or 5 then $\mathbf{Q}_p(\sqrt{15})$ is a ramified extension of $\mathbf{Q}_p$ but $\mathbf{Q}_p(\omega)$ is not.

If $p = 3$ then $h(z)$ is irreducible over $\mathbf{Q}_p$ by the Eisenstein criterion.

□

Theorem 7.2. *Let E_6 and P_6 be defined as in Section 5. Let $m \neq 0$ and let $P = mP_6 = (x, y)$. Assume that* $\mathbf{Gal}(K(P)/\mathbf{Q}) \simeq A_6$. *Define N as follows.*

If $h(z) = z^3 - 18z - 12$ has a root in $\mathbf{F}_p$ for every prime $p > 5$ with $\nu_p(x-1) > 0$, let $N = \mathbf{Q}$. Otherwise, let $N = \mathbf{Q}(\alpha)$, where α is a root of $h(z)$.

Then there exists a Galois extension M of N with $K(P) \subseteq M$ and $\mathbf{Gal}(M/L) \simeq 3A_6$.

Proof. Since $[N : \mathbf{Q}] \leq 3$ it follows that $\mathbf{Gal}(K(P)N/N) \simeq A_6$. By Theorem 6.2 it suffices to show that if for some valuation v of N, N_v contains ω, a primitive cube root of 1, then $9 \nmid |G_v|$.

Suppose that $\omega \in N_v$ and $9 \mid |G_v|$. Let v correspond to a divisor of the rational prime p. By Lemma 7.1, $p > 6$ as $\omega \in N_v$. Hence p does not ramify in N as the discriminant of $h(z)$ is $d = 2^4 \cdot 3^5 \cdot 5$. Furthermore a splitting field is tamely ramified. Thus $|G_v| = 9$ by Lemma 5.10. Hence p ramifies in $K(P)$.

Let $x - 4 = a/b$, where a, b are relatively prime integers. By Lemma 5.4 $p \nmid b$. Thus by (4.2) $p \mid c_j$ for some j with $2 \leq j \leq 6$. This value of j is unique since $p > 6$. By (4.1) $F_6(z, x-4) \equiv z^j f_0(z) \pmod{p}$ where $f_0(z) \in \mathbf{Q}_p[z] \subseteq N_v[z]$ with $f_0(0) \not\equiv 0 \pmod{p}$. By Hensel's lemma $F_6(z, x-4) = f_1(z) f_0(z)$, with $f_i(z) \in N_v[z]$ such that $f_1(z)$ has degree j. As $|G_v| = 9$, $j = 3$ or 6.

Suppose that $j = 3$. Thus $p \mid c_3 = x - 1$. By (4.1)

$$f_0(z) \equiv z^3 - 18z^2 + 90z - 120 \pmod{p}.$$

If $z_0 = z - 6$, this implies that $f_0(z) \equiv h(z_0) \pmod{p}$. By assumption $h(z_0)$ is reducible in the residue class field of N_v. As $p \nmid d$, $h(z_0)$ is reducible in $N_v[z]$. Thus $|G_v| \neq 9$.

Suppose that $j = 6$. As $c_6 = x + 2$, Lemma 5.3 implies that $\nu_p(c_6) = 2t$ is even. By (4.2) $\nu_p(\Delta_6(x-4)) = 10t$.

Assume that $3 \mid t$. Let $t = 3k$. Then $\nu_p(c_6) = 6k$. Thus if α is a root of $F_6(z, x-4)$ then $\alpha = p^k \alpha_0$, where α_0 is a root of a monic polynomial $f(z)$ whose coefficients are integers in N_v. Let $\Delta(f)$ denote the discriminant of f. Then

$$\nu_p(\Delta(P)) \leq \nu_p(\Delta(f)) = \nu_p(\Delta_6(x-4)) - 30k = 10t - 30k = 0$$

and p does not ramify in $K(P)$. Thus $3 \nmid t$.

The Newton polygon of $F_6(z, x-4)$ is the lower convex envelope of the points $(i, 2t)$, $0 \leq i \leq 5$ and $(6,0)$. This is a line of slope $-t/3$. As $3 \nmid t$ this implies that $\nu_p(\alpha) = 1/3$ for every root α of $F_6(z, x-4)$. Since $|G_v| = 9$ this implies that

$$F_6(z, x-4) = f_1(z) f_2(z), \tag{7.3}$$

where each $f_i(z)$ is a totally ramified cyclic cubic. Since $\omega \in N_v$, a splitting field of $f_i(z)$ is of the form $N_v(\pi_i)$, where $\pi_i^3 = u_i p$ with u_i a unit in N_v for $i = 1,2$.

Let α_i be a root of $f_i(z)$. Then $\alpha_i^6 \equiv 0 (\text{mod } \pi_i^{6t})$. Thus $\alpha_i = w_i \pi_i^t$ for $w_i \in N_v(\pi_i)$. Since $\nu_p(c_1 \cdots c_6) = \nu_p(c_6) = 2t$, it follows from (7.3) that w_i is a unit in $N_v(\pi_i)$. Thus

$$0 \equiv F_6(\alpha_i, x-4) \equiv \alpha_i^6 + c_6 \cdots c_1 (\text{mod } \pi_i^{6t+1}).$$

Therefore

$$-c_6 \cdots c_1 \equiv w_i^6 \pi_i^{6t} \equiv w_i^6 u_i^{2t} p^{2t} (\text{mod } p^{6t+1}).$$

Since $w_i \equiv \tilde{w}_i (\text{mod } \pi_i)$ for a unit $\tilde{w}_i \in N_v$, we see that

$$-c_6 \cdots c_1 \equiv \tilde{w}_i^6 u_i^{2t} p^{2t} (\text{mod } p^{2t+1}).$$

Thus

$$\tilde{w}_1^6 u_1^{2t} \equiv \tilde{w}_2^6 u_2^{2t} (\text{mod } p)$$

and so

$$(u_1 u_2^{-1})^{2t} \equiv (\tilde{w}_1^{-1} \tilde{w}_2)^6 (\text{mod } p).$$

As $3 \nmid t$ this implies that $u_1 u_2^{-1} = c^3$ for some $c \in N_v$. Therefore

$$(\pi_1 \pi_2^{-1})^3 = u_1 u_2^{-1} = c^3.$$

Hence $N_v(\pi_1) = N_v(\pi_2)$ is a splitting field of $F_6(z, x-4)$ and so $|G_v| \neq 9$. □

Theorem 7.4. *Let L be a finite extension of $\mathbf{Q}$ such that*

$$h(z) = z^3 - 18z - 12$$

has a root in L. Then for all but a finite number of even values of m, $\mathbf{Gal}(LK(mP_6)/L) \simeq A_6$ and there exists a Galois extension M of L with $K(mP_6) \subseteq M$ and $\mathbf{Gal}(M/L) \simeq 6A_6$.

Proof. Let $\hat{L}$ be the Galois closure of L over $\mathbf{Q}$. Let N be defined as in Theorem 7.2. Thus $N \subseteq L$. By Theorem 4.4

$$\mathbf{Gal}(\hat{L}K(mP_6)/\hat{L}) \simeq \mathbf{Gal}(NK(mP_6)/N) \simeq A_6,$$

for all but finitely many values of m. By Theorem 6.2 and 7.2 there exists a Galois extension M_0 of N with $K(mP_6) \subseteq M_0$ and $\mathbf{Gal}(M_0/N) \simeq 3A_6$. Hence

$$\mathbf{Gal}(M_0L/L) \simeq \mathbf{Gal}(M_0\hat{L}/\hat{L}) \simeq \mathbf{Gal}(M_0/N) \simeq 3A_6.$$

J.-F. Mestre [14], has shown that if m is even then there exists a Galois extension M_1 of $\mathbf{Q}$ with $K(mP_6) \subseteq M_1$ and $\mathbf{Gal}(M_1/\mathbf{Q}) \simeq 2A_6$. Thus

$$\mathbf{Gal}(M_1L/L) \simeq \mathbf{Gal}(M_1\hat{L}/\hat{L}) \simeq \mathbf{Gal}(M_1/\mathbf{Q}) \simeq 2A_6.$$

Let $M = M_0M_1$. Then M has the required properties. □

Theorem A is clearly a consequence of Theorem 7.4.

§8. The case $n = 7$

Lemma 8.1. *Let $N = \mathbf{Q}(\alpha)$ where α is a root of $h(z) = z^3 - 63z + 147$. Let v be a valuation of N.*

(I) *If v corresponds to a prime divisor of 2, 3 or 5, then N_v does not contain a primitive cube root of 1.*

(II) *If v corresponds to a divisor of 7 then $h(z)$ is reducible over N_v.*

Proof. Let ω be a primitive cube root of 1.

(I) If p = 2, 3 or 5 then $[\mathbf{Q}_p(\omega) : \mathbf{Q}] = 2$. Thus if N_v contains ω then $h(z)$ is reducible over $\mathbf{Q}_p$ and $\mathbf{Q}_p(\omega) = N_p$. The discriminant of $h(z)$ is $d = 3^5 \cdot 5 \cdot 7^3$. Hence

$$\mathbf{Q}_p(\omega) = \mathbf{Q}_p(\sqrt{d}) = \mathbf{Q}_p(\sqrt{105}).$$

If $p = 2$ or 5, then $\mathbf{Q}_p(\omega)$ is an unramified extension of $\mathbf{Q}_p$, while $\mathbf{Q}_p(\sqrt{105})$ is ramified. If $p = 3$, then $h(z)$ is irreducible over $\mathbf{Q}_p$ by the Eisenstein criterion.

(II) Let $K = \mathbf{Q}_7(\sqrt{7}) = N_v(\sqrt{7})$. Let $z = z_0\sqrt{7}$. Then

$$h(z) = 7\sqrt{7}\{z_0^3 - 9z_0 + 3\sqrt{7}\}.$$

The discriminant of $h_0(z) = z_0^3 - 9z_0 + 3\sqrt{7}$ is equal to $3^5 \cdot 5$ and $h_0(z) \equiv z_0(z_0^2 - 9)(\text{mod } \sqrt{7})$. Thus by Hensel's lemma $h_0(z)$ has 3 roots in K and so $h(z)$ is reducible over $\mathbf{Q}_7$. $\square$

Theorem 8.2. *Let E_7 and P_7 be defined as in Section 5. Let $m \neq 0$ and let $P = mP_7 = (x, y)$. Assume that* $\mathrm{Gal}(K(P)/\mathbf{Q}) \simeq A_7$. *Define N as follows.*

If $h(z) = z^3 - 63z + 147$ has a root in $\mathbf{F}_p$ for every prime $p > 7$ with $\nu_p(x - 1) > 0$, let $N = \mathbf{Q}$. Otherwise let $N = \mathbf{Q}(\alpha)$, where α is a root of $h(z)$.

Let v be a valuation of N which does not correspond to the prime $p = 7$. If N_v contains a primitive cube root ω of 1 the $9 \nmid |G_v|$.

Proof. Suppose that v is a counter example to the statement. Let v correspond to the rational prime p. By Lemma 8.1 $p \neq 2,3$ or 5. By assumption $p \neq 7$. Thus $p > 7$. Hence p does not ramify in N as the discriminant of $h(z)$ is $d = 3^5 \cdot 5 \cdot 7^3$. Since $9 \mid |G_v|$, p ramifies in $K(P)$. By Lemma 5.10 $|G_v| = 9$.

Let $x - 4 = a/b$, where a, b are relatively prime integers. By Lemma 5.4 $p \nmid b$. Thus by (4.2) $p \mid c_j$ for some j with $2 \leq j \leq 7$. This value of j is unique as $p > 6$. By (4.1) $F_7(z, x - 4) \equiv z^j f_0(z) \pmod p$ where $f_0(z) \in \mathbf{Q}_p[z] \subseteq N_v[z]$ with $f_0(0) \not\equiv 0 \pmod p$. By Hensel's lemma $F_7(z, x - 4) = f_1(z) f_0(z)$, with $f_i(z) \in N_v[z]$ such that $f_1(z)$ has degree j. As $|G_v| = 9$, $j = 3,4,6,7$. Each of these cases will be handled separately, though the arguments are quite similar.

Suppose that $j = 3$. Then $p \mid c_3 = x - 1$. By (4.1) $f_0(z) \equiv g(z) \pmod p$, where

$$g(z) = z^4 - 28z^3 + 252z^2 - 840z + 840.$$

If $z_0 = z - 7$ then $g(z) = z_0^4 - 42z_0^2 - 56z_0 + 105$. The resultant cubic of $g(z)$ is

$$h_0(z) = z^3 + 84z^2 + 1344z + 3136.$$

If $z_0 = z + 28$ then $h_0(z) = z_0^3 - 1008z_0 + 9408$. Thus if $z_0 = 4z_1$ then $h_0(z) = 64h(z_1)$. By assumption $h(z_1)$ is reducible in N_v. Thus 3 does not divide the order of the Galois group of $g(z)$ over N_v. Since $h(z)$ has discriminant $d = 3^5{\cdot}5{\cdot}7^3$, $h(z)$ and also $f_0(z)$ defines an unramified extension of N_v whose splitting field does not have degree divisible by 3. Hence $9 \not| |G|$.

Suppose that $j = 4$. By Lemma 5.3 $\nu_p(c_4) = 2s$ is even. Hence by (4.2) $\nu_p(\Delta_7(x-4)) = 6s$. Furthermore by (4.1) $F_7(z, x-4) = f_1(z)f_0(z)$, where $f_1(z) \equiv z^4 (\text{mod } p)$ and $f_0(z) \equiv g(z) (\text{mod } p)$ with

$$g(z) = z^3 - 21z^2 + 126z - 210.$$

The discriminant of $g(z)$ is $d = 2^3 \cdot 3^4 \cdot 7^2$ and so the splitting field of $g(z)$ and $f_0(z)$ is unramified over N_v.

The Newton polygon of $F_7(z, x-4)$ is the lower convex envelope of the points $(i, 2s)$, $0 \le i \le 3$ and $(i, 0)$, $4 \le i \le 7$. This has a segment of slope $-2s/4 = -s/2$. As the ramification index of a splitting field of $F_7(z, x-4)$ divides 9, $s = 2t$ is even.

Let $f_1(\alpha) = 0$. As $f_1(z) \equiv z^4 (\text{mod } p^{4t})$ it follows that $\nu_p(\alpha) \ge t$. Let Δ be the discriminant of the monic polynomial whose roots are $p^{-t}\alpha$ as α ranges over the roots of $f_1(z)$. By (4.2)

$$\nu_p(\Delta(P)) \le \nu_p(\Delta) = \nu_p(\Delta_7(x-4)) - 6s = 12t - 6s = 0$$

and p does not ramify in $K(P)$.

Suppose that $j = 7$. Let $\nu_p(c_7) = s$. The Newton polygon of $F_7(z, x-4)$ is the lower convex envelope of the points (i, s), $0 \le i \le 6$ and $(7, 0)$. This is a line of slope $-s/7$. As 7 does not divide the ramification index of a splitting field of $F_7(z, x-4)$, $s = 7t$ is divisible by 7.

Let $f_1(\alpha) = 0$. As $f_1(z) \equiv z^7 (\text{mod } 7t)$ it follows that $\nu_p(\alpha) \ge t$. Let Δ be the discriminant of the monic polynomial whose roots are $p^{-t}\alpha$ as α ranges over the roots of $f_1(z)$. By (4.2)

$$\nu_p(\Delta(P)) \le \nu_p(\Delta) = \nu_p(\Delta_7(x-4)) - 6s = 42t - 6s = 0$$

and p does not ramify in $K(P)$.

Suppose finally that $j = 6$. By Lemma 5.3 $\nu_p(c_6) = 2s$ is even. Hence $\nu_p(\Delta_7(x-4)) = 10s$ by (4.2). Furthermore

$$F_7(z, x-4) \equiv (z-7)f_0(z)f_1(z)(\text{mod } p^{2s})$$

where $f_i(z) \equiv z^3(\text{mod } p)$ for $i = 1,2$ and $f_0(z)f_1(z) \equiv z^6(\text{mod } p^{2s})$.

If $s = 3t$ is divisible by 3 then $\nu_p(c_6) = 6t$. Thus if α is a root of $f_1(z)f_2(z)$ then $\alpha = p^t\alpha_0$, where α_0 is a root of a monic polynomial $f(z)$ whose coefficients are integers in N_v. Let Δ denote the discriminant of f. Then

$$\nu_p(\Delta(P)) \le \nu_p(\Delta) = \nu_p(\Delta_7(x-4)) - 30t = 10s - 30t = 0$$

and p does not ramify in $K(P)$. Thus $3 \nmid s$.

The Newton polygon of $F_6(z, x-4)$ is the lower convex envelope of the points $(i, 2s)$ $0 \le i \le 5$ and $(i, 0)$ for $i = 6,7$. This has a segment of slope $-s/3$. Since $3 \nmid s$ it follows that $\nu_p(\alpha) = 1/3$ for every root of $f_1(z)f_2(z)$. Hence each $f_i(z)$ is a totally ramified cyclic cubic. Since a primitive cube root ω is in N_v, a splitting field of $f_i(z)$ has the form $N_v(\pi_i)$, where $\pi_i^3 = u_ip$ with u_i a unit in N_v for $i = 1,2$.

Let α_i be a root of $f_i(z)$. Then $\alpha_i^6 \equiv 0(\text{mod } \pi_i^{6s})$. Thus $\alpha_i = w_i\pi_i^s$ for $w_i \in N_v(\pi_i)$. Since $\nu_p(c_1 \cdots c_7) = \nu_p(c_7) = 2s$, it follows that w_i is a unit in $N_v(\pi_i)$. Thus

$$0 \equiv F_7(\alpha_i, x-4) \equiv -7c_7w_i^6\pi_i^{6s} - c_7 \cdots c_1(\text{mod } \pi_i^{6s+1}).$$

Therefore

$$-c_6 \cdots c_1 \equiv 7w_i^6\pi_i^{6s} \equiv 7w_i^6u_i^{2s}p^{2s}(\text{mod } \pi_i^{6s+1}).$$

Since $w_i \equiv \tilde{w}_i(\text{mod } \pi_i)$ for a unit $\tilde{w}_i \in N_v$, we get that

$$-c_6 \cdots c_1 \equiv 7\tilde{w}_i^6u_i^{2s}p^{2s}(\text{mod } p^{2s+1}).$$

Hence

$$\tilde{w}_1^6u_1^{2s} \equiv \tilde{w}_2^6u_2^{2s}(\text{mod } p)$$

and so

$$(u_1u_2^{-1}) \equiv (w_1^{-1}w_2)^6(\text{mod } p).$$

As $3 \not| s$ this yields that $u_1 u_2^{-1} = c^3$ for some $c \in N_v$. Therefore

$$(\pi_1 \pi_2^{-1})^3 = u_1 u_2^{-1} = c^3.$$

Hence $N_v(\pi_1) = N_v(\pi_2)$ is a splitting field of $F_7(z, x-4)$ and so $9 \not| |G_v|$.

□

Theorem 8.3. *Let m be an odd integer and let $P = mP_7 = (x, y)$. Assume that $\mathrm{Gal}(K(P)/\mathbf{Q}) \simeq A_7$ and $9 \mid |G_7|$. Let $\lambda = x - 4 = a/b$ where a, b are relatively prime integers. Then $7 \not| b$, $7 \not| c_j$ for $j \neq 6$ and $\nu_7(c_6) = \nu_7(x+2) > 1$.*

Proof. Since G_7 is a solvable subgroup of A_7, $63 \not| |G_7|$ and so $7 \not| |G_7|$. Hence $K(P)_7$ is a tamely ramified extension of $\mathbf{Q}_7$. Thus $|G_7| = 9$ by Lemma 5.10. Let $\nu = \nu_7$.

Suppose that $\nu(b) > 0$. then $\nu(c_j) = -\nu(b)$ for $1 \leq j \leq 7$. Since m is odd Lemma 5.3 implies that $\nu(c_6)$ is odd but $\nu(c_4)$ is even. Thus $\nu(b) = 0$. As $\nu(c_6)$ is odd it follows that $\nu(c_6) > 0$ but $\nu(c_j) = 0$ for $1 \leq j \leq 7, j \neq 6$.

If $\nu(c_6) = 1$ then $F_7(z, x-4)$ is irreducible by Eisenstein theorem. Hence $7 \mid |G_7| = 9$. This contradiction proves the result. □

Theorem 8.4. *Let m be an odd integer, $P_1 = mP_7$, $P_2 = (m+2)P_7$. Assume that $\mathrm{Gal}(K(P_i)/\mathbf{Q}) \simeq A_7$ for $i = 1, 2$. Then $9 \not| |G_7|$ for at least one of $i = 1, 2$.*

Proof. Let $P_i = (x_i, y_i)$ for $i = 1,2$. Direct computation shows that

$$2P_7 = (\frac{29^2}{420}, \frac{29.41}{840})$$

and

$$x_1 + x_2 = \frac{1}{420}\left[\frac{4(840y_1 - 29.41)^2}{(840x_1 - 2.29^2)^2} - 29^2\right]. \tag{8.5}$$

By Theorem 8.3 it suffices to show that $\nu_7(x_i + 2) = 1$ for $i = 1$ or 2. Suppose this is not the case. Then $x_i \equiv -2 \pmod{7^2}$ for $i = 1, 2$. By

Lemma 5.3 $x_i + 2 \equiv 0 \pmod{7^3}$ and so $y_i \equiv 0 \pmod{7^2}$. Thus (8.5) implies that

$$-4.420 \equiv \frac{4.29^2.41^2}{4(1681)^2} - 29^2 \equiv \frac{29^2}{41^2}(1-41^2) \pmod{7^3}$$

Therefore

$$1 \equiv 29^2/41^2 \pmod{7^2}.$$

Hence

$$12.70 \equiv 41^2 - 29^2 \equiv 0 \pmod{7^2}$$

which is not the case. □

Theorem 8.6. *Let L be a finite extension of $\mathbf{Q}$ such that*

$$h(z) = z^3 - 63z + 147$$

has a root in L. Then there exist infinitely many odd values m so that $\mathbf{Gal}(LK(mP_7)/L) \simeq A_7$ and there exists a Galois extension M of L with $K(mP_7) \subseteq M$ and $\mathbf{Gal}(M/L) \simeq 6A_7$.

Proof. Let $\hat{L}$ be the Galois closure of L over $\mathbf{Q}$. Let N be defined as in Theorem 8.2. Thus $N \subseteq L$. By Theorem 4.4

$$\mathbf{Gal}(\hat{L}K(mP_7)/\hat{L}) \simeq \mathbf{Gal}(NK(mP_7)/N) \simeq A_7$$

for all but finitely many values of m. By Theorems 6.2, 8.2 and 8.3, there exist infinitely many odd integers m and a Galois extension M_0 of N with $K(mP_7) \subseteq M_0$ and $\mathbf{Gal}(M_0/N) \simeq 3A_7$. Hence

$$\mathbf{Gal}(M_0L/L) \simeq \mathbf{Gal}(M_0\hat{L}/\hat{L}) \simeq \mathbf{Gal}(M_0/N) \simeq 3A_7.$$

By [5] Section 11 there exists a Galois extension M_1 of $\mathbf{Q}$ such that $K(mP_7) \subseteq M_1$ and $\mathbf{Gal}(M_1/\mathbf{Q}) \simeq 2A_6$. Thus

$$\mathbf{Gal}(M_1L/L) \simeq \mathbf{Gal}(M_1\hat{L}/\hat{L}) \simeq \mathbf{Gal}(M_1/\mathbf{Q}) \simeq 2A_7.$$

Let $M = M_0M_1$. Then M has the required properties. □

Theorem B is clearly a consequence of Theorem 7.4.

Theorem 8.7. *There exists a Galois extension M of $\mathbf{Q}$ such that*

$\mathbf{Gal}(M/\mathbf{Q}) \simeq 6A_7$.

Proof. By [16] p. 227 $\mathbf{Gal}(K(P_7)/\mathbf{Q}) \simeq A_7$. Since $P_7 = (5, 105)$, $\nu_7(x+2) = \nu_7(7) = 1$ and $x - 1 = 4$. Thus Theorems 6.2, 8.2 and 8.3 imply the existence of a Galois extension M_0 of $\mathbf{Q}$ with $K(P_7) \subseteq M_0$ and $\mathbf{Gal}(M_0/\mathbf{Q}) \simeq 3A_7$. By [5] Section 11 there exists a Galois extension M_1 of $\mathbf{Q}$ with $K(P_7) \subseteq M_1$ and $\mathbf{Gal}(M_1/\mathbf{Q}) \simeq 2A_7$. Then $M = M_1M_2$ has the required properties. □

References

[1] Belyi, G.V., On Galois extensions of a maximal cyclotomic field, *Math. USSR Izvestia AMS Translations*, **14** (1980), 247-256.

[2] Conway, J.H.,et al. *Atlas of finite groups*, Oxford: Clarendon Press 1985.

[3] Feit, W.,Rigidity and Galois groups *Proc. of the Rutgers group theory year 1983-1984*, Cambridge University Press 1984, 283-287.

[4] ———, Rigidity of $\mathrm{Aut}(PSL_2(p^2))$, $p \equiv \pm 2 \pmod 5$, $p \neq 2$, *Proc. of the Rutgers group theory year 1983-1984*, Cambridge University Press 1984, 351-356.

[5] ———, $\tilde{A}_5$ and $\tilde{A}_7$ are Galois groups over number fields, *J. Algebra* **104** (1986), 231-260.

[6] Feit, W., and Vojta, P., Examples of some **Q**-admissible groups, *J. of Number Theory*, **26** (1987), 210-226.

[7] Fried, M., Fields of definition of function fields and Hurwitz families-Groups as Galois groups, *Comm. Alg.* **5** (1977), 17-82.

[8] Hoyden-Siedersleben, G. and Matzat, B.H., Realisierung sporadischer einfacher Gruppen als Galoisgruppen uber Kreisteilungskörpern, *J. Algebra*, **101** (1986), 273-285.

[9] Hunt, D.C., Rational rigidity and the sporadic groups, *J. Algebra*, **99** (1986), 577-592.

[10] Lang, S., *Introduction to algebraic and abelian functions*, Addison -

Wesley, Reading, Mass. 1972.

[11] Matzat, B.H., Konstruction von Zahl - und Funktionenkörpern mit vorgegebener Galoisgruppe, *J. reine angew. Math.*, **349** (1984), 179-220.

[12] ———, Zum Einbettungsproblem der algebraischen Zahlentheorie mit nicht abelschen Kern, *Invent. Math.*, **80** (1985), 365-374.

[13] ———, *Konstructive Galoistheorie*, LNM 1284 (1987).

[14] Mestre, J.-F., Personal communication.

[15] Milne, J.S., *Arithmetic duality theorems*, Academic press, Boston...

[16] Schur, I., *Collected Works, Vol. III*, Springer-Verlag, Berlin-Heidelberg, New York, 1973.

[17] Serre, J.-P., *Groupes de Galois sur Q*, Seminaire Bourbaki 1987-88, 689.

[18] Sonn, J., $SL(2,5)$ and Frobenius Galois groups over Q, *Canada. J. Math.*, **32** (1980), 281-293.

[19] Thompson, J.G., Isomorphisms induced by automorphisms, *J. of the Australian Math. Soc.*, **16** (1973), 16-17.

[20] ———, Some finite groups which appear as $\mathbf{Gal}(L/K)$ where $K \subseteq Q(\mu_n)$, *J. Algebra*, **89** (1984), 437-499.

Department of Mathematics
Yale University
New Haven, Connecticut 06520
U. S. A.

Actions on lower central factors of free groups

B. Hartley

1. Introduction

Let F be a non-cyclic free group, $1 \neq R \triangleleft F$, and $G = F/R$. Let $R' = [R, R]$. Then G acts (on the right) on $\bar{R} = R/R'$ via conjugation in F, and in this way we obtain a *relation module* for G. These have been extensively studied by a number of authors. A general reference is [5]. The starting point for us is the following theorem of Passi.

Theorem 1.1 [13]. *$\bar{R}$ is a faithful $\mathbf{Z}G$-module.*

The proof has four main steps, which we briefly describe.

Step 1. $\bar{R}$ can be embedded in a free $\mathbf{Z}G$-module.

This follows from the "relation sequence"

$$0 \to \bar{R} \to (\mathbf{Z}G)^d \to \mathcal{G} \to 0. \qquad (1.1)$$

Here $d = d(F)$ is the rank of F, which may be infinite. $(\mathbf{Z}G)^d$ denotes the direct sum of d copies of $\mathbf{Z}G$, and $\mathcal{G}$ is the augmentation ideal of $\mathbf{Z}G$.

Step 2. Let J be the annihilator in $\mathbf{Z}G$ of $\bar{R} : J = \operatorname{ann}_{\mathbf{Z}G} \bar{R}$. Then J is an annihilator ideal, that is, the right annihilator of some right ideal of $\mathbf{Z}G$. The right ideal in question is the set of components of elements of $\bar{R}$, when $\bar{R}$ has been embedded in $(\mathbf{Z}G)^d$ as in (1.1).

Step 3. J is controlled by $\Delta^+(G)$, that is,

$$J = (J \cap \mathbf{Z}\Delta^+(G)).\mathbf{Z}G. \qquad (1.2)$$

Here $\Delta^+(G) = \langle D \lhd G : |D| < \infty \rangle$ is the torsion subgroup of the FC-centre of G, and (1.2) is a theorem of Martha Smith on annihilator ideals in semiprime group rings (see Passman [14], p.143).

Thus if $J \neq 0$, then $J \cap \mathbf{Z}D \neq 0$ for some finite normal subgroup D of G. Taking the inverse image of D in F allows us then to view $\bar{R}$ as a relation module, with non-zero annihilator, for the finite group D. We may easily reduce to the case $1 < d < \infty$ at this stage, if we wish. In this way, we reduce the proof of the theorem to the case when G and d are both finite.

Step 4.

$$\bar{R} \otimes \mathbf{Q} \cong (\mathbf{Q}G)^{d-1} \oplus \mathbf{Q} \quad (d \geq 1).$$

This theorem of Gaschütz can be proved by tensoring (1.1) with $\mathbf{Q}$, whereupon it splits. Theorem 1.1, for finite G, follows immediately.

Our aim is to show how this argument can be used as a model for similar proofs in which $\bar{R}$ is replaced by other sections of R on which G acts naturally.

2. Lower central factors

Let $\gamma_1(X) \geq \gamma_2(X) \geq \ldots$ be the lower central series of a group X. The higher relation modules $\gamma_c(R)/\gamma_{c+1}(R)$ come naturally to our attention. They are usually studied by Lie methods, going back to Magnus and Witt, as we now outline. Some of the exposition is based on [10]. In this section, F plays no part.

Let R be free on $\{x_i : i \in I\}$. We form the free associative ring A generated freely by the x_i, that is, the polynomial ring over $\mathbf{Z}$ in the non-commuting indeterminates x_i. Thus

$$A = \bigoplus_{c=0}^{\infty} A_c \tag{2.1}$$

where A_c is the homogeneous component of degree c. We can identify $\bar{R}$ with A_1, and we can equally well view A as the tensor algebra on $\bar{R}$. We have the Lie subring

$$\Lambda = \bigoplus_{c=1}^{\infty} \Lambda_c \tag{2.2}$$

of A, generated by the x_i under the Lie operation $(x, y) = xy - yx$, and we can identify $\gamma_c(R)/\gamma_{c+1}(R)$ with Λ_c by replacing cosets of group commutators by the corresponding Lie commutators. If ε is any additive endomorphism of A_1, then it extends to a ring endomorphism of A (linear substitution), preserving the homogeneous components A_c and Λ_c, and inducing on them an additive endomorphism, which we denote by $\varepsilon^{(c)}$. In particular, if θ is any group endomorphism of R, it acts on $A_1 = \bar{R}$ and hence on Λ_c; it also acts on $\gamma_c(R)/\gamma_{c+1}(R)$, and this action is compatible with the identification

$$\Lambda_c \cong \gamma_c(R)/\gamma_{c+1}(R) \tag{2.3}$$

just described. This means that if we wish to calculate the effect of endomorphisms of R on factors of the lower central series of R, we can work purely in the context of (2.1) and (2.2) and forget about the group R.

Let $\Gamma = \mathrm{Aut}_{\mathbf{Z}} A_1$ and $\Delta = \mathrm{End}_{\mathbf{Z}} A_1$ be respectively the group of all additive automorphisms of A_1 and the multiplicative semigroup of all additive endomorphisms of A_1. (We wish to ignore the additive structure of Δ). We note

Proposition 2.1. *The natural map* $\mathrm{Aut} R \to \Gamma$ *is surjective.*

This is a classical result of Nielsen in the finitely generated case. The infinitely generated case is due to Swan [2]; see also Macedońska [11].

Let $\Gamma^{(c)}$, $\Delta^{(c)}$ be the images of Γ and Δ respectively under the multiplicative (but not additive) map $\varepsilon \to \varepsilon^{(c)}$. In studying the actions of these, it is natural to tensor with the rationals and form $\mathbf{Q}A = \mathbf{Q} \otimes_{\mathbf{Z}} A$, $\mathbf{Q}A_c$, $\mathbf{Q}\Lambda_c$, etc. This brings $\Gamma_{\mathbf{Q}} = \mathrm{Aut}_{\mathbf{Q}} \mathbf{Q}A_1$ and $\Delta_{\mathbf{Q}} = \mathrm{End}_{\mathbf{Q}} \mathbf{Q}A_1$ into play. Also acting on $\mathbf{Q}A_c$ we have the symmetric group S_c, which acts by permuting variables and which commutes with the other actions described above. We now have the following rings of $\mathbf{Q}$-linear transformations of $\mathbf{Q}A_c$, where $\mathbf{Q}[X]$ denotes the rational span of X.

$$\begin{array}{ccccc} & & \mathbf{Q}[\Delta^{(c)}] & & \\ & \rotatebox{45}{$\leq$} & & \rotatebox{-45}{$\geq$} & \\ \mathbf{Q}[\Gamma^{(c)}] & & & & \mathbf{Q}[\Delta_{\mathbf{Q}}^{(c)}] \;\leq\; \mathrm{End}_{\mathbf{Q}S_c} \mathbf{Q}A_c . \\ & \rotatebox{-45}{$\leq$} & & \rotatebox{45}{$\geq$} & \\ & & \mathbf{Q}[\Gamma_{\mathbf{Q}}^{(c)}] & & \end{array} \tag{2.4}$$

If $|I| = n < \infty$, then Γ and Δ can be identified with $GL(n, \mathbf{Z})$ and the multiplicative semigroup $M_n(\mathbf{Z})$ of all $n \times n$ matrices over $\mathbf{Z}$, respectively, and $\Gamma_{\mathbf{Q}}$, $\Delta_{\mathbf{Q}}$ with $GL(n, \mathbf{Q})$ and $M_n(\mathbf{Q})$ respectively. It is then essentially classical that the rings (2.4) are all equal. In general we have the following result of Lichtman and the author [7].

Proposition 2.2. *Each of the rings* (2.4) *is dense with respect to each other.*

The meaning of this is as follows. If S and T are sets of linear transformations of a vector space V, we say that S is dense with respect to T if, given finitely many elements $v_1, \ldots, v_n \in V$ and $\tau \in T$, there exists $\sigma \in S$ such that $v_i\tau = v_i\sigma$ $(1 \leq i \leq n)$.

Now by Maschke's Theorem, $\mathbf{Q}A_c$ is completely reducible under $\mathbf{Q}S_c$. A general result of ring theory shows that it is completely reducible under $\mathrm{End}_{\mathbf{Q}S_c}\mathbf{Q}A_c$. Since mutually dense sets of linear transformations have the same invariant subspaces, we obtain

Proposition 2.3. $\mathbf{Q}A_c$ *is completely reducible under each of the rings* (2.4).

As a simple application of this, consider the characteristic subgroups of R lying between two successive terms $\gamma_c(R)$ and $\gamma_{c+1}(R)$ of the lower central series of R. By Proposition 2.1, these correspond precisely to the $\Gamma^{(c)}$-submodules of Λ_c, under the identification (2.3). Let M be such a submodule such that Λ_c/M is torsion-free. Then $\mathbf{Q}M$ is $\mathbf{Q}[\Gamma^{(c)}]$-invariant and hence is invariant under all the rings (2.4). In particular, $\mathbf{Q}M$, and hence $\mathbf{Q}M \cap \Lambda_c = M$ is $\Delta^{(c)}$-invariant. This gives the following, which is no doubt well known for other reasons.

Proposition 2.4 [7]. *If M is a characteristic subgroup of R lying between two successive terms of the lower central series of R and such that R/M is torsion free, then M is fully invariant in R.*

3. Relative higher relation modules

We return now to the situation of section 1, where R is a non-trivial normal subgroup of the non-cyclic free group F, and $G = F/R$. The higher relation modules $\gamma_c(R)/\gamma_{c+1}(R)$ have been studied from the above point of view. For example, Gupta, Laffey and Thomson calculated the character of the $\mathbf{Q}G$-module $\gamma_c(R)/\gamma_{c+1}(R) \otimes \mathbf{Q}$ [6]. More generally, Lichtman and the author [7] defined a *c-th relative higher relation module* to be a $\mathbf{Z}G$-module of the form $\gamma_c(R)/M$, where M is a characteristic (equivalently, fully invariant) subgroup of R such that $\gamma_{c+1}(R) \leq M < \gamma_c(R)$ and $\gamma_c(R)/M$ is torsion free. This is the last non-trivial term of the lower central series of the relatively free group R/M. The action of G, as before, is via conjugation in F. These modules were studied from the point of view of faithfulness for $\mathbf{Z}G$. The machinery of the last chapter shows that studying them amounts to dealing with the following situation.

(3.1) M is a $\Gamma^{(c)}$-submodule of Λ_c such that Λ_c/M is non-trivial and torsion free,

(3.2) $G \leq \Gamma$,

(3.3) $A_1 = \Lambda_1$ is a relation module for G.

We assume these for the moment. By Proposition 2.3,

$$\Lambda_c/M \hookrightarrow \mathbf{Q}\Lambda_c/\mathbf{Q}M \hookrightarrow \mathbf{Q}A_c = (\mathbf{Q}A_1)^{\otimes c}$$

By step 1 of the proof of Theorem 1.1, $\mathbf{Q}A_1$ can be embedded in a free $\mathbf{Q}G$-module, and hence, by general principles, so can $(\mathbf{Q}A_1)^{\otimes c}$. The following is now immediate when Λ_c/M has finite $\mathbf{Z}$-rank, and follows with a little more work in general [7].

Proposition 3.1. *Λ_c/M can be embedded in a free $\mathbf{Z}G$-module.*

Exactly as in Theorem 1.1, we can now apply Martha Smith's theorem to reduce the question of whether Λ_c/M is $\mathbf{Z}G$-faithful to the case when G is finite and $1 < d < \infty$. We can equally well discuss the $\mathbf{Q}G$-faithfulness of $\mathbf{Q}\Lambda_c/\mathbf{Q}M$. This embeds in $\mathbf{Q}A_c$ by Proposition 2.3, and as $\mathbf{Q}A_1 \cong (\mathbf{Q}G)^{d-1} \oplus \mathbf{Q}$ and $\mathbf{Q}[\Gamma^{(c)}] = \mathbf{Q}[\Gamma^{(c)}_{\mathbf{Q}}]$ (Proposition 2.2), we reach the following situation.

(3.1′) W is a non-zero $\Gamma^{(c)}_{\mathbf{Q}}$ (or $GL(t, \mathbf{Q})$)-submodule of $\mathbf{Q}\Lambda_c$,

(3.2′) $G \leq \Gamma \leq GL(t, \mathbf{Q})$, where $t = (d-1)|G| + 1$ is the rank of R,

(3.3′) $\mathbf{Q}A_1 \cong (\mathbf{Q}G)^{d-1} \oplus \mathbf{Q} \quad (d > 1)$.

We wish to know whether W is necessarily $\mathbf{Q}G$-faithful. Unfortunately, it need not be, which may make us wonder whether our time so far has been well spent. However, it usually is faithful, and this enables a theorem on faithfulness of our relative higher relation modules to be obtained. Before stating it, we note that the above is now a problem in the representation theory of the general linear group $GL(t, Q)$. We may as well take W to be irreducible; work of Kljačko [9] shows that the fact that $W \leq \mathbf{Q}\Lambda_c$ places almost no restriction on the possible W's that can occur. We have a finite subgroup G of $GL(t, \mathbf{Q})$ that operates in a certain way on the "natural" right module for $GL(t, \mathbf{Q})$, consisting of row vectors of length t over $\mathbf{Q}$, and we wish to know how G operates under other irreducible representations of $GL(t, \mathbf{Q})$. After analyzing this question, Lichtman and the author obtained the following [7].

Theorem 3.2. *The c-th relative higher relation module $V = \Lambda_c/M$ is $\mathbf{Z}G$-faithful unless all the following hold:*

(a) *G is finite of order $m > 1$,*

(b) *F has finite rank $d > 1$.*

(c) $c = \ell(1 + m(d-1)) \quad (\ell > 1)$ *is a non-trivial multiple of the rank of R.*

(d) $|G/C_G(V)| \leq 2$.

The exceptional cases can occur.

4. A property of representations of the general linear group

The representation theoretic question arising above was further studied in [8], where the following more general result was obtained.

Theorem 4.1. *Let K be an arbitrary field and V be a finite dimensional irreducible module for $GL(t, K)$ over K. If K is infinite, assume that V affords a polynomial representation of $GL(t, K)$. Let E be the natural $GL(t, K)$-module over K, consisting of row vectors of length t with entries from K, and let G be a non-trivial finite subgroup of $GL(t, K)$.*

Suppose that

$$E \cong KG \oplus E_1$$

as KG-modules, for some KG-module E_1. Then one of the following holds

(4.1a) *V contains a non-zero free KG-module,*

(4.1b) $\dim_K V = 1$,

(4.1c) *$E_1 = 0$, G is an elementary abelian 2-group of order > 2*

and

$$V \cong (\Lambda^t E)^{\otimes a} \otimes (\Lambda^m E)^{Fr^b}$$

for some $a, b \geq 0$. Here $m = 2$ or $t - 2$ and $b = 0$ if $\operatorname{char} K = 0$. In cases (b) *and* (c) *V contains no non-zero free KG-module.*

The exact nature of the exception (4.1c) need not concern us very much, but for completeness, we note that Fr denotes the Frobenius automorphism $\lambda \to \lambda^p$ when K has characteristic $p > 0$, and Λ is the exterior power. The term "polynomial representation" simply means that the entries of the representing matrices are polynomial functions of the entries of the matrices being represented. A similar but less precise result is available without the polynomial assumption.

The proof involves first showing that we may assume K to be infinite. This is because if K is finite and has algebraic closure $\bar{K}$, then $V \otimes_K \bar{K}$ is irreducible as a $\bar{K}[GL(t, K)]$-module, and can be extended to a $GL(t, \bar{K})$-module. In the infinite case, one follows the construction of the irreducible polynomial $GL(t, K)$- modules as given in Green [4], and constructs the required free module more or less explicitly.

We can think of Theorem 4.1 as saying that free modules have a certain "persistence" property. If a finite subgroup of $GL(t,K)$ has a free submodule under the natural action of $GL(t,K)$, then this usually persists under other irreducible K-representations of $GL(t,K)$. One wonders whether other types of module share this property.

5. Relation modules over fields of finite characteristic

The fact that Theorem 4.1 holds for fields of arbitrary characteristic makes it tempting to look for analogues of Theorem 3.2 for modules that are not necessarily additive by torsion-free, such as $\gamma_c(R)/\gamma_{c+1}(R)\gamma_c(R)^p$, where p is a prime, and perhaps suitable "relativized" versions of these. Some steps in this direction have been taken by A.D Gilchrist, a student of mine in Manchester, and have appeared in his Ph.D. thesis. However, there is probably much more to do in this area. We consider some of the problems that arise when one tries to proceed along the lines used above in the torsion-free case.

Write $\mathbf{F}_p$ for the field of order p, $\bar{R}_p$ for $\bar{R} \otimes_{\mathbf{Z}} \mathbf{F}_p$, and $\bar{R}_{p,c}$ for $\gamma_c(R)/\gamma_{c+1}(R)\gamma_c(R)^p \cong \Lambda_c \otimes_{\mathbf{Z}} \mathbf{F}_p$. Tensoring (1.1) with $\mathbf{F}_p$ we find

$$0 \to \bar{R}_p \to (\mathbf{F}_pG)^d \to \mathcal{G} \otimes \mathbf{F}_p \to 0 \tag{5.1}$$

whence in particular

Proposition 5.1. *$\bar{R}_p$ can be embedded in a free $\mathbf{F}_pG$-module.*

Since Λ_c is a $\mathbf{Z}$-direct summand of A_c, we have that $\bar{R}_{p,c}$ is an $\mathbf{F}_pG$-submodule of $A_c \otimes_{\mathbf{Z}} \mathbf{F}_p \cong (\bar{R}_p)^{\otimes c}$, so

Proposition 5.2. *$\bar{R}_{p.c}$ can be embedded in a free $\mathbf{F}_pG$-module.*

In proving the analogous result for our relative higher relation modules (Proposition 3.1) we used a complete reducibility result (Proposition 2.3). Such results can hardly be expected in non-zero characteristic, so this initially impels us to focus on *submodules* of $\bar{R}_{p,c}$, rather than quotient modules, though these may be of less group theoretical interest. The next problem occurs with Martha Smith's Theorem, which holds for annihilator ideals in semiprime group rings, but not in general. As a replacement, we have

Theorem 5.3 (Gilchrist [3]). *Let K be any field, G any group, and J any annihilator ideal in KG. If $J \neq 0$ then*

$$J \cap K\Delta^{+}(G) \cap Z(KG) \neq 0$$

where $Z(KG)$ is the centre of KG.

This enables one to reduce faithfulness questions for the modules considered, to the case of finite G as before. Here, two problems remain. Firstly, we can expect to encounter exceptional cases analogous to (d) of Theorem 3.2. We may try to avoid this by restricting to infinite G, but since the aim is to use Theorem 5.3 to reduce back to the finite case, the difficulty is likely to reappear. Now (d) of Theorem 3.2 corresponds to the situation when the W of (3.1′) affords a power of the determinant representation, and over an infinite field, polynomial degree considerations show that this can only occur when c is a multiple of the dimension t. This explains (c) of Theorem 3.2. Over a finite field we do not have a well defined polynomial degree to attach to our representations and the exceptional situation is less well under control. One is led to consider submodules U of $\bar{R}_{p,c}$ that "behave well" under extension of the field to the algebraic closure $\bar{\mathbf{F}}_p$, in the sense that $U \otimes \bar{\mathbf{F}}_p$ is invariant under the appropriate general linear group over $\bar{\mathbf{F}}_p$.

Finally, we would like to know that $\bar{R}_p$ is $\mathbf{F}_pG$-faithful, by analogy with Gaschütz's Theorem used in Step 4 of Theorem 1.1. This is not in general true, but it is if $d(F) = d > d(G)$, the minimum number of generators of the (finite) group G. For applying Schanuel's Lemma to the sequence

$$0 \to \bar{R}_p^* \to (\mathbf{F}_pG)^{d^*} \to \mathcal{G} \otimes \mathbf{F}_p \to 0$$

arising from a free presentation of G on $d^* = d(G)$ generators, and to (5.1), we have

$$\bar{R}_p^* \oplus (\mathbf{F}_pG)^d \cong \bar{R}_p \oplus (\mathbf{F}_pG)^{d*},$$

whence by the Krull-Schmidt Theorem $\mathbf{F}_pG$ is a direct summand of $\bar{R}_p$.

After taking these considerations into account, Gilchrist [3] obtains the following theorem.

Theorem 5.4. *Let $1 \neq R \lhd F$, where F is free non-cyclic, and suppose $G = F/R$ is infinite. Let M be a subgroup of R such that $\gamma_{c+1}(R)\gamma_c(R)^p < M \leq \gamma_c(R)$ for some prime p and integer $c \geq 1$, and suppose that M is generated modulo $\gamma_{c+1}(R)\gamma_c(R)^p$ by the set of R-values of a set of outer commutator words of weight c. Then $M/\gamma_{c+1}(R)\gamma_c(R)^p$ is a faithful $\mathbf{F}_pG$-module.*

Outer commutator words are defined as follows. Let Y be the free group freely generated by $y_1, y_2, \ldots$. Then y_1 is the unique outer commutator word of weight 1. If $w_1(y_1, \ldots, y_r)$ and $w_2(y_1, \ldots, y_s)$ are outer commutator words of weight r and s respectively, then $[w_1(y_1, \ldots, y_r), w_2(y_{r+1}, \ldots, y_{r+s})]$ is an outer commutator word of weight $r+s = n$, and all outer commutator words of weight n have this form, for suitable r, s, w_1 and w_2.

6. Some proofs

In this section we give the proof of Gilchrist's Theorem 5.3. This is based on a theorem of McCoy on annihilators in polynomial rings [12], and we thank Warren Dicks for bringing McCoy's Theorem to our attention. We first describe the connection between Theorem 5.1 and polynomial rings. The following, giving various characterizations of annihilator ideals, is trivial.

Lemma 6.1. *Let J be a subset of a ring R. Then the following conditions are equivalent.*

(i) *$J = r(A)$ is the right annihilator of some right ideal A of R*

(ii) *$J = r(A)$ for some two-sided ideal A of R*

(iii) *J is an ideal of R, and $J = r(A)$ for some subset A of R.*

The next result is marginally less trivial.

Lemma 6.2. *Let R be a ring, S be a subring of R, and assume that R is free as a right S-module. If J is an annihilator ideal of R, then $J \cap S$ is an annihilator ideal of S.*

Proof. Let $R = \bigoplus_{\lambda \in \Lambda} b_\lambda S$ be right S-free on the basis $(b_\lambda : \lambda \in \Lambda)$, and let J be the right annihilator of the subset A of R. For each $a \in A$, write $a = \sum_\lambda b_\lambda s(\lambda, a)$, with $s(\lambda, a) \in S$. Then $J \cap S$ is the right annihilator in S of $\{s(\lambda, a) : \lambda \in \Lambda, a \in A\}$.

We also require a result of Passman. Let K be an arbitrary field and G be an arbitrary group. Let $\Delta(G) = \{x \in G : |G : C_G(x)| < \infty\}$ be the FC-centre of the group G and let $\theta : KG \to K\Delta(G)$ be the projection, given by

$$\theta(\sum_{g \in G} \lambda_g g) = \sum_{g \in \Delta(G)} \lambda_g g \quad (\lambda_g \in K).$$

Let J be an annihilator ideal of KG, so that $J = r(A)$ for some two sided ideal A of KG. By the proof of [13, Theorem 4.2.9, p.128], we have $A\theta(J) = 0$. Hence $\theta(J) \leq J$. By elementary properties of controller subgroups [13,p.8] we have

Lemma 6.3. *If J is any annihilator ideal of KG, then*

$$J = (J \cap K\Delta(G))KG.$$

In particular, if $J \neq 0$, then $J \cap KH \neq 0$ for some finitely generated $H \triangleleft G$ such that $|G : C_G(H)| < \infty$. By Lemma 6.2, $J \cap KH$ is an annihilator ideal of KH. As is well known, we have $|H'| < \infty$, and so H has a finite characteristic subgroup T such that H/T is a free abelian group of finite rank. This allows us to view KH as a repeated skew Laurent polynomial extension of KT, hence demonstrating the connection with polynomial rings.

Skew polynomial rings are obtained as follows. Let R be a ring and σ an automorphism of R. The corresponding skew polynomial ring $R[X; \sigma]$ consists of all finite formal linear combinations $\sum_{n=0}^{\infty} r_n X^n$ with coefficients $r_n \in R$. These elements are added by adding coefficients of like powers of X, and multiplied using $Xr = r^\sigma X \quad (r \in R)$. Similarly we have the skew

Laurent polynomial ring $R[X^{\pm 1};\sigma]$, consisting of finite formal linear combinations $\sum_{n=-\infty}^{\infty} r_n X^n$, with similar operations. If we have an automorphism of $R[X,\sigma]$, we can form a further skew polynomial or skew Laurent extension, leading after finitely many iterations to repeated skew polynomial and skew Laurent extensions of R. We have the following extension of the theorem of McCoy already referred to.

Theorem 6.4. (Gilchrist [3]). *Let R be a ring and S be a repeated skew polynomial or skew Laurent extension of R. Let J be a non-zero annihilator ideal of S. Then $J \cap R$ is a non-zero annihilator ideal of R.*

Proof. By Lemma 6.2, the general case follows by induction from the case when $S = R[X;\sigma]$ or $S = R[X^{\pm 1};\sigma]$.

Suppose first that $S = R[X;\sigma]$. Let $J = r(A)$, where A is a right ideal of S. Suppose that $J \cap R = 0$, and choose a non-zero element $g(X)$ of minimal degree in J. Then

$$g(X) = a_0 + a_1 X + \ldots + a_r X^r \quad (a_i \in R)$$

where $a_r \neq 0$ and $r > 0$ by assumption. Since our assumption also implies that $a_r \notin r(A)$, we can choose $f = f(X) \in A$ such that $f(X)a_r \neq 0$. Write

$$f(X) = b_0 + b_1 X + \ldots + b_s X^s \quad (b_i \in R).$$

Since $f(X)a_r \neq 0$, we can choose t with $0 \leq t \leq s$ and $b_t a_r^{\sigma^t} \neq 0$.

Hence

$$b_t g(X)^{\sigma^t} \neq 0 \qquad (*)$$

where we allow σ to act on elements of S by acting on their coefficients. Choose t maximal subject to $(*)$. We have

$$0 = f(X)g(X) = \sum_{i=0}^{t} b_i g(X)^{\sigma^i} X^i.$$

Considering the leading coefficient, we have $b_t {a_r}^{\sigma^t} = 0$. Hence $b_t g(X)^{\sigma^t}$ is a non-zero element of S of degree $< r$, and $b_t^{\sigma^{-t}} g(X)$ is another such element, that furthermore belongs to J. This contradiction to the minimality of r establishes the result in the skew polynomial case.

Now let $S = R[X^{\pm 1};\sigma]$ and let $S_1 = R[X;\sigma] \leq S$. Let $J = r(A)$, where A is a two-sided ideal of S. We easily see that $A = S(A \cap S_1)$. Consequently, $J \cap S_1 = r(A \cap S_1) \cap S_1$, an annihilator ideal of S_1. If $J \neq 0$, then $J \cap S_1 \neq 0$, and the case already dealt with gives $J \cap R \neq 0$.

Taking account of Lemma 6.3 and the remarks after it, we deduce

Theorem 6.5. (Gilchrist [3]) *If K is any field, G is any group, and J is any non-zero annihilator ideal of KG, then $J \cap KH \neq 0$ for some finite normal subgroup H of G.*

For the rest of Theorem 5.3, we need a result on group rings of finite groups. This is taken from Gilchrist's thesis [3], but may have been known previously.

Theorem 6.6. *Let G be a finite group, K be any field, and I be any non-zero two-sided ideal of KG. Then $I \cap Z(KG) \neq 0$. Furthermore if I is invariant under some group Γ of automorphisms of G, then $I \cap Z(KG)$ contains a non-zero Γ-invariant element.*

Proof. If K has characteristic zero the claim follows from the fact that every two-sided ideal of KG is uniquely the direct sum of a number of such ideals, each generated by a centrally primitive idempotent, so we may suppose K has prime characteristic. Let $V_1, \ldots, V_s$ be a complete set of representatives for the irreducible (right) KG-modules, and let W_i be the sum of those minimal right ideals of KG that are isomorphic to V_i. Then $W_i \neq 0$ as is well known, and

$$W = W_1 \oplus \ldots \oplus W_s$$

is the socle of KG. Identify V_i with a minimal right ideal of KG (contained in W_i).

Let U be any right ideal of KG contained in W_i. If $\alpha \in KG$, then αU is either zero or a minimal right ideal of KG isomorphic to U. In any case, $\alpha U \leq W_i$. Therefore W_i is a two-sided ideal of KG. In fact, the W_i are exactly the minimal two sided ideals of KG. For any non-zero two sided ideal J of KG contains a minimal right ideal U. We have $U \leq W_i$ for some i. Let T be any minimal right ideal of KG contained in W_i. Then $U \cong T$ as right KG-modules. Let $\varphi : U \to T$ be a KG-isomorphism. Since KG is self-injective, φ can be extended to a right KG-map $\hat{\varphi} : KG \to KG$. We have $\hat{\varphi}(\beta) = \alpha\beta$ for all $\beta \in KG$, where $\alpha = \hat{\varphi}(1)$. Thus $T = \alpha U \leq J$. It follows that $W_i \leq J$, as claimed.

Now we show how to write down a canonical central element in W_i. Let $\rho_i : G \to (a_{\alpha\beta}(g))$ be the matrix representation of G obtained by a choice of basis $\{v_1, \ldots, v_n\}$ in V. Thus

$$v_\alpha g = \sum_{\beta=1}^{n} a_{\alpha\beta}(g) v_\beta \quad (1 \leq \alpha \leq n).$$

Let

$$w_{\alpha\beta} = \sum_{g \in G} a_{\alpha\beta}(g^{-1}) g.$$

If $h \in G$ then

$$\begin{aligned} w_{\alpha\beta} h = \sum_{g} a_{\alpha\beta}(g^{-1}) gh &= \sum_{g} a_{\alpha\beta}(hg^{-1}) g \\ &= \sum_{g,\gamma} a_{\alpha\gamma}(h) a_{\gamma\beta}(g^{-1}) g \\ &= \sum_{\gamma} a_{\alpha\gamma}(h) w_{\gamma\beta}. \end{aligned}$$

It follows that the K-linear map $v_\alpha \to w_{\alpha\beta} (1 \leq \alpha \leq n)$ is a KG-homomorphism. Hence $W_{i\beta} = \sum_{\alpha=1}^{n} K w_{\alpha\beta}$ is either zero or a minimal right ideal of KG isomorphic to V_i. We see that $W_{i\beta} \neq 0$ since $a_{\beta\beta}(1) = 1$. An easy calculation shows that $\sum_{\beta=1}^{n} W_{i\beta}$ is a two-sided ideal of KG, whence by what has already been proved,

$$\sum_{\beta=1}^{n} W_{i\beta} = W_i \tag{6.1}$$

When K is a splitting field for G, we can see by computing dimensions that the above sum is direct.

Now let

$$\xi_i = \sum_{\alpha=1}^{n} w_{\alpha\alpha}(g) = \sum_{g\in G}(\sum_{\alpha=1}^{n} a_{\alpha\alpha}(g^{-1}))g$$

$$= \sum_{g\in G} \chi_i(g^{-1})g \tag{6.2}$$

where $\chi_i(g^{-1})$ is the trace of $\rho_i(g^{-1})$, so that χ_i is the "natural character" associated to V_i.

Since χ_i is a class function, we see by trivial calculation that

$$\xi_i \in Z(KG) \cap W_i.$$

When K is a splitting field, the facts that $w_{\beta\beta} \neq 0$ and the sum (6.1) is direct, show that $\xi_i \neq 0$. In general this seems less obvious. We must show that the character χ_i is not identically zero on G.

Let E be the KG-endomorphism ring of V_i, let $u_1, \ldots, u_s$ be an E-basis of V_i, and let $w_1, \ldots, w_t$ be a K-basis of E. Thus $st = n$. Also E is a commutative field, a finite separable extension of K [1, VII. 1.10]. For $g \in G$, let

$$u_\alpha g = \sum_{\beta=1}^{s} \sigma_{\alpha\beta}(g)u_\beta \quad (\sigma_{\alpha\beta}(g) \in E)$$

Thus $\sigma : g \to (\sigma_{\alpha\beta}(g))$ is a homomorphism $G \to GL_s(E)$, and by the Jacobson density theorem, the matrices $\sigma(g)$ $(g \in G)$ span the full ring $M_s(E)$ of $s \times s$ matrices over E. Let S be the matrix trace map $M_s(E) \to E$ and T be the trace map from E to K. An easy calculation shows that $\chi_i = ST\sigma$. If $\chi_i(g) = 0$ for all $g \in G$, then ST vanishes on the K-span of matrices $\sigma(g)$, namely $M_s(E)$. However $T : M_s(E) \to E$ is trivially surjective, and $S : E \to K$ is surjective as E is a finite separable extension of K. When K is finite this is particularly clear. Thus χ_i is not identically zero on G, so $\xi_i \neq 0$, as required.

We have established that every non-zero two-sided ideal of KG contains at least one of the elements $\xi_i \in Z(KG)$.

If α is any automorphism of G, then α permutes the characters χ_i and hence permutes the central elements ξ_i, along with the direct summands W_i

of the socle of KG. If Γ is a group of automorphisms of KG leaving invariant a non-zero ideal of I of KG, and $\xi_i \in I$, then the sum of the elements in the Γ-orbit of ξ_i will be a non- zero Γ-invariant element in $Z(KG) \cap I$.

Proof of Theorem 5.3. This follows immediately from Theorems 6.5 and 6.6.

References

[1] N. Blackburn and B. Huppert, *Finite groups II*, Springer-Verlag, Berlin, 1982.

[2] J.M. Cohen. Aspherical 2-complexes, *J. Pure Appl. Algebra* **12** (1978), 101-110.

[3] A.D. Gilchrist, Ph.D. thesis, University of Manchester, 1987.

[4] J.A. Green, *Polynomial representations of* GL_n, Lecture Notes in Math. No. 830, Springer-Verlag, Berlin, 1980.

[5] K.W. Gruenberg, *Relation modules of finite groups*, CBMS Regional Conference series in Math., Vol. 25, American Math. Soc., Providence, R.I., 1976.

[6] N.D. Gupta, T. Laffey and M.W. Thomson, On the higher relation modules of a finite group, *J. Algebra* **59** (1979), 172-187.

[7] B. Hartley and A.I. Lichtman, Relative higher relation modules, *J. Pure Appl. Algebra* **22** (1981), 75-89.

[8] B. Hartley, Relative higher relation modules, and a property of irreducible K-representations of $GL_n(K)$, *J. Algebra* **104** (1986) 113-125.

[9] A.A. Kljačko, Lie elements in a tensor algebra , Sibirsk. Mat. Ž. **15** (1974), 1296-1304 (Russian); *Siberian Math. J.* **15** (1974). 914-921.

[10] L.G. Kovács, *Varieties of nilpotent groups of small class,* Lecture Notes in Math. No. 697, Springer-Verlag, Berlin, 1978.

[11] O. Macedónska-Nosalska, Note on automorphisms of a free abelian

group, *Canad. Math. Bull.* **23** (1980), 111-113.

[12] N. McCoy. Annihilators in polynomial rings. *Amer. Math. Monthly* **64** (1957), 28-29.

[13] I.B.S. Passi, Annihilators of relation modules II, *J. Pure Appl. Algebra* **6** (1975), 235-237.

[14] D.S. Passman, *The algebraic structure of group rings*, Wiley-Interscience, New York, 1977.

University of Manchester
Manchester M13 9PL
United Kingdom

Some countably free groups

Graham Higman

1. Introduction

We recall first some definitions. If κ is an infinite cardinal, a group is said to be *κ-free* if each of its subgroups of rank less than κ is free. A subgroup of a κ-free group is said to be *κ-pure* if it has rank less than κ, and is a free factor of every subgroup of rank less than κ containing it. A group is said to be *strongly κ-free* if it is κ-free, and every subgroup of rank less than κ is contained in a κ-pure subgroup. In this paper we shall be concerned with the cases $\kappa = \omega$ and $\kappa = \omega_1$, and we shall use the terms "locally free" and "countably free" instead of "ω-free" and "ω_1-free", because (at least to group theorists) they are more evocative. Notice that we rarely need to mention strongly locally free groups, because of the following lemma, which was proved, using different terminology, in [3].

Lemma 1.1. *A locally free group is countably free if and only if it is strongly locally free.*

In [3], I constructed a countably free group of cardinality ω_1 which is not free. In [6], among many other things, Alan Mekler showed that my group is not strongly countably free, and constructed 2^{ω_1} non–isomorphic strongly countably free groups of cardinality ω_1. Now one remarkable feature of these constructions is a certain economy of means. The groups are constructed as unions of ascending towers, and the basic building block is the pair (A, B) of free groups of countable rank, where A is freely generated by $\{a_i, i < \omega\}$ and B by $\{b_i, i < \omega\}$ and A is embedded in B by setting $a_i = b_i b_{i+1}^{-2}, i < \omega$. In my construction this block is used straightforwardly at every level, and in Mekler's it is used more subtly, in combination with free products, at a selected set of levels. But it remains the same block. The present paper has its origin in the observation that there are many pairs (A, B) sufficiently like the pair described above to substitute for it in the construction, and that the choice of which pairs to use is likely to influence the group constructed. We shall use the term *blocking pair* for a pair (A, B) which can be used in this way. We give a precise definition in section 2, and show that it works, besides developing a few further properties of blocking pairs. Each of sections 3 and 4 contains an application of the basic idea.

In section 3, we ask questions about $G/v(G)$, where G is the group constructed, and v is a verbal functor. This is suggested by an argument of Mekler, who proves in [6] that the group G constructed in [3] is not strongly countably free by observing that G/G^2 is countable. For the general verbal functor we ask two questions, which coincide for the functor $v(G) = G^2$, because G/G^2 is abelian. The first is : When is $G/v(G)$ countable? The second is : When is $G/v(G)$ countably normally generated (that is, generated by the elements of a countable set of conjugacy classes)? We obtain usable answers to both these questions, and we give examples to show the sort of thing that can happen. For instance, using the answer to the second question for the trivial functor $v(G) = 1$, we show how to construct a countably free group of cardinality ω_1, which is itself countably normally generated.

In section 4 we tackle more directly the question of the influence of the choice of blocking pairs on the result of the construction. This result is the union of a smooth tower $\{A_\alpha, \alpha < \omega_1\}$ of groups, where each pair $(A_\alpha, A_{\alpha+1})$ is a blocking pair. We call such a tower a *blocking tower*, and the sequence (of length ω_1) formed by the isomorphism types of the pairs $(A_\alpha, A_{\alpha+1})$ we call the *type sequence* of the tower. Thus the question is : How far can the type sequence of a blocking tower be reconstructed from the union of the tower? Now if Ω is a closed unbounded subset of ω_1, and $\{A_\alpha, \alpha < \omega_1\}$ is a blocking tower, so is $\{A_\alpha, \alpha \in \Omega\}$. Thus we can only hope to recover the type sequence if it consists of indecomposable types (where the blocking pair (A, B) is said to be indecomposable if there is no group C such that both (A, C) and (C, B) are blocking pairs), and even then we can only hope to recover it up to ultimate equality (where two sequences of length ω_1 are said to be ultimately equal if there is a countable ordinal ρ such that they agree at the α-th place for all $\alpha \geq \rho$). What we prove in section 4 is a weak theorem along these lines. It is weak first because we have to restrict severely the types of blocking pair that are allowed to occur, and secondly because we recover, up to ultimate equality, not the types (A, B), but only the quotient-types, that is the isomorphism types of the quotient groups B/A^B. Even with these weaknesses, the theorem is strong enough to have as a corollary the statement that there exist 2^{ω_1} non-isomorphic countably free groups of cardinality ω_1. However, in view of Mekler's result on strongly countably free groups, one wants to strengthen this statement, by showing that the groups differ in some stronger way than by mere non-isomorphism; and there is a natural way to do this. Let $\mathcal{L}_{\infty,\kappa}$, where κ is an infinite cardinal, be the language for group theory usually so denoted, so that the atomic formulae are those of first order group theory, but in addition to the usual logical connectives, we allow conjunctions and disjunctions of arbitrary sets of formulae, and quantification over sets of variables of cardinality less than κ. Then (see Mekler [6] and references there cited) a group of cardinality at least κ is $\mathcal{L}_{\infty,\kappa}$-equivalent to a free group of rank at least κ if and only if it is strongly κ-free. Thus (using Lemma 1.1) countably free groups form an $\mathcal{L}_{\infty,\omega}$-equivalence class. So it is natural to want to prove that the 2^{ω_1} countably free groups of cardinality ω_1 that we construct are $\mathcal{L}_{\infty,\omega_1}$-inequivalent, and this we do. (It should be said that Mekler [6] asserted without proof the existence of 2^{ω} $\mathcal{L}_{\infty,\omega_1}$-inequivalent countably free groups of cardinality ω_1, and that he has explained (*in litt.*) how his method would in fact give 2^{ω_1} such groups).

In the concluding section of the paper we bring together some rather obvious questions, raised by the considerations of Section 4, to which we do not know the answers.

2. Blocking pairs and blocking towers

We begin with the basic definition.

Definition 2.1. By a *blocking pair* we mean a pair (A, B) such that

(i) A, B are free groups of countably infinite rank;
(ii) A is a proper subgroup of B, but is not contained in a proper free factor of B;
(iii) every finitely generated free factor of A is a free factor of B.

We next prove two lemmas which establish closure properties of the relation of being a blocking pair.

Lemma 2.2. *If (A, B) and (B, C) are both blocking pairs, then so is (A, C).*

The only part of the definition which is not trivial is that A is not contained in a proper free factor of C. But if X is a free factor of C containing A then $B \cap X$ is a free factor of B containing A, and since (A, B) is a blocking pair, $B \cap X = B$. That is X is a free factor of C containing B, and since (B, C) is a blocking pair, $X = C$. The lemma follows. (Here, and throughout the paper, we use without mention the fact that if X is a free factor of a group G, then for any subgroup H of G, $X \cap H$ is a free factor of H).

Lemma 2.3. *Let the group G be the union of the chain $\{A_i, i \in I\}$ of subgroups without greatest element, and suppose that for i, j in I, whenever $A_i < A_j, (A_i, A_j)$ is a blocking pair. Then G is countably free; and if G is countable, (A_i, G) is a blocking pair for each i in I.*

Each A_i is a free group, and any finitely generated subgroup of G is contained in A_i for some i, so G is locally free. We show that if X is a finitely generated free factor of some A_i, then X is locally pure in G. To that end, suppose that $X < Y < G$, where Y is finitely generated. Then $X < Y < A_j$ for some j in I, where we may suppose $A_i < A_j$. Then (A_i, A_j) is a blocking pair. So X is a free factor of A_j, and hence of Y, as required. But any finitely generated subgroup of G is contained in some A_i, and hence in some finitely generated free factor of A_i. So G is strongly locally free, and so, by Lemma 1.1, countably free. Now suppose that G is countable. Then G is free, necessarily of infinite rank, because the chain $\{A_i\}$ has no greatest element. Thus (i) of Definition 2.1 holds for the pair (A_i, G) as does the first part of (ii). To check the second half of (ii), assume that X is a free factor of G containing A_i. Then, for all j such that $A_i < A_j, X \cap A_j$ is a free factor of A_j containing A_i, and since (A_i, A_j) is a blocking pair, $A_j \leq X$. Then $G = \cup A_j \leq X$, so that, as required, X is not a proper free factor. Finally (iii) is clear, since we have already seen that a finitely generated free factor of A_i is locally pure in G, and a locally pure subgroup of a free group is a free factor. This completes the proof of the lemma.

We next explain how blocking pairs can be used to construct countably free groups of cardinality ω_1 which are not strongly countable free.

Definition 2.4. A *blocking tower* is a set $\{A_\alpha, \alpha < \omega_1\}$ of groups indexed by the countable ordinals such that

(i) for each countable ordinal α, $(A_\alpha, A_{\alpha+1})$ is a blocking pair;
(ii) for each countable limit ordinal α, $A_\alpha = \bigcup_{\beta<\alpha} A_\beta$.

Lemma 2.5. *If $\{A_\alpha, \alpha < \omega_1\}$ is a blocking tower, then (A_α, A_β) is a blocking pair whenever $\alpha < \beta < \omega_1$.*

This follows immediately from Lemma 2.2 and 2.3 by transfinite induction.

Theorem 2.6. *The union of the groups in a blocking tower is a countably free group of cardinality ω_1 which is not strongly countably free.*

If $G = \cup A_\alpha$, where $\{A_\alpha, \alpha < \omega_1\}$ is a blocking tower, then by Lemma 2.5, G satisfies the hypothesis of Lemma 2.3, and so is countably free. The A_α are countable, and $A_\alpha < A_\beta$ if $\alpha < \beta$, so G has cardinality ω_1. If G is strongly countably free, A_0 is contained in a countably pure subgroup B. Then B is countable, so $A_\alpha \not\leq B$, for some α. The countable purity of B implies that B is a free factor of $\langle B, A_\alpha \rangle$, and hence that $B \cap A_\alpha$ is a proper free factor of A_α containing A_0, contrary to the fact that (A_0, A_α) is a blocking pair. The contradiction proves that G is not strongly countably free, and completes the proof of the theorem.

Next we want to show that many blocking towers exist. We say that blocking pairs (A_i, B_i) $i = 1, 2$ are *isomorphic* if there is an isomorphism $\theta : B_1 \to B_2$ with $A_1\theta = A_2$; and when we speak of the type of a blocking pair, we mean its isomorphism type. We shall also speak of the *quotient-type* of the pair (A, B), meaning the isomorphism type of the factor group B/A^B. We shall use the expression *type sequence* to mean a sequence of length ω_1 of types of blocking pairs. In particular, the type sequence of the blocking tower $\{A_\alpha, \alpha < \omega_1\}$ is the sequence whose α-th element is the type of $(A_\alpha, A_{\alpha+1})$ for all $\alpha < \omega_1$. We shall use the expression *quotient-type sequence* similarly.

Theorem 2.7. *Every type sequence is the type sequence of a blocking tower.*

We assume that $\{f(\alpha), \alpha < \omega_1\}$ is a type sequence, and construct the blocking tower by transfinite recursion. We first choose A_0, A_1 so that (A_0, A_1) is a blocking pair of type $f(0)$. Now assume that for some ordinal α in $2 \leq \alpha < \omega_1$ groups A_β have been chosen, for all $\beta < \alpha$, so that (i) for all $\beta < \gamma < \alpha$, (A_β, A_γ) is a blocking pair, (ii) for all β such that $\beta + 1 < \alpha$, the type of $(A_\beta, A_{\beta+1})$ is $f(\beta)$, and (iii) for limit ordinals β with $\beta < \alpha, A_\beta = \bigcup_{\gamma<\beta} A_\gamma$. We have to show that we can choose a group A_α so that these properties persist. If $\alpha = \beta + 1$, we choose a blocking pair (C, D) of type $f(\beta)$, identify A_β with C (as is possible, since both are free groups of countable infinite rank) and take $A_\alpha = D$. The only part of the extended conditions which is not automatic is that if $\gamma < \alpha$ (A_γ, A_α) is a blocking pair, and this follows from Lemma 2.2. If α is a limit ordinal, we set $A_\alpha = \bigcup_{\beta<\alpha} A_\beta$, as we must, and Lemma 2.3 assures us that the extended conditions holds. The theorem is proved.

In order to apply Theorems 2.6 and 2.7 we need a supply of blocking pairs, so we list some. In each case we are talking about a pair (A, B) where A is freely generated by $\{a_i, i < \omega\}$ and B by $\{b_i, i < \omega\}$; and Q is the factor group B/A^B, that is, the group generated by $\{b_i, i < \omega\}$ subject to the relations $a_i = 1, i < \omega$. Condition (i) of Definition 2.1 will be automatically satisfied, but conditions (ii) and (iii) will need to be checked.

Example 2.8 (i). Suppose first that A is embedded in B by setting $a_i = b_i b_{i+1}^{-n_i}, i < \omega$, where each n_i is an integer greater than 1. Then Q is a noncyclic torsion free abelian group of rank 1, and by choice of the integers n_i, can be taken to be any such group. Since $Q \neq 1$, A is a proper subgroup of B, and since Q has no cyclic free factor, A is not contained in a proper free factor of B. So part (ii) of Definition 2.1 holds. Moreover, directly from the equations, for any positive integer k,

$\langle a_0, a_1, \ldots, a_k, b_{k+1} \rangle = \langle b_0, b_1, \ldots, b_k, b_{k+1} \rangle$. Thus $\langle a_0, \ldots, a_k \rangle$ is a free factor of $\langle b_0, \ldots, b_{k+1} \rangle$, and hence of B. Since any finitely generated free factor of A is a free factor of $\langle a_0, \ldots, a_k \rangle$ for some k, part (iii) of Definition 2.1 holds, and we have a blocking pair.

Example 2.8 (ii). Suppose next that A is embedded in B by setting $a_i = b_i b_{i+1}^{-1} b_{i+2}^{-1} b_{i+1} b_{i+2}$, $i < \omega$. In this case Q is a nontrivial perfect group, and the argument goes through, with the minor modification that now $\langle a_0, \ldots, a_k, b_{k+1}, b_{k+2} \rangle = \langle b_0, \ldots, b_k, b_{k+1}, b_{k+2} \rangle$, to show that again we have a blocking pair.

Example 2.8 (iii). Suppose lastly that A is embedded in B by setting $a_i = b_{i+1}^{-1} b_i b_{i+1}$, $i < \omega$. In this case $Q = 1$, so we have to use a different method of proving that A is a proper subgroup of B. However, as words in the b_i, the elements a_i have the Nielsen property (see e.g., Magnus, Karrass and Solitar [5]), so that A contains no nontrivial element whose length as a reduced word in the b_i is less than 3. The remainder of the argument goes through as before, so again we have a blocking pair.

We next develop a little of the theory of blocking pairs and blocking towers, mainly for use in later sections.

Lemma 2.9. *If (A, B) is a blocking pair and X is a finitely generated free factor of B, then $A \cap X$ is a free factor of X.*

$A \cap X$ is certainly a free factor of A, so that a finitely generated free factor Y of $A \cap X$ is a free factor of A, and hence also of B, because (A, B) is a blocking pair. Since $Y \leq X$, Y is a free factor of X. Since Y was any finitely generated free factor of $A \cap X$, it follows that the rank of $A \cap X$ is at most the rank of X. But then $A \cap X$ is finitely generated, so that in the first part of the argument we can take Y to be $A \cap X$ itself, and obtain the desired result.

Recall next that a *retraction* of a group G is an endomorphism $\theta : G \to G$ such that $\theta^2 = \theta$, and that a *retract* of G is the image of G under a retraction. In particular, a free factor is a retract. We shall say that a subgroup H of a group G is a *local retract* of G if each finite subset of G is contained in a subgroup X such that $X \cap H$ is a retract of X.

Lemma 2.10. *If (A, B) is a blocking pair, A is a local retract of B.*

Any finite subset of B is contained in a finitely generated free factor X, and by Lemma 2.9, $X \cap A$ is a retract of X.

Corollary 2.11. *If G is the union of the blocking tower $\{A_\alpha, \alpha < \omega_1\}$ then each A_α is a local retract of G.*

A finite subset of G is contained in some A_β, where we may suppose $\alpha < \beta$. By Lemma 2.5, (A_α, A_β) is a blocking pair, so by Lemma 2.10 the required subgroup can be found within A_β.

We note next an important property that local retracts share with retracts.

Lemma 2.12. *If H is a local retract of G, and $N \triangleleft H$, then $H \cap N^G = N$.*

If $g \in H \cap N^G$ then $g \in H_0 \cap N_0^{G_0}$ for suitable finitely generated subgroups G_0, H_0, N_0 of G, H, N. Let K be a subgroup containing G_0, H_0 and N_0 such that there exists a retraction $\theta : K \to K \cap H$. Then $g = g\theta \in H_0\theta \cap (N_0\theta)^{G_0\theta} \leq H \cap N^H = N$, as required.

Corollary 2.13. *Let G be the union of the blocking tower $\{A_\alpha, \alpha < \omega_1\}$ and let N be a normal subgroup of A_0. Then $\{A_\alpha/N^{A_\alpha}, \alpha < \omega_1\}$ is a smooth tower of groups whose union is G/N^G.*

It follows from Lemma 2.12 and the fact that $(A_\alpha, A_{\alpha+1})$ is a blocking pair that A_α/N^{A_α} embeds in $A_{\alpha+1}/N^{A_{\alpha+1}}$, and the rest is obvious.

Lemma 2.14. *If (A, B) is a blocking pair, then $Q = B/A^B$ is locally free.*

Let $b_1, \ldots, b_r$ be elements of B, and suppose that $\langle b_1, \ldots, b_r\rangle A^B/A^B$ cannot be generated by fewer than r elements. We shall show that $b_1A^B, \ldots, b_rA^B$ are free generators of this group. If not, there is a nontrivial word w

such that $w(b_1, \ldots, b_r) \in A^B$, and we can clearly choose a finite generated free factor X of B so that $\langle b_1, \ldots, b_r\rangle \leq X$, and $w(b_1, \ldots, b_r) \in (A \cap X)^X$. By Lemma 2.9, X is the free product $Y * (A \cap X)$ for suitable Y. We denote by θ the retraction $X \to Y$, so that ker $\theta = (A \cap X)^X$. Then $b_i\theta = b_i\theta^2$, so that $b_i = b_i\theta \bmod (A \cap X)^X$ for each i, whence $\langle b_1, \ldots, b_r\rangle A^B = \langle b_1\theta, \ldots, b_r\theta\rangle A^B$. Now $\langle b_1\theta, \ldots, b_r\theta\rangle$ is a subgroup of a free group and therefore free, and $w(b_1\theta, \ldots, b_r\theta) = w(b_1, \ldots, b_r)\theta = 1$, since $w(b_1, \ldots, b_r) \in (A \cap X)^X$. But of course a free group of rank r cannot be generated by r elements between which there is a nontrivial relation. So $\langle b_1\theta, \ldots, b_r\theta\rangle$ has rank less than r, whence $\langle b_1, \ldots, b_r\rangle A^B/A^B = \langle b_1\theta, \ldots, b_r\theta\rangle\, A^B/A^B$ can be generated by fewer than r elements, contrary to hypothesis. The contradiction shows that $b_1A^B, \ldots, b_rA^B$ are indeed free generators of $\langle b_1, \ldots b_r\rangle A^B/A^B$, and establishes the lemma.

Corollary 2.15. *If G is the union of the blocking tower $\{A_\alpha, \alpha < \omega_1\}$ then, for each $\alpha, G/A_\alpha^G$ is locally free.*

The case $\alpha = 0$ follows from Corollary 2.13 and Lemma 2.14 (using Lemma 2.5). The general case follows, because $\{A_\beta, \alpha \leq \beta < \omega_1\}$ is also a blocking tower.

3. Verbal properties

Recall that a *verbal functor* v is defined by a set V of elements of a free group F, and that, for any group G, $v(G)$ is the subgroup generated by the images of elements of V in homomorphisms of F into G, the *instances*, as we say, of words in V in G. Typical verbal functors are the n-th power functor, assigning to G the subgroup G^n, and the commutator functor, assigning to G the derived subgroup G'. If F is an uncountable free group, and v is not the functor which assigns to each group G the group G itself, then $F/v(F)$ is a relatively free group of uncountable rank in a nontrivial variety, and so cannot be countable, or even countably normally generated; and the same holds for a strongly countably free group, because countability and countable normal generation can both be formulated in $\mathcal{L}_{\infty,\omega_1}$. For groups which are merely countably free the case is different, and in particular for the unions of blocking towers we have the following criteria :

Theorem 3.1. *Let G be the union of the blocking tower $\{A_\alpha, \alpha < \omega_1\}$ and write $Q_\alpha = A_{\alpha+1}/A_\alpha^{A_{\alpha+1}}, \alpha < \omega_1$. Let v be a verbal functor. Then*

(i) *$G/v(G)$ is countable if and only if, for all sufficiently large α, $A_{\alpha+1} = v(A_{\alpha+1})A_\alpha$.*

(ii) *$G/v(G)$ is countably normally generated if and only if, for all sufficiently large α, $v(Q_\alpha) = Q_\alpha$.*

Note that the condition $v(Q_\alpha) = Q_\alpha$ can also be written $A_{\alpha+1} = v(A_{\alpha+1})A_\alpha^{A_{\alpha+1}}$, so that the analogy between the two halves is stricter than might appear. The sufficiency of the condition requires only that the tower $\{A_\alpha\}$ be smooth, that is, that for limit ordinals α, $A_\alpha = \bigcup_{\beta<\alpha} A_\beta$. Suppose in fact that the tower is smooth, and that, for all $\alpha \geq \rho, A_{\alpha+1} = v(A_{\alpha+1})A_\alpha$. Then for any subgroup B of G, if $\alpha \geq \rho$, $A_\alpha \leq v(G)B$ implies $A_{\alpha+1} \leq v(G)B$. Using this at successor ordinals and smoothness at limits, we see, by transfinite induction, that $A_\rho \leq v(G)B$ implies $A_\alpha \leq v(G)B$ for all $\alpha \geq \rho$ and so implies $G = v(G)B$. In particular, $G = v(G)A_\rho$, so that $G/v(G)$ is countable. If we only assume that for $\alpha \geq \rho$, $A_{\alpha+1} = v(A_{\alpha+1})A_\alpha^{A_{\alpha+1}}$, the same argument works as long as B is normal in G, to give $G = v(G)A_\rho^G$, whence $G/v(G)$ is countably normally generated.

Turn now to the necessity of the conditions. If $G/v(G)$ is countable then, for all sufficiently large α, A_α contains a set of coset representatives for $v(G)$, so that $G = v(G)A_\alpha$. In particular, for g in $A_{\alpha+1}$.

(a) $g = v_1 \ldots v_r g_0$

where $v_1 \ldots, v_r$ are instances in G of the defining words of v, and g_0 belongs to A_α. By Corollary 2.11 $A_{\alpha+1}$ is a local retract of G, so that we can choose a subgroup H of G such that g and g_0 belong to H and each v_i is an instance in H of a defining word of v, and a retraction $\theta : H \to A_{\alpha+1} \cap H$. Then $g\theta = g, g_0\theta = g_0$, and each $v_i\theta$ is an instance in $H\theta = A_{\alpha+1} \cap H$ of a defining word of v. Thus applying θ to (a) gives $g \in v(A_{\alpha+1})A_\alpha$. That is, $A_{\alpha+1} = v(A_{\alpha+1})A_\alpha$, as required. If $G/v(G)$ is countably normally generated then, for all sufficiently large α, A_α meets each of the countable set of conjugancy classes which generate G *modulo* $v(G)$. Thus instead of (a) we have

(b) $g = v_1 \ldots v_r g_1^{x_1} \ldots g_s^{x_s}$

where $g_1, \ldots, g_s$ belong to A_α, and $x_1, \ldots, x_s$ to G. The argument goes through as before, provided H is taken large enough to contain $g_1,\ldots,g_r$ and $x_1,\ldots,x_r$, to give $g \in v(A_{\alpha+1})A_\alpha^{A_{\alpha+1}}$. The theorem follows.

Before giving applications of the theorem it is perhaps as well to note that its apparent generality is a little spurious, because the statements that $G/u(G)$ and $G/v(G)$ are countably normally generated may well be equivalent for different functors u and v. For instance $G/G^{n^r}, r > 1$, is countably normally generated if and only if G/G^n is; and a soluble group G is countably normally generated if and only if G/G' is countable. In the examples which follow, (A, B) is a blocking pair, $Q = B/A^B$, and $G(A, B)$ is the union of a blocking tower whose type sequence is constant, being equal to the type of (A, B) for all α.

Example 3.2. *There exists a countably free group of cardinality ω_1 which is countably normally generated.*

By Theorem 3.1 (ii), we may take $G(A, B)$ for any pair (A, B) such that $Q = 1$, and Example 2.8 (iii) gives such a pair.

Example 3.3. *There exists a countably free group G of cardinality ω_1 such that G/G' is countable but G is not countably normally generated.*

Of course, G/G' is countable if and only if it is countably normally generated. Thus by Theorem 3.1 (ii), we may take $G(A, B)$ for any pair (A, B) with $Q = Q' \neq 1$, and Example 2.8 (ii) gives such a pair.

Example 3.4. *If Π is any set of primes, there exists a countably free group G of cardinality ω_1 such that G/G^p, p prime, is countably normally generated if and only if $p \in \Pi$.*

There exists a noncyclic torsion-free abelian group X of rank 1 such that $X = X^p$ if and only if $p \in \Pi$; for if Π is nonempty X can be taken to be the additive group of rational numbers with denominators divisible only by primes in Π, and if Π is empty X can be taken to be the additive group of rationals with square free denominators. Then a blocking pair (A, B) can be chosen, of the kind described in Example 2.8 (i), so that $Q \cong X$, and

$G(A, B)$ has the required properties.

Example 3.5. *If G is the countably free group of cardinality ω_1 constructed in [3], then*
(i) *G/G^2 is countable;*
(ii) *G/G^{2^n} is countably normally generated for all $n \geq 1$; but*
(iii) *$G/(G^2)^2$ is uncountable.*

$G = G(A, B)$, where (A, B) is the blocking pair with $a_i = b_i b_{i+1}^{-2}$ for all $i < \omega$. Thus Q is the additive group of rationals with denominator a power of 2, and $Q = Q^{2^n}$ for all n. It follows at once that G/G^{2^n} is countably normally generated for all n, and that G/G^2, being abelian, is countable. This proves (i) and (ii). But (iii) requires a further calculation, which is most conveniently carried out in the free group of countably infinite rank in the variety $A_2 A_2$ of groups G with a normal subgroup K such that K and G/K are both elementary abelian 2-groups. These free groups are studied in satisfying detail in Kovács and Newman [4], and the methods of that paper can be used to establish the following fact.

Fact 3.6. *Let D be the free group in $A_2 A_2$ freely generated by $\{d_i, i < \omega\}$, and let C be the subgroup $\langle c_i, i < \omega\rangle$, where $c_i = d_i d_{i+1}^{-2}, i < \omega$. Then C is a proper subgroup of D.*

Because D is a (relatively) free group, there is an endomorphism θ such that $d_i\theta = c_i$, $i < \omega$, whence $D\theta = C$, so that what we have to prove is that $D\theta < D$. In [4] Kovács and Newman give an explicit basis for the derived group D', consisting of left normed commutators in the d_i, which includes in particular the elements k_i, $1 \leq i < \omega$, where $k_i = [d_i, d_0, d_1, \ldots, d_i]$. If we denote the basis by $\{h_i,\ i < \omega\}$ it is not hard, using the methods of calculation established in [4], to see (i) that $g\theta \in D'$ implies $g \in D'$, (ii) that for $i < \omega, h_i\theta = h_i x_i$, where x_i is product of basis elements weight greater than the weight of h_i; and (iii) in particular that $k_i\theta = k_i k_{i+1}, 1 \leq i < \omega$. From this it is clear that if there is an element g such that $g\theta = k_1$ then, for $1 \leq i < \omega, g \equiv k_1 k_2 \ldots k_i$ *modulo* terms of greater weight. This is impossible because g is uniquely a finite product of h_j's. Thus $k_1 \notin D\theta$, which proves the result.

Returning now to the blocking pair (A, B), there is a homomorphism of B to D with kernel $(B^2)^2$ in which the image of A is C. Taking inverse images under this homomorphism, Fact 3.6 translates into the statement that $(B^2)^2A$ is a proper subgroup of B. This is precisely what is needed to deduce from part (i) of Theorem 3.1 that $G/(G^2)^2$ is uncountable.

Example 3.7. Finally we mention briefly the situation when $G = G(A, B)$ where $a_i = b_i b_{i+1}^{-n}$, $i < \omega$, and v is the functor $v(G) = G^m$, n and m being arbitrary positive integers greater than 1. Then Q is the additive group of rationals with denominator a power of n, and there are three cases, according to the relationship between m and n. First, if m divides $n, B = B^m A$, so that G/G^m is countable and *a fortiori* countably normally generated. Secondly, if there is a prime dividing m which does not divide n, then $Q \neq Q^m$, so that G/G^m is not countably normally generated, and *a fortiori* not countable. Lastly there is the case when m divides a power of n, but does not divide n itself. In this case $Q = Q^m$, so that G/G^m is countably normally generated, but presumably a calculation along the lines of that establishing Fact 3.6 will show that $B \neq B^m A$, so that G/G^m is not countable. I have only carried out this calculation in case $n = p$ (prime) and $m = p^2$. In this case we get a result similar to Example 3.5, replacing 2 by the odd prime p, and allowing for the fact that G/G^p is no longer abelian. Specifically, we get a countably free group G of cardinality ω_1 such that G/G^{p^n} is countably normally generated for all n, but $G/(G^{\hat{p}})^{\hat{p}}$ is not countable, where $G^{\hat{p}} = G^p G'$.

4. Translucent blocking towers

In this section we consider a group G which is the union of a blocking tower $\{A_\alpha, \alpha < \omega_1\}$ in which each blocking pair $\{A_\alpha, A_{\alpha+1}\}$ is of the kind described in Example 2.8 (i), and ask how far G determines the type sequence of the tower. It is convenient to have a name, so we repeat the description.

Definition 4.1. A *translucent blocking pair* is a blocking pair (A, B) in which the generators $\{a_i,\ i < \omega\}$ of A and $\{b_i,\ i < \omega\}$ of B are related by $a_i = b_i b_{i+1}^{-n_i}$, where each n_i is an integer greater than 1. A *translucent blocking tower* is a blocking tower $\{A_\alpha, \alpha < \omega_1\}$ such that for all $\alpha < \omega_1$ $(A_\alpha, A_{\alpha+1})$ is a translucent blocking pair.

Notice that the set of quotient-types of translucent blocking pairs is the set T of isomorphism types of noncyclic torsion-free abelian groups of rank 1. T is, of course, a well understood set of cardinality 2^ω (see, for instance, Fuchs [2], Vol. II, pp. 145 ff.). Our first objective is to prove :

Theorem 4.2. *If the same group is the union of two translucent blocking towers, then the quotient-type sequences of these towers are ultimately equal.*

There are difficulties about a direct attack on this theorem. If $G = \cup A_\alpha = \cup C_\alpha$, where $\{A_\alpha, \alpha < \omega_1\}$ and $\{C_\alpha, \alpha < \omega_1\}$ are translucent blocking towers, a standard argument shows that there is a closed unbounded subset Ω of ω_1 such that for $\alpha \in \Omega$, $A_\alpha = C_\alpha$. But unless also $A_{\alpha+1} = C_{\alpha+1}$, it is not clear that we can conclude, even for such α, that the quotient-types of $(A_\alpha, A_{\alpha+1})$ and $(C_\alpha, C_{\alpha+1})$ are the same. We get round this difficulty by considering the normal closures A_α^G of the groups A_α in G. We make another definition.

Definition 4.3. Let f be a map from ω_1 to the set T of isomorphism types of noncyclic torsion-free abelian groups of rank 1. By a *translucent normal tower of type f for a group G* we mean a set $\{N_\alpha, \alpha < \omega_1\}$ of normal subgroups of G such that

(i) for each countable ordinal α, N_α is the normal closure of a countable subgroup of G;

(ii) for each countable ordinal α, the maximal abelian subgroups of G/N_α are all infinite cyclic, except for one conjugacy class of subgroups of type $f(\alpha)$; if X/N_α belongs to this conjugacy class, then $X^G = N_{\alpha+1}$;

(iii) for countable limit ordinals α, $N_\alpha = \bigcup_{\beta<\alpha} N_\beta$;

(iv) $G = \bigcup_{\alpha<\omega_1} N_\alpha$.

Then we prove :

Theorem 4.4. *If G is the union of the translucent blocking tower $\{A_\alpha, \alpha < \omega_1\}$, with quotient-type sequence f, and $N_\alpha = A_\alpha^G, \alpha < \omega_1$, then $\{N_\alpha, \alpha < \omega_1\}$ is a translucent normal tower of type f for G.*

Theorem 4.2 is almost a corollary of this, because it is easy to prove :

Theorem 4.5. *If f, g are maps from ω_1 to T, and the same group has a translucent normal tower of type f and a translucent normal tower of type g, then f and g are ultimately equal.*

We turn finally to the question of the logical language in which the differences between the groups can be expressed. The basic result is :-

Lemma 4.6. *If f is a map from ω_1 to T there is a sentence of $\mathcal{L}_{\omega_2,\omega_1}$ which is true of a group G if and only if G has a translucent normal tower of type f.*

This enables us to strengthen Theorem 4.2.

Theorem 4.7. *Groups which are unions of translucent blocking towers whose quotient-type sequences are not ultimately equal are not $\mathcal{L}_{\omega_2,\omega_1}$-equivalent.*

It is a corollary of Theorem 4.7 that there exist 2^{ω_1} $\mathcal{L}_{\omega_2,\omega_1}$-inequivalent countably free groups of cardinality ω_1. For it follows from the cardinal equality $\omega_1^2 = \omega_1$ that ω_1 can be expressed as the union of ω_1 disjoint unbounded subsets, say $\omega_1 = \bigcup_{\alpha<\omega_1} \Omega_\alpha$. There are 2^{ω_1} maps $\omega_1 \to T$ which are constant on each Ω_α, and two such maps are ultimately equal only if they are equal. In fact one can use the methods of Section 3 to impose further conditions on the groups. For instance, if Π is a set of primes with infinite complement, one can insist that, for each group G in the set, G/G^p is countably normally generated if and only if $p \in \Pi$. However, progress in this direction is restricted by the need to use only translucent blocking pairs.

Turning now to proofs, we begin with the technical lemma on which Theorem 4.4 depends.

Lemma 4.8. *Let $a_i, i < \omega$, be elements of a group A, and suppose that the subgroup $\langle a_i, i < \omega \rangle$ is not finitely generated. Let B be the group generated by A and elements $b_i, i < \omega$, subject to the defining relations $a_i = b_i b_{i+1}^{-n_i}, i < \omega$, where each n_i is an integer greater than 1. Let X be an abelian subgroup of B. Then* (i) *If $X \cap A \neq 1, X \leq A$ and* (ii) *if $X \cap A^g = 1$ for all g in B, X is free cyclic, or trivial.*

Let $B_i, i < \omega$, be the free product $A * \langle b_i \rangle$ where b_i has infinite order, and identify B_i with a subgroup of $B_{i+1}, i < \omega$, by putting $b_i = a_i b_{i+1}^{n_i}$. Then B is the union of the tower $\{B_i, i < \omega\}$. If $X \leq B_i$ for any i, Kurosh's subgroup theorem (see e.g. Magnus, Karrass and Solitar [5]) guarantees the truth of the lemma. Thus (i) is true in any case, and (ii) is true provided that, for any abelian subgroup X of B, the sequence $\{X \cap B_i, i < \omega\}$ is ultimately stationary. But we can also think of the step from B_i to B_{i+1} as being one of freely adjoining b_{i+1} as an n_i-th root of $a_i^{-1} b_i$, so that an abelian subgroup X of B_{i+1} satisfying $X > X \cap B_i > 1$ must be conjugate in B_{i+1} to a subgroup of $\langle b_{i+1} \rangle$. Thus it is sufficient to show that, for any i and for large enough j, no conjugate of $\langle b_i \rangle$ in B_j meets $\langle b_j \rangle$ nontrivially. To that end, we consider the normal form of b_i as an element of the free product $A * \langle b_j \rangle$. We have $b_{j-1} = a_{j-1} b_j^{n_{j-1}}$, and so $b_{j-1} = a_{j-2}(a_{j-1} b_j^{n_{j-1}})^{n_{j-2}}$, and so on. By induction on $j - i$, there is an expression for b_i of the form

$$(*) \qquad c_0 b_j^{m_0} c_1 b_j^{m_1} \dots c_r b_j^{m_r}$$

where (i) each m_k is a positive integer; (ii) each c_k is $a_s a_{s+1} \dots a_{j-1}$, for some s in $1 \leq s \leq j-1$; (iii) each such $a_s a_{s+1} \dots a_{j-1}$ occurs among the c_k; and (iv) in particular $c_0 = a_i a_{i+1} \dots a_j$. Because the m_k are positive we obtain a cyclically reduced expression for a conjugate of b_i by deleting those c_k which are 1, amalgamating powers of b_j where possible, and, if $c_0 = 1$, conjugating by a power of b_j. The result is still of the form $(*)$ with each m_k positive, but now the set $\{c_0, c_1, \dots, c_r\}$ consists precisely of those $a_s a_{s+1} \dots a_{j-1}, i \leq s \leq j-1$, which are not 1. The assumption that $\langle a_i, i < \omega \rangle$ is not finitely generated guarantees that for large enough j this set is nonempty. For such j, no conjugate of $\langle b_i \rangle$ in B_j meets $\langle b_j \rangle$ nontrivially, as required. The lemma follows.

We next prove Theorem 4.4. Assume, then, that G is the union of the translucent blocking tower $\{A_\alpha, \alpha < \omega_1\}$ with quotient-type sequence

f, and put $N_\alpha = A_\alpha^G, \alpha < \omega_1$. We have to show that $\{N_\alpha, \alpha < \omega_1\}$ is a translucent normal tower of type f for G. The only part of Definition 4.3 which requires an argument is part (ii), and here we may, without loss of generality, assume $\alpha = 0$. We restate what is required as another lemma.

Lemma 4.9. *Let G be the union of the translucent blocking tower $\{A_\alpha, \alpha < \omega_1\}$, and put $N_0 = A_0^G$. A maximal abelian subgroup of G/N_0 is either conjugate to $A_1 N_0/N_0$ or free cyclic.*

(Note that by Corollary 2.11 and Lemma 2.12, $A_1 N_0/N_0$ is isomorphic to $A_1/A_0^{A_1}$, so that it is a group of type $f(0)$). We write $\bar{G}$ for $G/N_0, \bar{A}_\alpha$ for $A_\alpha N_0/N_0$, and so on. We note first that, for any countable ordinal $\alpha > 0$, the pair $(\bar{A}_\alpha, \bar{A}_{\alpha+1})$ satisfies the conditions required of the pair (A, B) in Lemma 4.8. For $(A_\alpha, A_{\alpha+1})$ is a translucent blocking pair, and a homomorphic image $(\bar{A}_\alpha, \bar{A}_{\alpha+1})$ satisfies these conditions as long as $\bar{A}_\alpha$ is not finitely generated. But if α is a limit ordinal $\bar{A}_\alpha$ is the proper union $\bigcup_{\beta<\alpha} \bar{A}_\beta$, and if $\alpha = \beta + 1, \bar{A}_\alpha$ has the group $A_{\beta+1}/A_\beta^{A_{\beta+1}}$ of type $f(\beta)$ as homomorphic image. In neither case can $\bar{A}_\alpha$ be finitely generated; so we can apply Lemma 4.8 with $A = \bar{A}_\alpha, B = \bar{A}_{\alpha+1}$, for any $\alpha > 0$. Suppose now that X is a maximal abelian subgroup of $\bar{G}$, and let ρ be the smallest ordinal such that some conjugate of X has nontrivial intersection with $\bar{A}_\rho$. Without loss of generality, we may suppose that this conjugate is X itself. Of course, $\rho > 0$, since $\bar{A}_0$ is trivial. For $\alpha \geq \rho, X \cap \bar{A}_\alpha$ is nontrivial, and an application of Lemma 4.8 gives $X \cap \bar{A}_{\alpha+1} = X \cap \bar{A}_\alpha$. Using this at successor ordinals and smoothness at limits, we see that $X \cap \bar{A}_\alpha = X \cap \bar{A}_\rho$ for all α in $\rho \leq \alpha < \omega_1$, so that $X \leq \bar{A}_\rho$. Using smoothness again, it is clear that ρ is not a limit ordinal, so that $\rho = \sigma + 1$ for some σ. If $\sigma = 0, X \leq \bar{A}_1$, and by maximality $X = \bar{A}_1$. If $\sigma > 0$, then we can use Lemma 4.8 again; all conjugates of X meet $\bar{A}_\sigma$ trivially, and so X is free cyclic. This completes the proof of Lemma 4.9, and with it the proof of Theorem 4.4.

To prove Theorem 4.5 assume that the same group G is the union of the translucent normal towers $\{N_\alpha, \alpha < \omega_1\}$ and $\{P_\alpha, \alpha < \omega_1\}$ of type f and g respectively. A standard argument shows that there is a closed unbounded subset Ω of ω_1 such that $N_\rho = P_\rho$ for ρ in Ω. Indeed, the fact that N_α and P_α are normal closures of countable sets ensures the existence of increasing functions n and p on ω_1 such that $N_\alpha \leq P_{n(\alpha)}$ and $P_\alpha \leq N_{p(\alpha)}$ for all α; and, for any α, if ρ is the limit of the sequence $\alpha, n(\alpha), pn(\alpha), npn(\alpha), \ldots$, smoothness implies that $N_\rho = P_\rho$. But then $f(\rho) = g(\rho)$, since each is the type of any noncyclic maximal abelian subgroup X/N_ρ of G/N_ρ. Furthermore $N_{\rho+1} = P_{\rho+1} = X^G$. This, with smoothness, guarantees that $N_\alpha = P_\alpha$ for all $\alpha \geq \rho$, and hence $f(\alpha) = g(\alpha)$ for all $\alpha \geq \rho$. That is, f and g are ultimately equal, as required.

Since Theorem 4.2 is an immediate consequence of Theorem 4.4. and Theorem 4.5, there remains only Lemma 4.6 and its corollary Theorem 4.7. We discuss the proof of Lemma 4.6, rather than writing it out formally. We cannot directly assert the existence of a translucent normal tower by a formula of $\mathcal{L}_{\infty,\omega_1}$, because to describe such a tower requires ω_1 variables, (a countable set for each countable ordinal), and the assertion would involve quantifying over these variables simultaneously. We can, however, assert the existence of each countable initial segment of such a tower. Specifically, if f is a map $\omega_1 \to T$, as in the definition, and ρ is a countable ordinal, there is a formula $\Phi(f,\rho)$ of $\mathcal{L}_{\infty,\omega_1}$, with free variables a countable set X and a single variable y, whose meaning is that G has normal subgroups $N_\alpha, \alpha \leq \rho$, satisfying conditions (i) to (iii) of Definition 4.3 as far as they apply, such that N_0 is the normal closure of the set of elements represented by the variables in X, and the element represented by y belongs to N_ρ. (The only difficulty about this occurs in part (ii) of the definition, which refers to maximal abelian subgroups without bounding their cardinality. We can get round this by first requiring that G/N_α has no abelian subgroup of rank 2, and then referring only to countable maximal abelian subgroups). As the sentence asserting that G has a translucent normal tower of type f we may take

$$(**) \qquad \Psi \wedge (\exists X)(\forall y) \bigvee_{\alpha<\omega_1} \Phi(f,\alpha),$$

where Ψ is a sentence asserting that G is not countably normally generated. It is clear that if G has a translucent normal tower of type f then (**) holds in G. But conversely, if (**) holds in G, the second conjunctand guarantees the existence of a collection of countable initial segments of such a tower, all with the same base group N_0. The way in which the base group N_0 determines the whole tower guarantees that these initial segments unite to form a single tower, and the first conjunctand guarantees that this tower cannot be of countable height.

The discussion to this point has concentrated on producing a formula in $\mathcal{L}_{\infty,\omega_1}$. To show that the formula is in fact in $\mathcal{L}_{\omega_2,\omega_1}$ (that is, that the conjunctions and disjunctions are over sets of at most ω_1 formulae) requires for the most part only careful bookkeeping. There is, however, one point which should, perhaps, be mentioned. One ingredient, used many times in the construction, is a formula with a countable set X of free variables, whose meaning is that the elements these variables represent constitute a subgroup of given (countable) isomorphism type. One natural way of formalising this involves a disjunction over the permutations of X, which is inadmissible in $\mathcal{L}_{\omega_2,\omega_1}$, in the absence of the continuum hypothesis. But in fact, given any countable group H, there is a quantifier-free formula, involving only countable conjunctions and disjunctions, whose meaning is that the elements represented by the variables in X constitute a subgroup isomorphic to H. The easiest way to see this in general, perhaps, is to adapt the Scott sentence of H (see e.g. Barwise [1] or Scott [7]). But we need the formula mainly for torsion-free abelian groups of rank 1, and it is possibly more illuminating to describe a special construction for this case. If H is torsion free abelian of rank 1, choose in it an element $h \neq 1$, and let Π be the set of prime-powers q such that the equation $x^q = h$ is soluble in H. To say that a group K is isomorphic to H is to say that it is torsion-free abelian of rank 1, and contains an element $k \neq 1$ such that, for prime-powers q, the equation $x^q = k$ is soluble in K if and only if $q \in \Pi$. This translates directly into a formula involving only countable conjunctions and disjunctions. The upshot is that each formula $\Phi(f, \alpha)$, as well as Ψ, can be taken to belong to $\mathcal{L}_{\omega_1,\omega_1}$. It is only the explicit disjunction in $(**)$ that requires us to go up to $\mathcal{L}_{\omega_2,\omega_1}$.

Theorem 4.7 is, of course, a corollary of theorems 4.4 and 4.5, and Lemma 4.6, so we have now proved all the results stated at the beginning of the section.

5. Unanswered questions

In this final section I list, with brief comments, some questions raised by the considerations in Section 4, to which I do not know the answers.

First some terminology. If $\{A_\alpha, \alpha < \omega_1\}$ is a blocking tower, so is $\{A_\alpha, \alpha \in \Omega\}$, where Ω is any closed unbounded subset of ω_1, and we speak of the first tower as a *refinement* of the second. In the strict sense, no blocking tower is unrefinable, since we can always insert extra groups before A_0. But with slight inaccuracy we shall call a blocking tower *unrefinable* if this is the only refinement possible, so that its type sequence contains only indecomposable tyeps (as defined in Section 1). We use the terms *refinement* and *unrefinable* also, in the obvious sense, for segments of blocking towers, in particular for blocking pairs.

Question 1. Is it true that if the same group is the union of two unrefinable blocking towers, then the type sequences of these towers must be ultimately equal?

A translucent blocking pair is indecomposable, so a positive answer to this question would provide a satisfactory generalisation and deepening of Theorem 4.2, but there is very little evidence either way. Of course, the results of Section 3 show that Theorem 4.2 is not the whole truth, even for translucent blocking pairs. For instance, if (A_n, B_n) is the blocking pair such that $a_i = b_i b_{i+1}^{-n}$ for all $i, (A_2, B_2)$ and (A_4, B_4) have the same quotient-type, so that, in the notation of Examples 3.2 to 3.4, Theorem 4.2 does not distinguish between $G(A_2, B_2)$ and $G(A_4, B_4)$. However G/G^4 is uncountable for $G = G(A_2, B_2)$ and countable for $G = G(A_4, B_4)$, so that these groups are not isomorphic, or even $\mathcal{L}_{\omega_1, \omega_1}$-equivalent.

Note also that if the same group is the union of two blocking towers, these towers agree for the ordinals in a closed unbounded subset so that Question 1 is really about refinements of blocking pairs. There are many questions which ask for something stronger than Theorem 4.2 but weaker than a positive answer to Question 1. For instance, one might begin with the fairly concrete question :

Question 2. If (A, B) and (B, C) are both translucent blocking pairs, what are the other groups, D, say, such that (A, D) and (D, C) are both blocking pairs?

It is perhaps worth asking for even more than question 1 :

Question 3. Is it true that if the same group is the union of two blocking towers then these towers must have refinements which are ultimately equal?

This seems fairly unlikely, but I do not know of a counter example.

Clearly, one also has to ask :

Question 4. Does every blocking pair have an unrefinable refinement?

This also seems unlikely, on the face of it. But if we take a chain of refinements of a pair (A, B), and the set of subgroups in the union of these refinements is well ordered by inclusion its smooth closure is again a refinement. If not, we have a strictly descending chain of subgroups somewhere between A and B, which might give something to work on.

These questions are precise, but other questions can be asked, for which it is less clear when we have a satisfactory answer.

Question 5. Can one give a succinct description of (i) the set of isomorphism types or (ii) the set of quotient-types of (a) blocking pairs or (b) indecomposable blocking pairs?

For instance, I conjecture that the quotient type of a blocking pair is a countable locally free group with no cyclic free factor, and is subject to no further condition; which, if true, would be one answer to (ii) (a).

Question 6. Can one give a good characterization of the set of unions of (a) blocking towers or (b) unrefinable blocking tower?

It seems likely that any countably free group of cardinality ω_1 which is not strongly countably free is the union of a tower $\{A_\alpha, \alpha < \omega_1\}$ of countable free groups which in some sense sit awkwardly inside one another, but I do not know how precise one can make this.

Finally, notice that there is much choice built in to the proof of Theorem 2.7. (This occurs not when choice is mentioned, but when we say "identify A_β with C"). Thus the type sequence of a blocking tower no more determines the isomorphism type of the union than *vice versa*. So we may ask :

Question 7. If two groups are unions of blocking towers with the same type sequence, how different can they be?

6. Acknowledgements

It is a pleasure to record that the research embodied in this paper was done when I was enjoying the hospitality of the Department of Mathematics at Melbourne University. Some of the writing was also done there, but I finished the paper while enjoying similar hospitality in the Department of Mathematics at the National University of Singapore.

References

[1] K.J. Barwise, Back and forth through infinitary logic, in *Studies in Model Theory* (ed. Morley), Math. Association of America 1973.

[2] L. Fuchs, *Infinite abelian groups*, Academic Press 1973.

[3] G. Higman, Almost free groups, *Proc. London Math. Soc.* (3), **1** (1951).

[4] L.G. Kovács and M.F. Newman, On non-Cross varieties of groups, *J. Austral. Math. Soc.* **12** (1971).

[5] W. Magnus, A. Karrass and D. Solitar, *Combinatorial Group Theory*, Interscience Publications 1966.

[6] A.H. Mekler, How to construct almost free groups, *Canadian J. Math.* **32** (1980).

[7] D.S. Scott, Logic with denumerably long formulas and finite strings of quantifiers, in *The Theory of Models* (ed. Addision, Henkin and Tarski), North Holland 1965.

The Mathematical Institute
24-29 St. Giles
Oxford OX1 3LB
England

Automorphism groups of DRADs

Noboru Ito

§1. Introduction

Let v, k and λ be positive integers such that $v > 2k > 4\lambda$. Let $D = (P, B)$ be a symmetric 2-(v, k, λ) design, where P and B denote the sets of points and blocks of D respectively. Assume that D admits a bijection T from B to P such that (1) $T(\alpha) \notin \alpha$ and (2) if $T(\alpha) \in \beta$ then $T(\beta) \notin \alpha$. Then we define a digraph $D(T) = (V, \mathcal{A})$ as follows, where V and $\mathcal{A}$ denote the sets of vertices and arcs of $D(T)$ respectively: $V = P$ and $[a, b] \in \mathcal{A}$, where $a, b \in V$, if and only if $b \in T^{-1}(a)$.

Since D is a symmetric 2-(v, k, λ) design and T satisfies the conditions (1) and (2), $D(T)$ has the following properties: (i) $D(T)$ is asymmetric. (ii) $D(T)$ is regular of valency k. (iii) $D(T)$ is doubly regular of valency λ. Namely for any two distinct vertices a_1 and a_2 there exist λ vertices b such that $[a_i, b] \in \mathcal{A}$ for $i = 1, 2$ and there exist λ vertices c such that $[c, a_i] \in \mathcal{A}$ for $i = 1, 2$.

We call $D(T)$ a DRAD, which is an abbreviation of a doubly regular asymmetric digraph, associated with D.

The following is an open problem. If there exists a symmetric 2-(v, k, λ) design D_1, then does there exist a symmetric 2-design D_2 with the same parameters as D_1 which has a DRAD associated with it ?

Let $\alpha(a)$ be the set of vertices b of $D(T)$ such that $[a, b]$ is an arc, where a is a given vertex. Then $\alpha(a)$ is a block of D and $T(\alpha(a)) = a$. Let $\bar{\alpha}(a)$ be the set of vertices d of $D(T)$ such that $[d, a]$ is an arc, where a is a given vertex, and put $\bar{B} = \{\bar{\alpha}(a), a \in P\}$. Then $\bar{D} = (P, \bar{B})$ is also a symmetric 2-(v, k, λ) design which is isomorphic to the dual design of D. Furthermore let $\beta(a)$ be the set of vertices c of $D(T)$ such that c is not adjacent with a, where a is a given vertex.

For basic facts on association schemes, symmetric designs and permutation groups see [2], [8] and [17] respectively.

§2. Generalities

Let $D = (P, B)$ be a symmetric 2-(v, k, λ) design. Then we may assume that $P = \{1, 2, \ldots, v\}$ and $B = \{\alpha_1, \alpha_2, \ldots, \alpha_v\}$. Let A be an incidence matrix of D, where the (i,j)-entry $A(i,j)$ of A equals 1 if and only if $j \in \alpha_i$. If S and T are any permutation matrices, then SAT is also an incidence matrix of D by a different points and blocks labeling. Now it is easy to see the validity of the following proposition.

Proposition 1. *D admits a bijection T from B to P satisfying two conditions* (1) *and* (2) *in* §1 *if and only if D has a labeling such that the corresponding incidence matrix A satisfies*

$$(!) \qquad A(i,j)A(j,i) = 0 \text{ for } 1 \leq i, j \leq v.$$

We call an incidence matrix satisfying (!) an incidence matrix of DRAD form. It can be regarded as the adjacency matrix of the corresponding DRAD. We notice that there may arise many distinct DRADs from a given symmetric design.

Proposition 2. *Let G be the automorphism group of a* DRAD *$D(T)$ such that $T(\alpha_i) = i, 1 \leq i \leq v$. Then the permutation representations of G on P and on B are the same.*

Proof. It is obvious.

Proposition 2 implies that G can be regarded as the set of permutation matrices commuting with A, the adjacency matrix of the DRAD.

Proposition 3. *Let a be a vertex of $D(T)$ and G_a the stabilizer of a in G. Then G_a is faithful on $\alpha(a)$.*

Proof. Let s be an automorphism of $D(T)$ such that s is trivial on $\alpha(a)$. If $a \in \alpha(d)$, then, since $\alpha(d)$ contains λ vertices of $\alpha(a), \alpha(d)s = \alpha(d)$ and hence $ds = d$. So s is trivial on $\bar{\alpha}(a)$. Let c and d be vertices of $\beta(a)$ and $\bar{\alpha}(a)$ respectively and $\beta_1, \beta_2, \ldots, \beta_\lambda$ all blocks containing c and d. There exists a vertex d' in $\bar{\alpha}(a)$ not belonging to the intersection of $\beta_1, \beta_2, \ldots$, and β_λ. Let $\beta'_1, \beta'_2, \ldots, \beta'_\lambda$ be all blocks containing c and d'. Then $\{c\}$ is the intersection of $\beta_1, \ldots, \beta_\lambda, \beta'_1, \ldots, \beta'_\lambda$. Since s fixes each of the blocks β_i and β'_j, where $1 \leq i, j \leq l$, s fixes c. Hence s is trivial.

Furthermore any automorphism of $D(T)$ cannot interchange vertices a and b which are adjacent. If $D(T)$ is a DRAD associated with a Hadamard design, then every pair of vertices is adjacent. So G has odd order and by the theorem of Feit-Thompson G is solvable. Although regular asymmetric digraphs whose automorphism groups are non-solvable are abundant, the following question might be reasonable. Is the automorphism group of a DRAD solvable?

Remark. In our definition of a DRAD we did not use a relabeling of points. However, if $PAQ + Q^{-1}A^tP^{-1}$ is a $(0,1)$-matrix, where P and Q are permutation matrices, then $QPA + A^tP^{-1}Q^{-1}$ is also a $(0,1)$-matrix.

§3. DRADs with $\lambda = 1$

Let $\lambda = 1$. This is an interesting case of projective planes. Usually many DRADs arise from a given projective plane. Here we prove only the following.

Proposition 4. *Let $D(T)$ be a* DRAD *arising from a projective plane D of order n through a bijection T. Let s be an automorphism of $D(T)$ of order two. Then s is a Baer involution* (For this see [10], p.91). *In particular, n is a square.*

Proof. Let F be the set of vertices of $D(T)$ fixed by s. If F is not a Baer subplane, then F consists of a line α, all points on α, a point a outside α and all lines through a ([10], Theorem 4.4). Since $T(\alpha)$ is fixed by s and $T(\alpha) \notin \alpha$, we get that $T(\alpha) = a$. Let b be a point on α and consider the line β through a and b. Then since β is fixed by s and $T(\beta) \notin \beta$, $T(\beta)$ is fixed by s and $T(\beta) \in \alpha$. This violates the property (2) in §1.

In particular, if n is not a square, then the automorphism group has an odd order and hence it is solvable.

Acknowledgement. We owe Dr. T. Hiramine a simplification of the original proof.

§4. DRADs with rank four automorphism groups

Let $D = (P, B)$ be a symmetric 2-(16, 6, 2) design, where $P = \{1, 2, \ldots, 16\}$ and B is given as follows:

$$\begin{aligned}
\alpha_1 &= \{2,5,7,8,10,14\} \\
\alpha_2 &= \{3,5,6,8,11.15\} \\
\alpha_3 &= \{4,5,6,7,12,16\} \\
\alpha_4 &= \{1,6,7,8,9,13\} \\
\alpha_5 &= \{4,8,9,10,11,16\} \\
\alpha_6 &= \{1,5,10,11,12,13\} \\
\alpha_7 &= \{2,6,9,11,12,14\} \\
\alpha_8 &= \{3,7,9,10,12,15\} \\
\alpha_9 &= \{2,6,10,13,15,16\} \\
\alpha_{10} &= \{3,7,11,13,14,16\} \\
\alpha_{11} &= \{4,8,12,13,14,15\} \\
\alpha_{12} &= \{1,5,9,14,15,16\} \\
\alpha_{13} &= \{1,2,3,8,10,16\} \\
\alpha_{14} &= \{2,3,4,5,9,13\} \\
\alpha_{15} &= \{1,3,4,6,10,14\} \\
\alpha_{16} &= \{1,2,4,7,11,15\}.
\end{aligned}$$

It is easy to see that if we define a bijection T from B to P by $T(\alpha_i) = i$ for $i = 1, \ldots, 16$ then we get a DRAD $D(T)$.

Proposition 5. *Let G denote the automorphism group of the* DRAD *$D(T)$. Then G is generated by*

$$s = (1,2,3,4)(5,6,7,8)(9,10,11,12)(13,14,15,16)$$

and

$$t = (1)(2,14,7,10,8,5)(3,9,11)(4,6,13,12,16,15).$$

Proof. It is easy to check that s and t are indeed automorphisms of $D(T)$.

Furthermore it is not difficult to see that $G = \langle s, t\rangle$.

Obviously G has rank four. G contains a regular normal subgroup generated by $t^3 s = (1,2,11,12)(3,4,9,10)(5,8,15,14)(6,13,16,7)$ and $t^4 st^5 = (1,5,9,13)(2,8,10,16)(3,7,11,15)(4,6,12,14)$. It is a direct product of two cyclic groups of order four. G contains a regular non-normal subgroup generated by s and $t^4 st^5$. It is non-Abelian. Since the stabilizer of 1 in G equals $\langle t\rangle$, G has order 96. Obviously G is imprimitive.

The symmetric 2-$(16,6,2)$ design D in proposition 5 has an automorphism which is not an automorphism of $D(T)$, for instance, the following permutation on $P : (1,5,9,13)(2,6,10,14)(3,7,11,15)(4,8,12,16)$. So T is not uniquely determined by D. In general, let $D = (P, B)$ be a symmetric 2-design and T_1 and T_2 two bijections from B to P satisfying the conditions (1) and (2) in §1. We say that T_1 and T_2 are equivalent if there exists an automorphism z of D such that $T_2(\alpha z) = T_1(\alpha)z$ for any α in B.

Now by a theorem of W. Kantor [14] DRADs which have automorphism groups of rank three are Hadamard tournaments of quadratic residue type.

The next two sections will be devoted to proving the following result.

Theorem 1. *The above mentioned* $(16,6,2)$–DRAD $D(T)$ *is the only (up to equivalence)* DRAD *which has the automorphism group of rank four.*

§5. Primitive case

Let D be a symmetric 2-(v,k,λ) design and $D(T)$ a DRAD associated with D such that the automorphism group G of $D(T)$ has rank four. Let a be a vertex of $D(T)$ and G_a the stabilizer of a in G. Then $\alpha(a)$, $\beta(a)$ and $\bar{\alpha}(a)$ defined in §1 are the orbits of G_a together with $\{a\}$. The orbitals α and $\bar{\alpha}$ define digraph structures in D which coincide with $D(T)$ and its converse, and the orbital β defines a graph structure in D.

Proposition 6. $\alpha(a)$ *and* $\bar{\alpha}(a)$ *are the paired orbits of* G_a.

Proof. Let α' be paired to $\alpha(a)$. Then $\alpha' = \{at; a \in \alpha(a)t, t \in G\}$.

For any vertex d in $\bar{\alpha}(a)$ consider an element s of G whose cycle structure equals $(d,a,b,\ldots)\ldots$. Since $[d,a] \in \mathcal{A}$, $[a,b] \in \mathcal{A}$ and $b = as \in \alpha(a)$. Now $a = bs^{-1} \in \alpha(a)s^{-1}$ and $as^{-1} = d$. Hence we have that $\alpha' = \bar{\alpha}(a)$.

$\beta(a)$ is self-paired. So for any c in $\beta(a)$ there exists an element $s(c)$ of G whose cycle structure has the form $(ac)\ldots$ [17]. Apparently we may assume that $s(c)$ is an involution.

Proposition 7. (i) *G is imprimitive on V if and only if the β-diameter of D is infinite.* (ii) *G is primitive on V if and only if the β-diameter of D equals two.*

Proof. The α- and $\bar{\alpha}$-diameters of D are at most four [12] . Hence by a theorem of D. G. Higman [9] G is primitive on V if and only if the β-diameter of D is finite. So we have only to prove the only if part of (ii). Since we are assuming that the β-diameter of D is finite, there exists a vertex c in $\beta(a)$ such that $\beta(c)$ contains a vertex b in $\alpha(a)$ or a vertex d in $\bar{\alpha}(a)$. In the former case consider an element s of G whose cycle structure equals $(b,a,\ldots)\ldots$. Then since $[a,b] \in \mathcal{A}$. $(b,a,\ldots) = (b,a,d',\ldots)$ with $d' \in \bar{\alpha}(a)$. Hence $d' = as \in \beta(cs)$ and $cs \in \beta(bs) = \beta(a)$. We can treat the latter case similarly. So the β-diameter of D equals two.

After [13] let $\mu_1 = |\alpha(a) \cap \beta(b)|$, where $b \in \beta(a)$, $\mu_2 = |\alpha(a) \cap \bar{\alpha}(b)|$, where $b \in \beta(a)$, $\nu_1 = |\alpha(a) \cap \beta(b)|$, where $b \in \bar{\alpha}(a)$, $\nu_2 = |\alpha(a) \cap \bar{\alpha}(a)|$, where $b \in \bar{\alpha}(a)$, $\lambda' = |\beta(a) \cap \beta(b)|$, where $b \in \bar{\alpha}(a)$, and $\mu' = |\beta(a) \cap \beta(b)|$, where $b \in \beta(a)$. These are called intersection numbers.

There hold the following equations among them [13] :

$$(i) \qquad k\nu_1 = (v - 2k - 1)\mu_2,$$

$$(ii) \qquad k\lambda' = (v - 2k - 1)\mu_1,$$

$$(iii) \qquad \mu_1 + \mu_2 = k - \lambda,$$

$$(iv) \qquad \nu_1 + \nu_2 = k - \lambda,$$

$$(v) \qquad \nu_1 + \lambda' = v - 3k + 2\lambda,$$

(vi) $$2\mu_1 + \mu' = v - 2k - 2.$$

Furthermore the following fact holds [13] . G is primitive on V if and only if $\mu' \neq v - 2k - 2$ and $0 < \lambda' < v - 2k - l$.

Now let A, B and C be adjacency matrices of orbitals α, β and $\bar{\alpha}$ respectively. Then by Proposition 6, $C = A^t$, where t denotes transposition, $B = B^t$, and

$$A + B + C + I = J, \tag{1}$$

where I and J denote the identity and all-one matrices respectively.

Moreover we have the following equations:

$$A^2 = \lambda A + \mu_2 B + \nu_2 C, \tag{2}$$

$$AB = (k - 2\lambda - 1)A + \mu_1 B + \nu_1 C, \tag{3}$$

$$AV = kI + \lambda A + \lambda B + \lambda C = CA, \tag{4}$$

and

$$B^2 = (v - 2k - 1)I + \lambda' A + \mu' B + \lambda' C. \tag{5}$$

A has constant row sum and column sum both of which equal k, and B has constant row sum and column sum both of which equal $v-2k-1$. Hence by (1) and (4) A, B and C are pairwise commutative. Let $\tilde{A}= \langle I, A, B, C\rangle$ be an algebra over the field of complex numbers spanned by I, A, B and C. Then $\tilde{A}$ is a commutative algebra of dimension four. Since A is normal and not symmetric, two of its eigenvalues are non-real, say λ_1 and $\bar{\lambda}_1$, where $^-$ denotes complex conjugation. k is a simple eigenvalue. Let λ_2 be another real eigenvalue of A.

Let $E_0 = J/v$, E_1, E_2 and E_3 be the primitive idempotents of $\tilde{A}$. Then we have that

$$(6) \qquad I = E_0 + E_1 + E_2 + E_3.$$

Moreover, since $\tilde{A}$ has dimension four, we may put

$$(7) \qquad A = kE_0 + \lambda_1 E_1 + \lambda_2 E_2 + \bar{\lambda}_1 E_3$$

and

$$(8) \qquad B = (v - 2k - 1)E_0 + \lambda_3 E_1 + \lambda_4 E_2 + \lambda_3 E_3,$$

where λ_3 and λ_4 are real eigenvalues of B.

First we treat the case where G is primitive on V. Put $\lambda_1 = a + bi$ with non-zero b. Then $\lambda_1^2 = (a^2 - b^2) + 2abi$. On the other hand, from (2) we have that $\lambda_1^2 = \lambda\lambda_1 + \mu_2\lambda_3 + \nu_2\bar{\lambda}_1$. So we have that $a^2 - b^2 = \lambda a + \mu_2\lambda_3 + \nu_2 a$ and that $2a = \lambda - \nu_2 = \lambda_1 + \bar{\lambda}_1$. From (3) we have that $\lambda_1\lambda_3 = (k - 2\lambda - 1)\lambda_1 + \mu_1\lambda_3 + \nu_1\bar{\lambda}_1$ and hence that $\bar{\lambda}_1\lambda_3 = (k - 2\lambda - 1)\bar{\lambda} + \mu_1\lambda_3 + \nu_1\bar{\lambda}_1$. Adding last two equations we have that

$$(9) \qquad (\lambda - \nu_2 - 2\mu_1)\lambda_3 = (k - 2\lambda - 1 + \nu_1)(\lambda - \nu_2).$$

Now assume that $\lambda = \nu_2$. Then either $\mu_1 = 0$ in which case $\lambda' = 0$ by (ii) and hence G is imprimitive on V by [13] , or $\lambda_3 = 0$ in which case $a = 0 = b$ and hence we have a contradiction. Hence $\lambda \neq \nu_2$.

From (1) we obtain that $0 = 1 + \lambda_1 + \lambda_3 + \bar{\lambda}_1 = 1 + \lambda - \nu_2 + \lambda_3$. Hence from (9) we obtain that

$$(1 - \nu_2 - 2\mu_1)(\nu_2 - 1 - \lambda) = (k - 2\lambda - 1 - \nu_1)(\lambda - \nu_2),$$

which implies that $\mu_1 = \mu_1(\nu_2 - \lambda)$. Hence we have that $\nu_2 - \lambda = 1$, which implies that $\lambda_3 = 0$. So from (9) we obtain a contradiction that $k = 2\lambda + 1 - \nu_1$.

Thus G is imprimitive on V.

§6. Imprimitive case

From now on we assume that G is imprimitive on V.

Proposition 8. *Let L_a be the set of vertices of $D(T)$ whose β-distances from a is finite. Then $L_a = \{a\} \cup \beta(a)$.*

Proof. Let M be the global stabilizer of L_a in G. Then M is transitive on L_a. So $[M : G_a] = |L_a|$. Now deny the assertion. Then $|L_a| = v - k$. So we may put $v = m(v - k)$, where $[G : M] = m$. This implies that $v = k + (k/(m-1))$. Since $v > 2k + 1$, this is a contradiction.

Remark. L_a is a connected component of the β-graph of D containing the vertex a, and it is complete as the subgraph of the β-graph of D.

Now since $|L_a| = v - 2k$, we may put $v = m(v - 2k)$. This implies that $v = 2k + (2k/(m-1))$.

Proposition 9. *Let b be a vertex in $\alpha(a)$ and M_b the stabilizer of b in M. Then $[M : M_b] = 2k$.*

Proof. Deny the assertion. Then since $M \supset G_a$, we have that $[M : M_B] = k$. Let $G_{a,b}$ be the stabilizer of b in G_a. Then since $[G_a : G_{a,b}] = k$ and $[M : G_a] = v - 2k$, we have that $[M_b : G_{a,b}] = v - 2k$. Put $[G_a : M_b] = x$, where G_b is the stabilizer of b in G. Then we have that $k = (v-2k)x$. Since $[G : M_b] = [G : M][M : M_b] = [G : G_b][G_b : M_b]$, we have that $mk = vx$. Hence $m = 2x + 1$, $v = (v - 2k)(2x + 1)$ and $g.c.d.(v, k) = v - 2k$. From $\lambda(v-1) = k(k-1)$ it follows that $\lambda = \lambda v - k(k-1) = (v-2k)\{\lambda(2x+1) - x(k-1)\}$. Put $y = (2x+1) - x(k-1)$. Then again from $\lambda(v-1) = k(k-1)$ it follows that $y(v-2k)(2x+1) - y = (v-2k)x^2 - x$. Namely we have that $v - 2k = (x-y)/\{x(x-2y) - y\}$. Since $\lambda = (v-2k)y$ and $k = (v-2k)x$, we have that $x > 2y$. Hence we must have that $x - 2y = 1$ and $v - 2k = 1$, which is a contradiction.

Proposition 10. *There exists a positive integer z such that*
$$v = 4(z+1)^2, k = (z+1)(2z+1) \text{ and } \lambda = (z+1)z.$$

Remark. We recognize that these values coincide with those of parameters of symmetric designs associated with regular Hadamard matrices [16].

Proof. Using the notation in the proof of proposition 9 we have that $[M_b : G_{a,b}] = (v-2k)/2$, which implies that v is even, and that $2mk = xv$. Since $v = m(v-2k)$, we obtain that $m = x+1$ and $xv = 2(x+1)k$. Put $v = 2w$. Then we have the following equations:

$$\lambda(2w-1) = k(k-1), \tag{10}$$

$$k = x(w-k), \tag{11}$$

and

$$w = (x+1)(w-k). \tag{12}$$

By (10), (11) and (12) we may put

$$\lambda = z(w-k), \tag{13}$$

where z is a positive integer.

From (10) and (13) it follows that $z\{2(x+1)(w-k)-1\} = x\{x(w-k)-1\}$, which implies that

$$w-k = (x-z)/\{x(x-2z)-2z\}. \tag{14}$$

From (11) and (13) we have that $x > 2z$. If $x - 2z > 1$, then we get a contradiction that $1 \leq w-k < (1/2)$. Hence $x - 2z = 1$ and $w - k = z+1$. Therefore from (11), (12) and (13) we obtain that $v = 4(z+1)^2$, $k = (z+1)(2z+1)$ and $\lambda = (z+1)z$.

We also have that $k - \lambda = (z+1)^2$, $v - 2k - 1 = 2z+1$ and $m = 2(z+1)$.

Let Λ be the set of all connected components of the β-graph of D. Then by Proposition 8 $|\Lambda| = m = 2(z+1)$.

Proposition 11. *G is 2-transitive on Λ.*

Proof. By proposition 9 the global stabilizer M of L_a in G is transitive on $P - L_a = \alpha(a) \cup \bar{\alpha}(a)$.

Let K be the core of M in G, namely the largest normal subgroup of G contained in M.

Proposition 12. *K contains all the involutions of G. In particular, K is transitive on L_a.*

Proof. Let s be any involution of G and let $s = (u, v)\dots$ be the cycle structure of s. Then u and v belong to the same connected component of the β-graph of D. Namely s is trivial as a permutation on L. The last assertion is obvious, because K contains the involution $s(c)$ for any vertex c in $\beta(a)$.

Proposition 13. $\mu_1 = 0, \mu_2 = (z+1)^2$, $\nu_1 = z+1$, $\nu_2 = z(z+1)$, $\lambda' = 0$ *and* $\mu' = 2z$.

Proof. By Proposition 8 we have that $\lambda' = 0$ and $\mu_1 = 0$. These imply the rest.

Proposition 14. *M is 2-transitive on L_a.*

Proof. Let c be a vertex in $\beta(a)$. Then $[G_a : G_{a,c}] = 2z + 1$, where $G_{a,c}$ is the stabilizer of c in G_a.

Let L be the core of G_a in M, namely the largest subgroup of M contained in G_a.

Proposition 15. *KL is a proper subgroup of M.*

Proof. Otherwise $M = M_{L_b}L$, where b is a vertex in $\alpha(a)$, L_b is the β-connected component containing b and M_{L_b} is the global stabilizer of L_b in M. Then L is transitive on $\Lambda - \{L_a\}$ and $[L : L_{L_b}] = 2z + 1$, where L_{L_b} is the global stabilizer of L_b in L. Let p be a prime divisor of $2z + 1$ and L_p a Sylow p-subgroup of L. Now L_p leaves $\alpha(a) \cap \bar{\alpha}(c)$ invariant, where

$c \in \beta(a)$. Since $\mu_2 = |\alpha \cap \bar{\alpha}(c)| = (z+1)^2$ by Proposition 13, L_p must fix a vertex b' in $\alpha(a) \cap \bar{\alpha}(c)$. We may assume that $b' = b$. This implies that $L_p \subseteq L_{L_b}$, which is a contradiction.

Proposition 16. *M/L contains a regular normal 2-subgroup.*

Proof. Assume that M/L contains no regular normal subgroup. Then such 2-transitive permutation groups are known. Since M/L contains a proper subgroup whose index divides $2z+1$ by proposition 15, M/L is not a sproradic type. By Proposition 12, $KL \neq L$. Hence M/L is not simple. Therefore if M/L is either of Lie type or the alternating type, then by a theorem of Cooperstein [5] $z = 1$. This is a contradiction.

So we can put $2(z+1) = 2^m$. Then $v = 2^{2m}$, $k = 2^{m-1}(2^m - 1)$ and $\lambda = 2^{m-1}(2^{m-1} - 1)$; $\mu_1 = 0$, $\mu_2 = 2^{2(m-1)}$, $\nu_1 = 2^{m-1}$, $\nu_2 = 2^{m-1}(2^{m-1} - 1)$, $\lambda' = 0$ and $\mu' = 2(2^{m-1} - 1)$.

We notice that $\lambda = \nu_2$. So as in §5 we have that $\lambda_1 + \bar{\lambda}_1 = \lambda - \nu_2 = 0$ Since the trace of A equal to zero, from (7) we have that

$$k + \lambda_2 \mu_2 = 0, \tag{15}$$

where μ_2 is the rank of E_2.

From (1) and (7) we have that

$$1 + 2\lambda_2 + \lambda_4 = 0. \tag{16}$$

From (3) and (8) we have that

$$\lambda_2 \lambda_4 = (2^{m-1} - 1)\lambda_2 = 2^{m-1}\lambda_2. \tag{17}$$

From (16) and (17) we have that $\lambda_4 = 2^m - 1$ and $\lambda_2 = 1 - 2^{m-1}$. Now from (15) it follows that $k = 2^{m-1}(2^m - 1) = (2^{m-1} - 1)\mu_2$, which implies that $m = 2$.

Now it is known that there exist three non isomorphic symmetric 2-(16,6,2) designs which are customarily denoted by $B6$, $B7$ and $B8$. $B6$ may also be called the Kummer design. (See [1], [4] and [15]). The automorphism group of $B8$ has rank larger than four. The automorphism group of $B7$

has rank three and there exist DRADs associated with it. However, the automorphism groups of associated DRADs are intransitve. Our $(16,6,2)$-DRAD comes from $B6$. The checking for uniqueness is tedious, because there are many DRADs associated with $B6$.

Remark. We remark that Hadamard matrices of order 16 corresponding to $B6, B7$ and $B8$ are of type I, type II and type III respectively in the notation of Hall [7]. For an intimately related result see [6].

§7. DRADs coming from regular Hadamard matrices

Let A be a regular Hadamard matrix of order n so that the number of 1's in each of the rows and columns of A are constant. Then we say that A is in a DRAD form if and only if there exist no i and j such that the (i,j)- and (j,i)-entries of A equal one. In particular, the diagonal entries of A equal minus one.

Proposition 17. *Let A be a regular Hadamard matrix of order n in a* DRAD *form. Put*

$$B = \begin{pmatrix} A & -A & A & A \\ A^t & A^t & -A^t & A^t \\ A & A & A & -A \\ -A^t & A^t & A^t & A^t \end{pmatrix}.$$

Then B is a regular Hadamard matrix of order $4n$ of a DRAD *form.*

Proof. The $(i,n+j)$- and $(n+j,i)$-entries of B, where $1 \leq i,\ j \leq n$, equal $-A(i,j)$ and $A(i,j)$ respectively. So it is straightforward to check that B is a regular Hadamard matrix of a DRAD form.

Remark. In the array of Proposition 17, if A is a regular Hadamard matrix of order n then B is a regular Hadamard matrix of order $4n$.

If we change -1 in A to 0, then we get an adjacency matrix C of a DRAD D of order n. If we do the same thing to B, then it turns into an adjacency matrix C'' of a DRAD D'' of order $4n$. Therefore Proposition

17 gives us a quadrupling procedure of a DRAD coming from a regular Hadamard matrix. If we put

$$C = \begin{pmatrix} 0 & 1 & 0 & 0 \\ 0 & 0 & 1 & 0 \\ 0 & 0 & 0 & 1 \\ 1 & 0 & 0 & 0 \end{pmatrix}.$$

then C'' corresponds to our rank four (16, 6, 2)-DRAD in §4.

Proposition 18. *Let C be the adjacency matrix of a* DRAD *D of order n.*

(i) *Let S be a permutation matrix corresponding to an automorphism of D. Then*

$$\begin{pmatrix} S & 0 & 0 & 0 \\ 0 & S & 0 & 0 \\ 0 & 0 & S & 0 \\ 0 & 0 & 0 & S \end{pmatrix} \text{ and } \begin{pmatrix} 0 & 0 & I & 0 \\ 0 & 0 & 0 & I \\ I & 0 & 0 & 0 \\ 0 & I & 0 & 0 \end{pmatrix}$$

are permutation matrices corresponding to automorphisms of the DRAD *D'', where I denotes the identity matrix of order n.*

(ii) *Let X be a permutation matrix of order two such that $CX = XC^t$. Then*

$$\begin{pmatrix} 0 & X & 0 & 0 \\ 0 & 0 & X & 0 \\ 0 & 0 & 0 & X \\ X & 0 & 0 & 0 \end{pmatrix}$$

is an automorphism of the DRAD *D''.*

Proof. (i) S commutes with C. Then S commutes with C^t and $J - C$. (ii) X is symmetric and commutes with J.

Remark. For our rank four (16, 6, 2)-DRAD let

$$X = \begin{pmatrix} x & 0 & 0 & 0 \\ 0 & 0 & 0 & y \\ 0 & 0 & x & 0 \\ 0 & y & 0 & 0 \end{pmatrix},$$

$$\text{where } x = \begin{pmatrix} 1 & 0 & 0 & 0 \\ 0 & 0 & 0 & 1 \\ 0 & 0 & 1 & 0 \\ 0 & 1 & 0 & 0 \end{pmatrix} \text{ and } y = \begin{pmatrix} 0 & 0 & 1 & 0 \\ 0 & 1 & 0 & 0 \\ 1 & 0 & 0 & 0 \\ 0 & 0 & 0 & 1 \end{pmatrix}.$$

Then X satisfies the assumption in Proposition 18, (ii). Therefore the corresponding $(64, 28, 12)$-DRAD has the transitive automorphism group G''. Moreover G'' contains a regular subgroup generated by

$r = (1,2,3,4)(5,6,7,8)(9,10,11,12)(13,14,15,16)(17,18,19,20)(21,22,23,24)(25,26,27,28)(29,30,31,32)(33,34,35,36)(37,38,39,40)(41,42,43,44)(45,46,47,48)(49,50,51,52)(53,54,55,56)(57,58,59,60)(61,62,63,64)$,

$s = (1,8,9,16)(2,7,10,15)(3,6,11,14)(4,5,12,13)(17,24,25,32)(18,23,26,31)(19,22,27,30)(20,21,28,29)(33,40,41,48)(34,39,42,47)(35,38,43,46)(36,37,44,45)(49,56,57,64)(50,55,58,63)(51,54,59,62)(52,53,60,61)$

and

$t = (1,17,33,49)(2,20,34,52)(3,19,35,51)(4,18,36,50)(5,31,37,63)(6,30,38,62)(7,29,39,61)(8,32,40,64)(9,25,41,53)(10,28,42,60)(11,27,43,59)(12,26,44,58)(13,23,44,55)(14,22,46,54)(15,21,47,53)(16,24,48,56)$,

where $s^{-1}rs = r^{-1}$, $t^{-1}rt = r^{-1}$ and $t^{-1}st = s^{-1}$.

§8. Hadamard tournaments

Hadamard tournaments are DRADs arising from Hadamard designs.

Proposition 19. *Let A be an adjacency matrix of* DRAD *form of a Hadamard tournament H of order $v = 4\lambda + 3$. Put*

$$A' = \begin{pmatrix} A^t & 0 & \bar{A} \\ j & 0 & 0 \\ A^t & j^t & A \end{pmatrix}$$

where j is an all-one row vector of size v and $\bar{A} = J - A$. Then A' also is an adjacency matrix of DRAD *form of a Hadamard tournament H' of order $2v + 1$.*

Proof. Let a_i be the i-th row vector of A $(1 \leq i \leq v)$ and let

$\cdot$ denote the inner product in the space of row vectors of size v. Then $a_i \cdot a_i = k = 2\lambda + 1$ and $a_i \cdot a_j = \lambda$ for $i \neq j$, where $1 \leq i,\ j \leq v$. Hence $a_i \cdot (j - a_j) = k - \lambda = \lambda + 1$ for $i \neq j$.

Now let $D = (P, B)$ be a Hadamard (v, k, λ)-design and α a block of D. Further let $K(\alpha)$ be the number of 2-subsets $\{\beta, \gamma\}$ of $B - \{\alpha\}$ such that $\alpha \cap \beta = \alpha \cap \gamma$. If $\{\beta, \gamma\}$ is such a 2-subset, then $\alpha \cup \beta \cup \gamma = P$. So if $\{\beta_1, \gamma_1\}$ and $\{\beta_2, \gamma_2\}$ are distinct 2-subsets counted for $K(\alpha)$, then they are disjoint. Hence we have that $0 \leq K(\alpha) \leq k$. We call $K(\ )$ the Kimberley function of D. If D has the transitive automorphism group, then the Kimberley function of D is a constant function.

Proposition 20. *Let $H = (D, T)$ be a (v, k, λ)-Hadamard tournament with $\lambda \geq 2$. Assume that H has the transitive automorphism group so that the Kimberley function of D takes a constant value c. Then $c < k$.*

Proof. Assume that $c = k$. Then for any two distinct blocks α and β there exists a unique block $\gamma \neq \alpha, \beta$ such that $\alpha \cap \gamma = b \cap \gamma$. Put $a = T(\alpha)$, $b = T(\beta)$ and $c = T(\gamma)$. Then we may assume that a, b, c is a 3-cycle. In fact, if $[a, b]$ and $[c, b]$ are arcs, then $b \in \alpha \cap \gamma$. Since $b \notin \beta$, this is a contradiction. Moreover if $[a, b]$ and $[a, c]$ are arcs, then either $[a, b]$ and $[c, b]$, or $[a, c]$ and $[b, c]$ are arcs. Now $[a, b, c]$ is dominant in the following sense. For any vertex $d \neq a, b, c$ either $[a, d]$ or $[b, d]$, or $[c, d]$ is an arc. Furthermore, $[a, b, c]$ is dominant if and only if $P = \alpha \cup \beta \cup \gamma$. Now take two distinct vertices d and e from $\alpha \cap \beta$, which is possible because of the assumption $\lambda \geq 2$. Put $d = T(\delta)$ and $e = T(\epsilon)$. Then as above, we may assume that there exists a dominant 3-cycle $[d, e, f]$. But, then, all of $[f, a]$, $[f, b]$ and $[f, c]$ are arcs, which is a contradiction.

Proposition 21. *In the notation of Proposition* 19 *let S be a permutation matrix corresponding to an automorphism of the tournament H. Then*

(i)

$$\begin{pmatrix} S & 0 & 0 \\ 0 & 1 & 0 \\ 0 & 0 & S \end{pmatrix}$$

is a permutation matrix corresponding to an automorphism of the tourna-

ment H'.

(ii) *The automorphism group of H'' is not transitive.*

Proof. (i) It is obvious. (ii) Let α_i, β_i and $\bar{\alpha}_i$ be the i-th row vectors of A, A^t and $\bar{A}$ respectively, $1 \leq i \leq v$. We use the same notation for blocks, for instance, α_i is a block of H whose incidence vector is α_i. Here we notice that A^t and $\bar{A}$ are incidence matrices of the dual and complement designs of H respectively. Let γ_i denote the i-th row vector of A', $1 \leq i \leq 2v+1$. When γ_i is considered as a block, we have that $\gamma_i = \beta_i \cup \bar{\alpha}_i$ for $1 \leq i \leq v$; $\gamma_{v+1} = P$, the set of points of H; $\gamma_i = \beta_i \cup \alpha_i$ for $v+2 \leq i \leq 2v+1$. Let $K(\ )$ be the Kimberley function of the Hadamard design from which H' arises. Then it is clear that $K(\gamma_{v+1}) = v$. So Proposition 20 applies.

Remark. It can be shown that $K(\gamma_1) \neq v$.

Proposition 22. *Let G be the automorphism group of the Hadamard tournament of order $q = p^a$ of quadratic residue type. If a is divisible by P, then G contains a regular non-Abelian subgroup.*

Proof. Let $GF(q)$ be a Galois field of q elements. Put $a = p^b c$, where $b \geq 1$ and c is relatively prime to p. Then the invariant subfield of an automorphism s of $GF(q)$ of order p equals $GF(p^{p^{b-1}c})$. Now G contains the split extension E of the additive group of $GF(q)$ by $\langle s \rangle$; $E = GF(q)\langle s \rangle$, and $GF(q) \cap \langle s \rangle = \{e\}$, where $GF(q)$ is the regular normal subgroup of G and $\langle s \rangle$ is the stabilizer of a vertex in E. Take a normal subgroup N of E of index p^2 such that E/N is elementary Abelian. The center of E equals $GF(p^{p^{b-1}c})$ and $[GF(q) : GF(p^{p^{b-1}c})] = p^{p^{b-1}c(p-1)}$. So N contains elements of $GF(q) - GF(p^{p^{b-1}c})$. Hence $\langle as, N \rangle$ is a required non-Abelian regular subgroup of G, where $E/N = \langle N+a, Ns \rangle$, $a \in GF(q)$.

Remark. Proposition 22 shows that there are many non-Abelian difference sets associated with Hadamard tournaments of quadratic residue type.

§9. Difference sets and DRADs

Let $D = (P, B)$ be a symmetric design arising from a difference set α in a multiplicative group P of order v which is identified with the set of points of D and $B = \{\alpha a, a \in P\}$. Now a DRAD (D,T) is called a P-DRAD if $T(\alpha a) = T(\alpha)a$ for every $a \in P$.

Proposition 23. *D admits a P-DRAD if and only if there exists an element $b \in P$ such that $b \notin \alpha$ and $b\alpha^{-1}b \cap \alpha = \emptyset$.*

Proof. Suppose that we have a P-DRAD (D,T) and put $b = T(\alpha)$. Then $b \notin \alpha$. If $bb_1^{-1}b = b_2$, where $b_1, b_2 \in \alpha$, then put $b_1^{-1}b = b^{-1}b_2 = c$. Thus $b = b_1c$, $bc = b_2$, $bc \in \alpha$ and $b \in \alpha c$. Since $T(\alpha c) = bc = b_2$, both $[b, bc]$ and $[bc, b]$ are arcs, which is a contradiction.

Conversely, if an element b of P satisfies the assumption in our proposition, then define T by $(T\alpha a) = ba$ for every $a \in P$. If we have that $ba_1 \in \alpha a_2$ and $ba_2 \in \alpha a_1$, then $ba_2a_1^{-1} = ba_2a_1^{-1}b^{-1}b$ belongs to α and $b\alpha^{-1}b$.

Proposition 24. *Let $P = Z/(v)$ be a cyclic group of order v. Let β be a difference set in P with parameters $v = 4x^2+1$, $k = x^2$ and $\lambda = (x^2-1)/4$, where x is odd and v is a prime, consisting of biquadratic residues modulo v.* (For this see [3]). *Then there exists a P-DRAD associated with the symmetric design $D = (P, B)$ arising from β.*

Proof. Since x is odd, -1 is not a biquadratic residue. So the element $b = 0$ satisfies the condition in Proposition 23.

Remark. The automorphism groups of P-DRADs in Proposition 24 have rank 5.

References

[1] Assmus, Jr. E.F. and Salwach C.J. The (16,6,2) designs, *Internat. J. Math. & Math. Sci.* **2** (1979), 261-281.

[2] Bannai, E. and Ito, T. *Algebraic combinatorics I.* Menlo Park: Benjamin, 1984.

[3] Baumert, L.D. *Cyclic difference Sets.* LNM 182, Berlin-Heidelberg-New York: Springer, 1971.

[4] Burau, W. Über die zur Kummerkonfiguration analogen Schemata von 16 Punkten und 16 Blöcken und ihre Gruppen, *Hambg. Abh.* **26** (1963), 129-144.

[5] Cooperstein, B.N. Minimal degree for a permutation representation of a classical group, *Israel J. Math.* **30** (1978), 213-235.

[6] Enomoto, H. and Mena R.A. Distance-regular digraphs of girth 4, to appear.

[7] Hall, Jr. M. Hadamard matrices of order 16, *J.P.L. Research summary* No. 36-10. **1** (1961), 21-26.

[8] Hall, Jr. M. *Combinatorial theory.* Boston: Blaisdell, 1967.

[9] Higman, D.G. Intersection matrices for finite permutation groups, *J. Alg.* **6** (1967), 22-42.

[10] Hughes, D.R. and Piper, F.C. *Projective planes.* New York-Heidelberg-Berlin: Springer, 1973.

[11] Ito, N. Hadamard tournaments with transitive automorphism groups, *Europ. J. Combinatorics*, **5** (1984), 37-42.

[12] Ito, N. Doubly regular asymmetric digraphs, to appear in *Discrete Math.*

[13] Iwasaki, S. On finite permutation groups of rank 4, *J. Math. Kyoto Univ.* **13** (1973), 1-20.

[14] Kantor, W.M. Automorphism groups of designs, *Math. Z.* **109** (1969), 246-252.

[15] Lander, E.S. *Symmetric designs: an algebraic approach.* London Math. Soc. LNS 74. Cambridge: University Press, 1983.

[16] Wallis, W.D., Street, A.P. and Seberry Wallis, J. Combinatorics: *Room*

squares, sum-free sets, Hadamard matrices. LNM 292. New York - Heidelberg - Berlin: Springer, 1972.

[17] Wielandt, H. *Finite permutation groups*. New York: Academic Press, 1964.

Department of Applied Mathematics
Konan University
Kobe 658
Japan

Invariant lattices in Lie algebras and their automorphism groups

A.I. Kostrikin

1. Introduction

Now that the classification of finite simple groups is supposed to be completed, the different natural realizations of the various types of simple groups become more significant. This sort of question is traditional in finite group theory. For example, many important finite groups arose in connection with geometrical configurations.

For simple groups, however, the most significant realizations or constructions occur in the context of Lie algebras. In fact, almost the whole historical development of the theory of simple groups (finite and compact) is inseparable from that of simple Lie algebras over $\mathbf{C}$ and their analogues over $\mathbf{R}$, $\mathbf{Z}$ and $\mathbf{F}_q$. Only the sporadic simple groups provide a challenge to the group-theorists since they do not yield to any single doctrine. However, even here the Lie algebra point of view is becoming important. A significant event of recent years was the realization of the largest sporadic group, the Monster F_1, using vertex operators and the representation theory of affine Lie algebras [1]. This result is indirectly related to 20 of the 26 sporadic groups contained in F_1. However, there is, of course, no reason to believe that these will all be able to be given a "Lie interpretation".

The aim of my lecture is to describe another approach to the realization problem, which turns out to be very fruitful. The basic idea is that of obtaining an orthogonal decomposition of a certain complex simple Lie algebra L . I remind you of certain known facts. In the papers [2-5], for each complex simple Lie algebra L of type A_{p^m-1} (p prime), B_n, C_{2^m}, D_n, G_2, F_4, E_6, E_7, E_8, there is given a decomposition

$$L = H_0 \oplus H_1 \oplus \ldots \oplus H_h \tag{1}$$

(h the Coxeter number) as a direct sum of Cartan subalgebras H_k pairwise mutually orthogonal with respect to the Killing form $K(*,*)$; $K(H_i, H_j) = 0$ for $i \neq j$. Such decompositions are called Orthogonal decompositions $(\mathcal{OD})$.

An algebra L has a multiplicative orthogonal decomposition ($\mathcal{MOD}$), if for every i and j there exist a k such that $[H_i, H_j] \subseteq H_k$. An $\mathcal{MOD}$ for a complex simple Lie algebra L *exists* if and only if L has one of the following types:

$$A_1;\ B_{2^m-1},\ m \geq 2;\ D_{2^m},\ m \geq 2;\ G_2;\ E_8.$$

All $\mathcal{MOD}$'s of a given Lie algebra are conjugate (see in this connection [4, 6, 7, 8]).

Now, it is natural to study the finite group $\mathrm{Aut}_{\mathcal{OD}}(L)$ of those automorphisms from the complex group $\mathrm{Aut}(L)$ which preserve an $\mathcal{OD}$ ($\mathrm{Aut}_{\mathcal{MOD}}(L)$ in the case of an $\mathcal{MOD}$), that is $\Phi(H_i) = H_j$, $\forall \Phi \in \mathrm{Aut}_{\mathcal{OD}}(L)$, $0 \leq i, j \leq h$. A natural restriction is imposed on the decomposition (1): transitivity of the action of $\mathrm{Aut}_{\mathcal{OD}}(L)$ on the set $\{H_i\}_0^h$ of Cartan subalgebras. Although far from every $\mathcal{OD}$ is transitive, all the $\mathcal{MOD}$'s are transitive, and denoting by $\mathcal{E}_k$ an elementary abelian group of order 2^k, we have the following list for $\mathrm{Aut}_{\mathcal{MOD}}(L)$:

$$\begin{aligned}
A_1 &: \ \mathcal{E}_2 \cdot SL_2(2);\\
B_{2^m-1} &: \ \mathcal{E}_k \cdot SL_{m+1}(2),\ k = 2^{m+1} - 2;\\
D_{2^m} &: \ \mathcal{E}_k(\mathcal{E}_{m+1} \cdot SL_{m+1}(2)),\ k = 2^{m+1} - 1, m > 2;\\
D_4 &: \ \mathcal{E}_7(\mathcal{E}_3 \cdot SL_3(2))\langle T\rangle,\ T \text{ is an outer automorphism of order 3;}\\
G_2 &: \ \mathcal{E}_3 \cdot SL_3(2), \quad \text{a nonsplit extension;}\\
E_8 &: \ \mathcal{E}_5 \cdot (\mathcal{E}_{10} \cdot SL_5(2)), \quad \text{a nonsplit extension.}
\end{aligned}$$

Thus let (1) be a transitive $\mathcal{OD}$ of a Lie algebra L and let G be a subgroup of $\mathrm{Aut}_{\mathcal{OD}}(L)$ transitive on the set $\{H_i\}_0^h$. Usually we shall take as G the full group $\mathrm{Aut}_{\mathcal{OD}}(L)$ or its commutator subgroup. Choosing in one of the subalgebras H_k, for example in H_0, an integer lattice Λ_0 which is spanned on the basis dual to the system of simple roots of the algebra L, and then translating it by means of the G-action on L, we obtain a G-invariant positive integer lattice $\Lambda = \bigoplus_0^h \Lambda_k$. Its metric is induced from the Killing form on L: $(x|y) \in \mathbf{Z}$ for all x, $y \in \Lambda$, $||x||^2 = (x|x) > 0$ for $x \neq 0$.

The next restriction on $\mathrm{Aut}_{\mathcal{OD}}(L)$ and on G is imposed in order that Λ contains only a finite number of similarity classes of G-invariant sublattices Γ. The transitivity of $\mathcal{OD}$ does not suffice for this. Usually $\Lambda \supset \Gamma \supset m\Lambda$

for some prime power m. We then forget about the original Lie algebra L and consider the isometry groups $\mathrm{Aut}_{\mathbf{Z}}(\Gamma)$ of all its G-invariant sublattices Γ. Most of them including Λ itself, do not have interesting isometry groups. For instance we have the wreath product $\mathrm{Aut}_{\mathbf{Z}}(\Lambda) \cong H \wr S_{p+1}$ of the group $H = \mathrm{Aut}_{\mathbf{Z}}(\Lambda_0)$ and the symmetric group S_{p+1}.

If there were not very many isometry groups $\mathrm{Aut}_{\mathbf{Z}}(\Gamma)$ close to being simple, then of course my theme would have little interest for the specialist in group theory. However, fortunately, there *do* exist isolated gems, and it is these that justify the above approach via sublattices of Lie algebra.

I will now state the problem precisely. In Λ we consider the complete $\mathbf{Z}G$-submodules Γ, that is, satisfying $\mathbf{Q}\Lambda = \mathbf{Q}\Gamma$. These submodules fall into similarity classes: $\Gamma' \sim \Gamma''$ if $\Gamma'' = \lambda\Gamma'$ for some $\lambda \in \mathbf{Q}$. Thus an integer G-invariant lattice $\Gamma \subset \Lambda$ is called indivisible if there is no $m \in \mathbf{Z}$, $|m| > 1$, such that $\Gamma \subset m\Lambda$. We then have:

Theorem (D. N. Ivanov). *Let G be any finite group. A $\mathbf{Q}G$-module V has a finite number of similarity classes of complete $\mathbf{Z}G$-submodules iff V is absolutely irreducible.*

I shall formulate again the basic problems.

Problem A. Find all $\mathcal{OD}$ of any given complex simple Lie algebra L and compute the corresponding group $\mathrm{Aut}_{\mathcal{OD}}(L)$.

Problem B. For each $\mathcal{OD}$ with absolutely irreducible action of $\mathrm{Aut}_{\mathcal{OD}}(L)$ on $\mathbf{Z}$-lattice Λ, classify all indivisible invariant sublattices $\Gamma \subset \Lambda$ (or more generally, invariant under a subgroup $G \subset \mathrm{Aut}_{\mathcal{OD}}(L)$ with absolutely irreducible action on Λ).

Problem C. Compute the isometry groups $\mathrm{Aut}_{\mathbf{Z}}(\Gamma)$ for "interesting" invariant lattices.

Problem D. The same questions for R-lattices where R is a maximal order in some algebraic number field.

None of these problems is solved in full generality but for some types of algebras there has been significant progress.

2. Results

2.1 The exceptional types. Here we encounter very interesting situations.

The type E_8 is the first one for which there was found a special $\mathcal{MOD}$ (called a Dempwolf decomposition by J. Thompson). Using a computer, P. E. Smith [9] was able to find a $\mathcal{E}_5 \cdot SL_5(2)$-invariant lattice Γ corresponding to this $\mathcal{MOD}$ with isometry groups isomorphic to $\mathbf{Z}_2 \times F_3$, where F_3 is Thompson's sporadic simple group. (This is how this group's existence was first established.) Since the algebra E_8 is involved in the moonshine associated with the Monster, it is appropriate to study all its transitive $\mathcal{OD}$'s, and the corresponding integer lattices.

The set of all distinct $\mathcal{OD}$'s for the Lie algebra L of type G_2 is still unknown. However, the transitive ones have all been found (by K. Chakrijan [10]), as have all the invariant lattices in L, compatible with some $\mathcal{MOD}$(see [11]).

Theorem. *The lattice Λ in L contains up to similarity, duality and isometry, just five proper G-invariant sublattices Γ_i, $o \leq i \leq 4$. Exactly one lattice is interesting, namely Γ_0, which does not contain 2Λ but contains 4Λ. Its isometry group $\mathrm{Aut}_{\mathbf{Z}}(\Gamma_0)$ is isomorphic to $\mathbf{Z}_2 \times G_2(3)$, where $G_2(3)$ is the Dickson-Chevalley simple group of order $2^6 3^6 \cdot 7 \cdot 13$.*

For the Lie algebras of types F_4 and E_6, some $\mathcal{OD}$'s have been proposed in [5]; however it's rather difficult to see if they are transitive or not. Happily, my student V. P. Burichenko [12], using quite different ideas, has very recently constructed, for these Lie algebras, certain $\mathcal{OD}$'s with transitive (and probably even absolutely irreducible) actions on them of the groups $\mathrm{Aut}_{\mathcal{OD}}(L)$. His construction is interesting in that he starts with exactly the groups $3^3 \cdot SL_3(3)$ (for F_4) and $3^{3+3} \cdot SL_3(3)$ (for E_6). (Compare this with the construction of Thompson and Smith [6, 9]). Although I cannot give the details, I would like to mention an analogy with the following result.

Theorem (A. V. Borovick [13]). *Let G be a simple algebraic group* *defined over a finite field* $\mathbf{F}_q$. *Let* X *be a finite subgroup of* $G(\mathbf{F}_q)$, *which does not normalize any connected, closed* $\mathbf{F}_q$*-defined subgroup* $H \subset G$, $H \neq \{1\}$. *Then one of the following holds:*

(1) $Y \subseteq X \subseteq \mathrm{Aut}(Y)$ *for some simple group* Y;

(2) $X \subseteq \mathbf{N}_G(J)$ *for certain known* J *and* $\mathbf{N}_G(J)$.

For the exceptional types we have the following table:

G	J	$\mathbf{C}_G(J)/J$	$\mathbf{N}_G(J)/\mathbf{C}_G(J)$
G_2	$\mathbf{Z}_2^3$	1	$SL_3(2)$
F_4	$\mathbf{Z}_3^3$	1	$SL_3(3)$
E_6	$\mathbf{Z}_3^3$	$\mathbf{Z}_3^3$	$SL_3(3)$
E_7	—	—	—
E_8	$\mathbf{Z}_2^5$	$\mathbf{Z}_2^{10}$	$SL_5(2)$

The analogy with orthogonal decompositions is immediately clear. V. P. Burichenko has found in the complex Lie group $F_4(\mathbf{C})$ a subgroups N, $N \cong \mathbf{Z}_3^3$, with the property that the character of its natural representation on L (where L is the Lie algebra of type F_4) is equal to $\sum \chi$, where χ runs through all the non-trivial linear characters of N. It follows that $L = \bigoplus_x L_x$, $[L_{x_1}, L_{x_2}] \subseteq L_{x_1 x_2}$ where L_x is the eigenspace for N corresponding to the character χ ($\dim L_x = 2$). It is easy to see that $[L_x, L_{\bar{x}}] = 0 \ \forall x$, and $(L_{x_1} | L_{x_2}) = 0$ if $x_1 x_2 \neq 1$. Hence $H_x = L_x \oplus L_{\bar{x}}$ is a nilpotent subalgebra on which the Killing form of L is non-degenerate. Consequently each H_x is a Cartan subalgebra, and moreover $H_{x_1} = H_{x_2} \Leftrightarrow x_1 = x_2$. In this way we obtained 13 pairwise orthogonal Cartan subalgebras. It can be shown that $\mathrm{Aut}_{OD}(L) = 3^3 \cdot SL_3(3)$, and the action of this finite group on $\{H_x\}$ is transitive.

The argument in the case of the Lie algebra of type E_6 is analogous, though more complicated. From a combinatorial point of view, the decompositions in [5] are essentially the same as those of V. P. Burichenko.

The invariant $\mathbf{Z}$-lattices in these two algebras have not yet been investigated; however it is, in principle at least, possible that they will yield integral realizations of the groups $L_4(3)$ (with $3^3 \cdot L_3(3)$ as a maximal subgroup and with 52 as a degree of a $\mathbf{Q}$-valued character), $B_3(3)$ (having $3^{3+3} L_3(3)$ as a maximal subgroup), $C_3(3)$ or Fi_{22}. This is pure conjecture based on observation of the atlas of finite groups [14]. Quite recently, V.P. Burichenko classified the invariant **Z**-lattices in F_4 and realized the group $2 \cdot L_4(3)$.

For the Lie algebra of type E_7, the transitive $\mathcal{OD}$'s remain unknown, if indeed there are any. (Compare the corresponding assertion of A. V. Borovick). All the same, various non-transitive $\mathcal{OD}$'s have been constructed (see [4]).

2.2 For the types B_n and D_n there is an abundance of $\mathcal{OD}$'s, as the following result shows.

Theorem ([5]). *To each positive number $n \geq 4$ and each group A of all odd order $2n-1$ there corresponds a transitive orthogonal decomposition of the Lie algebra L of type D_n with automorphism groups*

$$\mathrm{Aut}_{\mathcal{OD}}(L) \cong \mathcal{E}_{2n-1} \cdot A \cdot B$$

where $\mathcal{E}_{2n-1}$ is a normal elementary abelian subgroup of order 2^{2n-1} and B is the group of all bijective mappings $\pi : A \to A$ such that $\pi(uvu) = \pi(u)\pi(v)\pi(u)$ for any $u,\ v \in A$.

An analogous assertion is valid for transitive $\mathcal{OD}$'s of the algebras of type B_n, $n \geq 2$.

The component B of the factorization $\mathcal{E}_{2n-1} \cdot A \cdot B$ contains Aut A as a subgroup, and coincides with it only in the case of an abelian group A.

To date not many results have been obtained concerning the invariant lattices in the cases B_n and D_n. However V. P. Burichenko has proved that in the cases B_{2^m-1}, $m > 2$, and D_{2^m}, $m > 2$, all invariant lattices compatible with a $\mathcal{MOD}$ necessarily contain roots. On the other hand in the cases B_3 and D_4 there exist invariant lattices without roots (K. Chakrijan, V. P. Burichenko).

2.3 The known examples of $\mathcal{OD}$'s in the case of the Lie algebra of type C_{2^m} are not interesting as far as the associated lattices Γ are concerned.

2.4 Type A_{p^m-1}. This is probably the most thoroughly investigated case K^2U, A. I. Bondal, Fam Hiu Tiep). In the case $m = 1$ the construction of an $\mathcal{OD}$ is very simple. Let ε be a primitive p-th root of 1, and let

$$D = \begin{pmatrix} 1 & & & 0 \\ & \varepsilon & & \\ & & \ddots & \\ 0 & & & \varepsilon^{p-1} \end{pmatrix}, \quad P = \begin{pmatrix} 0 & 0 & \dots & 0 & 1 \\ 1 & 0 & \dots & 0 & 0 \\ . & . & \dots & . & . \\ 0 & 0 & \dots & 1 & 0 \end{pmatrix}.$$

Write $V = \mathbf{F}_p^2$ and set $J_v = D^a P^b$ for $v = (a,b) \in V \backslash \{0\}$. (Here the elements of $\mathbf{F}_p$ are identified with $0, 1, \dots, p-1$.) We denote by $\mathbf{P}V$ the projective line with points $\sigma, \sigma_0, \dots$. We shall use the notation $(a,b) \in \sigma$ to mean that σ is determined by the vector $(a,b) \neq (0,0)$. Set $H_\sigma = \langle J_{(a,b)} \mid (a,b) \in \sigma \rangle_{\mathbf{C}}$. For the Lie algebra L of type A_{p-1}, we have the $\mathcal{OD}$, $L = \bigoplus_{\sigma \in \mathbf{P}V} H_\sigma$, which we call *standard*. It is known that $\mathrm{Aut}_{\mathcal{OD}}(L) = \mathrm{Int}_{\mathcal{OD}}(L)\langle \psi \rangle$, where $\psi(x) = -{}^t x$ for $x \in L$. In this case we take

$$G = \mathrm{Int}_{\mathcal{OD}}(L) = \{\hat{u}\,\hat{\varphi} \mid u \in V, \varphi \in SL(V)\},$$

where $\hat{u}(J_v) = \varepsilon^{u \cdot v} J_v$, $u \cdot v = ac + bd$ for $u = (a,b)$, $v = (c,d)$. Setting $B(u,v) = -bc$, we have, under the condition $p > 2$,

$$\hat{\varphi}(J_u) = \varepsilon^{\frac{1}{2}\{B(\varphi(u),\varphi(u)) - B(u,u)\}} J_{\varphi(u)}.$$

Observe that G acts on 2-transitively on $\{H_\sigma\}$ and absolutely irreducibly on L. An analogous standard $\mathcal{OD}$ exists for each m: one simply considers the matrices

$$J_u = J_{(a_1,b_1)} \otimes \cdots \otimes J_{(a_m,b_m)}; \; u = (a_1, b_1, \dots, a_m, b_m).$$

In the cases A_1, A_2, A_3, A_4, all $\mathcal{OD}$'s are conjugate. However, as has been shown by D. N. Ivanov [15], for $m > 2$ there exist non-standard $\mathcal{OD}$'s. Their construction depends on the properties of non-desarguesian projective planes.

The complete description given in [16] of all invariant sublattices (under $G = \mathrm{Int}_{\mathcal{OD}}(L)$) in the case L of type A_{p-1}, is based on a reformulation of the problem in terms of a ring of functions on the algebraic variety

$$M = \{(\varepsilon^a, \varepsilon^b) \mid (a,b) \in V \backslash (0,0)\}.$$

I shall not give the details (which may be found in [16]); however, there occur in that paper the symbols

$$\Gamma^{k,l}, 0 \leq k \leq l \leq p-1,\ k \neq p-1;\ \Delta^{k,r}, 1 \leq k \leq \frac{p-3}{2}, r \in \mathbf{F}_p^*.$$

(whose meaning I shall not give), which are used to denote certain (nearly all) of the invariant sublattices. There are $\frac{(p-1)(p-2)}{2}$ of the symbols $\Gamma^{k,l}$, and $\frac{(p-1)(p-3)}{2}$ of the $\Delta^{k,r}$; there are in addition two lattices which together with the $\Gamma^{k,l}$ and the $\Delta^{k,r}$, give all invariant lattices in this case.

The lattices $\Gamma^{k,l}$ are invariant not only under G but also under $\hat{\varphi}$, where $\varphi \in GL_2(p)$. As to the lattices $\Delta^{k,r}$, we have

$$\hat{\varphi}(\Delta^{k,r}) = \Delta^{k,r(det\varphi)^k},$$

whence it follows that the set $\{\Delta^{k,r}\}$ is partitioned into orbits. In particular when $p = 5$, there is just one such orbit, and when $p = 7$, there are three orbits with representatives $\Delta^{1,1}$, $\Delta^{2,1}$, $\Delta^{2,3}$.

Recall that the *dual* of a lattice $(\Delta, (*|*))$, $\Delta \subset \mathbf{R}^n$, is the lattice

$$\Delta^* = \{x \in \mathbf{R}^n \mid (x|y) \in \mathbf{Z}, \forall y \in \Delta\}.$$

A lattice Δ is *unimodular* if $\Delta^* = \Delta$, and is *even* if $||x||^2 = (x|x) \in 2\mathbf{Z}, \forall x \in \Delta$, and is decomposable if

$$\Delta = \Delta' \oplus \Delta'',\ \Delta' \neq 0,\ \Delta'' \neq 0,\ (\Delta'|\Delta'') = 0.$$

(a lattice is then indecomposable if it is not decomposable).

Theorem 1 ([16]). *The following statements are valid:*

(i) *To within a scalar factor the lattices*

$$\Gamma^{k,p-k}, 1 \leq k \leq \frac{p-1}{2};\ \Delta^{k,r}, 1 \leq k \leq \frac{p-3}{2}, k \equiv 1 (mod\, 2), r \in \mathbf{F}_p^*$$

are even and unimodular.

(ii) *These lattices are all indecomposable and, except for* $\Gamma^{1,p-1}$, $\Gamma^{\frac{p-1}{2},\frac{p-1}{2}}$, *have no roots.*

Theorem 2 ([16]). (i) *For every* $r \in \mathbf{F}_p^*$, $\Delta^{1,r}$ *contains a minimal vector the square of whose length is* 4.

(ii) *The even unimodular lattice* $\Gamma^{1,p-1}$ *contains as a sublattice the root sublattice* $A_{p^2-1}, p > 3$, *and coincides with the root lattice* E_8 *in the case* $p = 3$.

(iii) *The even unimodular lattice* $\Gamma^{\frac{p-1}{2},\frac{p+1}{2}}$ *contains the root lattice*

$$(p+1)A_{p-1}, \quad p \geq 3.$$

Corollary. (i) *Among the invariant lattices in* A_2 $(p = 3)$ *there is exactly one even, unimodular lattice* $\Gamma^{1,2} \cong E_8$ *with the Weyl group* $W(E_8)$ *as group of automorphisms.*

(ii) *Among the invariant lattices in* A_4 $(p = 5)$ *there are 6 even, unimodular lattices, namely:*

$\Gamma^{1,4}(\sim A_{24})$ *with automorphism group* $\mathbf{Z}_2 \times S_{25}$;

$\Gamma^{2,3}(\sim 6A_4)$ *with automorphism group* $(S_5)^6 \ltimes (GL_2(5)/\mathbf{Z}_2)$

and 4 *lattices* $\Delta^{1,r}$, $r \in \mathbf{F}_5^*$, *isomorphic to the Leech lattice* Λ; *it is known that* $\mathrm{Aut}(\Lambda)/\mathbf{Z}_2 \cong \cdot 1$ (*the first of the Conway sporadic simple groups*).

Thus one obtains one "interesting" invariant lattice in each of the two Lie algebras of types $A_{p-1} : p = 3, 5$, where the lattices are respectively $\Gamma^{1,2} \cong E_8$ and $\Delta^{1,1} \cong \Lambda$.

Hence we have the agreeable outcome whereby the Conway group $\cdot 1$ is realized in terms of the completely unremarkable complex simple Lie algebra of type A_4. What might one expect when $p > 5$?. The lattice $\Gamma^{k,l}$ are not useful in this context, since they do not yield realizations of groups close to being simple. There remain the lattices of the form $\Delta^{k,r}$, which my student Fam Hiu Tien is at present investigating. He has obtained several interesting results, not as yet, however, in final form.

Finally, mention must be made of the result of K. S. Abduhalikov [17], giving an algorithm for describing all invariant lattices in the Lie algebra of type A_{p^m-1} for $m > 1$, $p > 2$. Quite recently, K.S. Abduhalikov proved that

$$\mathrm{Aut}_{\mathbf{Z}}\,(\Delta^{1,r}) \cong \mathbf{Z}_2 \times q^2 \cdot Sp_{2m}(p)$$

for every $q = p^m > 5$.

2.5 Behind the scenes, so to speak, of my story, there is one further character, namely the all-seeing Winnie-the-Pooh, who has formulated (in Russian! – see in this connection the explanation in [2] by the translator L. Queen) the above problem for the Lie algebra of type A_5. The authors of the work [2] forgot to mention that in the formulation of their problem B. Zahoder, the talented translator into Russian of Milne's "Winnie-the-Pooh", had a part to play.

The generalized problem of Winnie-the-Pooh.

Do there exist orthogonal decompositions of the complex simple Lie algebras of type A_n, $n \neq p^m - 1$, and C_n, $n \neq 2^m$?

Many have tried to solve this problem, so far without success. Even the innocent case A_5 of Winnie-the-Pooh seems to require some new idea.

References

[1] I.B. Frenkel, J. Lepowsky and A. Meurman, A natural representation of the Fischer-Griess Monster with the modular function J as character, *Proc. Nat. Acad. Sci. U.S.A.* **81** (1984), 3256 - 3260.

[2] K^2U (A.I. Kostrikin, I.A. Kostrikin and V.A. Ufnarovskii), Orthogonal decompositions of simple Lie algebras (type A_n), *Trudy Mat. Inst. Steklov* **158** (1981), 105 - 120; English transl. in *Proc. Steklov Inst. Math.* 1983, Issue 4 (**158**).

[3] K^2U, Orthogonal decompositions of simple Lie algebras, *Dokl. Akad. Nauk SSSR* **260** (1981), 526 - 530; English transl. in *Soviet Math. Dokl.* **24** (1981).

[4] K^2U, Multiplicative decompositions of simple Lie algebras, *Dokl. Akad. Nauk SSSR* **262** (1982), 29 -33; English transl. in *Soviet Math. Dokl.* **25** (1982).

[5] K^2U, On decompositions of classical Lie algebras, *Trudy Mat. Inst. Steklov* **166** (1984), 107 - 122; English transl. in *Proc. Steklov Inst. Math.* 1986, Issue 1 (**166**).

[6] J.G. Thompson, A conjugacy theorem for E_8, *J. Algebra* **38** (1976), 525 - 530.

[7] W.H. Hesselink, Preprint, Univ. Göttingen, 1980.

[8] D. Jackson, Ph.D. Dissertation, Cambridge Univ., 1981.

[9] P.E. Smith, Ph.D. Dissertation, Cambridge Univ., 1975.

[10] K. Chakrijan, Transitive $\mathcal{OD}$ for Lie algebras of type G_2, *Proc. Acad. Sci. Bulgaria*, to be published.

[11] K^2U, Invariant lattices of type G_2 and their automorphism groups, *Trudy Mat. Inst. Steklov* **165** (1984), 79 - 97; English transl. in *Proc. Steklov Inst. Math.* 1985, Issue 3 (165).

[12] V.P. Burichenko, Transitive orthogonal decompositions of the simple complex Lie algebras of type F_4 and E_6, XIX *All Union Algebraic Conference*, Abstracts, Lvov, 1987, Part I, p. 46.

[13] A.V. Borovick, Jordan subgroups and orthogonal decompositions, in print.

[14] J.H. Conway, R.T. Curtis, S.P. Norton, R.A. Parker and R.A. Wilson, *Atlas of finite groups*, Clarendon Press, Oxford, 1985.

[15] D.N. Ivanov, Orthogonal decompositions of Lie algebras of type A_{p^n-1} and D_n, XIX *All Union Algebraic Conference*, Abstracts, Lvov, 1987, Part I, p. 114.

[16] A.I. Bondal, A.I. Kostrikin and Fam Khyu T'ep, Invariant lattices, the Leech lattice and its even unimodular analogues in the Lie algebras A_{p-1}, *Mat. Sbornik* **130** (172) (1986), 435 - 464; English transl. in *Math. USSR Sbornik* **58** (1987), No. 2.

[17] K.S. Abduhalikov, On invariant lattices in Lie algebras of type A_{q-1}, *Uspehii Mat. Nauk* **43** (1988), 187 - 188.

Moscow University
U.S.S.R.

Yet more on finite groups with few defining relations

In memoriam K.A. Hirsch

B.H. Neumann

Summary. This talk gave a brief survey of some of the finite groups with "zero -deficiency" (or "balanced") presentations that the author has studied in the recent past. Some results obtained in joint work with J. L. Mennicke are to be published in the *J. Austral. Math. Soc.*, Ser. A. Some new results are also presented.

0. Introduction and acknowledgements

This expanded version of the final invited talk of the conference has, inevitably, some overlap with my paper [9], and with a joint paper with J. L. Mennicke [8] that is in course of publication. Some new results are presented in the final section. It is a pleasure to acknowledge the mathematical help I have received from Jens L. Mennicke, the friendly hospitality he and his colleagues and their spouses have shown me while I was at Bielefeld, the support of the DAAD (Deutscher Akademischer Austauschsdienst) and of the National University of Singapore. Special thanks are also due, and gladly rendered, to the organisers of the conference, especially Peng Tsu Ann, Malcolm J. Wicks, Cheng Kai Nah, Leong Yu Kiang, Tan Sie Keng, not only for making it such an excellent conference, but also for many kindnesses.

1. Known facts

We use the following notation for group presentations:

$$\mathcal{G} = gp(a_1, a_2, \ldots, a_d;\ u_1 = v_1, u_2 = v_2, \ldots, u_e = v_e).$$

(This notation was introduced by Philip Hall to deal also with other algebraic systems than groups.) Here the u_i and v_i are group words in the generators $a_1, \ldots, a_d$. The "deficiency" of the presentation is defined to be

$$d - e.$$

This is not a group invariant: counter-examples are the Baumslag-Solitar [1] group

$$gp(a, b;\ a^2 b = b a^3),$$

which, as remarked by Graham Higman [5], can also be generated by a^4 and b, but then needs 2 defining relations; and the "clover-knot" group

$$gp(a, b;\ a^2 = b^3),$$

(Dunwoody and Petrowski [3]).

The following fact is trivial: If $\mathcal{G}$ is finite, then

$$d - e \leq 0.$$

Presentations with $d = e$ are called "zero-deficiency" or "balanced" presentations. Many finite groups with zero-deficiency presentations (called "interesting" groups by David L. Johnson [6, p. 63]) are known: examples are the finite cyclic groups (with $d = e = 1$)

$$C_n = gp(a;\ a^n = 1),$$

the symmetric group of degree 3

$$S_3 = gp(a, b;\ b^{-1}ab = a^2,\ b^2 = 1),$$

the quaternion group

$$Q_8 = gp(a, b;\ b^{-1}ab = a^{-1},\ a^{-1}ba = b^{-1}).$$

More examples with $d = e = 2$ will come later. Examples of finite groups with (minimally) $d = e = 3$ were made by J. L. Mennicke [7], by J. W. Wamsley [12], and by D. L. Johnson [6].

There are no known examples of finite groups with (minimally) $d = e = 4$. Finite nilpotent groups with minimally $d = 4$ generators require $e \geq 5$

defining relations. Such groups were constructed by G. Havas and M. F. Newman [4], of orders $2^{16}, 2^{17}, 2^{18}, 2^{19}$.

A classical fact is that finite groups with zero-deficiency presentations have trivial Schur multiplicator: this goes back to Issai Schur [10], [11] himself. As a reminder: If $\mathcal{N}$ is a (necessarily normal) subgroup of the group $\mathcal{G}$ contained in both the centre and the derived group of $\mathcal{G}$, then $\mathcal{N}$ is isomorphic to a subgroup of the Schur multiplicator of the factor group $\mathcal{G}/\mathcal{N}$. If $\mathcal{G}$ is written as a factor group $\mathcal{F}/\mathcal{R}$ of a free group, then the Schur multiplicator can be obtained from $\mathcal{R}/[\mathcal{F},\mathcal{R}]$ by cutting out a free abelian factor of rank equal to that of $\mathcal{F}$. That the result does not depend on the choice of $\mathcal{F}$ and $\mathcal{R}$ is also a (non-obvious) fact.

Another classical fact, due also to Schur [10], is that the multiplicator of a finite group is finite. It is abelian, isomorphic to a subgroup of the direct product of the multiplicators of the Sylow subgroups of the given group. Thus, for example, it is trivial if the given group has only cyclic Sylow subgroups. Such groups are easily seen to have zero-deficiency presentations, with $d = e = 2$.

2. More known results

More (known) examples of finite groups with $d = e = 2$ are

$$\mathcal{B}(p,q,r) = gp(a,b;\ a^p = b^q = (ab)^r).$$

This is finite if, and only if,

$$\frac{1}{|p|} + \frac{1}{|q|} + \frac{1}{|r|} > 1. \tag{1}$$

Without loss of generality

$$0 < p \le |q| \le |r|.$$

The case $p = 1$ is uninteresting. More interesting cases are

(D) $p = 2, \quad q = \pm 2, \quad |r| \ge 2,$ arbitrary;
(T) $p = 2, \quad q = \pm 3, \quad r = \pm 3;$
(O) $p = 2, \quad q = \pm 3, \quad r = \pm 4;$

(I) $\qquad p = 2, \quad q = \pm 3, \quad r = \pm 5.$

These are various extensions of the dihedral (D), tetrahedral (T), octahedral (O), icosahedral (I) groups; they were completely determined by H. S. M. Coxeter [2], though some of them occur much earlier; see [9, pp 233-234] for the relevant references. These presentations I generalised in [9] to

$$\mathcal{G}(p,q,r,s,t,u) = gp(a,b,c;\ a^p = c^s, b^q = c^t, (ab)^r = c^u).$$

Necessary for finiteness is

$$\frac{1}{|p|} + \frac{1}{|q|} + \frac{1}{|r|} > 1; \tag{1}$$

$$\Delta := -pqu + prt + qrs \neq 0. \tag{2}$$

(Δ is the index in $\mathcal{G}$ of its derived group $\mathcal{G}'$). Sufficient for finiteness is (1) and (2) and

$$(s,t) = (s,u) = (t,u) = 1. \tag{3}$$

(One proves without difficulty that c is in the centre, and a power of c in the derived group.) The groups $\mathcal{G}(p,q,r,s,t,u)$ are certain central extensions of the polyhedral groups

$$\mathcal{P}(p,q,r) := gp(a,b;\ a^p = b^q = (ab)^r = 1),$$

of order

$$|\mathcal{G}| = 2n|\mathcal{P}|,$$

where n is a function of the parameters p,q,r,s,t,u. (Observe that the groups $\mathcal{B}$ mentioned earlier are obtained from the polyhedral groups $\mathcal{P}$ by omitting the final "$= 1$" from the defining relations.) By suitable choice of s,t,u, one can give n any arbitrary integer value.

The groups $\mathcal{G}$ can be generated by a and b alone. Can they be defined by two relations in a and b? In [9] I conjectured that the answer is negative in general, especially in the icosahedral case with larger n. This conjecture was demolished by J. L. Mennicke: see our joint paper [8] for the details.

3. New results

In a search for groups that could possibly be the fundamental groups of compact closed 3-dimensional manifolds, J. L. Mennicke had also looked at the group

$$\mathcal{M} := gp(a, b;\ (ab)^2 = a^{10}, b^{-3}ab^3 = a^{-1}),$$

and had found that it is finite, of order

$$|\mathcal{M}| = 120.$$

This can be generalised to the zero-deficiency presentation

$$\mathcal{M}(p,q) := gp(a, b;\ (ab)^2 = a^{2p}, b^{-q}ab^q = a^{-1}),$$

where p and q are arbitrary odd numbers. Mennicke's group is

$$\mathcal{M} = \mathcal{M}(5,3).$$

The cases that p and q are not both odd can also be treated, but is less interesting.

Theorem 3.1. *If p and q are odd, then $\mathcal{M}(p,q)$ is finite, of order*

$$|\mathcal{M}(p,q)| = 8|pq|.$$

Proof. The defining relations are, for future reference

$$abab = a^{2p}, \tag{3.11}$$

$$b^{-q}ab^q = a^{-1}. \tag{3.12}$$

Observe first that a^{2p} is permutable with ab, and, being also (trivially) permutable with a, it is permutable with b, and thus central. Now

$$a^{2p} = b^{-q}a^{2p}b^q = a^{-2p},$$

and thus

$$a^{4p} = 1. \tag{3.13}$$

Also $(ab)^2$, being equal to a^{2p}, is central. Transformation by b^q gives

$$abab = b^{-q}ababb^q = a^{-1}ba^{-1}b,$$

and

$$a^2ba^2 = b. \tag{3.14}$$

Now as p is assumed odd, $p-1$ is even, and repeated application of (3.14) gives

$$a^{p-1}ba^{p-1} = b,$$

$$a^{-p}ba^p = a^{-2p+1}ba = a^{-2p}aba = b^{-1},$$

using (3.11). This,

$$a^{-p}ba^p = b^{-1}, \tag{3.15}$$

is analogous to (3.12); and using it to transform the (central) element $abab$ by a^p, one obtains

$$abab = a^{-p}ababa^p = ab^{-1}ab^{-1},$$

$$b^2ab^2 = a, \tag{3.16}$$

in analogy to (3.14). Now q is also assumed odd, so repeated application of (3.16) gives

$$b^{q-1}ab^{q-1} = a,$$

$$a^{-1} = b^{-q}ab^q = b^{-2q+1}ab,$$

and thus also

$$b^{2q} = baba \ (= abab), \tag{3.17}$$

in analogy to (3. 11). This shows that b^{2q} is central - which could also have been inferred from (3. 12) directly. It also shows that

$$\mathcal{G}(p,q) \cong \mathcal{G}(q,p).$$

To continue with the proof, observe that

$$ba = a^{2p-1}b^{-1} \quad \text{(from (3.11))},$$

$$b^2 a = ab^{-2} \quad \text{(from (3.16))},$$

and thus inductively

$$b^j a = a^{2p-1}b^{-j} \quad \text{when } j \text{ is odd},$$

$$b^j a = ab^{-j} \quad \text{when } j \text{ is even}.$$

Similarly, and more generally, one shows that

$$b^j a^i = a^{f(i,j)} b^{g(i,j)},$$

where f and g are (easily computable) functions of i, j. Thus all elements of the group can be written in the form

$$a^f b^g,$$

with

$$0 \leq f < 4p, \quad 0 \leq g < 2q,$$

because of (3.13) and (3.17), which implies that

$$b^{2q} = a^{2p}.$$

It follows that the order of $\mathcal{G}$ can not exceed $8|pq|$. It is not difficult to see that the group has, in fact, precisely this order. Its structure is also easily described: It is an extension of an abelian group, the direct product of the cyclic group of order p generated by a^4 and the cyclic group of order q generated by b^4 by a quaternion group generated by a^p and b^q - I omit the details.

A similar type of finite group, that arose from an attempt to generalise the above, is

$$\mathcal{N}(q,r) := gp(a,b;(ab)^2 = a^{r+1}, b^{-q}ab^q = a^r),$$

with q and r again odd integers, $r \neq \pm 1$.

Theorem 3.2. *If q and $r \neq \pm 1$ are odd, then $\mathcal{N}(q,r)$ is finite, of order*

$$|\mathcal{N}(q,r)| = 2|q(r^2-1)|.$$

Proof. The defining relations are repeated for reference:

$$abab = a^{r+1}, \tag{3.21}$$

$$b^{-q}ab^q = a^r. \tag{3.22}$$

Again a^{r+1} is easily seen to be central; hence

$$a^{r+1} = abab = b^{-q}ababb^q = a^r ba^r b,$$

$$ba^r b = a. \tag{3.23}$$

Now, as (3.21) can be simplified to

$$bab = a^r,$$

(3.23) gives also here

$$b^2ab^2 = a. \tag{3.16}$$

More generally,

$$b^j ab^j = a \quad \text{when } j \text{ is even},$$

$$b^j ab^j = a^r \quad \text{when } j \text{ is odd}. \tag{3.24}$$

With $j = q$ this gives, as q is assumed odd,

$$b^q ab^q = a^r = b^{-q} ab^q,$$

and thus

(3.25) $$b^{2q} = 1.$$

To find the order of a, again transform the central element a^{r+1} by b^q, to get

$$a^{r+1} = b^{-q} a^{r+1} b^q = a^{r(r+1)},$$

which gives

(3.26) $$a^{r^2-1} = 1.$$

As in the proof of Theorem 3.1, one goes on to show that

$$b^j a = ab^{-j} \quad \text{when } j \text{ is even,}$$

$$b^j a = a^r b^{-j} \quad \text{when } j \text{ is odd,}$$

and then, by an easy induction, that

$$b^j a^i = a^{f(i,j)} b^{g(i,j)},$$

with f and g easily computed functions of i, j. Thus every element of the group can be written in the form

$$a^f b^g.$$

with

(3.27) $$0 \le f < r^2 - 1, \quad 0 \le g < 2|q|.$$

This shows that the order of the group can not exceed $2|q(r^2-1)|$. To show that this is in fact the exact order, one considers vectors

$$\begin{pmatrix} f \\ g \end{pmatrix}$$

with f and g integers modulo $r^2 - 1$ and $2|q|$, respectively, and makes a and b act as permutations on the set of these vectors by putting

$$\begin{pmatrix} f \\ g \end{pmatrix} a = \begin{pmatrix} f+1 \\ -g \end{pmatrix} \quad \text{if } g \text{ is even,}$$

$$\begin{pmatrix} f \\ g \end{pmatrix} a = \begin{pmatrix} f+r \\ -g \end{pmatrix} \quad \text{if } g \text{ is odd,}$$

$$\begin{pmatrix} f \\ g \end{pmatrix} b = \begin{pmatrix} f \\ g+1 \end{pmatrix}.$$

It is not difficult to see that these permutations generate a transitive permutation group on the set of $2|q(r^2 - 1)|$ vectors; and one also readily verifies that these permutations satisfy the defining relations (3.21) and (3.22). The structure of $\mathcal{N}(q,r)$ can also be described, as an extension of a dihedral group of order $2|r+1|$ generated by a^{r-1} and b^q by another metacyclic group with generators a^* and b^* (corresponding to a and b), with defining relations

$$a^{*\,r-1} = b^{*\,q} = 1, \quad a^{*\,-1} b^* a^* = b^{*\,-1},$$

this latter group being a central extension of a dihedral group of order $2|q|$ by a cyclic group generated by a^{*2}. Alternatively $\mathcal{N}(q,r)$ can be described as an extension of the direct product of two cyclic groups, one of order $\frac{1}{2}(r^2 - 1)$ generated by a^2, the other of order $|q|$ and generated by b^2, by a four-group, whose action can be described without difficulty.

Some or all of the parameters that occur in these presentations can also be chosen even, but the groups that turn up appear (to me) less interesting.

References

[1] Gilbert Baumslag and Donald Solitar, Some two-generator one-relator non-Hopfian groups, *Bull. Amer. Math. Soc.* **68** (1962), 199-201.

[2] H. S. M. Coxeter, The binary polyhedral groups, and other generalizations of the quaternion group, *Duke Math. J.* **7** (1940), 367-379.

[3] M. J. Dunwoody and A. Pietrowski, Presentations of the trefoil group, *Canad. Math. Bull.* **16** (1973), 517-520.

[4] George Havas and M. F. Newman, Minimal presentations for finite groups of prime-power order, *Comm. Algebra* **11** (1983), 2267-2275.

[5] Graham Higman, unpublished.

[6] D. L. Johnson, *Topics in the theory of group presentations* [London Mathematical Society Lecture Notes Series, 42]. Cambridge University Press, Cambridge, 1980.

[7] Jens Mennicke, Einige endliche Gruppen mit drei Erzeugenden und drei Relationen, *Arch. Math.* (Basel) **10** (1959), 409-418.

[8] J. L. Mennicke, and B. H. Neumann, More on finite groups with few defining relations, *J. Austral. Math. Soc.*, Ser. A (to be published).

[9] B. H. Neumann, Some finite groups with few defining relations, *J. Austral. Math. Soc.*, Ser. A **38** (1985), 230-240.

[10] J. Schur, Über die Darstellung der endlichen Gruppen durch gebrochene lineare Substitutionen, *J. reine angew. Math.* **127** (1904), 20-50.

[11] J. Schur, Untersuchugen über die Darstellung der endlichen Gruppen durch gebrochene lineare Substitutionen, *J. reine angew. Math.* **132** (1907), 85-117.

[12] J. W. Wamsley, A class of three-generator, three-relator, finite groups, *Canad. J. Math.* **22** (1970), 36-40.

Department of Mathematics
Institute of Advanced Studies
Australian National University
Canberra, ACT 2601
Australia

and

Division of Mathematics and Statistics
Commonwealth Scientific and Industrial Research Organization
Canberra, ACT 2601
Australia

Elementary proof of the simplicity of sporadic groups

Michio Suzuki

1. Introduction

According to the classification theorem, a finite simple group is isomorphic to one of the simple groups in the following list:

(1) cyclic groups of prime order;
(2) alternating groups of degree at least five;
(3) simple groups of Lie type;
(4) 26 sporadic simple groups.

Of these groups, each of the 26 sporadic groups is defined individually, and each group has been proved to be simple by a separate argument.

The purpose of this note is to prove the simplicity of each sporadic group in an elementary manner using as few properties of the given group as possible, preferably using only the definition of the group. We also seek a method which is applicable to many sporadic groups, not just one of them. Since sporadic groups are defined in many different ways, it will not be possible (at present) to give a uniform treatment. We will divide sporadic groups into four families depending on how these groups are defined, and for each family we will give a method of proving the simplicity of the groups in that family.

2. Four families of sporadic groups

The first family contains sporadic groups which are defined as a primitive transitive extension. The second family consists of those sporadic groups which are defined by giving the structure of the centralizer of an involution. The other two families are the family of Conway groups, and the family of Fischer groups related to groups generated by the conjugacy class of 3–transpositions.

It is true that each sporadic group can be defined as a primitive transitive extension and can also be defined by giving the structure of the centralizers of involutions. So, the distribution of sporadic groups into families is rather arbitrary. However, we will use the basic original definition for each group.

3. First family

The first family consists of five Mathieu groups and five rank three permutation groups. The following list contains the degree and the structure of the stabilizer of one point in the defining permutation representation of each group.

M_{11}	11	A_6 2
M_{12}	12	M_{11}
M_{22}	22	$PSL(3,4)$
M_{23}	23	M_{22}
M_{24}	24	M_{23}
J_2	100	$PSU(3,3)$
HS	100	M_{22}
S	$2\ 3^4 11$	$G_2(4)$
Mc	$5^2 11$	$PSU(4,3)$
Ru	$2^2 5\ 7\ 29$	the Tits group

We note that all the stabilizers in the last column are simple except the first. The notation X 2 denotes a group Y having a normal subgroup X of index 2 such that every element of $Y - X$ induces an outer automorphism of order 2 in X.

A (well-known) proof of the simplicity of the groups in this family is based on the following (known) lemma.

Lemma 1. *Let G be a primitive transitive permutation group of degree n, and let H be the stabilizer of a point. Let N be a normal subgroup $\neq \{1\}$ of G. Then, the order of N is of the form mn where n is the degree of G and m is the order of a normal subgroup of H. If H is simple, then G is simple except possibly when n is a power of the order of some simple group.*

For example, the Higman-Sims group is simple because the stabilizer M_{22} is simple and the degree n is not a power of $|X|$ for any simple group X.

This is the usual proof of the simplicity of each group in this family. Lemma 1 is easy to prove and elementary. Applications of the lemma to individual groups are routine. Since the method is uniformly applicable to every group in the family, we do not need to say more, except perhaps to remark that for the last three groups, applications of Sylow's theorem for $p = 11$ or 29 will prove the nonexistence of a simple group of order n.

4. Second family

Many of the sporadic groups in the second family are defined as simple groups having prescribed centralizer of involutions. Thus, for example, the first Janko group J_1 is defined as a simple group with abelian S_2-subgroups which contains an involution t such that the centralizer of t is isomorphic to $\langle t\rangle \times A_5$. In this case, we do not need to prove the simplicity of J_1. On the other hand, the Fischer group F_2 is defined as a group generated by the conjugacy class of a $\{3,4\}$-transposition with prescribed centralizer. Thus, it is necessary to prove that F_2 is indeed simple.

We define a sporadic group of the second kind as a finite group G which contains an involution t such that

(i) G is generated by a conjugacy class of t, and
(ii) the structure of $\mathbf{C}_G(t)$ is given as one of the groups in the second column of the following list :

J_2, J_3	$2^{4+1}A_5$
J_4	$2^{12+1}(3\ M_{22}\ 2)$
He	$2^{6+1}PSL(3,2)$
Ly	$2\ A_{11}$
ON	$4\ PSL(3,4)\ 2$
F_2	$2\ ({}^2E_6(2))\ 2$
F_3	$2^{8+1}A_9$
F_5	$2\ (HS)2$
F_1	$2\ F_2$

The first column lists the sporadic group defined by the centralizer $\mathbf{C}_G(t)$ given in the second column.

In the above list, we used the following notation. The symbol $2^{2n+1}X$ denotes a group C with a normal subgroup E such that

(i) E is an extra-special 2-group of order 2^{2n+1},
(ii) $C/E = X$, and
(iii) $\mathbf{F}^*(C) = E$.

Here, $\mathbf{F}^*(C)$ is the generalized Fitting subgroup of C.

The symbol $m\,Y$ for a simple group Y denotes a central extension D of Y such that

(i) The center $\mathbf{Z}(D)$ is a cyclic group of order m,
(ii) $D/\mathbf{Z}(D) = Y$, and
(iii) $[D, D] \supset \mathbf{Z}(D)$.

And, finally, we set $m\,Y\ 2 = (m\,Y)\ 2$.

To prove the simplicity of the groups of the second family, we need the following lemma.

Lemma 2. *Let N be a normal subgroup of even order of a finite group G. Then, for any involution t of G, we have $N \cap \mathbf{C}_G(t) \neq \{1\}$.*

Proof. This is known, but for the convenience of the readers we will sketch a proof. Let $H = \mathbf{C}_G(t)$, and let Q be an S_2-subgroup of H. By Sylow's theorem, Q is contained in an S_2-subgroup R of G. Since $N \lhd G$, $N \cap R \lhd R$ and $N \cap R$ is an S_2-subgroup of N. By assumption, N is of even order so $N \cap R \neq \{1\}$. By a fundamental property of p-groups, we have $N \cap R \cap \mathbf{Z}(R) \neq \{1\}$. Thus, we have $\{1\} \neq N \cap \mathbf{Z}(R) \subset N \cap \mathbf{C}_G(t)$. □

We will now prove the following proposition.

Proposition 3. *Let G be a finite group of order divisible by 4. Suppose that G contains an involution t satisfying the following two conditions :*

(i) *any normal subgroup $\neq \{1\}$ of $\mathbf{C}_G(t)$ contains t;*

(ii) *G is generated by the conjugacy class of t.*
Then, G is simple.

Proof. Let $H = \mathbf{C}_G(t)$. Suppose that G is not simple, and let N be a proper normal subgroup of G.

First, we will prove that N is of odd order . Suppose that N is of even order. By Lemma 2, we have $N \cap H \neq \{1\}$. Since $N \cap H \lhd H$, condition (i) gives us $t \in N \cap H$. Hence, the normal subgroup N contains t, and therefore all conjugates of t. By (ii), we get $N = G$ which contradicts the assumption that N is a proper subgroup of G. This proves that N is of odd order.

It follows that $N \cap H$ is also of odd order. Hence, by (i), we have $N \cap H = \{1\}$. Thus, the element t acts on N without fixed points. By Burnside's lemma ([4], II p.131), t inverts every element of N.

Let us consider the centralizer $M = \mathbf{C}_G(N)$. Then, $M \lhd G$, but $M \neq G$ because $t \notin M$. It follows that M is of odd order. We will show that $\langle M, t\rangle$ is a normal subgroup of G. Set $G_0 = \langle M, t\rangle$. If $g \in G$, then $g^{-1}Mg = M$. We need to show that

$$g^{-1}tg \in G_0.$$

If $x \in N$, then $x^t = x^{-1}$ and $(gxg^{-1})^t = gx^{-1}g^{-1}$. So, we get

$$g^{-t}g = t^{-1}g^{-1}tg \in \mathbf{C}_G(N) = M.$$

This implies that $g^{-1}tg \in \langle M, t\rangle = G_0$. Hence, we have $G_0 \lhd G$. Clearly, the order of G_0 is $2|M|$, twice an odd number. It follows that we have $G_0 = G$. However, this contradicts the assumption that the order of G is divisible by 4. Thus, G must be simple. □

The simplicity of the groups in the second family will be proved by the following two lemmas.

Lemma 4. *Let $C = \mathbf{C}_G(t)$ be a centralizer of the form $2^{2n+1}X$. Then, C satisfies the condition (i) of the proposition.*

Proof. By assumption, $E = \mathbf{F}^*(C)$ is an extra-special 2-group of order 2^{2n+1}. Then, the center of E is $\langle t\rangle$.

Let H be a normal subgroup of C. Suppose that $H \cap E \neq \{1\}$. Then, $H \cap E \lhd E$. By a fundamental property of p-groups, we have $H \cap E \cap \mathbf{Z}(E) \neq \{1\}$. Since $\mathbf{Z}(E) = \langle t \rangle$, we get $t \in H$.

Suppose that $H \cap E = \{1\}$. We will prove that $H = \{1\}$. We have $[H, E] \subset H \cap E = \{1\}$. It follows that $H \subset \mathbf{C}_C(E)$. By a fundamental property of the generalized Fitting subgroup $\mathbf{F}^*(C)$ ([4], II p.452), we have $\mathbf{C}_C(E) \subset E$. Hence, we get $H \subset E$. This gives us $\{1\} = H \cap E = H$. Thus, the group C satisfies the condition (i) of the Proposition. □

The proposition, together with Lemma 4, gives us the simplicity of J_2, J_3, J_4, He, and F_3.

Remark. The Fischer-Griess group F_1 can be defined by using the centralizer of involutions in the second conjugacy class. In this case, the structure of the centralizer is $2^{24+1}(\cdot 1)$ where $\cdot 1$ is the Conway group. So, the simplicity of F_1 can be proved by Lemma 4.

The remaining groups of the second kind can be handled by a similar argument.

Lemma 5. *Let $C = \mathbf{C}_G(t)$ be a centralizer of the form $2X$ or $mX2$. Then, C satisfies the condition* (i) *of Proposition* 3.

Proof. In all cases, X is a simple group, and $\langle t \rangle$ is the unique subgroup of order 2 in the center $Z = \mathbf{Z}(C)$. Suppose that C does not satisfy the condition (i) of Proposition 3, and let H be a normal subgroup $\neq \{1\}$ of C such that $t \notin H$. Then, $Z \cap H = \{1\}$. Set $K = ZH$. Then, K is a normal subgroup of C which contains Z.

Let L be the normal subgroup of C corresponding to $m\,X$. Then, L is an irreducible central extension of X. Suppose that K contains L. In this case, we have $L = (L \cap H)Z = (L \cap H) \times Z$. This contradicts the irreducibility of L. So, K does not contain L.

Since $L/Z = X$ is simple, we get $L \cap K = Z$. It follows that

$$[L, H] \subset L \cap H = L \cap K \cap H = Z \cap H = \{1\}.$$

This contradicts the structure of $(m\,X)2$. Thus, the group C of the form $2\,X$ or $m X\,2$ satisfies the condition (i) of Proposition 3. □

Remark. The first Janko group J_1 may be defined as a group G having an abelian S_2-subgroup such that $G = G'$ and such that G contains an involution t with $\mathbf{C}_G(t) = \langle t \rangle \times A_5$. It can be proved that G is simple. But, it seems to require some fusion argument in proving the conjugacy of involutions. In the presentation of J_1 due to Livingstone [3], J_1 is defined as a primitive transitive extension of $PSL(2,11)$ with degree 266. In this form, the simplicity of J_1 can be proved without difficulty by using Lemma 1.

5. Fischer groups

Each of the Fischer groups is defined as a group generated by a conjugacy class D of 3-transpositions. One of the main theorems of Fischer is the assertion that, as a permutation group on the elements of D, the Fischer group is a primitive permutations group of rank 3 [2]. We will list for each Fischer group the dgree $|D|$ and the structure of the stabilizer of an element t of D.

$M(22)$	$3510{=}2\ 3^3 5\ 13$	$2\ PSU(6,2)$
$M(23)$	$31671{=}3^4 17\ 23$	$2\ M(22)$
$M(24)$	$306,936{=}2^3 3^3 7^2 29$	$Z_2 \times M(23)$

The information contained in the above table was given by Fischer. We note that the last column uses the same notation as in the previous section and that the stabilizer is the centralizer of the involution t of D.

Since Fischer groups are generated by D, Proposition 3, together with Lemma 5, proves that $M(22)$ and $M(23)$ are simple.

For the group $M(24)$, Proposition 3 does not apply. We need more information about $M(24)$. Fischer proved that the commutator subgroup of $M(24)$ is a subgroup of index 2. (This is proved by a fusion argument.) With this information at hand, we can prove the simplicity of $M(24)'$ as follows. The group $M(24)$ is a primitive transitive extension of a direct product $H = \langle t \rangle \times M(23)$. The group H is the centralizer of t. It is easy to see (given the simplicity of $M(23)$) that the normal subgroups of H are $\{1\}$, $\langle t\rangle$, $M(23)$, and H. Let N be a proper normal subgroup of $M(24)$. Then, the order of N is even by Lemma 1. Thus, $N \cap H \neq \{1\}$by Lemma 2. As in the proof of Proposition 3, $t \in N \cap H$ would imply the contradiction that $N = M(24)$. So, the only possibility is $N \cap H = M(23)$. By Lemma 1, the order of N is half the order of $M(24)$. Hence the commutator subgroup $M(24)'$ is the only proper normal subgroup of $M(24)$. The order $M(24)'$ is obviously not a square, so $M(24)'$ is simple (cf. the proof of Theorem 2.11 in vol. I [4], p.295).

Remark. The group $M(24)'$ is a primitive transitive extension of $M(23)$ (and of rank 3). Thus, Lemma 1 gives us the following. Either $M(24)'$ is simple, or there is a simple group of order $2^3 3^3 7^2 29$. It is certainly possible to prove that there is no simple group of this order and finish the proof of the simplicity of $M(24)'$ based on Lemma 1. However, it seems that we need more than just Sylow's theorems to prove the nonsimplicity of a group of order $2^3 3^3 7^2 29$.

We can also prove the simplicity of $M(22)$ and $M(23)$ using Lemma 1. But, it seems more complicated than the proof based on Proposition 3.

6. Conway groups

The Conway group $\cdot 0$ is the group of automorphisms of the Leech lattice Λ in R^{24}, the 24-dimensional Euclidean space. Let $\{e_i\}$ be an orthonormal basis of R^{24}. Then, each element of Λ is a certain linear combination of the basis elements with integral coefficients. For example,

$$v_2 = -3e_1 + e_2 + \ldots + e_{24} \quad \text{and}$$
$$v_3 = 5e_1 + e_2 + \ldots + e_{24}$$

are elements of Λ. If $x \in \Lambda$, it is proved that $(x, x) = 16n$ for some integer n. Let

$$\Lambda_n = \{x \in \Lambda | (x, x) = 16n\}.$$

Then, Λ_1 is empty, $v_2 \in \Lambda_2$, and $v_3 \in \Lambda_3$. It can be proved that $\cdot 0$ acts transitively on the set Λ_2 as well as on Λ_3 (and Λ_4).

It is easy to see that $-1 \in \cdot 0$. The Conway groups are defined as follows:

$$\begin{aligned}
\cdot 1 &= \cdot 0/\langle -1 \rangle \\
\cdot 2 &= \text{the stabilizer of } v_2, \quad \text{and} \\
\cdot 3 &= \text{the stabilizer of } v_3.
\end{aligned}$$

The orders of these groups are

$$\begin{aligned}
|\cdot 0| &= 2^{22}3^9 5^4 7^2 11\ 13\ 23, \\
|\cdot 1| &= |\cdot 0|/2, \\
|\cdot 2| &= 2^{18}3^6 5^3 7\ 11\ 23, \quad \text{and} \\
|\cdot 3| &= 2^{10}3^7 5^3 7\ 11\ 23.
\end{aligned}$$

If we let $\sigma \in M_{24}$ act on the 24 unit vectors $\{e_i\}$, then σ is an element of $\cdot 0$. Thus, we may consider M_{24} as a subgroup of $\cdot 0$. In particular, if $\alpha = (2, 3, \dots, 24)$ is a 23-cycle, then α may be considered as an element of $\cdot 0$. As an element of $\cdot 0$, α fixes e_1 but moves $e_2, \dots, e_{24}$ cyclically.

The proof of the simplicity of the Conway groups is based on the following lemmas.

Lemma 6. *Let G be the Conway group $\cdot 0$, and let α be the element of order 23 described above. Then, for $A = \langle \alpha \rangle$, we have*

$$\mathbf{C}_G(\alpha) = \langle -1 \rangle \times A, \text{ and } \mathbf{N}_G(A) = \langle -1 \rangle \times \langle \alpha, \beta \rangle$$

where $\langle \alpha, \beta \rangle \subset M_{23}$ and $\beta^{-1} \alpha \beta = \alpha^2$.

Lemma 7. *Let G be any finite group, and let H be a normal subgroup of G. For any Sylow subgroup P of H, we have*

$$G = H\mathbf{N}_G(P).$$

If p is a prime divisor of $|G|$, then p divides either $|H|$ or $|\mathbf{N}_G(P)|$.

Lemma 7 is trivial and known. Lemma 6 is a property of the Conway group which is fairly easy to prove, but nonetheless, it depends on fundamental properties of the Conway group.

Proposition 8. *Let G denote the Conway group $\cdot 0$. Let H be a normal subgroup of G such that $-1 \in H$. Then, either $H = G$ or $H = \langle -1 \rangle$. The Conway group $\cdot 1$ is simple.*

Proof. We divide the proof into two parts. First, we consider the case when the prime number 23 divides the order of H. We will prove that $H = G$. Since 23 divides $|H|$, H contains a subgroup of order 23. This subgroup is an S_{23}-subgroup, so it is conjugate to $A = \langle \alpha \rangle$ where α is the element of order 23 defined before. Since $H \triangleleft G$, we have $A \subset H$. Thus, A is an S_{23}-subgroup of H. By Lemma 7, we get

$$G = H\mathbf{N}_G(A).$$

On the other hand, Lemma 6 gives us

$$\mathbf{N}_G(A) = \langle -1 \rangle \times \langle \alpha, \beta \rangle \subset \langle -1, M_{23} \rangle.$$

Since M_{23} is simple, M_{23} is generated by conjugates of α. We have $\alpha \in H$ and $H \triangleleft G$. Hence, H contains all conjugates of α. It follows that $M_{23} \subset H$. By definition, $-1 \in H$. Therefore,

$$\mathbf{N}_G(A) \subset \langle -1, M_{23} \rangle \subset H.$$

Thus, we have $G = H\mathbf{N}_G(A) = H$.

Suppose that the order $|H|$ is not divisible by 23. Let p be a prime factor of $|H|$, and let P be an S_p-subgroup of H. Then, by Lemma 7, $|\mathbf{N}_G(P)|$ is divisible by 23. Thus $\mathbf{N}_G(P)$ contains an element α' of order 23. Since $\langle \alpha' \rangle$ is conjugate to A, we have

$$\mathbf{C}_G(\alpha') = \langle -1 \rangle \times \langle \alpha' \rangle$$

by Lemma 6. Thus, if $p \neq 2$, the element α' induces an automorphism of P such that each nonidentity element belongs to an orbit of length 23. It follows that

$$|P| \equiv 1 \pmod{23}.$$

However, this is not possible as we can see from the order of G. If $p = 2$, the same argument shows

$$|P| \equiv 2 \pmod{23}.$$

It follows that H is a 2-group of order 2 or 2^{12}.

Suppose that $|H| = 2^{12}$. Then, $\mathbf{C}_G(H) \neq G$. The preceding part of the proof shows that $\mathbf{C}_G(H)$ is a 2-group. On the other hand, we will show that $\mathbf{C}_G(H)$ is not a 2-group. To show this, let ρ be an element of order 13 in G. If $|\mathbf{C}_H(\rho)| = 2^a$, then $1 \leq a \leq 12$ and

$$2^{12} \equiv 2^a \pmod{13}.$$

This gives us $a = 12$. Thus, $\mathbf{C}_G(H)$ contains ρ and is not a 2-group. This contradicts the assertion that $\mathbf{C}_G(H)$ is a 2-group.

We have shown that $|H| = 2$. Hence, $H = \langle -1 \rangle$ and Proposition 8 is proved. □

Proposition 9. *The Conway groups ·2 and ·3 are simple.*

Proof. Set $G = \cdot 2$ (or $\cdot 3$). It follows easily from definition that

$$\langle \alpha, \beta \rangle \subset M_{23} \subset G.$$

We remark that $-1 \notin G$ and so, for $A = \langle \alpha \rangle$, we have

$$\mathbf{N}_G(A) = \langle \alpha, \beta \rangle.$$

Let H be a normal subgroup of G. If $|H|$ is divisible by 23, the proof of Proposition 8 can be repeated in this case. We conclude that $H = G$.

Suppose that $|H|$ is not divisible by 23. If P is an S_p-subgroup of H, then we get

$$|P| \equiv 1 \pmod{23}.$$

If $G = \cdot 3$, this is impossible unless $|P| = 1$. (Note that $2^{11} \nmid |G|$.) This completes the proof for ·3.

If $G = \cdot 2$, we have a possibility that $|H| = 2^{11}$. Suppose that this is the case, and let $K = \mathbf{C}_G(H)$. Then, we get $K \neq G$, so $K = H$. It follows that H is an elementary abelian 2-group and that G/H is isomorphic to a subgroup of $GL(11,2)$. This gives us the contradiction that 5^3 divides the order of G/H while 5^3 does not divide the order of $GL(11,2)$. This completes the proof that the Conway group ·2 is simple. □

References

[1] J.H. Conway, Three lectures on exceptional groups. In : *Finite Simple Groups. Proceedings of an Instructional Conference*, pp. 215-247. London, New York, Academic Press 1971.

[2] B. Fischer, Finite groups generated by 3-transpositions, *Invent. math.* **13** (1971), 232-246.

[3] D. Livingstone, On a permutation representation of the Janko group, *J. of Algebra* **6** (1967), 43-55.

[4] M. Suzuki, *Group Theory I and II*, New York, Berlin, Heidelberg, Springer-Verlag 1982 and 1986.

Department of Mathematics
University of Illinois at Urbana-Champaign
1409 West Green Street
Urbana, IL 61801
U. S. A.

Fricke, free groups and $\mathbf{SL_2}$

J.G. Thompson

This note was stimulated by a paper of Horowitz [1], more especially by his Theorem 8.1.

Definition. Suppose $\chi : G \to S$ is a class function on the group G with values in a set S. If F is a group and $a, b \in F$, we say that a and b are χ-equivalent if and only if $\chi(\rho(a)) = \chi(\rho(b))$ for all $\rho \in \mathrm{Hom}(F, G)$.

Since χ is a class function, the above definition implies that each χ-equivalence class in F is a union of conjugacy classes.

Definition. If χ and F are as in the previous definition, we say that χ *almost separates* the conjugacy classes of F if and only if each χ-equivalence class is the union of a finite number of conjugacy classes.

Theorem 8.1 of Horowitz [1] proves that if $G = SL(2, \mathbf{C})$ and $\chi = \mathrm{tr} : G \to \mathbf{C}$, then for each free group F, tr almost separates the conjugacy classes of F.

Moreover, and I find this very interesting, Horowitz shows that if the rank of the free group F is > 1, then for each integer m, there is a tr-equivalence class which contains more than m conjugacy classes. If E is a tr-equivalence class, denote by $m(E)$ the number of F-conjugacy classes contained in E. The determination of $m(E)$ seems to be an interesting and difficult problem. The only contribution of this note is to adapt Horowitz's argument and to show that we can embed F in a subgroup $\bar{F}$ of $SL(2, \mathbf{C})$ such that elements of F are tr-equivalent if and only if they are conjugate in $\bar{F}$. This is worth recording, since the construction of $\bar{F}$ suggests that some Galois theory is involved in the determination of the numbers $m(E)$.

As an example of the interest in the $m(E)$, knowledge about it would necessarily imply some information concerning the number of $SL(2, \mathbf{Z})$-conjugacy classes of elements of a given trace, a problem whose connection with indefinite binary quadratic forms is well known. This is exploitable. Namely, the principal congruence subgroup $\Gamma(2)$ of $SL(2,\mathbf{Z})$ is the direct product of a group of order 2 and a group Γ generated by $\begin{pmatrix} 1 & 2 \\ 0 & 1 \end{pmatrix}$ and $\begin{pmatrix} 1 & 0 \\ 2 & 1 \end{pmatrix}$, which is free. For each integer t, let T_t be the set of elements of Γ of trace t. Then T_t is the union of a finite number of Γ-conjugacy classes, and this set of classes is itself the disjoint union of a certain number of tr-equivalence classes (where we take $\Gamma = F$, the free group of rank 2). In this way, the classes of indefinite binary quadratic forms of discriminant $t^2 - 4$ are linked to the tr-equivalence classes defined above. I hope that someone, possibly myself, will pursue this matter.

Returning for a moment to the definition of tr-equivalence, it is also of interest to take for F the fundamental group of a compact Riemann surface, a case which presents subtlety when the genus exceeds 1. This topic is not treated here either.

The underlying idea in the work of Horowitz depends on the identity

$$u + u^{-1} = (\operatorname{tr} u) \cdot I, \quad u \in SL(2,\mathbf{C}),$$

whence

$$\operatorname{tr} uv + \operatorname{tr} uv^{-1} = \operatorname{tr} u \cdot \operatorname{tr} v. \tag{1}$$

Recognition that (1) is exploitable goes back to Fricke or possibly even earlier to Vogt [2].

Suppose r is an integer ≥ 1. If $1 \leq i_1 < \ldots < i_s \leq r$, introduce an indeterminate $x_{i_1 i_2 \cdot \ldots \cdot i_s}$, obtaining in this fashion $2^r - 1$ indeterminates, one for each non empty ordered (in the usual ordering) subset of $\{1,\ldots,r\}$. Call X the set of these $2^r - 1$ indeterminates, and form the polynomial ring $\mathbf{Z}[X]$. Possibly Fricke, and certainly Horowitz, has given a rigorous proof, combinatorial in nature, that if F is a free group of rank r on $u_1,\ldots,u_r$, then for each $u \in F$, there is $P_u \in \mathbf{Z}[X]$ with the following nice property:

For each $\rho \in \operatorname{Hom}(F, SL(2,\mathbf{C}))$, set

$$t_{i_1 \cdot \ldots \cdot i_s} = \operatorname{tr} \rho(u_{i_1} \cdot \ldots \cdot u_{i_s}).$$

Then tr $\rho(u)$ is obtained by specializing the $x_{i_1\cdot\ldots\cdot i_s}$ to $t_{i_1\cdot\ldots\cdot i_s}$ and evaluating P_u, so that

$$(2) \qquad \operatorname{tr} \rho(u) = P_u(t_{i_1\cdot\ldots\cdot i_s}).$$

If $r = 2$, the polynomial P_u is uniquely determined by (2), but for $r \geq 3$, this is not so. When $r = 3$, for example, the 7 "indeterminates" $t_{i_1\cdot\ldots\cdot i_s}$ satisfy an algebraic relation:

$$(3) \qquad t_{123}^2 - \alpha t_{123} + \beta = 0, \quad (\text{all } \rho \in \operatorname{Hom}(F_3, SL(2, \mathbf{C})))$$

where

$$\begin{aligned} \alpha &= t_{12}t_3 + t_{13}t_2 + t_{23}t_1, \\ \beta &= t_1^2 + t_2^2 + t_3^2 + t_{12}^2 + t_{13}^2 + t_{23}^2 \\ &\quad - t_1t_2t_{12} - t_1t_3t_{13} - t_2t_3t_{23} + t_{12}t_{13}t_{23} - 4. \end{aligned}$$

This relation led Magnus to the notion of a Fricke ring [2].

The aim of this note is to record an explicit construction of a subgroup G_r of $SL(2,k)$ which contains a free subgroup F_r of rank r such that elements of F_r are tr-equivalent if and only if they are conjugate in G_r. Moreover, the field k is explicitly constructed as the algebraic closure of a finitely generated extension field of $\mathbf{Q}$. This result is not in itself of much moment, but the group G_r is intriguing.

Lemma 1. *Suppose k is an algebraically closed field, $u, v \in SL(2,k)$,* tr u = tr v, tr $uv \neq 2$. *Then there is j in $SL(2,k)$ such that*

$$(4) \qquad \operatorname{tr} j = 0, \quad j^{-1}uj = v.$$

If char $k = 2$, *j is unique. If* char $k \neq 2$, *j and $-j$ are the only solutions to* (4).

Proof. First, suppose that tr $u \neq 2, -2$. We may then assume that

$$u = \begin{pmatrix} \varsigma & 0 \\ 0 & \varsigma^{-1} \end{pmatrix}.$$

Let

$$v = \begin{pmatrix} a & b \\ c & d \end{pmatrix},$$

so that

$$ad - bc = 1, \quad a + d = \varsigma + \varsigma^{-1}.$$

We seek $\alpha, \beta, \gamma, \delta$ such that

$$\begin{pmatrix} \varsigma & 0 \\ 0 & \varsigma^{-1} \end{pmatrix}\begin{pmatrix} \alpha & \beta \\ \gamma & \delta \end{pmatrix} = \begin{pmatrix} \alpha & \beta \\ \gamma & \delta \end{pmatrix}\begin{pmatrix} a & b \\ c & d \end{pmatrix},$$

$$\alpha\delta - \beta\gamma = 1, \qquad \alpha + \delta = 0.$$

So we want

$$\begin{aligned} \varsigma\alpha &= \alpha a + \beta c, \\ \varsigma\beta &= \alpha b + \beta d, \\ \varsigma^{-1}\gamma &= \gamma a + \delta c, \\ \varsigma^{-1}\delta &= \gamma b + \delta c. \end{aligned}$$

Since tr $uv \neq 2$, we have $\alpha \neq \varsigma^{-1}$.
Hence

$$\gamma(\varsigma^{-1} - a) = \delta c.$$

Since $\delta = -\alpha$, we get

$$\gamma = \frac{-\alpha c}{\varsigma^{-1} - a},$$

and since $\alpha\delta - \beta\gamma = 1$, we also have

$$\beta = \frac{-\alpha^2 - 1}{\gamma} = (\frac{\alpha^2 + 1}{\alpha c})(\varsigma^{-1} - a).$$

So j is uniquely determined by α. As for α, we have

$$\varsigma\alpha = \alpha a + (\frac{\alpha^2 + 1}{\alpha c})(\varsigma^{-1} - a) \cdot c,$$

and so

$$\varsigma\alpha^2 = \alpha^2 a + (\alpha^2 + 1)(\varsigma^{-1} - a),$$

$$\alpha^2 = \frac{\varsigma^{-1} - a}{\varsigma - \varsigma^{-1}}.$$

If μ ia a square root of $\frac{\varsigma^{-1}-a}{\varsigma-\varsigma^{-1}}$ in k, and we set

$$\alpha=\mu,\quad \beta=\frac{\mu^2+1}{\mu c}(\varsigma^{-1}-a),$$

$$\gamma=\frac{-\mu c}{\varsigma^{-1}-a},\quad \delta=-\mu,$$

we check that $j=\begin{pmatrix}\alpha & \beta\\ \gamma & \delta\end{pmatrix}$ satisfies the conditions of the lemma. So the proof is complete in case tr $u\neq 2,-2$.

Suppose tr $u=2$. Since tr $uv\neq 2$, we conclude that $u\neq I$, and we may assume that

$$u=\begin{pmatrix}1 & 1\\ 0 & 1\end{pmatrix}.$$

Again using tr $uv\neq 2$, we may assume that

$$v=\begin{pmatrix}1 & 0\\ c & 1\end{pmatrix}$$

for some $c\in k^{\times}$. We seek $\begin{pmatrix}\alpha & \beta\\ \gamma & \delta\end{pmatrix}$ such that $\alpha\delta-\beta\gamma=1, \alpha+\delta=0$, and

$$\begin{pmatrix}1 & 1\\ 0 & 1\end{pmatrix}\begin{pmatrix}\alpha & \beta\\ \gamma & \delta\end{pmatrix}=\begin{pmatrix}\alpha & \beta\\ \gamma & \delta\end{pmatrix}\begin{pmatrix}1 & 0\\ c & 1\end{pmatrix},$$

that is,

$$\alpha+\gamma=\alpha+\beta c,\quad \beta+\delta=\beta,$$

$$\gamma=\gamma+\delta c,\quad \delta=\delta,$$

whence

$$\delta=0,\quad \alpha=0,\quad \beta c=\gamma,\quad \beta\gamma=-1=\beta^2 c,$$

so $\beta^2=-c^{-1}$, and the lemma holds in this case, too. If tr $u=-2$, replace u,v by $-u,-v$, respectively; the proof is complete.

Lemma 2. *Suppose r is an integer ≥ 1. Let*

$$A=\mathbf{Q}[x_i,y_i,z_i,t_i|1\leq i\leq r]$$

be the polynomial ring in $4r$ variables. Let

$$D_i = x_i t_i - y_i z_i - 1, \quad 1 \leq i \leq r,$$

and let P be the A-ideal generated by $\{D_1, \ldots, D_r\}$. Then P is a prime.

Proof. We first check that for each i, x_i is not a zero divisor $(\bmod P)$. For suppose $f \in A$ and $x_i f \in P$. Write

$$x_i f = a_1 D_1 + \ldots + a_r D_r,$$

and write

$$a_j = x_i a_{ij} + b_j,$$

where b_j does not involve the variable x_i, $j = 1, \ldots, r$.
Set $\tilde{f} = f - \sum_{j=1}^{r} a_{ij} D_j$, so that

$$x_i \tilde{f} = b_1 D_1 + \ldots + b_r D_r,$$

$$f \equiv \tilde{f} \pmod{P},$$

$$b_1, \ldots, b_r \text{ do not involve } x_i.$$

This forces

$$x_i \tilde{f} = b_i x_i t_i,$$

$$(5) \qquad 0 = -b_i x_i t_i + b_1 D_1 + \ldots + b_r D_r;$$

so

$$\tilde{f} = b_i t_i.$$

We may write (5) as

$$0 = \sum_{\substack{j=1 \\ j \neq i}}^{r} b_j D_j - b_i x_i t_i + b_i (x_i t_i - y_i z_i - 1),$$

and so

$$b_i (y_i z_i + 1) = \sum_{\substack{j=1 \\ j \neq i}}^{r} b_j D_j.$$

By induction, the ideal generated by any proper subset of $\{D_1, \ldots, D_r\}$ is prime, and so we get

$$b_i \in \text{ideal generated by } D_1, \ldots, D_i^{\wedge}, \ldots, D_r,$$

so $\tilde{f} \in P$, $f \in P$.

Now suppose $f \in A$. Then there are g in $\mathbf{Q}[x_i, y_i, z_i | 1 \le i \le r] = B$, say, and a large integer N such that

$$(x_1 \cdot \ldots \cdot x_r)^N f \equiv g \pmod{P}.$$

Since $B \cap P = \{0\}$, we see that P is prime.

Let K be the field of fractions of A/P, and let k be an algebraic closure of K. Let a_i, b_i, c_i, d_i be the images in k of x_i, y_i, z_i, t_i, respectively, $i = 1, \ldots, r$, and set

$$u_i = \begin{pmatrix} a_i & bi \\ c_i & d_i \end{pmatrix}.$$

By construction, $F_r = \langle u_1, \ldots, u_r \rangle$ is free of rank r and $F_r \subseteq SL(2, k)$.

Suppose $u, v \in F_r - \{1\}$ and $uv \neq 1$. Then by Horowitz [1], $\mathrm{tr}_k uv \neq 2$; so if u, v are tr-equivalent, there is $j = j(u, v) \in SL(2, k)$ such that $j^2 = -I$ and $j^{-1} u j = v$. Let G_r be the group generated by F_r, together with all the $j(u, v)$ where u, v range over all pairs of tr-equivalent elements of F_r such that $uv \neq 1$. Then G_r has the property that each conjugacy class of G_r meets F_r either in the empty set or in a tr-equivalence class.

Finally, let K_0 be the field generated by K, together with all the matrix entries of all the $j(u, v)$ as above. Then by the construction of j in Lemma 1, we see that K_0/K is a Galois extension whose Galois group has exponent 2. How to exploit this group G_r is left as an open problem.

Bibliography

[1] Horowitz, R. D., Characters of Free Groups Represented in the Two-Dimensional Special Linear Groups, *Comm. Pure & Appl. Math.*, vol. **XXV**, 634-649 (1972).

[2] Magnus, W., Rings of Fricke Characters and Automorphism Groups of Free Groups, *Math. Z.*, **170**, 91-103 (1980).

University of Cambridge
United Kingdom

National University of Singapore
Singapore

Hecke operators and non congruence subgroups

J.G. Thompson

This paper stems from unpublished work of A.O.L. Atkin, who told me of his results in 1982. These results suggest a conjecture which I have taken the liberty to call

Atkin's conjecture. *If G is of finite index in $SL(2,\mathbf{Z})$, k is an integer ≥ 1, and f is a modular form of weight k on G, then for each prime p, $f \circ T_p^G$ is on $\bar{G}$.*

In order to clarify the terms appearing in the conjecture, some well known notation and some definitions are appropriate.

If n is an integer ≥ 1, $\Gamma(n)$ is the principal congruence subgroup of $SL(2,\mathbf{Z})$ of level n consisting of those $\left(\begin{smallmatrix} a & b \\ c & d \end{smallmatrix}\right)$ in $SL(2,\mathbf{Z})$ such that

$$a-1 \equiv d-1 \equiv c \equiv d \equiv 0 \pmod{n}.$$

A congruence subgroup of $SL(2,\mathbf{Z})$ is any subgroup which contains $\Gamma(n)$ for some n. A non congruence subgroup is a subgroup of finite index in $SL(2,\mathbf{Z})$ which is not a congruence subgroup. If G is any subgroup of $SL(2,\mathbf{Z})$ of finite index, then $\bar{G}$ denotes the intersection of all the congruence subgroups which contain G.

If R is a subring of the field of real numbers, $GL_2^+(R)$ denotes the set of elements of $GL_2(R)$ of positive determinant.

Let F be the set of all complex-valued functions on the upper half plane

$$h = \{x+iy \mid x, y \in \mathbf{R},\ y > 0\}.$$

If k is an integer ≥ 1, we get an action $|_k$ of $GL_2^+(\mathbf{Q})$ on F by defining, for $\left(\begin{smallmatrix} a & b \\ c & d \end{smallmatrix}\right) \in GL_2^+(\mathbf{Q})$ and $f \in F$,

$$(f|_k A)(\tau) = (ad-bc)^{k/2}(c\tau+d)^{-k} f\left(\frac{a\tau+b}{c\tau+d}\right), \quad (\tau \in h).$$

One checks directly that $f|_k AB = (f|_k A)|_k B$; that is, we do indeed have an action. If G is a subgroup of $GL_2^+(\mathbf{Q})$, F^G denotes the set of fixed points of G on F. If $f \in F^G$, we say that f is on G, an abuse of language which should cause no confusion.

If $G_1 \subseteq G_2 \subseteq GL_2^+(\mathbf{Q})$ and $|G_2 : G_1|$ is finite we have a trace map

$$\begin{aligned} \mathrm{tr}_{G_1,G_2} : F^{G_1} &\to F^{G_2} \\ f &\mapsto \sum_R f|_k R, \end{aligned}$$

where R ranges over a set of representatives for the cosets $G_1 R$ of G_1 in G_2.

Now suppose G is of finite index in $SL(2, \mathbf{Z})$, and $A \in GL_2^+(\mathbf{Q})$. We define, for $f \in F^G$

$$f \circ T_A^G = \sum_R (f|_k A)|_k R,$$

where R ranges over a set of representatives for the cosets of $G \cap A^{-1}GA$ in G. In case n is an integer ≥ 1, we set

$$T_n^G = T_M^G \quad \text{with} \quad M = \begin{pmatrix} n & 0 \\ 0 & 1 \end{pmatrix}.$$

The T_n^G are called Hecke operators (in their action on F^G). For the purposes of this paper, it is enough to know that modular forms of weight k on G lie in F^G.

Lemma 1. *If $G \triangleleft \Gamma$, $A \in GL_2^+(\mathbf{Q})$ and f is an element of F^G, then $f \circ T_A^G$ is fixed by each element of $\Gamma \cap G^A$.*

Proof. For convenience, set $|_k = |$. By definition,

$$f \circ T_A^G = \sum_i f|A|R_i,$$

where $\{R_i\}$ is a transversal to $G \cap G^A$ in G. Since $G \triangleleft \Gamma$, $\Gamma \cap G^A$ normalizes both G and G^A. This implies that $\Gamma \cap G^A$ permutes by conjugation the family of all transversals to $G \cap G^A$ in G. So for each $S \in \Gamma \cap G^A$ and each R_i, there is an equation

$$(*) \qquad S^{-1} R_i S = H(S, i) R_j,$$

where $H(S,i) \in G \cap G^A$, and where, for a fixed S, the map $R_i \mapsto R_j$ is a permutation of $\{R_i\}$. Thus,

$$\begin{aligned}
(f \circ T_A^G)|S &= \sum_i f|A|R_i S \\
&= \sum_i f|A|S|S^{-1} R_i S \\
&= \sum_i f|A|S^{-1} R_i S \quad (\text{as } S \in G^A \text{ and } f|A \text{ is on } G^A) \\
&= \sum_i f|A|H(S,i) R_j \quad (\text{by } (*)) \\
&= \sum_i f|A|R_j \quad (\text{as } H(S,i) \in G \cap G^A \text{ and } f|A \text{ is on } G \cap G^A) \\
&= f \circ T_A^G \quad (\text{since } R_i \mapsto R_j \text{ is a permutation}).
\end{aligned}$$

Corollary. *If $\bar{G} \subseteq G \cdot (\Gamma \cap G^A)$, then $f \circ T_A^G \in F^{\bar{G}}$, for all $f \in F^G$.*

Proof. $f \circ T_A^G$ is certainly in F^G, so the lemma implies the corollary.

Remark. In practice, we work with subgroups G of Γ which are not necessarily normal in Γ. In order to apply the preceding discussion to such subgroups, set

$$G_0 = \bigcap_{g \in \Gamma} G^g.$$

If $f \in F^G$, then since $G_0 \subseteq G$, we also have $f \in F^{G_0}$. If $A \in GL_2^+(\mathbf{Q})$, then both T_A^G and $T_A^{G_0}$ may be applied to f. It is not difficult to check that if $m = |G \cap G^A : G_0 \cap G_0^A|$, then setting

$$F = f \circ T_A^G$$

$$F_0 = f \circ T_A^{G_0}$$

we have

$$F = \frac{1}{m} \sum_j F_0 | S_j, \tag{**}$$

where $\{S_j\}$ is a transversal to G_0 in G. An appropriate picture for the verification of $(**)$ is

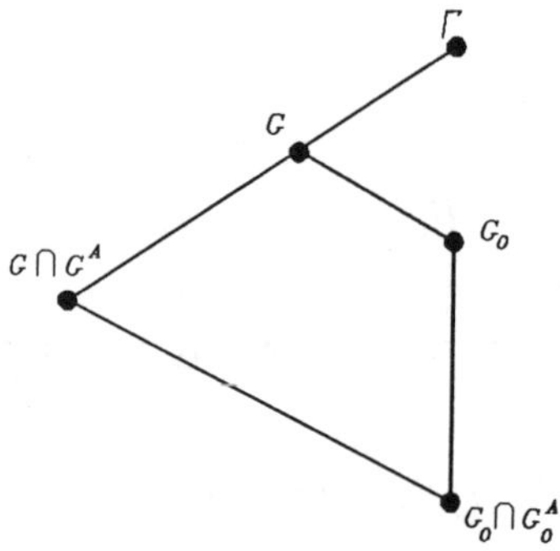

To verify $(**)$, we use the transparent commutativity

$$\begin{aligned}\operatorname{tr}_{G_0\cap G_0^A,G} &= \operatorname{tr}_{G\cap G^A,G} \circ \operatorname{tr}_{G_0\cap G_0^A,G\cap G^A} \\ &= \operatorname{tr}_{G_0,G} \circ \operatorname{tr}_{G_0\cap G_0^A,G_0}.\end{aligned}$$

Finally, we are using the even more transparent fact that since $f|A$ is invariant under $G\cap G^A$, $\operatorname{tr}_{G_0\cap G_0^A,G\cap G^A}$, when applied to $f|A$, is simply multiplication by $m = |G\cap G^A : G_0\cap G_0^A|$. Thus, we have proved

Lemma 2. *Suppose G is of finite index in Γ,*

$$G_0 = \bigcap_{g\in\Gamma} G^g,$$

and $\bar{G}_0 = \Gamma$. Suppose p is a prime and

$$f\circ T_p^{G_0} \quad \textit{is on} \quad \Gamma \quad \textit{for all} \quad f\in F^{G_0}.$$

Then $f\circ T_p^G$ is on Γ for all $f\in F^G$.

It is both a pleasure and a duty to record a result, the proof of which I received just a few days after the end of the Singapore Conference.

Theorem (Serre). *If $G\lhd\bar{G}=\Gamma$, then Atkin's conjecture is true for G and all primes p.*

The proof is contained in the following letter, the clarity of which justifies its inclusion in an intact form.

Paris, June 24, 1987

Dear Thompson,

This conjecture of Atkin, about T_p in the non-congruence case, can indeed be proved by using the theorems of Ihara and Mennicke.

Here is the proof:

1. Notation

It is the same as yours, except that I do not divide by $\{\pm 1\}$. Hence $\Gamma = SL_2(\mathbf{Z})$, $A = \left(\begin{smallmatrix} p & 0 \\ 0 & 1 \end{smallmatrix}\right)$ and G is a normal subgroup of finite index of Γ with $\bar{G} = \Gamma$ (i.e. no proper congruence subgroup of Γ contains G).

As usual, I write $\Gamma_0(p) = \Gamma \cap A^{-1}\Gamma A$.

2. Theorems used in the proof

There are two of them:

Theorem 1 (Ihara) - *The group $SL_2(\mathbf{Z}[1/p])$ is the amalgamated sum of Γ and $A^{-1}\Gamma A$.*

See e.g. my Tree book (Astérisque 46, p.110, cor.2 - it is p.80 of the English translation).

Theorem 2 (Mennicke) - *Every subgroup of finite index of $SL_2(\mathbf{Z}[1/p])$ is a congruence subgroup.*

See J. Mennicke, *On Ihara's modular group*, Invent. Math. 4 (1967), p.202-228, or my paper "*Le problème des groupes de congruence pour SL_2*", Ann.of Math. 92 (1970), p.489-527.

3. The main statement

Call S the quotient Γ/G, and $f : \Gamma \to S$ the natural projection of Γ onto the finite group S.

Define two homomorphisms, f_1 and f_2, of $\Gamma_0(p)$ into S by:

$$f_1(x) = f(x) \qquad \text{and} \qquad f_2(x) = f(AxA^{-1}) \quad \text{for } x \in \Gamma_0(p).$$

(This makes sense since $x \in \Gamma_0(p) \Rightarrow AxA^{-1} \in \Gamma$.)

Theorem 3 - *The homomorphism*

$$(f_1, f_2) : \Gamma_0(p) \to S \times S$$

is surjective.

I shall prove this below, as an application of th. 1 and th. 2. Note that the kernel of f_1 is $G \cap \Gamma_0(p) = G \cap A^{-1}\Gamma A$ and the kernel of f_2 is $A^{-1}GA \cap \Gamma$. Hence the kernel of (f_1, f_2) is

$$G \cap A^{-1}\Gamma A \cap A^{-1}GA \cap \Gamma = G \cap A^{-1}GA,$$

and th. 3 gives:

Corollary - *The index of $G \cap A^{-1}GA$ in Γ is $(p+1)s^2$, where $s = |S| = (\Gamma : G)$.*

Equivalently: the index of $G \cap A^{-1}GA$ in G is $(p+1)s$.

4. A trivial lemma

Lemma 1 - *Let X, X_1, X_2 be groups, and let*

$$g_1 : X \to X_1 \qquad \textit{and} \qquad g_2 : X \to X_2$$

be surjective homomorphisms. Suppose $(g_1, g_2) : X \to X_1 \times X_2$ is not surjective. Then, there exists a group $X_{12} \neq \{1\}$ and surjective homomorphisms $h_i : X_i \to X_{12}$ $(i = 1, 2)$, such that $h_1 \circ g_1 = h_2 \circ g_2$.

Proof. If $N_1 = \operatorname{Ker}(g_i)$, define X_{12} as $X/N_1 \cdot N_2$, and define h_i in an obvious way. The hypothesis that (g_1, g_2) is not surjective means that $N_1 \cdot N_2 \neq X$, hence $X_{12} \neq \{1\}$. Hence the lemma.

5. Proof of theorem 3

(a) Let us first check that each $f_i : \Gamma_0(p) \to S$ is surjective. Call π the natural projection $\Gamma \to SL_2(\mathbf{Z}/p\mathbf{Z})$.

Lemma 2 - *The map* $(f, \pi) : \Gamma \to S \times SL_2(\mathbf{Z}/p\mathbf{Z})$ *is surjective.*

Proof. If not, apply lemma 1, and call N the kernel of the corresponding map: $\Gamma \to X_{12}$. The group N would be a proper congruence subgroup of Γ containing G, contrary to the assumption $\bar{G} = \Gamma$.

Let now s be any given element of S. By lemma 2, there exists $x \in \Gamma$ with $f(x) = s$ and $\pi(x) = 1$, i.e. $x \equiv 1 \pmod{p}$. We then have $x \in \Gamma_0(p)$ and $f_1(x) = s$, which proves that f_1 is surjective. Similarly, we have $A^{-1}xA \in \Gamma_0(p)$ and $f_2(A^{-1}xA) = s$, which proves that f_2 is surjective.

(b) We may now apply lemma 1 to

$$(f_1, f_2) : \Gamma_0(p) \to S \times S.$$

Hence, if (f_1, f_2) is not surjective, there are a non trivial group T, and surjective homomorphisms $h_i : S \to T$ $(i = 1, 2)$, such that

$$h_1 \circ f_1 = h_2 \circ f_2.$$

Call $h : \Gamma_0(p) \to T$ the homomorphism $h_1 \circ f_1 = h_2 \circ f_2$. Define:

$$F_1 : \Gamma \to T \text{ by } F_1(x) = h_1(f(x)) \text{ for } x \in \Gamma,$$
$$F_2 : A^{-1}\Gamma A \to T \text{ by } F_2(A^{-1}xA) = h_2(f(x)) \text{ for } x \in \Gamma.$$

If x belongs to $\Gamma_0(p) = \Gamma \cap A^{-1}\Gamma A$, we have

$$F_1(x) = h_1(f_1(x)) = h(x),$$
$$F_2(x) = h_2(f(AxA^{-1})) = h_2(f_2(x)) = h(x),$$

hence the restrictions of F_1 and F_2 to $\Gamma_0(p)$ coincide with h. By theorem 1, this implies that there is an homomorphism

$$F : SL_2(\mathbf{Z}[1/p]) \to T$$

whose restriction to Γ (resp. to $A^{-1}\Gamma A$) is F_1 (resp. F_2). By theorem 2, the kernel N of F is a congruence subgroup of $SL_2(\mathbf{Z}[1/p])$ (note that T is finite, since it is a quotient of S). Hence $N \cap \Gamma$ is a congruence subgroup of $SL_2(\mathbf{Z})$. It is clear that $N\cap\Gamma$ contains G. On the other hand, the restriction of F to Γ is $F_1 = h_1 \circ f_1$, which is surjective; hence the index of $N \cap \Gamma$ in Γ is equal to $|T|$, which is > 1; this contradicts the hypothesis $\bar{G} = \Gamma$, qed.

6. Application to Hecke operators

Let $GL_2^+(\mathbf{Z}[1/p])$ be the subgroup of $GL_2(\mathbf{Z}[1/p])$ made up of the elements with positive determinant. This group contains $\Gamma = SL_2(\mathbf{Z})$ and $A = \left(\begin{smallmatrix} p & 0 \\ 0 & 1 \end{smallmatrix}\right)$ (it is even *generated* by Γ and A).

Let E be a right module over $GL_2^+(\mathbf{Z}[1/p])$, and let E^G be the subgroup of E fixed by G. The Hecke operator T_p^G relative to G is the homomorphism

$$T_p^G : E^G \to E^G$$

defined (as in your lecture) by:

$$E^G \xrightarrow{i} E^{G'} \xrightarrow{t} E^G \quad (\text{here } G' = G \cap A^{-1}GA),$$

where

$$i : E^G \to E^{G'} \text{ is the map } e \mapsto e \cdot A,$$

and

$$t : E^{G'} \to E^G \text{ is the trace map } e \mapsto \sum e \cdot x_i,$$

where the x_i are representatives of the right cosets of G mod G'.

Theorem 4 (Atkin's conjecture) - *The image of T_p^G is contained in E^Γ.*

I shall prove this in a more precise form:

Theorem 4′ - *For every $e \in E^G$, one has*

$$T_p^G(x) = T_p^\Gamma(\mathrm{Tr}(e)),$$

where Tr *is the trace map $E^G \to E^\Gamma$, and $T_p^\Gamma : E^\Gamma \to E^\Gamma$ is the standard Hecke operator relative to Γ.*

Proof. Put $G_p = G \cap A^{-1}\Gamma A = G \cap \Gamma_0(p)$. We have $G \supset G_p \supset G'$. From what we have seen above, (esp. cor. to th.3) we know that $(G : G_p) = p+1$ and $(G_p : G') = (\Gamma : G) = s$. More precisely, the natural maps

$$G_p\backslash G \to \Gamma_0(p)\backslash\Gamma \quad \text{and} \quad G'\backslash G_p \to G\backslash\Gamma$$

are bijective (indeed, they are obviously injective, and, if they were not surjective, the index of G' in G would be strictly smaller than $(p+1)s$).

Let y_j ($j = 0, \ldots, p$) be representatives of G mod G_p, and z_ℓ ($\ell = 1, \ldots, s$) be representatives of G_p mod G'. We may take for representatives x_i of G mod G' the products $z_\ell y_j$, hence:

$$T_p^G(e) = \sum_{\ell,j} e \cdot A z_\ell y_j .$$

Put $e' = \sum_\ell e \cdot A z_\ell A^{-1}$. Since the $A z_\ell A^{-1}$ are representatives of Γ mod G, e' is equal to the trace $\mathrm{Tr}(e)$ of e relative to the inclusion $G \subset \Gamma$. The formula above gives

$$T_p^G(e) = \sum_j e' \cdot A y_j .$$

But the y_j are representatives of Γ mod $\Gamma_0(p) = \Gamma \cap A^{-1}\Gamma A$. Hence we have

$$\sum_j e' \cdot A y_j = T_p^\Gamma(e').$$

Since $e' = \mathrm{Tr}(e)$, this proves Theorem 4′.

Conclusion: the Hecke operators do not give anything new on non-congruence groups. Too bad!

Yours,

J-P. Serre

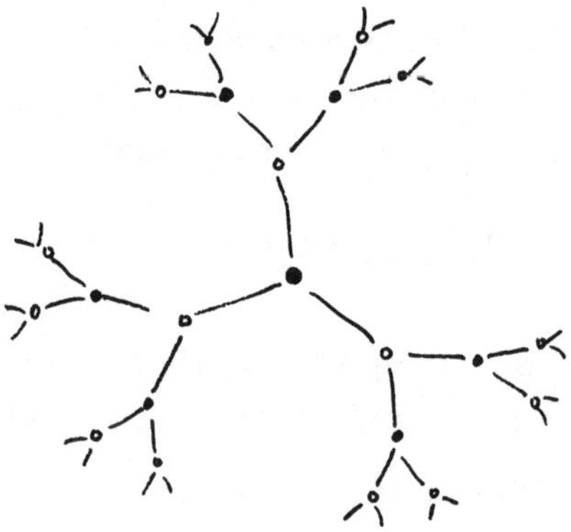

Concluding remarks. The preceding result proves that Atkin's conjecture is true in the important special case just treated. Whether or not the conjecture is true for an arbitrary subgroup of finite index and an arbitrary prime remains an open problem. In any case, it is now clear that any theory of Hecke operators on spaces of modular forms for non congruence subgroups is radically different from the rich theory for congruence subgroups, and the most likely outcome is that the theory in the non congruence case is vacuous.

University of Cambridge
United Kingdom

National University of Singapore
Singapore

Contributed papers

Soluble groups which are the product of a nilpotent and a polycyclic subgroup

Bernhard Amberg, Silvana Franciosi and Francesco de Giovanni

1. Introduction

It was shown by the first author in [1], Theorem A and Theorem C, that every soluble group $G = AB$ which is the product of two polycyclic subgroups A and B, one of which is nilpotent, is likewise polycyclic, and for the torsion-free rank of G the following rank formula holds:

$$r_0(G) = r_0(A) + r_0(B) - r_0(A \cap B).$$

Lennox and Roseblade in [7] and independently Zaicev in [12] have shown that every soluble product of two polycyclic subgroups is polycyclic, and the rank formula also holds in this case (see for instance [2], Satz 5.2). Here we use the above result in [1] to prove the following

Theorem A. *Let the soluble group $G = AB$ with derived length n be the product of a nilpotent subgroup A and a polycyclic subgroup B. Then G is nilpotent-by-polycyclic and, if G is not abelian, for the torsion-free rank of the Fitting factor group the following inequality holds:*

$$r_0(G/\mathbf{Fit}G) \leq (2n-3)r_0(B).$$

When A and B are abelian and B is finitely generated, this result was obtained by Heineken in [5] using different arguments; note that, as a product of two abelian subgroups, $G = AB$ is metabelian in this case by a famous theorem of Itô [6].

Theorem A becomes false for non-soluble groups in general, as can be seen from the alternating group A_5 of degree 5, which is the product of the alternating group A_4 of degree 4 and a cyclic group of order 5. - If B is the additive group of rational numbers and α is the inversion of B, the semidirect product $G = \langle\alpha\rangle B$ shows that the group G in Theorem A need not be polycyclic-by-nilpotent.

A soluble group $G = AB$ which is the product of a nilpotent subgroup A and a Černikov subgroup B is not always nilpotent-by-Černikov or even an extension of a nilpotent group by a minimax group, as the following example shows.

Example 1. Let p be any prime and let G be the holomorph of a quasicyclic group B of type p^∞. Then $G = AB$ where $A = \mathrm{Aut}B$ is an uncountable abelian group and $B = \mathbf{Fit}G$ is the Fitting subgroup of G. In particular $G/\mathbf{Fit}G$ is not a Černikov group and not even a minimax group.

Every soluble group $G = AB$, which is the product of two minimax subgroups A and B, one of which is hypercentral, is likewise a minimax group, and for the minimax rank and the p^∞-rank of G the following rank formulas hold:

$$m(G) = m(A) + m(B) - m(A \cap B),$$
$$m_p(G) = m_p(A) + m_p(B) - m_p(A \cap B) \quad \text{for every prime } p$$

(see for instance [11], Theorem 1, or [3], Corollary and [10], Theorem 1 and Theorem 3). Here we use this result to establish the following

Theorem B. *Let the soluble group $G = AB$ with derived length n be the product of a hypercentral subgroup A and a minimax subgroup B. Then G is an extension of a Gruenberg group by a minimax group and, if G is not abelian, for the minimax rank and the p^∞-rank of the Gruenberg factor group the following inequalities hold:*

$$m(G/\mathbf{K}(G)) \leq (2n-3)m(B),$$
$$m_p(G/\mathbf{K}(G)) \leq (2n-3)m_p(B) \quad \textit{for every prime } p.$$

If, in addition, B is a Černikov group, then G is an extension of a Gruenberg group by a Černikov group.

Clearly, in Theorem B the Gruenberg radical $\mathbf{K}(G)$ can be replaced by the Hirsch-Plotkin radical $\mathbf{R}(G)$. It should be observed that in Theorem

B there is no bound for the torsion-free rank of the Hirsch-Plotkin factor group $G/\mathbf{R}(G)$, as the following example shows.

Example 2. For any positive integer n let $\pi(n)$ be the set of the first n prime numbers, and let B_n be the additive group of rational numbers whose denominators are $\pi(n)$-numbers. Then B_n is a torsion-free abelian minimax group of rank 1. For each p in $\pi(n)$, let α_p be the automorphism $b \mapsto pb$ of B_n. The group of automorphisms A_n of B_n which is generated by all the α_p for $p \in \pi(n)$ is a free abelian group of rank n. Also $B_n = \mathbf{R}(G_n)$ is the Hirsch-Plotkin radical of the semidirect product $G_n = A_n B_n$. Then $r_0(G_n/\mathbf{R}(G_n)) = n$, but $r_0(B_n) = 1$ for all n.

It follows in particular from Theorem B that a soluble product of a hypercentral subgroup A and a polycyclic subgroup B is an extension of a Gruenberg group by a polycyclic group. However, statements like Theorem B do not hold when the minimax condition is replaced by the condition that B has only finite Prüfer rank or only finite abelian section rank. This can be seen from the following example.

Example 3. For each odd prime p let B_p be a cyclic group of order p and let $B = \mathop{\mathrm{Dr}}_{p>2} B_p$ be their direct product. For every such p let α_p be the automorphism of B which inverts the elements of B_p and acts trivially on every B_q for $q \neq p$. The group of automorphisms A of B which is generated by all the α_p is an infinite elementary abelian 2-group. Further $B = \mathbf{R}(G)$ is the Hirsch-Plotkin radical of the semidirect product $G = AB$. Obviously the factor group $G/\mathbf{R}(G)$ has infinite Prüfer rank. Note that G is even a torsion group.

Notation

The notation is standard and can be found in [8]. In particular we note:

The *Gruenberg radical* $\mathbf{K}(G)$ of a group G is the subgroup of G generated by all its abelian ascendant subgroups. A group G is a *Gruenberg group* if G equals its Gruenberg radical.

Fit$G$ denotes the *Fitting subgroup* of the group G and $\mathbf{R}(G)$ its *Hirsch-Plotkin radical.*

A soluble group G is a *minimax group* if it has a finite series whose factors are finite or infinite cyclic or quasicyclic of type p^∞; the number of infinite cyclic factors in such a series is the *torsion-free rank* $r_0(G)$ of G, the number of factors of type p^∞ for the prime p is the *p^∞-rank* $m_p(G)$ of G, and the minimax rank of G is

$$m(G) = r_0(G) + \sum_p m_p(G).$$

2. Automorphism groups of torsion-free abelian minimax groups

In the proofs of the "rank inequalities" essential use will be made of the following facts about abelian automorphism groups of torsion-free abelian minimax groups.

Lemma 2.1. *Let Γ be an abelian and rationally irreducible group of automorphisms of the torsion-free abelian minimax group $M \neq 1$. Then Γ is finitely generated and $r_0(\Gamma) < m(M)$.*

Proof. By Folgerung 3.2 of Baer [4] the group Γ is finitely generated and so $r_0(\Gamma) < m(M)$ by Lemma 5 of Wilson [10].

We use this result to obtain the following more general information about hypercentral groups of automorphisms of torsion-free abelian minimax groups.

Lemma 2.2. *Let Γ be a hypercentral group of automorphisms of the torsion-free abelian minimax group $M \neq 1$. Then there exists a normal subgroup Λ of Γ which stabilizes a series of finite length of M such that Γ/Λ is a finitely generated nilpotent group with $r_0(\Gamma/\Lambda) < m(M)$.*

Proof. There exists in M a Γ-invariant series of finite length

$$1 = M_0 \leq M_1 \leq \ldots \leq M_t = M,$$

whose factors are torsion-free groups on which Γ acts rationally irreducibly (see e.g. [4]), Hilfssatz 3.4). Put $L = M_{i+1}/M_i$ and $C_i = \mathbf{C}_\Gamma(L)$. The factor group $\Gamma_i = \Gamma/C_i$ is an irreducible linear group over the field of rational numbers and hence it is central-by-finite (see [9], p.106 and p.47).

Let $\mathbf{Z}(\Gamma_i) = \Sigma_i/C_i$. Then there exists in L a Σ_i-invariant series of finite length

$$1 = L_0 \leq L_1 \leq \ldots \leq L_r = L,$$

whose factors are torsion-free groups on which Σ_i acts rationally irreducibly. Denote by $\Lambda_{i,j}$ the centralizer of L_{j+1}/L_j in Σ_i. By Lemma 2.1 the group $\Sigma_i/\Lambda_{i,j}$ is finitely generated and $r_0(\Sigma_i/\Lambda_{i,j}) < m(L_{j+1}/L_j)$. Put $\Lambda_i = \cap_{j=0}^{r-1}\Lambda_{i,j}$. Then Σ_i/Λ_i is finitely generated and

$$r_0(\Sigma_i/\lambda_i) \leq \sum_{j=0}^{r-1} r_0(\Sigma_i/\Lambda_{i,j}) < \sum_{j=0}^{r-1} m(L_{j+1}/L_j) = m(M_{i+1}/M_i).$$

Therefore also Γ/Λ_i is finitely generated and

$$r_0(\Gamma/\Lambda_i) < m(M_{i+1}/M_i).$$

It follows that $\Lambda = \cap_{i=0}^{t-1}\Lambda_i$ is a normal subgroup of Γ which stabilizes a series of finite length of M, such that Γ/Λ is a finitely generated nilpotent group and

$$r_0(\Gamma/\Lambda) \leq \sum_{i=0}^{t-1} r_0(\Gamma/\Lambda_i) < \sum_{i=0}^{t-1} m(M_{i+1}/M_i) = m(M).$$

As an application of Lemma 2.2 we obtain the following

Lemma 2.3. *Let N be an abelian minimax normal subgroup of a group G with hypercentral factor group G/N. Then the factor group $G/\mathbf{K}(G)$ is finitely generated and*

$$r_0(G/\mathbf{K}(G)) \leq m(N).$$

If N is finitely generated and G/N is nilpotent, then $G/\mathbf{Fit}G$ is finitely generated and

$$r_0(G/\mathbf{Fit}G) \leq r_0(N).$$

Proof. Let C_1 be the centralizer in G of the (finite) socle of the torsion subgroup T of N and let C_2 be the centralizer in G of N/T. Then clearly G/C_1 is finite and G/C_2 is hypercentral. By Lemma 2.2 there exists a normal subgroup C/C_2 of G/C_2 such that G/C is finitely generated, $r_0(G/C) \leq m(N/T)$, and C stabilizes a series of finite length of N/T. If $E = C \cap C_1$, then G/E is also finitely generated,

$$r_0(G/E) = r_0(G/C) \leq m(N/T),$$

and E stabilizes an ascending series of N. Then N is hypercentrally embedded in E, so that E is hypercentral and hence is contained in $\mathbf{K}(G)$.

A similar argument proves the second statement of the lemma.

3. Proof of Theorem A

We first prove that G is nilpotent-by-polycyclic. Assume that this is false and choose a counterexample $G = AB$ with minimal derived length. If K is the last non-trivial term of the derived series of G, then G/K is nilpotent-by-polycyclic. Let L/K be a nilpotent normal subgroup of G/K with polycyclic factor group G/L.

Assume first that $A \cap K = 1$. Then $A \cap BK \simeq (A \cap BK)K/K \leq BK/K$ is polycyclic. By Theorem A of [1] also $BK = B(A \cap BK)$ is polycyclic. In particular K is polycyclic, and hence the soluble group of automorphisms $G/\mathbf{C}_G(K)$ of K is polycyclic (see [8], Part 1, p.82). It follows that also $G/(L \cap \mathbf{C}_G(K))$ is polycyclic. Since K is contained in the centre of $L \cap \mathbf{C}_G(K)$ and $(L \cap \mathbf{C}_G(K))/K$ is nilpotent, then $L \cap \mathbf{C}_G(K)$ is nilpotent and so G is nilpotent-by-polycyclic. This contradiction shows that $A \cap K$ is non-trivial.

The group $AK = A(B \cap AK)$ is a factorized subgroup of $G = AB$, which contains $A \cap K$ as a normal subgroup. It follows from the first case that $AK/(A \cap K)$ is nilpotent-by-polycyclic. If c is the nilpotency class of A, then

$$A \cap K = \mathbf{Z}_c(A) \cap K \leq \mathbf{Z}_c(AK).$$

Therefore also AK is nilpotent-by-polycyclic.

Put $H = AK \cap L$. Since $K \leq H \leq L$ and L/K is nilpotent, H is a nilpotent-by-polycyclic subnormal subgroup of G. Let

$$H = H_0 \triangleleft H_1 \triangleleft \ldots \triangleleft H_t = G$$

be the standard series of H in G. Since A normalizes H, it is well-known that A also normalizes each H_i, so that $\mathbf{N}_G(H_i) = A(B \cap \mathbf{N}_G(H_i))$ for every $i \leq t$. The factor group $\mathbf{N}_G(H_i)/H_i$ has the factorization

$$\mathbf{N}_G(H_i)/H_i = (AH_i/H_i)((B \cap \mathbf{N}_G(H_i))H_i/H_i),$$

where A_iH_i/H_i is an epimorphic image of

$$AH/H \simeq A/(A \cap H) = A/(A \cap L) \simeq AL/L \leq G/L,$$

so that AH_i/H_i is polycyclic. By Theorem A of [1] it follows that $\mathbf{N}_G(H_i)/H_i$ is polycyclic.

Suppose now that H_i is nilpotent-by-polycyclic for some $i < t$. Then $\mathbf{N}_G(H_i)$ is nilpotent-by-polycyclic, so that also its subgroup H_{i+1} is nilpotent-by-polycyclic. It follows by induction that also G is nilpotent-by-polycyclic. This contradiction proves the first part of Theorem A.

Assume now that the rank inequality in Theorem A is false. Among all the counterexamples with minimal derived length choose one $G = AB$ for which the torsion-free rank of B is minimal. Consider first the case when G is metabelian.

Assume that $A \cap G' = 1$. Then $A \cap BG' \cong (A \cap BG')G'/G' \leq BG'/G'$ is polycyclic. By Theorem A of [1] also $BG' = B(A \cap BG')$ is polycyclic. Moreover by Theorem C of [1] it follows that

$$r_0(BG') = r_0(B) + r_0(A \cap BG') - r_0(A \cap B).$$

Clearly we have also that

$$r_0(G') = r_0(BG') - r_0(B) + r_0(B \cap G')$$

and

$$r_0(A \cap BG') \leq r_0(B) - r_0(B \cap G').$$

Hence we obtain that

$$r_0(G') = (r_0(B) + r_0(A \cap BG') - r_0(A \cap B)) - r_0(B) + r_0(B \cap G') \leq r_0(B).$$

Now Lemma 2.3 yields that

$$r_0(G/\mathbf{Fit}G) \leq r_0(G') \leq r_0(B).$$

This contradiction shows that $A \cap G' \neq 1$.

If c is the nilpotency class of A, then $A \cap G'$ is contained in $\mathbf{Z}_c(AG')$, so that also the normal closure $N = (A \cap G')^G$ is contained in $\mathbf{Z}_c(AG')$. If $\bar{G} = G/N$. then $\bar{A} \cap \bar{G}' = 1$ and hence by the first case

$$r_0(\bar{G}/\mathbf{Fit}\bar{G}) \leq r_0(\bar{B}) \leq r_0(B).$$

Let $F/N = \mathbf{Fit}(G/N)$ and $F_0 = F \cap AG'$. then $N \leq \mathbf{Z}_c(F_0)$, so that F_0 is a nilpotent normal subgroup of G. We have that

$$r_0(G/F_0) \leq r_0(G/F) + r_o(G/AG') \leq r_0(B) + r_0(G/AG').$$

If G/AG' is finite, then

$$r_0(G/\mathbf{Fit}G) \leq r_0(G/F_0) \leq r_0(B).$$

This contradiction shows that $G/AG' \cong B/(B \cap AG')$ is finite. Then $AG' = A(B \cap AG')$ is a factorized subgroup of $G = AB$ with $r_0(B \cap AG') < r_0(B)$. Therefore it follows that

$$r_0(AG'/\mathbf{Fit}(AG')) \leq r_0(B \cap AG').$$

Then also

$$r_0(G/\mathbf{Fit}(AG')) = r_0(G/AG') + r_0(AG'/\mathbf{Fit}(AG')) \leq$$

$$\leq r_0(B/(B \cap AG')) + r_0(B \cap AG') = r_0(B).$$

Since $\mathbf{Fit}(AG') \leq \mathbf{Fit}G$, it follows that $r_0(G/\mathbf{Fit}G) \leq r_0(B)$. Hence the rank inequality holds when G is metabelian.

Consider now the general case and let K be the last non-trivial term of the derived series of G. If F/K is the Fitting subgroup of G/K, then we have that

$$r_0(G/F) \leq (2(n-1)-3)r_0(B).$$

If c is the nilpotency class of A, then

$$A \cap K = \mathbf{Z}_c(A) \cap K \leq \mathbf{Z}_c(AK).$$

If E is the Fitting subgroup of AK, then $E/(A \cap K)$ is the Fitting subgroup of $AK/(A \cap K)$. Put $H = AK \cap F$ and $L = \mathbf{Fit}H$. Clearly $r_0(H/L) \leq r_0(AK/E)$. Since F/K is nilpotent, H is a normal subgroup of G. let

$$H = H_0 \lhd H_1 \lhd \ldots \lhd H_t = G$$

be the standard series of H in G. Since G is nilpotent-by-polycyclic, the Fitting subgroup of H_i is a nilpotent normal subgroup of H_{i+1} and hence it is contained in the Fitting subgroup of H_{i+1}. Therefore L is contained in the Fitting subgroup of G, and so the normal closure L^G is nilpotent.

From

$$(AL^G \cap F)/L^G = L^G(A \cap F)/L^G \cong (A \cap F)/(A \cap L^G)$$

and

$$(A \cap F)/(A \cap L) \cong (A \cap F)L/L \leq (AK \cap F)/L = H/L$$

it follows that $(AL^G \cap F)/L^G$ is polycyclic and $r_0((AL^G \cap F)/L^G) \leq r_0(AK/E)$. The factor group $AL^G/(AL^G \cap F) \cong AF/F \leq G/F$ is polycyclic and

$$r_0(AL^G/(AL^G \cap F)) \leq r_0(G/F) \leq (2(n-1)-3)r_0(B).$$

By Theorem A of [1] also the factor group $G/L^G = (AL^G/L^G)(BL^G/L^G)$ is polycyclic and by Theorem C of [1]

$$\begin{aligned} r_0(G/L^G) &\leq r_0(AL^G/L^G) + r_0(BL^G/L^G) \\ &\leq r_0((AL^G/(AL^G \cap F)) + r_0((AL^G \cap F)/L^G) + r_0(B) \\ &\leq (2(n-1)-3)r_0(B) + r_0(AK/E) + r_0(B). \end{aligned}$$

This last inequality shows that, in order to obtain a contradiction, it is enough to prove that $r_0(AK/E) \leq r_0(B)$.

Since $E/(A \cap K)$ is the Fitting subgroup of $AK/(A \cap K)$, we may assume without loss of generality that $A \cap K = 1$. Now as in the first part of the proof it can be shown that $r_0(AK/E) \leq r_0(B)$. This contradiction proves Theorem A.

4. Proof of Theorem B

We first prove that G is an extension of a Gruenberg group by a minimax group. Assume that this is false, and let $G = AB$ be a counterexample with minimal derived length. If K is the last non-trivial term of the derived series of G, then there exists a normal Gruenberg subgroup L/K of G/K such that G/L is a minimax group.

Assume first that $A \cap K = 1$. Then $A \cap BK \cong (A \cap BK)K/K \leq BK/K$ is a minimax group. Therefore $BK = B(A \cap BK)$ is also a minimax group (see [11], Theorem 1 or [3], Corollary). In particular K is a minimax group, so that its torsion subgroup T is a Černikov group. Let C_1 be the centralizer in L of the (finite) socle of T and let C_2 be the centralizer in L of K/T. Then $K \leq C_1 \cap C_2 \leq L$. Since K is hypercentrally embedded in $C_1 \cap C_2$ and $(C_1 \cap C_2)/K \leq L/K$ is a Gruenberg group, also $C_1 \cap C_2$ is a Gruenberg group. The soluble group of automorphisms L/C_2 of the torsion-free abelian minimax group K/T is a minimax group (see [4]). Since L/C_1 is finite, it follows that the factor group $G/(C_1 \cap C_2)$ is also a minimax group. This contradiction shows that $A \cap K \neq 1$.

Now the proof that G is an extension of a Gruenberg group by a minimax group is very similar to the proof of Theorem A. To see this observe that, if S is the Gruenberg radical of $H = AK \cap L$, then S^G is a Gruenberg group. In fact, if x is any element of S, the subgroup $\langle x \rangle$ is ascendant in $\langle x, K \rangle \leq H$, and $\langle x, K \rangle$ is ascendant in G, since L/K is a Gruenberg group. Therefore S is contained in the Gruenberg radical of G and hence S^G is a Gruenberg group. Using the above arguments it is now easy to show that G/S^G is a minimax group. If, in particular, B is a Černikov group, the proof shows that G/S^G is a Černikov group.

The proof of the inequalities for the minimax rank and the p^∞-rank in Theorem B is the same as that for the torsion-free rank in Theorem A. Here one has again to use Lemma 2.3 as well as Theorem 1 and Theorem 3 of Wilson [10]. This observation completes the proof of Theorem B.

5. Some concluding remarks

If the soluble group $G = AB$ is the product of a nilpotent subgroup A

and a finite subgroup B, then clearly G is nilpotent-by-finite and the order of $G/\mathbf{Fit}G$ is bounded by $m!$, where m is the order of B. This bound can be attained as the example of the semidirect product $G = S_3 \ltimes Q_8$ shows when we consider a factorization $G = AB$ where A has order 2^4 and B has order 3. - Under the stated hypothesis inequalities like those in Theorem A and Theorem B do not hold, as the following example shows.

Example 4. Let B be an elementary abelian group of order 3^4 and A a Sylow 2-subgroup of $GL(4,3)$ and consider the semidirect product $G = AB$. Then

$$\mathbf{Fit}G = B \quad \text{and} \quad |G/\mathbf{Fit}G| = |A| = 2^9 \not\leq (2n-3)|B|,$$

where n is the derived length of G. Observe that the group G is abelian-by-nilpotent.

However, for nilpotent-by-abelian groups the following extension of Satz 1 of Heineken [5] holds.

Proposition. *If the nilpotent-by-abelian group $G = AB$ is the product of a nilpotent subgroup A and a finite subgroup B, then $|G/\mathbf{Fit}G| \leq B$.*

Proof. Let N be a nilpotent normal subgroup of G with abelian factor group G/N. Then $\mathbf{Fit}(G/N') = (\mathbf{Fit}G)/N'$ by a result of P. Hall (see [8], Part 1, p. 56), and hence it is enough to prove the result for metabelian groups.

Assume now that the proposition is false, and let $G = AB$ be a counterexample such that B has minimal order.

Assume first that $A \cap G' = 1$. Then $|G'| = |AG' : A| \leq |B|$. If $AG' = G$, then $|G/\mathbf{Fit}G|/|G'| \leq |B|$ by Heineken's result. Therefore $AG' < G$. Since $AG' = A(B \cap AG')$, it follows that $|AG'/\mathbf{Fit}(AG')| \leq |B \cap AG'|$. Hence

$$|G/\mathbf{Fit}(AG')| = |G/AG'| + |AG'/\mathbf{Fit}(AG')| \leq |B/(B \cap AG')| + |B \cap AG'| = |B|.$$

This contradiction shows that $A \cap G' \neq 1$.

If c denotes the nilpotency class of A, then $A \cap G'$ is contained in $\mathbf{Z}_c(AG')$. Hence, if F is the Fitting subgroup of AG', then $F/(A \cap G')$ is the Fitting subgroup of $AG'/(A \cap G')$. Therefore, by the first part of the proof, we have that

$$|AG'/F| \leq |B \cap AG'|$$

and so

$$|G/F| = |G/AG'| + |AG'/F| \leq |B/(B \cap AG')| + |B \cap AG'| = |B|.$$

This contradiction proves the proposition.

References

[1] B. Amberg, Factorizations of infinite soluble groups, *Rocky Mountain J. Math.* **7** (1977), 1-17.

[2] B. Amberg, Produkte von Gruppen mit endlichem torsionfreiem Rang, *Arch. Math.* (Basel) **45** (1985), 398-406.

[3] B. Amberg, S. Franciosi and F. de Giovanni, Groups with an *FC*-nilpotent triple factorization, *Ricerche Mat.*, to appear.

[4] R. Baer, Polyminimaxgruppen, *Math. Ann.* **175** (1968), 1-43.

[5] H. Heineken, Produkte abelscher Gruppen und ihre Fittinggruppe, *Arch. Math.* (Basel) **48** (1987), 185-192.

[6] N. Itô, Über das Produkt von zwei abelschen Gruppen, *Math. Z.* **62** (1955), 400-401.

[7] J.C. Lennox and J.E. Roseblade, Soluble products of polycyclic groups, *Math. Z.* **170** (1980), 153-154.

[8] D.J.S. Robinson, *Finiteness Conditions and Generalized Soluble Groups*, Springer, Berlin, 1972.

[9] B.A.F. Wehrfritz, *Infinite Linear Groups*, Springer, Berlin 1973.

[10] J.S. Wilson, On products of soluble groups of finite rank, *Comment. Math. Helv.* **60** (1985), 337-353.

[11] J.S. Wilson, Soluble groups which are products of minimax groups, *Arch. Math.* (Basel) **50** (1988), 193-198.

[12] D.I. Zaicev, Factorizations of polycyclic groups, *Mat. Zametki* **29** (1981), 481-490 = Math. Notes **29** (1981), 247-252.

Fachbereich Mathematik
Universität Mainz
Saarstraße 21
D-6500 Mainz
Federal Republic of Germany

Istituto di Matematica
Facoltà di Scienze
Università di Salerno
I-84100 Salerno
Italy

Dipartimento di Matematica
Università di Napoli
Via Mezzocannone 8
I-80134 Napoli
Italy

The tame range of automorphism groups and GL_n

S. Bachmuth and *H.Y. Mochizuki*

I. Background

This paper is intended as a survey, with a discussion of open problems centered about automorphism groups of relatively free groups. In recent work on the automorphism group $\mathrm{Aut}(G)$ of a group G, the "rank" of G has played an unexpectedly sensitive role. Analogous phenomena have occurred recently in other settings as well. We begin by listing these results in the following two theorems.

Theorem A.

(1) *Let $M(n)$ be the free metabelian group of rank n. Then* $\mathrm{Aut}(M(n))$ *is finitely generated (f.g.) for $n = 2$ and $n \geq 4$, but* $\mathrm{Aut}(M(3))$ *is not f.g.*

(2) *Let $J(n)$ be the free group of rank n in the Jacobi variety. Then* $\mathrm{Aut}(J(n))$ *is f.g. for $n = 2$ and $n \geq 4$, but* $\mathrm{Aut}(J(3))$ *is not f.g.*

(3) *Let $C(n)$ be the free group of rank n in the variety of center-by-metabelian groups. Then* $\mathrm{Aut}(C(n))$ *is f.g. for $n \geq 4$, but* $\mathrm{Aut}(C(2))$ *and* $\mathrm{Aut}(C(3))$ *are not f.g.*

(4) *Let $t_1, \ldots, t_k$ be indeterminates over the ring $\mathbf{Z}$ of integers. Then, if $k \geq 2$, $GL_n(\mathbf{Z}[t_1, t_1^{-1}, \ldots, t_k, t_k^{-1}])$ is f.g. for $n = 1$ and $n \geq 3$, but $GL_2(\mathbf{Z}[t_1, t_1^{-1}, \ldots, t_k, t_k^{-1}])$ is not f.g.*

(5) *Let $T(n)$ be the Torelli group of genus n. Then $T(n)$ is f.g. for $n = 1$ and $n \geq 3$, but $T(2)$ is not f.g.*

Proof. The proofs for the $\mathrm{Aut}(M(n))$ can be found in [2], [6] and [7] for the cases $n = 2$, $n = 3$ and $n \geq 4$ respectively. ($M(n) = F(n)/F(n)''$ where $F(n)$ is the free group of rank n and $F(n)$, $F(n)'$, $F(n)''$, $\ldots$, $F(n)^{(d)}$, $\ldots$ are the terms of the derived series for $F(n)$.) The proofs for $\mathrm{Aut}(C(n))$, due to Elena Stöhr, are in [15]. The result for GL_n, $n \geq 3$ is due to Suslin [16] and the result for GL_2 appears in [5]. The result for $T(n)$, $n \geq 3$ is due to D. Johnson [11] and the result for $T(2)$ can be found in McCullough and Miller [13] and also in G. Mess [14], where he furthermore proves that $T(2)$ is actually free. ($T(n)$ is the subgroup of the mapping class group of the closed orientable surface of genus n consisting of the mapping classes that act trivially on homology.) Thus we need only consider $\mathrm{Aut}(J(n))$.

The Jacobi variety is defined by the law $[x,y,z][z,x,y][y,z,x] = 1$. In [3] it was shown that Jacobi variety is 3-metabelian (i.e. every 3 generator group in the variety is metabelian), and hence $J(n) = M(n)$ for $n = 2$ or $n = 3$. Thus, to complete the proof of Theorem A, we need to show that $\mathrm{Aut}(J(n))$ is f.g. for $n \geq 4$.

It is known (see part (1) of Theorem B below) that for $n \geq 4$ the following commutative diagram has an exact column:

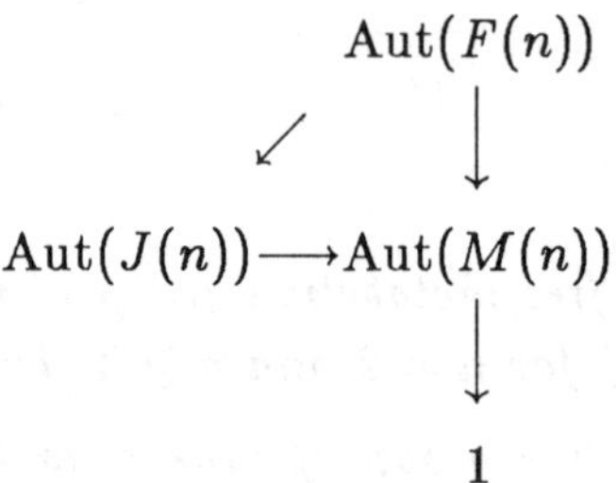

If $K(n)$ denotes the kernel of $\mathrm{Aut}(J(n)) \to \mathrm{Aut}(M(n))$, then

$$1 \to K(n) \to \mathrm{Aut}(J(n)) \to \mathrm{Aut}(M(n)) \to 1$$

is exact, and since $\mathrm{Aut}(M(n))$, $n \geq 4$ is f.g., it is sufficient to show that $K(n)$ is f.g. We in fact show that $K(n)$ is a finite group of exponent 2 and order $2^{n\binom{n}{4}}$. In [3] it was shown that $J(n)$ is center-by-metabelian. More precisely, for $n \geq 4$, $J(n)/Z(J(n)) \cong M(n)$, where $Z(J(n))$, the center of $J(n)$, is an elementary 2-group of order $2^{\binom{n}{4}}$. The elements of $K(n)$ are automorphisms of the form

$$\begin{aligned} x_1 &\longrightarrow x_1 c_1 \\ &\vdots \\ x_n &\longrightarrow x_n c_n \end{aligned}$$

where $c_i \in Z(J(n))$, and all such maps define automorphisms of $J(n)$. Moreover the elements of $K(n)$ induce the identity map on $J(n)' = [J(n), J(n)]$.

From these facts we conclude that $K(n)$ is an abelian group of exponent 2 and order $2^{n\binom{n}{4}}$. □

If a family $G(n)$ of groups satisfies a strong enough finiteness condition, such as being polycyclic, then the finite generation of the $\mathrm{Aut}(G(n))$ is known. (n is the rank or the minimum number of generators of $G(n)$.) This observation together with Theorem A raises the question of whether there can be a naturally related family $G(n)$ for which $\mathrm{Aut}(G(n)$ is not f.g. for more than a finite number of values of n, for example, if the $G(n)$ are relatively free groups. One purpose of this article is to make conjectures about the behaviour of $\mathrm{Aut}(G(n))$ as $n \to \infty$. There is an example [8] of a solvable variety whose free groups $G(n)$ of rank n have the property that $\mathrm{Aut}(G(n)$ is not f.g. for all $n \geq 2$ (see Case 2 of Problem (4) in section III of this paper), but we conjecture that this phenomenon can never happen if the $G(n)$ are torsion-free or, more generally, if the variety does not satisfy a power law.

It turns out that at least for parts (1) and (4) of Theorem A a much stronger result holds. In order to state this strengthened version we define the analogue of the "tame" elements of matrix groups for the automorphism groups.

Definition. (a) Let R be a ring with 1. Then, $\alpha \in GL_n(R)$ is called *tame* if $\alpha \in GE_n(R)$, the subgroup of $GL_n(R)$ generated by the set of elementary and diagonal matrices in $GL_n(R)$. An elementary matrix is a matrix which (possibly) differs form the identity matrix by having exactly one off-diagonal entry which is (possibly) nonzero.

(b) Let $G(n)$ be a relatively free group of rank n in a variety. $\alpha \in \mathrm{Aut}(G(n))$ is defined to be *tame* if α is in the image of the natural map $\mathrm{Aut}(F(n)) \to \mathrm{Aut}(G(n))$, i.e. α is induced by an automorphism of $F(n)$.

J. Nielsen (see [12], Section 3.5) gave a finite set of easily described generators for $\mathrm{Aut}(F(n))$. These are reminiscent of the elementary and diagonal elements in $GL_n(R)$. Thus the tame elements should be viewed as the obvious elements that $GL_n(R)$, respectively $\mathrm{Aut}(G(n))$. possesses.

Theorem B.

(1) $\mathrm{Aut}(M(n))$ *consists only of tame elements for* $n = 2$ *and* $n \geq 4$

but any set of generators for Aut($M(3)$) *contains infinitely many non-tame elements.*

(2) $GL_n(\mathbf{Z}[t_1, t_1^{-1}, \ldots, t_k, t_k^{-1}])$ *consists entirely of tame elements for* $n = 1$ *and* $n \geq 3$. *But for* $k \geq 2$, *any set of generators of* $GL_n(\mathbf{Z}[t_1, t_1^{-1}, \ldots, t_k, t_k^{-1}])$ *contains infinitely many non-tame elements.*

(3) *The subgroup of tame elements of* Aut($J(n)$) *has finite index if* $n \geq 4$.

Proof. The references given after Theorem A contain the proof of parts (1) and (2) of Theorem B. The proof of part (3) follows from the proof of Theorem A. □

II. The tame range

Theorem B implies the following mysterious situation:

Suppose $M(4)$ has generators x,y,z,w so that $M(3)$ with generators x,y,z is embedded in $M(4)$. Theorem B says that almost any automorphism α of $M(3)$ that one might write down

$$\alpha : \begin{array}{l} x \to \alpha(x) \\ y \to \alpha(y) \\ z \to \alpha(z) \end{array}$$

cannot be lifted to an automorphism of $F(3)$, yet the *same* automorphism extended to $M(4)$

$$\alpha^* : \begin{array}{l} x \to \alpha(x) \\ y \to \alpha(y) \\ z \to \alpha(z) \\ w \to w \end{array}$$

always lifts to an automorphism of $F(4)$.

There is no algorithm known to the authors which decides whether or not a specific $\alpha \in$ Aut($M(3)$) lifts to an automorphism of $F(3)$. Thus, although virtually all automorphisms of $M(3)$ do not lift, if one writes down even a mildly complicated automorphism, it may be either very difficult or impossible to decide whether or not it does lift. Simple examples of elements of $\mathrm{Aut}(M(3))$ which cannot be lifted can be found among those which fix all but one generator. For example, for each $i = 1, 2, \ldots,$ define

$$\alpha_i: \begin{array}{l} x \to x[y, z, \underbrace{x, \ldots, x}_{i+1 \text{ times}}] \\ y \to y \\ z \to z\,. \end{array}$$

Each α_i does not lift to an automorphism of $F(3)$.

The above discussion is motivation for the following definition.

Definition. Let $\chi(n) = GL_n(R)$ or $\chi(n) = \mathrm{Aut}(G(n))$. We say $\{\chi(n), n \geq 1\}$ has *tame range* equal to d, denoted $TR\{\chi(n)\} = d$, if $\chi(k)$ consists entirely of tame elements for all $k \geq d$, and d is the least such positive integer. If no such integer $d \geq 1$ exists, we define $TR\{\chi(n)\} = \infty$.

Thus, a consequence of Theorem B is the following:

Theorem C. (1) $TR\{\mathrm{Aut}(M(n))\} = 4$.
(2) $TR\{GL_n(\mathbf{Z}[t_1, t_1^{-1}, \ldots, t_k, t_k^{-1}]\} = 3$ *if* $k \geq 2$.

III. Problems, conjectures, and further results

(1) Determine $TR\{GL_n(\mathbf{Z}[t, t^{-1}])\}$.

Suslin's paper [16] mentioned earlier establishes that the tame range is at most three, and thus $TR\{GL_n(\mathbf{Z}[t, t^{-1}])\}$ is 1 or 3. Although this is a very tantalizing problem, it has thus far eluded the efforts of many people. We are unaware of any progress made since the publication of [5], and we refer to [5] for a discussion of what we know concerning this problem. There is some evidence for either answer, so that we are reluctant to venture an opinion on the eventual outcome. If the value is 3, then this would give a simplified proof that $TR\{\mathrm{Aut}(M(n))\} \geq 4$ since $GL_2(\mathbf{Z}[t, t^{-1}])$ is a homomorphic image of the kernel of the natural map $\mathrm{Aut}(M(3)) \to \mathrm{Aut}(M(3))_{ab}$.

If K is a field, then the same remarks apply for $TR\{GL_n(K[t_1, t_1^{-1}, t_2, t_2^{-1}])\}$ as for $TR\{GL_n(\mathbf{Z}[t, t^{-1}])\}$. This should be kept in mind in the discussion of Example (4) below.

On the other hand, it was shown in [5] that $TR\{GL_n(K[t_1, t_2, t_2^{-1}])\} = 3$. One should note the similarity between the rings $\mathbf{Z}[t, t^{-1}]$, $K[t_1, t_1^{-1}, t_2, t_2^{-1}] = K[t_1, t_1^{-1}][t_2, t_2^{-1}]$ and $K[t_1, t_2, t_2^{-1}] = K[t_1][t_2, t_2^{-1}]$. Namely, the coefficient rings $\mathbf{Z}$, $K[t_1, t_1^{-1}]$ and $K[t_1]$ are all Euclidean domains.

(2) Determine $TR\{\mathrm{Aut}(J(n)\}$.

As observed in Theorem B, for $n \geq 4$, $\mathrm{Aut}(J(n))$ contains a subgroup of finite index consisting entirely of tame elements. One could say that $VTR\{\mathrm{Aut}(J(n))\} = 4$, the *virtual tame range* of $\mathrm{Aut}(J(n))$ is four. The tame range of $\mathrm{Aut}(J(n))$ is an open problem, and all can be said with certainty is that $TR\{\mathrm{Aut}(J(n))\} \geq 4$. The problem can be reduced to the study of a single automorphism of $J(4)$. Namely, suppose $J(4) = gp\langle x, y, z, w\rangle$ and α_4 is defined by

$$\alpha_4 : \begin{array}{l} x \to x[[x,y],[z,w]] \\ y \to y \\ z \to z \\ w \to w. \end{array}$$

α_4 can be extended to an element α_k in $\mathrm{Aut}(J(k))$, $k \geq 5$, in the obvious way. If α_k is the image of an element in $\mathrm{Aut}(F(k))$, where k is the minimum possible value, then one can show that $TR\{\mathrm{Aut}(J(n))\} = k$; if no such k exists, then by definition $TR\{\mathrm{Aut}(J(n))\} = \infty$.

We know of no reason which rules out any values larger than three, but we suspect the answer to be 5. Showing that α_4 is not in the image of $\mathrm{Aut}(F(4))$ so that $TR\{\mathrm{Aut}(J(n))\} \geq 5$ would be a very interesting first step.

(3) Determine $TR\{\mathrm{Aut}(C(n))\}$.

This problem is closely related to the preceding problem of determining $TR\{\mathrm{Aut}(J(n))\}$. In fact, during the course of showing that $TR\{\mathrm{Aut}(C(n))\} \geq 4$, E. Stöhr showed that $TR\{\mathrm{Aut}(C(n))\} = 4$ if and only if the automorphism α_4 defined above (considered now as an automorphism of $C(4)$ rather than $J(4)$) lifts to an automorphism of $F(4)$. In a letter, she pointed out that her calculations lead her to believe that $TR\{\mathrm{Aut}(C(n))\}$ is in fact five. Her conjecture should be compared with our discussion of the free (abelian-by-nilpotent) groups in (6) below.

(4) Determine $TR\{\mathrm{Aut}(\chi(n))\}$, where $\chi(n) = F(n)/F(n)''[F(n), F(n)]^k$.

This problem naturally breaks up into two distinct cases according to whether k is square free or not. G. Baumslag, J. Dyer, and the authors [1] have already shown that $\mathrm{Aut}(\chi(2))$ is infinitely generated if k is not square free and f.g. if k is a square free integer. We will deal with each case separately.

Case 1. k is a square free integer

Not only is $\mathrm{Aut}(\chi(2))$ f.g., but it is possible to show that $\mathrm{Aut}(\chi(2))$ consists entirely of tame elements. Furthermore, it $n \geq 4$, it seems clear that a proof that $\mathrm{Aut}(F(n)) \to \mathrm{Aut}(\chi(n))$ is also surjective should be possible along the same lines of the proof in [7] that for $n \geq 4$, $\mathrm{Aut}(F(n)) \to \mathrm{Aut}(M(n))$ is surjective. This leaves the case $n = 3$, and here there exists serious problems in deciding whether or not $\mathrm{Aut}(\chi(3))$ contains only tame elements, or even if $\mathrm{Aut}(\chi(3))$ is f.g.. We cannot go into detail without introducing a large amount of background material, so we merely remark that the nature of the problems with $\mathrm{Aut}(\chi(3))$ is reminiscent of and analogous with those of $GL_2(K[x, x^{-1}, y, y^{-1}])$ for a field K, specifically $K = \mathbf{Z}_p$ where p is a prime, with the added difficulties that group automorphisms possess compared to module automorphisms. We do suspect the same answers in both cases - namely, $\mathrm{Aut}(\chi(3))$ and $GL_2(K[x, x^{-1}, y, y^{-1}])$ are both f.g. or both infinitely generated, and both consist only of tame elements or both contain non-tame elements.

In summary, it appears feasible to establish that $TR\{\mathrm{Aut}(\chi(n)\}$ is either 1 or 4. We believe that a complete solution to this problem would of necessity require techniques that would have as a by-product a solution of the tame range of $GL_n(\mathbf{Z}[t, t^{-1}])$ and $GL_n(K[x, x^{-1}, y, y^{-1}])$, just as the techniques developed to show that $\mathrm{Aut}(M(3))$ is infinitely generated were used in [5] to show the same results for $GL_2(\mathbf{Z}[x, x^{-1}, y, y^{-1}])$ and $GL_2(K[x, x^{-1}, y, y^{-1}, z, z^{-1}])$.

Case 2. k is not square free

As already observed, $\mathrm{Aut}(\chi(2))$ is not f.g. This means in particular that almost all elements of $\mathrm{Aut}(\chi(2))$ are not induced by elements of $\mathrm{Aut}(F(2))$. One might guess that $\mathrm{Aut}(\chi(n))$ is not f.g. for all $n \geq 2$, and indeed the authors have succeeded in showing this to be the case [8]. Thus, as a corollary, we have the only known example where $TR\{\mathrm{Aut}(\chi(n))\} = \infty$.

(5) Determine $TR\{\mathrm{Aut}(H(n,d))\}$, where $H(n,d) = F(n)/F(n)^{(d)}$ the free solvable group of class d.

This is far and away the most intractable problem discussed in this section and represents a goal to strive for. It does not seem practical to attack this problem directly; rather it makes sense to first acquire experience via intermediate families as in Example 6 below. But because this is the most interesting of all the problems, one cannot help but speculate as to what may lie ahead.

For a fixed d, put $f(d) = TR\{\mathrm{Aut}(H(n,d))\}$. The basic question is whether $f(d)$ is a finite number for each d, and if yes, whether $f(d) \to \infty$ as $d \to \infty$ or $f(d) = 4$ (or possibly a constant from some point on).

Finiteness of the tame range of the automorphism groups of free solvable groups would have enormous implications as well as afford untold aesthetic pleasure. But is finiteness a reasonable expectation? History certainly counsels against it. Results involving higher solvable groups are often in the nature of (ingenious) counterexamples to what would be suggested by the abelian and metabelian cases. (Harlampovich's [10] example of a f.g. solvable group with unsolvable word problem is a spectacular recent such illustration.) Thus, considering the complexity of $H(n,d)$, experience dictates that the most reasonable expectation is for $\mathrm{Aut}(H(n,d))$ to be infinitely generated for each $d \geq 3$ and $n \geq 3$.

Having said this, let us point out that there are several heuristic, yet compelling arguments which lead one to expect that $f(d)$ is finite for each d and moreover $f(d) \to \infty$ as $d \to \infty$. For example, one such argument is a varietal induction that predicts $f(d) = 2^d$. The reasoning goes roughly as follows:

We know that for $d = 2$, $\mathrm{Aut}(F(n) \to \mathrm{Aut}(H(n,d))$ is surjective for each $n \geq 4$. In other words, any automorphism of $H(n,d)$ can be lifted modulo $F'' = F(n)'' = [F(n)', F(n)']$ to a free automorphism as soon as n reaches 4. One reason advanced as to why 4 is the magic number is that F'' is defined by a law involving 4 variables, and that all of the variables must be utilized with independent choices of elements in order to be able to lift any conceivable automorphism. By this reasoning, F''' requires 8 variables, ... , $F^{(d)}$ requires 2^d variables, ... and thus $f(d) = 2^d$ is the correct generalization of the known result $f(2) = 4$. (Notice that if $d = 1$, $f(d) = 1$ so the formula $f(d) = 2^d$ begins with $d \geq 2$. But the above reasoning is valid for $d = 1$; i.e. one can lift modulo F' beginning with 2 generators.)

As a final observation we remark that the fact that for every d, the map $\mathrm{Aut}(F(2)) \to \mathrm{Aut}(H(2,d))$ is a surjection does not invalidate the above reasoning. The surjection for $n = 2$ is a consequence of a result of the authors and E. Formanek [2] to the effect that the kernel of $\mathrm{Aut}(H(2,d)) \to \mathrm{Aut}(F(2)/F(2)')$ consists entirely of inner automorphisms. In our context, this latter result should be interpreted to say that there are no (non-trivial) automorphisms of $H(2,d)$, and thus the problem of lifting to free automorphisms does not effectively appear until the 3 generator case. With the appearance of more complicated mappings beginning with 3 generators, $\mathrm{Aut}(H(n,d))$ becomes infinitely generated until n reaches 2^d, and then ... Certainly a lovely thought and food for the mathematical soul.

(6) Determine $TR\{\mathrm{Aut}(G(n,c))\}$, where $G(n,c) = F(n)/[F(n)_{c+1}]'$, the free (abelian-by-nilpotent of class c) group.

Here $F(n)_1 = F(n)$ and $F(n)_k$ is the kth term of the lower central series of $F(n)$. For a fixed c, put $f(c) = TR\{\mathrm{Aut}(G(n,c))\}$. As in the previous problem, the basic question is whether $f(c)$ is finite for each c and if $f(c) \to \infty$ as $c \to \infty$. We speculate that $f(c) = c + 3$. (The result $TR\{\mathrm{Aut}(M(n))\} = 4$ is the case $c = 1$.)

For a fixed n, let us write $F = F(n)$, and $\mathbf{Z}(F/F_{c+1})$ the integral group ring of the free nilpotent group of class c (and rank n). Then $F_{c+1}/(F_{c+1})'$ is a $\mathbf{Z}(F/F_{c+1})$-module, and each automorphism of $G(n,c)$ which acts trivially on F/F_{c+1} induces a module automorphism. Technical considerations indicate that each increase in the nilpotency class increases by one the dimension when everything becomes tame. On the other hand a varietal argument as in Example 4 would indicate that $f(c) = 2(c+1)$. There is no hard evidence for either of these tame range values - only heuristic argument. But this problem may be accessible and is perhaps the most important next step for an understanding of the automorphism groups of relatively free groups - particularly the higher solvable groups.

(7) Determine $TR\{GL_n(R_c)\}$, where R_c is the integral group ring of a free nilpotent group of class c and finite rank larger than one.

R_1 is the group ring of a free abelian group and hence as noted in

Section I, $TR\{GL_n(R_1)\} = 3$. For $c > 1$, the tame behavior of $GL_n(R_c)$ and $\text{Aut}(G(n+1,c))$ of Problem 6 are probably the same for all n - as is indeed the case for $GL_n(R_1)$ and $\text{Aut}(M(n+1))$. In fact the techniques developed for handling one group may turn out to be just what is needed for the other group. To illustrate this possibility, the methods developed in [4], which was a major step in the eventual proof that$\text{Aut}(M(3))$ is infinitely generated, were subsequently used in establishing that $GL_2(R_1)$ (and also the Torelli group of genus 2) is infinitely generated. Thus the answer to $TR\{GL_n(R_c)\}$ could be of major importance for the insight or even the inspiration it might bear on $TR\{\text{Aut}(G(n,c))\}$ of problem 6.

The very interesting results in Brown, Lenagen and Stafford [9] show that $TR\{GL_n(R_c)\}$ is finite and less than or equal to 3 plus the Hirsh number of the free nilpotent group. This is very promising start on this problem.

References

[1] S. Bachmuth, G. Baumslag, J. Dyer, H. Y. Mochizuki, Automorphism groups of 2-generator metabelian groups, to appear in *J. London Math Soc.*

[2] S. Bachmuth, E. Formanek and H.Y. Mochizuki, IA-automorphisms of two-generator torsion-free groups, *J. Algebra* **40** (1976), 19-30.

[3] S. Bachmuth and J. Lewin, The Jacobi identity in groups, *Math. Z.* **83** (1964), 170-176.

[4] S. Bachmuth and H.Y. Mochizuki, IA-automorphisms of the free metabelian group of rank 3, *J. Algebra* **55** (1978), 106-115.

[5] ———, $E_2 \neq SL_2$ for most Laurent polynomial rings, *Amer. J. Math.* **104** (1982), 1181-1189.

[6] ———, The non-finite generation of $AUT(G)$, G free metabelian of rank 3, *Trans. Amer. Math. Soc.* **270** (1982), 693-700.

[7] ———, $AUT(F) \to AUT(F/F'')$ is surjective for free group F of rank ≥ 4, *Trans. Amer. Math Soc.* **292** (1985), 81-101.

[8] ———, Automorphism groups which are infinitely generated in every dimension, in preparation.

[9] K.A. Brown, T.H. Lenagen and J.T. Stafford, K-theory and stable structure of some noetherian group rings, *Proc. London Math. Soc.* (3) **42** (1981), 193-230.

[10] O.G. Harlampovich, A finitely presented solvable group with unsolvable word problem, *Izvestya Akad. Nauk USSR*, ser. Mat. Tom **45**, No 4 (1981), 852-873 (Russian).

[11] D. Johnson, The structure of the Torelli group I: A finite set of generators for $\mathcal{T}$, *Annals of Math.* **118** (1983), 423-442.

[12] W. Magnus, A. Karrass, and D. Solitar, *Combinatorial group Theory*, New York: Wiley, 1966.

[13] D. McCullough, and A. Miller, The genus 2 Torelli group is not finitely generated, *Topology and its Applications* **22** (1986), 43-49.

[14] G. Mess, Thesis, University of California, Berkeley.

[15] E. Stöhr, On automorphisms of free centre-by-metabelian groups, *Arch. der Math.* **48** (1987), 376-380.

[16] A.A. Suslin, On the structure of the special linear group over polynomial rings, *Isv. Akad. Nauk.* **11** (1977), 221-238.

University of California
Santa Barbara
Ca 93106
U. S. A.

Universal groups, binate groups and acyclicity

A.J. Berrick

In [3 p85] it is stated that Lemma 11.11, used to prove the acyclicity (triviality of homology with trivial integer coefficients) of a certain universal group in algebraic K-theory, "serves to establish the acyclicity of many other groups besides, but that, as they say, is another story". This work presents that other story.

The key property of groups under consideration is a *binate* ("arranged in pairs") structure conforming to the hypotheses of [3 (11.11)]. We begin with a theoretical consideration of such structures and then offer a list of examples. Almost all the examples were previously known to be acyclic groups; however the arguments given in the literature tend to be of an ad hoc nature, with analogies on occasion noted though not explored. A number of these groups are universal in some sense. One newcomer to the list (in that it has not previously been described as acyclic) is P. Hall's countable universal locally finite group.

There is another way in which universality enters into this discussion. We begin by considering groups which are universal among binate groups.

1. Binate Groups

A binate-filtered group is to be thought of as the limit of a binate tower. This is an ascending sequence of groups built up in a prescribed manner from a base group. In the case of universal binate tower the base group actually determines the whole tower.

Definition 1.1. Let H be an arbitrary non-trivial group. The *universal binate tower* or *filtration* with *base* H is the tower of groups

$$H = H_0 \leq H_1 \leq \ldots \leq H_i \leq H_{i+1} \leq \cdots$$

where H_{i+1} is the HNN-extension $(H_i \times H_i)*_\psi$ resulting from the monomorphism

$$\psi : 1 \times H_i \rightarrowtail H_i \times H_i \quad (1,h) \mapsto (h,h).$$

That is,

$$H_{i+1} = gp\langle H_i \times H_i, t_i | t_i^{-1}(1,h)t_i = (h,h)\rangle.$$

The embedding $H_i \hookrightarrow H_{i+1}$ is given by $h \mapsto (h,1)$.

Note that in the above setup for each i $H_i \le H_{i+1} \le \dots$ is the universal binate tower with base H_i.

Definition 1.2. Now suppose that H is a subgroup of a group G. Any non-trivial quotient of the universal binate tower with base H under a homomorphism $\pi : \cup H_i \to G$ such that $\pi|_H = id$ will be called a *binate tower* or *filtration* (in G) with *base* H and *limit* $(\cup H_i)\pi$. A limit of a binate tower is a *binate-filtered* group.

We can characterise binate-filtered groups in terms of the centraliser apparatus of [3].

Lemma 1.3. *$G_0 \le G_1 \le \dots$ is a binate tower in G if and only if for each $i \ge 0$ there is a monomorphism $\phi_i : G_i \to \mathbf{C}_{G_{i+1}}(G_i)$ and $u_i \in G_{i+1}$ such that for all $g \in G_i$*

$$g = [g\phi_i, u_i]. \tag{1.4}$$

Proof. Given the stated ascending sequence in G, let $H = G_0$ and inductively define the requisite $\pi : H_i \to G_i$ from the universal binate tower with base H as follows. If $\pi : H_i \to G_i$ is already defined then let $t_i\pi = u_i$ and put

$$(h,1)\pi = h\pi, \quad (1,h)\pi = h\pi\phi_i.$$

Observe that π respects the extra relation $t_i^{-1}(1,h)t_i = (h,h)$ because $u_i^{-1}(1,h)\pi u_i = h\pi\phi_i h\pi = (h,h)\pi$.

For the converse argument, the above equations serve to define each u_i and ϕ_i so as to satisfy relation (1.4). □

In Lemma (1.3) the hypothesis of injectivity is actually a consequence of (1.4). Note also that, for all $g_1, g_2 \in G_i$,

$$[g_1 g_2 \phi_i, u_i] = [g_1 \phi_i, u_i][[g_1 \phi_i, u_i], g_2 \phi_i][g_2 \phi_i, u_i].$$

This means that, so long as we require that $[G_i \phi_i, u_i] \leq G_i$, then, precisely when ϕ_i has its image in $C_{G_{i+1}}(G_i)$, we have a homomorphism

$$G_i \to G_i \qquad g \mapsto [g\phi_i, u_i].$$

It will be clear from the proof of the fundamental technical result, immediately below, that acyclicity is obtained even when condition (1.4) is weakened to the hypothesis that this endomorphism when composed with the inclusion $G_i \hookrightarrow G_{i+j}$, some $j > 0$, is homologous to the inclusion $G_i \hookrightarrow G_{i+j}$. However this generalisation behaves less well on the passage to quotients. Moreover, I know of no groups which I would wish to add to the list in §3 below that satisfy only the more general property.

Lemma 1.5. *Any binate-filtered group is an infinitely generated acyclic group.*

Proof. In the previous notation, let $G = \cup G_i$ be the limit of the binate filtration $G_0 \leq G_1 \leq \ldots$. Since any finite subset of G lies in some G_i, if G is finitely generated then for some k,

$$G = G_k = G_{k+1} = \ldots .$$

So ϕ_k sends G to $\mathbf{C}_{G_{k+1}}(G_k) = \mathbf{Z}(G)$ (the centre of G) and (1.4) reveals that $G = [\mathbf{Z}(G), G] = 1$, a contradiction. Thus any generating set for G is indeed infinite.

We now establish acyclicity. (Although this is already in the literature, it seems appropriate to present the argument here.) By the universal coefficient theorem it is enough to prove triviality of homology with trivial coefficients in an arbitrary field; we therefore work with such homology groups. Since homology preserves direct limits, the following assertion will suffice.

For $k \geq 0, n \geq 1, H_q(G_k) \to H_q(G_{k+n})$ is trivial whenever $1 \leq q < 2^n$.

First note that by relabelling as necessary we need only consider the case $k = 0$. The proof goes by induction on n, being clear for $n = 1$ since (1.4) makes each element of G_0 trivial in the abelianisation of G_1. Assume therefore the triviality of the map induced by the inclusion $i : G_0 \hookrightarrow G_{n-1}$ on homology in dimensions between 1 and $2^{n-1} - 1$. We have to show that $j_* \circ i_* = 0 : H_q(G_0) \to H_q(G_n)$ where $j : G_{n-1} \hookrightarrow G_n$ and $2^{n-1} \leq q < 2^n$. To do this, define homomorphisms

$$\begin{aligned} \Delta_k &: G_k \to G_k \times G_k & g &\mapsto (g,g) \\ in_L &: G_{n-1} \to G_{n-1} \times G_{n-1} & g &\mapsto (g,1) \\ in_R &: G_{n-1} \to G_{n-1} \times G_{n-1} & g &\mapsto (1,g) \end{aligned}$$

and examine the effect of the inductive hypothesis on the (natural) Künneth isomorphism $H_q(G_k \times G_k) \cong \sum_{0 \leq p \leq q} H_p(G_k) \otimes H_{q-p}(G_k)$. Because

$$\Delta_{0*} x_q = x_q \otimes 1 + 1 \otimes x_q + \sum_{0 \leq p \leq q} x_p \otimes x_{q-p} \qquad x_i \in H_i(G_k)$$

under this correspondence, composition with i_* now annihilates all summands after the first two, leaving

$$i_* \circ \Delta_{0*} x_n = i_* x_n \otimes 1 + 1 \otimes i_* x_n,$$

or in other words

$$i_* \circ \Delta_{0*} = in_{L*} \circ i_* + in_{R*} \circ i_*.$$

With the further definitions

$$\eta : G_{n-1} \times \mathbf{C}_{G_n}(G_{n-1}) \to G_n \qquad g \mapsto gj \cdot g\phi$$

$$\nu : G_n \to G_n \qquad g \mapsto u_n^{-1} g u_n$$

(so that, for example, $\eta \circ in_L = j$ and $\eta \circ in_R = \phi$), we rewrite (1.4) in the form

$$\eta \circ (id \times \phi) \circ \Delta_{n-1} = \nu \circ \phi.$$

On composing with i and passing to homology we obtain (noting that $\Delta_{n-1} \circ i = i \circ \Delta_0$ and that conjugation induces the identity map on homology)

$$\eta_* \circ (id \times \phi_*) \circ (in_{L*} \circ i_* + in_{R^*} \circ i_*) = \phi_* \circ i_* .$$

That is,

$$\eta_* \circ in_{L^*} \circ i_* + \eta_* \circ in_{R^*} \circ \phi_* \circ i_* = \phi_* \circ i_* ,$$

or

$$j_* \circ i_* + \phi_* \circ i_* = \phi_* \circ i_* ,$$

which forces the induction through. □

Definition 1.6. A group G is *binate* if every finitely generated subgroup of G is contained in a binately-filtered subgroup of G.

There are various similar properties by which binate groups may be recognised; these are recorded below.

Proposition 1.7. *Let G be a group. The following are equivalent.*
(i) *G is binate.*
(ii) *Every finitely generated subgroup of G is the base of a binate tower in G.*
(iii) *For each finitely generated subgroup H of G there exist $u_H \in G$ and $\phi_H : H \to G$ such that for all $h \in H$ $h = [h\phi_H, u_H]$.*
(iv) *$G = \text{dirlim} G_\lambda$ where for each λ there exist $\mu \geq \lambda$, $u_\lambda \in G_\mu$ and $\phi_\lambda : G_\lambda \to G_\mu$ such that for all $g \in G_\lambda$ $g = [g\phi_\lambda, u_\lambda]$.*

Since any group is the direct limit of its finitely generated subgroups and homology respects direct limits, we straightaway obtain the following important consequence of Lemma 1.5.

Theorem 1.8. *All binate groups are infinitely generated and acyclic.*

2. Quotient-binate groups

There is a class of binate groups which boasts some surprising group-theoretic properties, which are explored in this section.

Definition 2.1. A *quotient-binate* group is one all of whose non-trivial quotients are binate.

A *normally binate* group is one with the property that for each normal subgroup N and finitely generated subgroup H the map ϕ_H (see (1.7)(iii)) sends $N \cap H$ into $\mathbf{C}_N(N \cap H)$.

Note that normal subgroups of normally binate groups need not be binate since the definition in no way constrains the behaviour of the elements u_H.

Evidently normally binate groups are quotient-binate (indeed, non-trivial quotients must also be normally binate), since each ϕ_H respects the identifications involved in passage to quotients. However, the converse is quite likely false: the family of homomorphisms ϕ_H for a given quotient-binate group need bear no relation to the families of homomorphisms in its quotients. Our interest lies in the behaviour of quotient-binate groups. The reason for the introduction of normally binate groups is that they permit construction of quotient-binate groups.

Example 2.2. A binate group G will be normally binate if for each finitely generated subgroup H the map ϕ_H assumes one of the following forms (which form may vary with H).

(i) There exists $s_H \in G$ such that for each $g \in H$

$$g\phi_H = s_H^{-1} g s_H.$$

Groups each of whose maps ϕ_H is of this form are called *mitotic* and further referred to in (3.1), (3.4), (3.6) below.

(ii) Let $\alpha : H \to H$ be an endomorphism that restricts to an endomorphism of $N \cap H$ (for example, conjugation by an element of $\mathbf{N}_G(H)$). There exists $r_H \in G$ such that

$$g\phi_H = [g\alpha, r_H].$$

(As hitherto, the fact that ϕ_H has its image in $\mathbf{C}_G(H)$ makes the map a homomorphism.)

Proposition 2.3. *Quotient-binate groups have no proper subgroups of finite index; every non-normal subgroup has infinitely many conjugates.*

Proof. Each subgroup of finite index contains a normal one and so gives rise to a finite quotient - but by (1.8) any such quotient must be trivial. The second assertion is of course obtained from the first by application to the normaliser of a given subgroup. □

The next theorem highlights one of the most striking features of quotient-binate groups. It uses a lemma which is of independent interest.

Lemma 2.4. *A quotient-binate group acts trivially on any finitely generated, residually finite group.*

Proof. This follows from Corollary 2.3; for, any subgroup of the automorphism group of a finitely generated residually finite group is also residually finite, however quotient-binate groups have no non-trivial finite nor residually finite images. □

Theorem 2.5. *Let $N \hookrightarrow G \twoheadrightarrow Q$ be an extension of groups. If N and Q are acyclic, then so is G. Conversely, if N is finitely generated and G is quotient-binate, then both N and Q are acyclic.*

Proof. The former assertion is well-known; it simply follows from the evanescence of all terms in the Lyndon-Hochschild-Serre spectral sequence for the extension. For the converse, Q will be binate, hence acyclic, even without the hypothesis on N. Now suppose that N is finitely generated, whence its homology groups are too. Therefore Q, being quotient binate, acts trivially on $H_*(N)$ (by the lemma) and the spectral sequence of the extension produces a Serre homology sequence

$$\ldots \to H_{q+1}(Q) \to H_q(N) \to H_q(G) \to H_q(Q) \to \ldots$$

in which each term $H_q(N)$ vanishes because all others do. □

For the continuation, let $\mathrm{Out}(N)$ denote the group of outer automorphisms of N.

Theorem 2.6. *Let N be a non-trivial normal subgroup of a quotient-binate group G. Then at most one of the following holds.*
(i) *N is finitely generated.*
(ii) *N is residually finite.*
(iii) *Any acyclic subgroup of* $\mathrm{Out}(N)$ *is residually finite.*

Proof. If both (i) and (ii) hold, then the theorem makes N acyclic and in particular non-abelian. On the other hand Lemma 2.4 forces N to be central in G.

Now suppose that (iii) holds. By Corollary 2.3 the group $Q = G/N$ admits only trivial images in $\mathrm{Out}(N)$. Since Q is acyclic a universal coefficient theorem shows that the group $H^2(Q; \mathbf{Z}(N))$ is zero. The exact sequence (of pointed sets) [4]

$$H^2(Q; \mathbf{Z}(N)) \to \mathcal{E}xt(Q, N) \to \mathrm{Hom}(Q, (\mathrm{Out}(N)))$$

forces the extension to be trivial. Hence N is a quotient of G, a fact incompatible with each of finite generation and residual finiteness of N. □

A more general investigation of the normal subgroups of acyclic groups is to be found in [6]. In [5] there is a construction of an acyclic group in which a given abelian group centrally embeds. By the above such acyclic groups cannot be quotient-binate, at least when the abelian group is finitely generated. However, the proof of acyclicity presented in [5] involves a Künneth formula argument not so very different from the one in Lemma 1.5 above.

What makes Theorem 2.6 especially pertinent is the following, strikingly contrasting fact. Its proof is deferred to (3.5).

Theorem 2.7. *Any group may be embedded* 2*-step subnormally in a binate group.*

3. Applications

This section justifies the attention paid to theory above by compiling a list of well-known groups which can be shown to be binate. The groups are itemised in order of their appearance in the literature.

3.1. Philip Hall's countable universal locally finite group [8]

Since this group is unique up to isomorphism we are at liberty to vary the construction slightly so as to render the binateness more apparent. We appeal to (iv) of Proposition 1.7 above. Let G_0 be a finite group of order greater than 2. For each $i \geq 0$, given the finite group G_i let $G_{i+1} =$

$\mathrm{Symm}(G_i \times G_i)$ with $G_i \times G_i$ embedded in $\mathrm{Symm}(G_i \times G_i)$ by means of the (right) regular representation ρ on the first factor and G_i in $G_i \times G_i$ by inclusion as the first factor. Then, by the proof of Lemma 1 of [8] any isomorphism between subgroups of $G_i \times G_i$ is expressible as conjugation in G_{i+1}. ϕ_i will be the embedding of G_i in $G_i \times G_i$ (and thence in G_{i+1}) by inclusion as the second factor. Since this produces a subgroup of $G_i \times G_i$ isomorphic to the one obtained from the diagonal inclusion (the isomorphism is $(id \times \bar{g}) \mapsto (\bar{g} \times \bar{g})$), we take u_i to be an element of G_{i+1}, conjugation by which restricts to this isomorphism. Thus we have the formula

$$(g \times g)\rho = u_i^{-1} \cdot (1 \times g)\rho \cdot u_i$$

which evidently reduces to the form (1.4). Likewise one sees that (3.1)(i) holds. The direct limit of this system is clearly a mitotic, countable, locally finite group satisfying P. Hall's criteria for universality (namely, every finite group is embedded, and any two isomorphic finite subgroups are conjugate).

3.2. The group G of compact-support self-homeomorphisms of $\mathbf{R}^n$ [13]

This is the doyen among examples with previously published proofs of acyclicity, and indeed its proof has served as the model for all others. A proof of binateness (by implication) for this group is presented in [14] (see (3.9) below). This can be seen more directly by observing that, for the filtration $\ldots \leq G_i \leq G_{i+1} \leq \ldots$ of G_n obtained by setting as G_i those homeomorphisms supported on the open ball of radius i centred at the origin, Mather's ψ_1 and k can be chosen as our ϕ_i and u_i respectively.

3.3. The general linear group of the cone on a ring [15]

To study the algebraic k-theory of a ring R, [15] embeds R in another ring, its cone CR, whose direct limit general linear group $GL(CR)$ is then shown to be acyclic. This is proved by checking that $GL(CR)$ is a *flabby* group and demonstrating, by means of a Künneth argument similar to the above, that all flabby groups are acyclic. A remarkable feature of flabby groups is that, provided one considers the generalised version of binateness discussed after Lemma 1.3 above, the homomorphism ϕ may be globally defined (that is, each ϕ_H can be chosen independently of H). Here are the details. In the notation of [15], a ***flabby*** group G is a direct sum group (in other words, admits a homomorphism $\oplus : G \times G \to G$) which has an endomorphism $\tau : G \to G$ and, corresponding to each finitely generated subgroup H of G, elements b, c and d of G such that whenever $h \in H$ we have

$$c(h \oplus 1)c^{-1} = d(1 \oplus h)d^{-1} = h,$$

These quickly yield the formula

$$c^{-1}hc = [1 \oplus hr, d^{-1}b].$$

Thus, because the map $h \mapsto c^{-1}hc$ is homologous to the identity, we have the generalised binate structure incorporating the global endomorphism $\phi : h \mapsto 1 \oplus hr$. Alternatively, we may proceed strictly within the binate framework by setting $h\phi_H = c(1 \oplus hr)c^{-1}$ and $u_H = cd^{-1}bc^{-1}$. It is evident that from the point of view of binateness and so acyclicity the structure of a flabby group allows redundancies (appropriately enough, given the terminology); however this situation has served as the prototype for some subsequent proofs of the acyclicity of groups (e.g. [9]). In the specific case of $GL(CR)$, an alternative, "tighter" proof of acyclicity through binateness is presented in [3, ch.11], where it gives rise to our opening quotation.

3.4. Further universal locally finite groups [11]

Given a locally finite group L of arbitrary infinite cardinality, Theorem 6.5 of [11] shows how to embed L in a locally finite group of the same cardinality satisfying the same universality criteria as in (3.1). Unlike the countable case (3.1) such universal groups are not unique up to isomorphism [12]. A slight modification of the construction of [11] forces the universal group to be mitotic as well. Specifically, let $G_1 = L \times S$ where S is the restricted symmetric group on some countable set. For each i form the supergroup G_{i+1} of G_i by taking the subgroup of the constricted symmetric group $S_i = \overline{\mathrm{Symm}}(G_i \times G_i)$ generated by $(G_i \times G_i)\rho$ and those elements of S_i which serve to conjugate pairs of isomorphic finite (= finitely generated) subgroups of $(G_i \times G_i)\rho$ (one element per pair). Then dirlim G_i is the group we seek. (Its existence is alluded to, without details, in [1].) Incidentally, all such universal groups, whether binate or not, are simple [11 Theorem 6.1.d] (so none of the alternatives of Theorem 2.6 holds!) Note that by [1 Theorem 4.3] all existentially closed groups are mitotic.

3.5. The cone on a group [10]

Let G be an arbitrary group. Its cone CG is defined in [10] as the semi-direct product of $G^{\mathbf{Q}}$ by $\mathrm{Aut}(\mathbf{Q})$. Hence $G^{\mathbf{Q}}$ is the set of functions from the rationals $\mathbf{Q}$ to G which map all numbers outside some finite interval to the identity; the group structure on G determines that on $G^{\mathbf{Q}}$. Likewise $\mathrm{Aut}\mathbf{Q}$ denotes the restricted symmetric group on $\mathbf{Q}$ comprising those permutations which act as the identity outside some finite interval $I_r = [-r, r]$. Thus a permutation a acts (on the right) on any $b \in G^{\mathbf{Q}}$ by $b \circ a(x) = b(a(x))$, $x \in \mathbf{Q}$, and the product structure of CG is given by

$$(a_1, b_1) \cdot (a_2, b_2) = (a_1 \circ a_2, (b_1 \circ a_2) \cdot b_2).$$

The acyclicity of $K = CG$ is not proved explicitly in [10] (the reader is instead referred to [13]), so it seems worth specifying the binate structure here. To accord with (1.7)(iv) we put

$$K_n = \{(a, b) \in C \mid \mathrm{supp}(a) \cup \mathrm{supp}(b) \subseteq I_n\},$$

where supp denotes the smallest closed set in $\mathbf{Q}$ outside which a given function acts as (or maps to) the identity. As in [13] we use a permutation k_n such that $\mathrm{supp}(k_n) \subseteq I_{n+1}$ and for each $j \in \mathbf{N}$ $k_n^j I_n \cap I_n = \emptyset$. The homomorphism $\phi_n : K_n \to C_K(K_n)$ is now defined by

$$(a, b) \mapsto [(a, b)\phi_n : \mathbf{Q} \to \mathbf{Q} \times G]$$

where

$$((a, b)\phi_n)(x) = \begin{cases} (k_n^j \circ a \circ k_n^{-j}(x), b \circ k_n^{-j}(x)) & x \in k_n^j I_n, \\ (x, 1) & \text{otherwise.} \end{cases}$$

This leads to the required formula $(a, b) = [(a, b)\phi_n, (k_n, 1)]$.

It is now an easy matter to prove Theorem 2.7. For, the arbitrary group G normally embeds in $G^{\mathbf{Q}}$ by means of $g \mapsto b_g$, where $b_g : \mathbf{Q} \to G$ sends 0 to g and non-zero numbers to $1 \in G$. We have already noted that $G^{\mathbf{Q}}$ is normal in the binate group CG. A somewhat sharper result is true. Let G^0 denote the group of all finite-support functions from $\mathbf{Q} - \{0\}$ to G. If we embed G^0 in $G^{\mathbf{Q}}$ as those functions that are trivial on 0, then we obtain CG as isomorphic to the semidirect product $(G \times G^0) \rtimes \mathrm{Aut}\mathbf{Q}$.

3.6. Mitotic groups [1]

These groups were first used to provide combinatorial, finitely generated and, later [2], finitely presented results analogous to those of [10]. However, the alternative construction suggested by (2.1)(ii) (or (i) and (ii) in tandem) would also work. The fact that mitotic groups have mitotic quotients is stated in [1], and a proof of acyclicity given.

3.7. The centraliser apparatus of [3]

This is the immediate origin of the present work. [3 (11.11)] is stated in terms of our (1.7)(iv). (Incidentally, note that the proof given for the more general Lemma 3.11 of [3] is debugged in [7] and in the special case of interest reduces to the one presented above.)

3.8. Examples of [9]

Numerous examples of flabby groups are exhibited in [9], as variations of the following prototype, itself modelled on (3.3) above. Let

$$V = S_0 \supset S_1 \supset S_2 \supset \dots$$

be a nested sequence of closed subspaces of an infinite-dimensional Hilbert space V (over $\mathbf{R}$, $\mathbf{C}$ or $\mathbf{H}$) such that for all $i \in \mathbf{N}$, S_{i-1}/S_i is isomorphic to V. Then the direct limit of the groups $\{g \in GL(V) \mid g(S_i^\perp) = S_i^\perp\}$ as $i \to \infty$ is shown to be flabby. Here $GL(V)$ denotes the group of all continuous linear automorphisms of V, although the group of invertible isometries would work just as well. Likewise invertible or unitary elements in a properly infinite von Neumann algebra may be considered, as may measure-preserving automorphisms of a Lebesgue measure space, or simply the permutations on an arbitrary infinite set. Moreover, examples of non-flabby acyclic groups are also given; in the light of our (2.1) and (2.3) these have the interesting property of admitting non-trivial quotients only of uncountable order.

3.9. Pseudo-mitotic groups [14]

In [14] a generalisation is given of the mitotic group structure of [1], which is readily seen to be equivalent to (1.7)(iii) when ψ_1 is taken to be our ϕ_H. The article also shows that the example of [13] ((3.2) above) conforms to this pattern.

References

[1] G. Baumslag, E. Dyer and A. Heller: The topology of discrete groups, *J. Pure Appl. Algebra* **16** (1980), 1-47.

[2] G. Baumslag, E. Dyer and C.F. Miller III: On the integral homology of finitely presented groups *Topology* **22** (1983), 27-46.

[3] A.J. Berrick: *An Approach to Algebraic K-Theory*, Research Notes in Math. 56 (Pitman, London, 1982).

[4] A.J. Berrick: Group extensions and their trivialisation, *l'Enseign. Math.* **31** (1985), 151-172.

[5] A.J. Berrick: Two functors from abelian groups to perfect groups, *J. Pure Appl. Algebra* **44** (1987), 35-43.

[6] A.J. Berrick: Remarks on the structure of acyclic groups, in preparation.

[7] A.J. Berrick and M.E. Keating: *K*-theory of triangular matrix rings, in *Applications of Algebraic K-Theory to Algebraic Geometry and Number Theory (Proc. Boulder CO, 1983)*, Contemporary Math. **55** (A M S, Providence RI, 1986), 69-74.

[8] P. Hall: Some constructions for locally finite groups, *J. London Math. Soc.* **34** (1959), 305-319.

[9] P. de la Harpe and D. MacDuff: Acyclic groups of automorphisms, *Comment. Math. Helvetici* **58** (1983), 48-71.

[10] D.M. Kan and W.P. Thurston: Every connected space has the homology of a $K(\pi, 1)$, *Topology* **15** (1976), 253-258.

[11] O.H. Kegel and B.A.F. Wehrfritz: *Locally Finite Groups* (North Holland, Amsterdam, 1973).

[12] A. Macintyre and S. Shelah: Uncountable universal locally finite groups, *J. Algebra* **43** (1976), 168-175.

[13] J.N. Mather: The vanishing of the homology of certain groups of homeomorphisms, *Topology* **10** (1971), 297-298.

[14] K. Varadarajan: Pseudo-mitotic groups, *J. Pure Appl. Algebra* **37** (1985), 205-213.

[15] J.B. Wagoner: Delooping classifying spaces in algebraic K-theory, *Topology* **11** (1972), 349-370.

Department of Mathematics
National University of Singapore
Kent Ridge 0511
Singapore

Residual solvability of the unit groups of group algebras

Ashwani K. Bhandari and *I.B.S. Passi*

1. Introduction

Let $U(KG)$ denote the unit group of the group algebra KG of a finite group G over a field K. It is known that, in case the characteristic of K is zero, $U(KG)$ is solvable if and only if G is Abelian [3]. In this paper, which is a sequel to our earlier paper [5], we characterize the residual solvability of $U(KG)$ when K is either (i) a $\mathcal{P}$-adic field of characteristic zero or (ii) an algebraic number field. It turns out that, in particular, $U(\mathbf{Q}G)$ is residually solvable if and only if either (i) G is Abelian or (ii) G is a Hamiltonian group of order $2^n m$, m odd, such that the multiplicative order of 2 modulo m is odd. Finally, we note that for the field $\mathbf{R}$ of real numbers $U(\mathbf{R}G)$ is residually solvable if and only if G is Abelian.

Our approach leads us to study the behaviour of the derived series of the unit groups of quaternion algebras.

2. The multiplicative group of a central division algebra

Let $\hat{K}$ be a $\mathcal{P}$-adic field (i.e. $\hat{K}$ is complete with respect to a discrete valuation and has residue class field finite) of characteristic zero. Such a field is necessarily a finite extension of $\mathbf{Q}_p$, the field of p-adic numbers, for some rational prime p ([16], §5, p.36). Let $\hat{R}$ be the ring of integers of $\hat{K}$ and $\hat{P}$ be its unique maximal ideal. Let D be a central division algebra over $\hat{K}$. For an element $\alpha \in D$, we denote by $N(\alpha)$ the reduced norm of α. Note that D has a unique maximal order $\Gamma = \{\alpha \in D \mid N(\alpha) \in \hat{R}\} = \{\alpha \in D \mid \alpha \text{ is integral over } \hat{R}\}$. Further, Γ has a unique maximal 2-sided ideal $\underline{P} = \{\alpha \in \Gamma \mid N(\alpha) \in \hat{P}\}$ which is a principal left ideal, and $\Gamma/\underline{P}$ is a finite field containing $\hat{R}/\hat{P}$ as a subfield. We refer the reader to [13] for details of these facts.

The following theorem is essentially available in [12].

2.1. Theorem. *Let $\hat{K}$ be a $\mathcal{P}$-adic field of characteristic zero. Let D be a central division algebra over $\hat{K}$ and let D^* be the unit group of D. Then the second derived group of D^* is residually nilpotent. In particular, D^* is residually solvable.*

Proof. Let $H_o = SL(1, D) = \{\alpha \in D \mid N(\alpha) = 1\}$. Then H_0 is a subgroup of the unit group of Γ. Let

$$H_n = H_0 \cap (1 + \underline{\mathcal{P}}^n), \quad n \geq 0.$$

Then we observe that for $x \in H_m$, $y \in H_n$, we have $x^{-1}y^{-1}xy \in H_{n+m}$, $m, n \geq 0$. This implies that H_{n+1} is normal in H_n, $n \geq 0$, and H_1 is residually nilpotent, since $\cap_n \underline{\mathcal{P}}^n = (0)$. The quotient H_0/H_1 embeds as a subgroup in the multiplicative group of $\Gamma/\underline{\mathcal{P}}$ which is commutative. Therefore, the derived group H_0' of H_0 is contained in H_1. Consequently, H_0' is residually nilpotent. Clearly, the derived group $(D^*)'$ of D^* is contained in H_0. Hence the second derived group of D^* is residually nilpotent.

For a field K we denote by $\mathbf{H}_K$ the usual quaternion algebra over K, i.e.

$$\mathbf{H}_K = \{a_1 + a_2 i + a_3 j + a_4 k \mid a_t \in K, 1 \leq t \leq 4, i^2 = j^2 = -1, ij = -ji = k\}.$$

For $\alpha = a_1 + a_2 i + a_3 j + a_4 k$, the reduced norm $N(\alpha) = \sum_{t=1}^{4} a_t^2$. We denote by $\mathbf{H}_K^*$ and K^* the unit groups of $\mathbf{H}_K$ and K respectively. The stufe of a field K, denoted by $S(K)$, is defined to be the least integer n such that $-1 = \alpha_1^2 + \alpha_2^2 + \ldots + \alpha_n^2$ for $\alpha_1, \alpha_2, \ldots, \alpha_n \in K$, if such expressions exist and is defined to be ∞ otherwise. Recall that, if K is a field of characteristic $\neq 2$, then $\mathbf{H}_K$ is a division algebra in case $S(K) > 2$ (equivalently, if the quadratic form $X_1^2 + X_2^2 + X_3^2$ is anisotropic over K), and is isomorphic to $M_2(K)$, the total matric algebra of 2×2 matrices over K, otherwise [see e.g. ([1], §10, pp. 145-146)].

The following corollary is an immediate consequence of Theorem 2.1.

2.2 Corollary. *The multiplicative group of a quaternion division algebra over a $\mathcal{P}$-adic field of characteristic zero is residually solvable.*

3. Residual solvability of the unit groups of group algebras

Let K be a field of characteristic zero. Suppose $U(KG)$ is residually solvable. Then, in particular, $U(\mathbb{Q}G)$ is residually solvable and, therefore, by ([5], Theorem 1.1)

either

$$G \text{ is Abelian}, \quad \text{(Type I)}$$

or

$$\left.\begin{array}{l} G \text{ is a Hamiltonian group of order } 2^n m,\ m \text{ odd},\\ \text{such that the multiplicative order of 2 modulo } m \text{ is odd.}\end{array}\right\} \quad \text{(Type II)}$$

Thus for studying the residual solvability of unit groups of group algebras over fields of characteristic zero, we need only to examine groups of type II.

Let G be a finite Hamiltonian group. Then

$$G = A \times E \times Q_8$$

where A is an Abelian group of odd order, E an elementary Abelian 2-group and

$$Q_8 = \langle x, y \mid x^4 = 1, x^2 = y^2, x^{-1}yx = y^{-1}\rangle$$

is the quaternion group of order 8 ([6], Theorem 7.8, p.308).

The quaternion algebras play a role in view of the following

3.1. Lemma. *If K is a field of characteristic $\neq 2$, then*

$$KQ_8 \simeq K \oplus K \oplus K \oplus K \oplus \mathbf{H}_K .$$

Proof. The homomorphism induced by

$$x \longmapsto (1, 1, -1, -1, i)$$
$$y \longmapsto (1, -1, 1, -1, j)$$

is an isomorphism.

The following result reduces our investigation to the residual solvability of the unit groups of quaternion algebras. For a field K, let $K_d = K(\varsigma_d)$, where ς_d is a primitive dth root of unity.

3.2. Lemma. *Let K be a field of characteristic zero, and let $G = A \times E \times Q_8$ be a finite Hamiltonian group, where E is an elementary Abelian 2-group and A an Abelian group of odd exponent e. Then $U(KG)$ is residually solvable if and only if $\mathbf{H}^*_{K_d}$ is residually solvable for every $d|e$.*

Proof. The group algebra $K(A \times E)$ is isomorphic to a direct sum of fields of the type K_d, $d|e$, and each such field occurs with a non-zero multiplicity [9]. Therefore, KG is isomorphic to a direct sum of the group algebras K_dQ_8 whose structure is given by Lemma 3.1 and the assertion follows.

We now consider $\mathcal{P}$-adic fields. Let $\mathbf{F}_q$ denote a finite field of order q.

3.3. Theorem. *Let $\hat{K}$ be a $\mathcal{P}$-adic field of characteristic zero and residue class field $\mathbf{F}_q$ with $q = p^f$, p prime. Let G be a finite non-Abelian group. Then $U(\hat{K}G)$ is residually solvable if and only if $p = 2$, $[\hat{K} : \mathbf{Q}_2]$ is odd and G is of type* II.

Proof. If $p > 2$ or if $p = 2$ and $[\hat{K} : \mathbf{Q}_2]$ is even, then $S(K) < 2$ {([15], Theorem 1, p.20) and ([11], Proposition 5), ([13], Theorem 6.17, p.239) or ([17], Theorem 1.3, p.33)}. Hence $\mathbf{H}^*_{\hat{K}}$ is isomorphic to $GL(2, \hat{K})$ and is not residually solvable. Therefore, by Lemma 3.2, in these cases $U(\hat{K}G)$ is not residually solvable.

Suppose now that $p = 2$ and $[\hat{K} : \mathbf{Q}_2]$ is odd and G is a group of type II. Let d be a divisor of the exponent of the odd order summand of G and let r be the order of q modulo d. Then $[\hat{K}_d : \hat{K}] = r$ ([16], Corollary 1, p.77). Since the order of 2 modulo d is odd, r must be odd. Hence the degree $[\hat{K}_d : \mathbf{Q}_2]$ is odd and the quadratic form $X_1^2 + X_2^2 + X_3^2$, which is anisotropic over $\mathbf{Q}_2$ ([5], Theorem 1, p.20), remains anisotropic over $\hat{K}_d$ by a theorem of Springer ([7], Theorem 2.3, p.198). Consequently, the quaternion algebra $\mathbf{H}_{K_d}$ is a division algebra (see also [17], Theorem 1.3, p.33) and, by Corollary 2.2, $\mathbf{H}^*_{\hat{K}_d}$ is residually solvable. Hence by Lemma 3.2, $U(\hat{K}G)$ is residually solvable and the proof is complete.

An immediate consequence is

3.4. Corollary. *Let G be a finite non-Abelian group. Then $U(\mathbf{Q}G)$ is residually solvable if and only if G is of type* II.

We next consider global fields. Let K be an algebraic number field. It is well known that $S(K) = 1, 2, 4$ or ∞.

The following theorem reduces the residual solvability of the unit groups of group algebras over global fields to that over local fields.

3.5. Theorem. *Let K be an algebraic number field and let G be a finite non-Abelian group. Then $U(KG)$ is residually solvable if and only if there exists a finite prime $\mathcal{P}$ in K such that the completion of K at $\mathcal{P}$ has stufe* 4 *and G is of type* II.

Proof. If $U(KG)$ is residually solvable, then, in particular, by Lemma 3.2, $\mathbf{H}_K^*$ is residually solvable. The derived group of $\mathbf{H}_K^*$ being equal to $SL(1, \mathbf{H}_K) = \{\alpha \in \mathbf{H}_K \mid N(\alpha) = 1\}$ ([18], Lemma 5), it follows from ([10], Theorem 1) that there exists a finite prime $\mathcal{P}$ in K such that the quaternion algebra over the completion of K at $\mathcal{P}$ is a division algebra and the necessity follows. The sufficiency is clear from Theorem 3.3.

We present the following characterization with indebtedness to the referee.

3.6. Corollary. *Let K be an algebraic number field with ring of integer $\mathcal{O}$. Let $(2) = \Pi_i \mathcal{P}_i^{e_i}$, where $\mathcal{P}_i$ are distinct prime ideals in K and let $|\mathcal{O}/\mathcal{P}_i| = 2^{f_i}$. Let G be a finite non-Abelian group. Then $U(KG)$ is residually solvable if and only if at least one $e_i f_i$ is odd and G is of type* II.

Proof. Observe that K has a finite prime $\mathcal{P}$ such that the completion $\hat{K}_{\mathcal{P}}$ of K at $\mathcal{P}$ has stufe 4 if and only if $\mathcal{P} = \mathcal{P}_i$ for some i and $[\hat{K}_{\mathcal{P}} : \mathbf{Q}_2] = e_i f_i$ ([16], Theorem 1, p.30) is odd.

For an algebraic number field K, if $S(K) = 1$ or 2, then clearly K

cannot have a completion with stufe 4. If $S(K) = 4$, then K has no real Archimedean prime. Therefore, by the Hasse-Minkowski Principle ([14], Theorem 6.5, p.223), there exists a finite prime $\mathcal{P}$ in K such that the completion $\hat{K}_{\mathcal{P}}$ of K at $\mathcal{P}$ has stufe 4. Suppose $[K : \mathbf{Q}]$ is odd. With the notation of Corollary 3.6, since $[K : \mathbf{Q}] = \sum_i e_i f_i$ ([16], Proposition 10, p.14) at least one $e_i f_i$ must be odd. Thus we have

3.7. Corollary. *Let K be an algebraic number field such that either $S(K) < \infty$ or $[K : \mathbf{Q}]$ is odd (and $S(K) = \infty$). Let G be a finite non-Abelian group. Then $U(KG)$ is residually solvable if and only if $S(K) > 2$ and G if of type II.*

3.8. Remarks. (i) The algebraic number fields K which are not covered by Corollary 3.6 are those for which $S(K) = \infty$ and $[K : \mathbf{Q}]$ is even. It is possible for such a field to possess or not to possess a completion $\hat{K}_{\mathcal{P}}$ at a finite prime $\mathcal{P}$ with $S(\hat{K}_{\mathcal{P}}) = 4$, as may be seen by considering, for example, $\mathbf{Q}(\sqrt{17})$ and $\mathbf{Q}(\sqrt{5})$ respectively.

(ii) From ([4], Theorem 3.1) it follows that if G is a finite group whose derived group is a p-group, then $U(\mathbf{Z}G)$ is residually solvable. Thus the behaviour of the derived series of $U(\mathbf{Q}G)$ is, as one would expect, in sharp contrast to that of $U(\mathbf{Z}G)$, where $\mathbf{Z}$ is the ring of rational integers (see also [8], p.530).

We conclude by considering the case of the real group algebra.

3.9. Theorem. *For a finite group G, $U(\mathbf{R}G)$ is residually solvable if and only if G is Abelian.*

Proof. Consider the real quaternion algebra $\mathbf{H}_{\mathbf{R}}$. Let

$$H_0 = SL(1, \mathbf{H}_{\mathbf{R}}) = \{\alpha \in \mathbf{H}_{\mathbf{R}} \mid N(\alpha) = 1\}.$$

Then we have an exact sequence

$$1 \to \{\pm 1\} \to H_0 \to SO(3) \to 1,$$

where $SO(3)$ is the special orthogonal group over $\mathbf{R}$ of degree 3 ([7], Chap. III, p.66). The group $SO(3)$ being a simple group ([2], Theorem 5.3, p.178), we get

$$H_0' = H_0,$$

since $-1 \in H_0'$. Thus $\mathbf{H}_{\mathbf{R}}^*$ is not residually solvable and consequently, by Lemma 3.2, $U(\mathbf{R}G)$ can be residually solvable only if G is Abelian.

References

[1] A.A. Albert, Structure of Algebras, *American Math. Society Colloquium Publications*, Volume XXIV (1961).

[2] E. Artin, *Geometric Algebra*, Interscience Publisher, Inc., New York (1957).

[3] J.M. Bateman, On the solvability of the unit groups of group algebras, *Trans. Amer. Math. Soc.*, **157** (1971), 73-86.

[4] A.K. Bhandari, Some remarks on the unit groups of integral group rings, *Arch. Math.*, **44** (1985), 319-322.

[5] A.K. Bhandari and I.B.S. Passi, Residual solvability of the unit groups of rational group algebras, *Arch. Math.*, **48** (1987), 213-216.

[6] B. Huppert, *Endliche Gruppen I*, Springer-Verlag (1967).

[7] T.Y. Lam, *The Algebraic Theory of Quadratic Forms*, W.A. Benjamin (1973).

[8] I. Musson and A. Weiss, Integral group rings with residually nilpotent unit groups, *Arch. Math.* **38** (1982), 514-530.

[9] S. Perlis and G.L. Walker, Abelian group algebras of finite order, *Trans. Amer. Math. Soc.*, **68** (1950), 420-426.

[10] V.P. Platonov and A.S. Rapinčuk, On the group of rational points of three-dimensional groups, *Dokl. Acad. Nauk SSSR* **247** (1979), 279-282.

[11] Y. Pourchét, Sur la represéntation en somme de carrés des polynomes a une indéterminéé sur un corps de nombres algébriques, *Acta Arith.*, **19** (1971), 89-107.

[12] C. Riehm, The norm 1 group of $\mathcal{P}$-adic division algebras, *Amer. J. Math.*, **92** (1970), 499-524.

[13] K.W. Roggenkamp and V. Huber-Dyson, *Lattices over orders* I, LNM 115, Springer-Verlag (1970).

[14] W. Scharlau, *Quadratic Forms and Hermitian Forms*, Springer-Verlag (1985).

[15] J.-P. Serre, *A course in Arithmetic*, Springer-Verlag (1973).

[16] J.-P. Serre, *Local Fields*, Springer-Verlag (1979).

[17] M.F. Vignéras, *Arithmétique des Algébres de Quaternions*, LNM 800, Springer-Verlag (1980).

[18] S. Wang, On the commutator group of a simple algebra, *Amer. J. Math.*, **72** (1950), 323-334.

Centre for Advanced Study in Mathematics
Panjab University
Chandigarh 160014
India

Stabilisers of injective modules over nilpotent groups

C.J.B. Brookes

§1. A successful method much used in the study of automorphisms of finitely generated groups G is to consider the canonical homomorphism

$$\pi_{ab} : \operatorname{Aut} G \to \operatorname{Aut} G_{ab}$$

where G_{ab} is the abelianisation G/G' of G. Under this, an automorphism α of G is mapped to the automorphism α_{ab} it induces on G_{ab}. The kernel of π_{ab}, called the IA-automorphism group IA(G) of G, contains the group Inn G of inner automorphisms.

A recent paper [2] of Bachmuth, Baumslag, Dyer and Mochizuki uses this approach to study the automorphisms of a 2-generator metabelian group $G = \langle g, h\rangle$. One of the questions they consider is whether $\operatorname{Aut} G$ is finite generated. They prove that $\operatorname{Im} \pi_{ab}$ is finitely presented and deduce that $\operatorname{Aut} G$ is finitely generated if and only if IA(G) is. Their paper is primarily an investigation of IA(G).

To prove that $\operatorname{Im} \pi_{ab}$ is finitely presented one considers G' as a $\mathbf{Z}G_{ab}$-module. It is cyclic, with generator $[g, h]$, and is therefore isomorphic to $\mathbf{Z}G_{ab}/I$ for some ideal I. Any $\beta \in \operatorname{Aut} G_{ab}$ extends to an automorphism of $\mathbf{Z}G_{ab}$. This extension will also be called β. Lemma 6 of [2] shows that $\operatorname{Im} \pi_{ab}$ consists of those β which stabilise I. The problem is therefore reduced to one about an ideal I of the integral group ring of a finitely generated abelian group A and the stabiliser Stab I of I in $\operatorname{Aut} A$. Farkas and Snider [10] considered these stabilisers in depth, conjecturing them to be arithmetic subgroups of $GL(n, \mathbf{Z})$ in the case where $A \cong \mathbf{Z}^n$. Corollary 12 of [10] says that Stab I is indeed arithmetic when I is semiprime. The proof relies on an important theorem of Roseblade [21, Theorem D] concerning the structure of prime ideals in $\mathbf{Z}A$. In [10] other ideals are also shown to have stabilisers arithmetic in $GL(n, \mathbf{Z})$. These are the I for which Stab I and all its subgroups of finite index are irreducible subgroups of $GL(n, \mathbf{Q})$. Here the deduction is more complex, using a linearisation process introduced by Auslander and Baumslag [1] and an earlier variant of Roseblade's theorem due to Bergman [3].

Returning to the proof of the finite presentability of $\operatorname{Im}\pi_{ab}$ for 2-generator groups, observe that if G_{ab} is not free abelian of rank 2 then $\operatorname{Aut} G_{ab}$ is finite. On the other hand, when $G_{ab} \cong \mathbf{Z}^2$ there are two cases; either Stab I is arithmetic as above or some subgroup of finite index in Stab I is reducible. In the former case we know from Borel and Harish-Chandra [7] that all arithmetic groups are finitely presented, and in the latter the reducible subgroup is triangularisable, polycyclic and therefore finitely presented.

The purpose of this note is to suggest a similar approach to the study of automorphisms of more complicated groups. As an illustration we shall consider finitely generated abelian-by-nilpotent groups but often the questions can be less restrictively formulated. These generalisations are left to the reader's imagination. The focal point will be the quest for analogues to the Farkas and Snider stabiliser results.

Throughout, knowledge of the structure of finitely generated nilpotent groups will be assumed. The development of the theory of isolated subgroups can be found in Hall's Edmonton notes [14].

§2. Let M_0 be a characteristic abelian subgroup of a group G and let $H = G/M_0$. Then M_0 is a $\mathbf{Z}H$-module and there is a homomorphism $\pi_H : \operatorname{Aut} G \to \operatorname{Aut} H$ mapping an automorphism α of G to the automorphism α_H it induces on H. In other words Aut G acts on H. More generally, suppose that a group Γ acts on H and let M be a $\mathbf{Z}H$-module. For $\gamma \in \Gamma$ one can define a $\mathbf{Z}H$-module structure on $M\gamma$ via $(m\gamma)h = (mh^{\gamma^{-1}})\gamma$. For example, if M is cyclic being isomorphic to $\mathbf{Z}H/I$ for some right ideal I of $\mathbf{Z}H$, then $M\gamma \cong \mathbf{Z}H/I^\gamma$ where the action of γ on H is extended to one on $\mathbf{Z}H$ by linearity. Note that when Γ is taken to be the group H acting on itself via conjugation, the $\mathbf{Z}H$-module structure defined above is identical to the usual one.

When H was G_{ab} and M_0 was a cyclic $\mathbf{Z}G_{ab}$-module, we wanted to describe those $\beta \in \operatorname{Aut} G_{ab}$ for which the (right) ideals I and I^β were the same, or alternatively, for which the modules $\mathbf{Z}G_{ab}/I$ and $\mathbf{Z}G_{ab}/I^\beta$ were isomorphic. For non-abelian H these two questions are no longer equivalent, but it is clearly the latter alternative in which we should be interested. On the other hand, even when H is free abelian of finite rank we have no answer as yet for arbitrary I. Instead we shall concentrate on a slightly different question which is dealt with in the free abelian case by the work of Farkas and Snider.

The idea is to set up an equivalence relation on $\mathbf{Z}H$-modules weaker than that given by isomorphism, and then to consider stabilisers for the action of our group Γ induced on the equivalence classes.

When A is a finitely generated abelian group and P is a prime ideal of $\mathbf{Z}A$, a non-zero cyclic submodule of $\mathbf{Z}A/P$ is isomorphic to $\mathbf{Z}A/P$. In our generalisation we want this isomorphism appearing as an equivalence. On the other hand we would not want $\mathbf{Z}A/P$ to be equivalent to $\mathbf{Z}A/P \oplus \mathbf{Z}A/Q$ for different primes P and Q. To help distinguish between these two cases we define a non-zero submodule of a module M to be *essential* (in M) if its intersection with any other non-zero submodule is also non-zero. Now we say that M_1 and M_2 are *similar* if they have essential submodules M_1' and M_2' respectively, with $M_1' \cong M_2'$. It is well known (see [23] for example) that any module M has a minimal injective extension, unique up to isomorphism, called its injective hull $E(M)$. Moreover M is essential in $E(M)$. Thus M_1 and M_2 are similar if and only if they have isomorphic injective hulls, and there is a bijection between the similarity classes of arbitrary modules and the isomorphism classes of injective modules. For ease of notation we shall identify the two sets of classes.

Γ permutes the classes via $[M]\gamma = [M\gamma]$, where $[M]$ denotes the similarity class of M and $\gamma \in \Gamma$. Likewise $E(M)\gamma = E(M\gamma)$. In the particular case where $\Gamma = \operatorname{Aut} H$, the modules M and $M\gamma$ are isomorphic when γ is inner. Thus the group Inn H of inner automorphisms fixes each similarity class and we have an action of the outer automorphism group Out $H \cong \operatorname{Aut} H/\operatorname{Inn} H$ induced on the set of similarity classes.

Now suppose H is finitely generated nilpotent and so $\mathbf{Z}H$ is a Noetherian ring (Hall [12]). If M is a finitely generated $\mathbf{Z}H$-module it is Noetherian and any direct sum of non-zero modules embedded as a submodule in M can have only a finite number of summands. Defining a module to be *uniform* if it is non-zero and all its non-zero submodules are essential, we see that M has some essential submodule of the form $M_1 \oplus \ldots \oplus M_r$ with each M_i uniform. Thus $E(M) \cong E(M_1 \oplus \ldots \oplus M_r) \cong E(M_1) \oplus \ldots \oplus E(M_r)$. Uniformity is inherited by $E(M_i)$ and so each $E(M_i)$ is an indecomposable injective. The Krull-Schmidt theorem says that the isomorphism classes $[E(M_i)]$ and their multiplicities are unique in any such decomposition of $E(M)$ into indecomposable injectives. In particular the number r is well-defined and is known as the *uniform dimension* (or Goldie rank) of M [11]. An element γ of our group Γ stabilises $[E(M)]$ if and only if it permutes the finite set of classes $[M_i] = [E(M_i)]$. Thus $\bigcap_i \mathrm{Stab}_\Gamma[M_i]$ has finite index in $\mathrm{Stab}_\Gamma[M]$, which means that we need really only consider indecomposable injectives.

In the case where $A \cong \mathbf{Z}^n$, any indecomposable injective contains a submodule $\mathbf{Z}A/P$ for some prime ideal P and therefore is isomorphic to $E(\mathbf{Z}A/P)$. Also $\mathrm{Stab}_{GL(n,\mathbf{Z})}[\mathbf{Z}A/P] = \{\beta \in GL(n,\mathbf{Z}) : P^\beta = P\}$. We have already seen this to be an arithmetic, and therefore finitely presented, subgroup of $GL(n,\mathbf{Z})$.

Question 1. *Is* $\mathrm{Stab}_{\mathrm{Out}H}[M]$ *finitely presented, or even isomorphic to an arithmetic group, for any indecomposable injective* $\mathbf{Z}H$*-module* M*?*

The automorphism group of a finitely generated nilpotent group is isomorphic to an arithmetic group (see Segal [22, Chapters 5 and 6]).

Question 2. *Is there a canonical homomorphism* $\mathrm{Aut}H \to GL(n,\mathbf{Z})$ *with the images of both* $\mathrm{Aut}H$ *and* $\mathrm{Stab}_{\mathrm{Aut}H}[M]$ *arithmetic in* $GL(n,\mathbf{Z})$ *for some* n*?*

If the above are too optimistic we could consider images under the canonical homomorphism $\pi_{ab} : \mathrm{Aut}H \to \mathrm{Aut}H_{ab}$.

Question 3. *Is* $\pi_{ab}(\mathrm{Stab}_{\mathrm{Aut}H}[M])$ *finitely presented?*

Question 4. *When* $H_{ab} \cong \mathbf{Z}^n$*, is* $\pi_{ab}(\mathrm{Stab}_{\mathrm{Aut}H}[M])$ *arithmetic in* $GL(n,\mathbf{Z})$*?*

§**3.** The definitive reference on the subject of injective modules over Noetherian rings is Jategaonkar's book on localisation [15]. For our purposes we can rely on definitions about the existence of certain submodules, ignoring the structure of the injectives as a whole. It is perhaps worth mentioning though that the endomorphism ring of an indecomposable injective is a local ring, the unique maximal ideal consisting of the non-invertible endomorphisms, all of which have non-zero kernels.

If an indecomposable injective $\mathbf{Z}H$-module M is $\mathbf{Z}$-torsion-free then $M \otimes_{\mathbf{Z}} \mathbf{Q}$ may be viewed as a $\mathbf{Q}H$-module and as such it is also an indecomposable injective. On the other hand, if M is not $\mathbf{Z}$-torsion-free it contains a non-zero submodule M_1 annihilated by some rational prime p. This prime is uniquely determined since any submodule annihilated by a different prime would have zero intersection with M_1. In fact if M_1 is taken to be the submodule of all elements of M annihilated by p, then M_1 is an indecomposable injective $\mathbf{F}_p H$-module. Moreover, for γ acting on H, we have $M_1\gamma \cong M_1$ if and only if $M\gamma \cong M$. Thus in both cases we need only consider indecomposable injective kH-modules for a suitable field k. In fact from now on, k will denote an arbitrary field.

The key result behind proving arithmeticity in the free abelian case is

Theorem 1 (Roseblade [21, Theorem D]). *Let P be a prime ideal of kA where A is a free abelian group of rank n, and k is a field. Then there are isolated subgroups $C \leq B \leq A$ invariant under* $\mathrm{Stab}_{GL(n,\mathbf{Z})}(P)$ *satisfying*

(i) $kB/P \cong (kB/(P \cap kB)) \otimes_{kB} kA$,
(ii) $f.f.(kC/(P \cap kC))$ *is algebraic over* k,
(iii) $\mathrm{Stab}_{GL(n,\mathbf{Z})}(P)$ *induces a finite automorphism group on* B/C.

Here $f.f.(\ )$ denotes the fraction field of the domain in question. An alternative way of expressing (i) is to stipulate that $f.f.(kA/P)$ is purely transcendental over $f.f.(kB/(P \cap kB))$. Clearly B and C may be taken to be unique subgroups respectively minimal with respect to (i) and maximal with respect to (ii).

The original proof was by induction. Recently Bieri and Groves [4] have produced a more geometric approach based on the complement to the invariant Σ_M originally defined by Bieri and Strebel [6] to describe those metabelian groups that are finitely presented. To prove any generalisation for nilpotent groups and indecomposable injectives we should perhaps be looking to define an invariant $\Sigma_{[M]}$, say, along the same lines. Certainly it is interesting to note that in the commutative case $\Sigma_{M_1} = \Sigma_{M_2}$ if M_1 and M_2 are similar finitely generated kA-modules.

The initial step is to produce a subgroup of H corresponding to the subgroup B in the theorem. One can certainly choose an isolated subgroup L of H minimal with respect to the property that our indecomposable injective kH-module M has a submodule of the form $V \otimes_{kL} kH$ for some non-zero kL-module V. But how special is this choice? It is not necessarily unique. For example, the pair (V, L) may be replaced by its conjugate (Vh, L^h) for any $h \in H$. In fact the injective hull $E(V)$ of V may be used instead of V. In other words $E(V) \otimes_{kL} kH$ may be embedded in M. As a prelude to proving this it is easy enough to demonstrate that $E(V)$ embeds in M. This can be deduced immediately from the fact that M is injective as a kL-module. This in turn only requires the following manipulation of definitions.

Recall that a module M over a ring R is injective if and only if, given R-modules M_1 and M_2, an R-module monomorphism $\psi : M_1 \to M_2$ and an R-module homomorphism $\phi : M_1 \to M$, there exists an R-module homomorphism $\rho : M_2 \to M$ such that $\phi = \rho\psi$, or in other words, such that the following diagram commutes

$$\begin{array}{ccc} M_1 & \overset{\psi}{\hookrightarrow} & M_2 \\ \big\downarrow \phi & \swarrow_{\rho} & \\ M & & \end{array}$$

Now let M be our injective kH-module and suppose that there are kL-modules V_1 and V_2, a kL-module monomorphism $\psi' : V_1 \to V_2$ and a kL-module homomorphism $\phi' : V_1 \to M$. These maps extend respectively to give a kH-module monomorphism $\psi : V_1 \otimes_{kL} kH \to V_2 \otimes_{kL} kH$ and a kH-module homomorphism $\psi : V_1 \otimes_{kL} kH \to M$. The kH-injectivity of M means we can complete the commutative diagram

$$\begin{array}{ccc} V_1 \otimes_{kL} kH & \overset{\psi}{\hookrightarrow} & V_2 \otimes_{kL} kH \\ \big\downarrow \phi & \swarrow_{\rho} & \\ M & & \end{array}$$

The restriction ρ' of ρ to V_2 satisfies $\phi' = \rho'\psi'$. Note for future reference that this argument applies for any subgroup L of H.

The proof that $E(V) \otimes_{kL} kH$ embeds in M is a little harder. Since $V \otimes_{kL} kH$ lies in the uniform module M we know that V, and therefore

$E(V)$, is uniform. Between L and H there lies a subnormal series of isolated subgroups with each factor infinite cyclic. An induction argument on this series using Lemma 3.4 of [9] shows that $E(V) \otimes_{kL} kH$ is a uniform kH-module. Hence $[M] = [V \otimes_{kL} kH] = [E(V) \otimes_{kL} kH]$. The injectivity of M now ensures that $E(V) \otimes_{kL} kH$ embeds in M.

The minimality of L means that the module V is *impervious*, that is, it has no non-zero submodule of the form $V_1 \otimes_{kL_1} kL$ for any proper isolated subgroup L_1 of L and kL_1-module V_1. Also, we have already noted that V inherits uniformity from M and conversely, that $E(V) \otimes_{kL} kH$ does so from V.

Proposition 1. *There is a one-one correspondence between isomorphism classes of indecomposable injective kH-modules M and H-orbits of pairs $([V], L)$ with L an isolated subgroup of H and V an impervious indecomposable injective kL-module. Under this correspondence*

$$[M] \leftrightarrow \{([Vh], L^h) : h \in H\}$$

if and only if $V \otimes_{kL} kH$ embeds in M.

When seeking stabilisers, this correspondence means that one need only consider actions on classes of impervious modules.

Corollary. *With the notation as above,*

$$\mathrm{Stab}_{\mathrm{Aut}H}[M] = \mathrm{Stab}_{N_{\mathrm{Aut}H}(L)}[V].\,\mathrm{Inn}\,H.$$

Proof. Let $\beta \in \mathrm{Aut}H$. Under the correspondence

$$[M] \leftrightarrow \{([Vh], L^h) : h \in H\}$$

and

$$[M\beta] \leftrightarrow \{([(V\beta)h], (L^\beta)^h) : h \in H\}.$$

Here the kL^β-module structure of $V\beta$ is the obvious one, obtained by considering $V\beta$ as a submodule of $(V \otimes_{kL} kH)\beta$. Now $\beta \in \mathrm{Stab}_{\mathrm{Aut}H}[M]$ if and only if there is some $h \in H$ for which $L^\beta = L^h$ and $V\beta \cong Vh$. But $L^\beta = L^h$ precisely when $\beta = \beta_1\beta_h$ for some $\beta_1 \in N_{\mathrm{Aut}H}(L)$ where β_h denotes the inner automorphism induced by h. In such circumstances $V\beta \cong Vh$ if and only if $V\beta_1 \cong V$, or $\beta_1 \in \mathrm{Stab}_{N_{\mathrm{Aut}H}(L)}[V]$.

§4. The proof of Proposition 1 relies on Mackey's theorem [16] and the following strengthened version of Lemma 3.5 of [9].

Lemma. *Let L be an isolated subgroup of H and V be a kL-module. Then every non-zero kH-submodule of $V \otimes_{kL} kH$ contains a non-zero submodule of the form $V' \otimes_{kL} kH$. Furthermore, if V is impervious and uniform, then V' may also be taken to be so, with $[V'] = [V]$.*

Proof. The argument is an induction on a subnormal series

$$L = H_1 \triangleleft \ldots \triangleleft H_r = H$$

of isolated subgroups. There is nothing to prove when $H = L$. Suppose the lemma is true on replacing H by H_i, and consider a non-zero kH_{i+1}-submodule U of $V \otimes_{kL} kH_{i+1}$. By Lemma 3.5 of [9] it contains a non-zero kH_{i+1}-submodule of the form $W \otimes_{kH} kH_{i+1}$ for some kH_i-module W.

Now $V \otimes_{kL} kH_{i+1}$ may be viewed as a direct sum $\bigoplus_{t\in T} (V \otimes_{kL} kH_i)t$ of kH_i-modules, where T is a transveral to the cosets of H_i in H_{i+1}. Clearly W may be taken to be finitely generated and uniform, and therefore to lie in the direct sum of finitely many of the summands $(V \otimes_{kL} kH_i)t$. Moreover we may assume that V is also finitely generated.

$$E(W) \leq E(\bigoplus_j (V \otimes_{kL} kH_i)t_j) \cong \bigoplus_j E((V \otimes_{kL} kH_i)t_j)$$

for some $t_j \in T$. Decomposing each $E((V \otimes_{kL} kH_i)t_j)$ into a finite direct sum of indecomposable injectives and applying the Krull-Schmidt theorem we know that $E(W)$ is isomorphic to one of the indecomposable summands and therefore $E(W)$ may be embedded in $E((V \otimes_{kL} kH_i)t_j)$ for some j.

In fact what we have done here is to prove that a non-zero submodule of any direct sum has, in turn, a non-zero submodule which may be embedded in one of the summands. This is really the crux of the Krull-Schmidt theorem in this context.

Replacing our original choice of W by a suitable kH_i-submodule of Wt_j^{-1}, we may assume W to be isomorphic to a submodule of $V \otimes_{kL} kH_i$. The inductive hypothesis now gives us a non-zero submodule $V' \otimes_{kL} kH_i$ of W for some kL-module V'. Moreover, if V is impervious and uniform one may take V' to be impervious with $[V'] = [V]$. The proof is completed by observing that $V' \otimes_{kL} kH_{i+1}$ embeds in $W \otimes_{kH_i} kH_{i+1}$, and thence in U.

Proof of Proposition 1. Suppose that $[V_1 \otimes_{kL_1} kH] = [V_2 \otimes_{kL_2} kH]$ for isolated subgroups L_i of H and impervious indecomposable injective kL_i-modules V_i. The aim is to show that $L_2 = L^t$ and $[V_2] = [V_1 t]$ for some $t \in H$.

There is some non-zero kH-module which embeds in both $V_1 \otimes_{kL_1} kH$ and $V_2 \otimes_{kL_2} kH$. From the lemma we know there to be an impervious kL_2-module V_2' with $[V_2'] = [V_2]$, for which $V_2' \otimes_{kL_2} kH$ embeds in this kH-module. As a consequence V_2' embeds in $V_1 \otimes_{kL_1} kH$.

Mackey's theorem describes $V_1 \otimes_{kL_1} kH$ as a kL_2-module; it is a direct sum $\bigoplus_{t \in T} V_t$ where T is a set of double coset representatives for L_1, L_2 in H. Thus $H = \dot{\bigcup}_{t \in T} L_1 t L_2$. Moreover, each kL_2-module V_t is induced up from a $k(L_2 \cap L_1^t)$-module. The result used during the proof of the lemma says that some non-zero, necessarily impervious, submodule of V_2' is isomorphic to a submodule of some V_t. But the lemma applied to V_t with $L_2 \cap L_1^t$ and L_2 in the roles of L and H respectively, says that no kL_2-submodule of V_t can be impervious, unless $L_2 = L_2 \cap L_1^t$. We deduce that $L_2 \le L_1^t$ and so the torsion-free rank of L_1 is at least that of L_2. By symmetry these ranks are the same and so $L_2 = L_1^t$. Mackey's theorem, in its complete form, says that $V_t = V_1 t$ whenever $L_2 = L_2 \cap L_1^t$. Thus $[V_2] = [V_2'] = [V_1 t]$ and we are done.

§5. In the light of the correspondence established in Proposition 1 we concentrate on impervious modules for the rest of the paper. These basic building blocks are also important in their own right, being the only sort of module which can underlie a finitely presented abelian-by-nilpotent group. Going back to our original group G with abelian characteristic subgroup M_0 for which $H = G/M_0$ is nilpotent, we have

Proposition 2. *If M_0 contains a non-zero $\mathbf{Z}H$-submodule of the form $V \otimes_{\mathbf{Z}L} \mathbf{Z}H$ for some proper isolated subgroup L of H and $\mathbf{Z}L$-module V, then G is not finitely presented.*

For metabelian G this follows immediately from the Bieri and Strebel classification [6]. The proof of Proposition 2 also uses other results of Bieri and Strebel; for example, a soluble group is finitely presented only if all its metabelian images are [6]. Since we can clearly assume G to be finitely generated we need only consider finitely generated M_0 and V. We shall produce a $\mathbf{Z}H$-submodule M_* of M_0 so that G/M_* has a nilpotent-by-cyclic normal subgroup G_0/M_* of finite index, which is not of finite Prüfer rank. Theorem C of [5] implies that such a subgroup has a metabelian image which is not finitely presented. Thus G_0 is not finitely presented. This suffices to prove the proposition since G/G_0 is finite (see Robinson [19, 2.2.4]). It remains to define M_*.

There is an isolated normal subgroup H_1 of H, containing L, with H/H_1 infinite cyclic. Let $U = V \otimes_{kL} kH_1$. The $\mathbf{Z}H$-module $M_1 = U \otimes_{\mathbf{Z}H_1} ZH$ is residually finite (Hall [13]), so we can let I_1 be the annihilator in $\mathbf{Z}H_1$ of some finite $\mathbf{Z}H$-module image of M_1. Thus $\mathbf{Z}H_1/I_1$ is finite and $I_1^h = I_1$ for every $h \in H$. Let $I = I_1\mathbf{Z}H$, an ideal of $\mathbf{Z}H$. Because $UI_1 < U$ we have $M_1 > M_1I$. Indeed I annihilates the finite image of M_1 that we started with.

A key attribute of all ideals in the integral group ring of a finitely generated nilpotent group is the weak Artin-Rees property (Nouazé, Gabriel [17], Roseblade [20]). In our context this says that there is some $r \in \mathbf{N}$ with $M_0I^r \cap M_1 \leq M_1I$. We can now introduce our candidate for M_*, namely $M_0I^r + M_1I$, and check that M_0/M_* does not have finite (Prüfer) rank. To do this we look at $(M_*+M_1)/M_*$, which is isomorphic to $M_1/(M_* \cap M_1)$. But $M_* \cap M_1 = M_1I$, since $M_0I^r \cap M_1 \leq M_1I$, and so M_0/M_* has a submodule isomorphic to M_1/M_1I, which certainly does not have finite rank, being isomorphic to $(U/UI_1) \otimes_{\mathbf{Z}H_1} ZH$.

The final step is to produce the normal subgroup G_0 of finite index in G, containing M_0. In fact for various i we define subgroups G_i which are the pre-images of certain subgroups H_i of H. Thus $H_i = G_i/M_0$. First we observe that $\mathbf{Z}H_1/I_1^r$ is a finite set acted upon by H_1 via right multiplication. The kernel H_2 of this action is a normal subgroup of finite index in H_1. In fact it is normal in H, since $H_2 = (I_1^r+1)\cap H_1$ and $(I_1^r)^h = I_1^r$ for $h \in H$. The finite-by-infinite cyclic group H/H_2 has an infinite cyclic normal subgroup H_0/H_2 of finite index, yielding our subgroup G_0 of finite index in G. It remains to check that G_0/M is nilpotent-by-cyclic. But G_2 is nilpotent because H_2 acts trivially on M_0/M_0I^r, and hence on M_0/M_*.

§**6.** The next problem to consider is that of the analogue to the subgroup C in Roseblade's theorem. Here there seems to be a choice of candidate. The version of the theorem as stated is weaker than the original, but is the form that arises naturally from the consideration of valuations. Bieri and Groves present this, leaving the deduction of the stronger version as an appendix to their paper [4]. Theorem D of [21], when translated into our language, actually defines C to be the unique subgroup maximal with respect to $C/(C \cap (P+1))$ being finite. The choice is therefore between using finite actions and algebraic ones for the definition of C. Whichever we settle upon, we are still going to have the same problem as we encountered when considering the analogue to the subgroup B, namely uniqueness. Plumping for the algebraic option we can, given an imperivous indecomposable injective kH-module M, choose an isolated subgroup K of H maximal with respect to M having a kK-submodule that is finite dimensional as a k-vector space.

Question 5. *Is K unique up to conjugacy in H?*

Question 6. *Is K normal in H?*

The reason for thinking that normality might hold here is bound up with the analysis of impervious modules using a Clifford-style theory that we shall come to shortly (see Theorem 3).

In the meantime let us simplify matters and consider the 'generic' case, where $K = 1$. In other words we are assuming M to be an impervious indecomposable injective kH-module in which no finite dimensional k-vector space is stabilised by a non-trivial subgroup of H. As a consequence of this assumption H is torsion-free.

Question 7. *Is* $\mathrm{Stab}_{\mathrm{Out}H}[M]$ *finite?*

An affirmative answere here would be a first step towards dealing with Questions 1 and 2. A proof would require a new method, since valuations and the geometric invariant associated with them definitely rely on a commutative framework. On the other hand there must be grounds for hope of answering the corresponding question about the abelianisation positively.

Question 8. *Is* $\pi_{ab}(\mathrm{Stab}_{\mathrm{Aut}H}[M])$ *finite?*

The theme of [8] was that in this context a reduction to an (almost) commutative situation can be made in a fairly controlled fashion. In particular, given a finitely generated $\mathbf{Z}H$-module M one can find an ideal I of $\mathbf{Z}H$ so that H acts on M/MI as a finite-by-abelian group, while M/MI retains a structure closely allied to that of M. Can one produce a geometric invariant using these images?

§7. There is another application for this work on automorphisms. Let H_1 be an isolated normal subgroup of H. Then H acts on H_1 via conjugation and a kH-module M may be analysed using its kH_1-module structure. Suppose M is injective. Then it is also injective as a kH_1-module and therefore contains an indecomposable injective kH_1-submodule W. In fact it is possible to pick W to be *minimal* in M. By this we mean that W contains a non-zero submodule W_1 with the property that any non-zero kH_1-module homomorphism $W_1 \to M$ is actually a monomorphism. In the case where H_1 is abelian, we would choose a non-zero element $m \in M$ with largest possible annihilator P in kH_1 and set $W_1 = mkH_1 \cong kH_1/P$. The ideal P is necessarily prime, and so the injective hull $W = E(mkH_1) \cong E(kH_1/P)$ is indecomposable. The image of m under any map $W_1 \to M$ which is not a monomorphism would have annihilator in kH_1 strictly larger than P. In the non-commutative case we mimic this argument, this time reserving the letter W for the injective hull of a critical kH_1-submodule W_1 of minimal Krull dimension. This dimension due to Rentschler and Gabriel [18] is designed for Noetherian rings. It suffices here to say that it takes ordinal values and that any module over a Noetherian ring has a (necessarily uniform) submodule, all of whose proper quotients have strictly smaller dimension. Such a submodule is *critical*.

A version of Clifford's theorem is the basis of all that follows.

Proposition 3 (Brookes, Brown [9, Lemma 3.2], Brookes [8, Lemma 1]). *With the above notation,* $W_1 kH \cong W_1 kS \otimes_{kS} kH$ *where* $S = \mathrm{Stab}_H[W]$.

Note that this is saying that $W_1 kH = \bigoplus_{t \in T} W_1 kSt$ where T is a transversal to the right cosets of S in H. Furthermore, the kH_1-modules $W_1 kSt$ are the homogeneous components, in that any uniform kH_1-submodule of $W_1 kH$ is similar to some $W_1 t$ and lies in $W_1 kSt$.

The significance of this is that if M is impervious we deduce that $Stab_H[W]$ is of finite index in H, for any isolated normal subgroup H_1 and minimal indecomposable injective kH_1-module W in M. To tie this in with Questions 7 and 8 we need to be dealing with stabilisers of impervious modules. Using the correspondence established in Proposition 1 we know $[W] = [V \otimes_{kL_1} kH_1]$ for some isolated subgroup L_1 of H_1 and impervious indecomposable injective kL_1-module V. In fact V is also minimal in M. To see this let W_1 be a kH_1-submodule used to define the minimality of W. Any non-zero kH_1-submodule W_2 of W_1 would also fulfil this purpose since any non-zero homomorphism $W_2 \to M$ extends, using the kH_1-injectivity of M, to one from W_1 to M. That the extension is a monomorphism implies that the original map is also one. Passing to a submodule if necessary, the Lemma ensures that we may assume W_1 to be of the form $V_1 \otimes_{kL_1} kH$ for some kL_1-module V_1 with $[V_1] = [V]$. Any non-zero kL_1-module homomorphism $\phi' : V_1 \to M$ extends to a non-zero kH_1-module homomorphism $\phi : V_1 \otimes_{kL_1} kH_1 \to M$, which necessarily is a monomorphism. Thus $\ker\phi' = 0$ and V is minimal in M.

An inductive argument on a subnormal series of isolated subgroups between L_1 and H_1 now yields the following generalisation of Proposition 3.

Theorem 2 (Brookes, Brown [9, Theorem A]). *With the above notation* $V_1 kH \cong V_1 kS_1 \otimes_{kS_1} kH$ *where* $S_1 = \mathrm{Stab}_{N_H(L_1)}[V]$.

The key observations in the proof are illustrated by the case where L_1 is normal in H_1. The argument used to deduce the Corollary from Proposition 1 yields that $\mathrm{Stab}_H[W] = \mathrm{Stab}_N[V] \,.\, H_1$ where $N = N_H(L_1)$. Under the assumption of normality we have $H_1 \leq N$ and so $\mathrm{Stab}_H[W] \leq N$.

It follows from Proposition 3 that $W_1kH \cong W_1kN \otimes_{kN} kH$ and hence $V_1kH \cong V_1kN \otimes_{kN} kH$. We can apply Proposition 3 again, this time with L_1 and N in the roles of H_1 and H, and deduce that $V_1kN \cong V_1kS_1 \otimes_{kS_1} kN$, completing the proof in this special case.

When M is impervious it follows that $\mathrm{Stab}_{N_H(L_1)}[V]$, and hence $N_H(L_1)$, is of finite index in H. But in a finitely generated nilpotent group isolated subgroups having only finitely many conjugates are actually normal. Summing up we have

Theorem 3. *Let M be an impervious indecomposable injective kH-module and H_1 be any isolated normal subgroup. Let W be an indecomposable injective kH_1-submodule minimal in M and V be the corresponding impervious indecomposable injective kL_1-submodule with $[W] = [V \otimes_{kL_1} kH_1]$, for some isolated subgroup L_1 of H_1. Then L_1 is normal in H and V is minimal in M. Moreover* $\mathrm{Stab}_H[V]$ *is of finite index in H.*

What implications for the structure of impervious modules would affirmative answers to Questions 7 and 8 have? There we were concerned only with modules M in which no finite dimensional k-vector space was stabilised by a non-trivial subgroup of H.

The answer 'Yes' to Question 7 would imply that $|\mathrm{Stab}_H[V] : L_1C_H(L_1)|$ is finite for any impervious kL_1-module V arising in Theorem 3 from such a module M. So $L_1C_H(L_1)$ would have finite index in H. Thus, with this assumption, the subgroups of L that could play the role of L_1 would be distinctly limited and there would be a very close relationship between the structure of impervious modules and the abelian subgroups of H.

On the other hand, for torsion-free H let $H_1 = H'$ and assume only that $\mathrm{Stab}_H[V]$, and hence H, acts finitely on the abelianisation L/L'. Since centralisers in the torsion-free nilpotent group H/L' are isolated, we actually have H fixing L/L' pointwise. Therefore H' acts trivially via conjugation on $L/\gamma_3(L)$, where $\gamma_i(L)$ denotes the ith term of the lower central series of L. But L is a subgroup of H' and so L acts trivially on $L/\gamma_3(L)$. In other words $L/\gamma_3(L)$ is abelian. Thus $\gamma_2(L) = \gamma_3(L)$ and L itself is an abelian group. We have therefore deduced that $[W]$ corresponds to the pair $([V], L)$ with L lying in the centre Z of H. If M is kZ-torsion-free, say, then the only possible pair is with $L = 1$ and $[V] = [k]$. In this case $[W] = [kH']$. Thus we deduce from our assumption about stabilisers that M is kH'-torsion-free if it is kZ-torsion-free.

As a result of Proposition 2, these techniques applied with $k = \mathbf{Q}$ or $\mathbf{F}_p$ for some prime p shed light on the structure of finitely presented abelian-by-nilpotent groups. In fact there is no example known of such a group where H is torsion-free of nilpotent class 3 or more, and where the underlying $\mathbf{Z}H$-module M_0 contains a non-zero $\mathbf{Z}H'$-submodule of the form $V \otimes_{\mathbf{Z}L} \mathbf{Z}H'$ for some $\mathbf{Z}L$-module V with $L \leq Z \cap H'$. When the class of H is less than 3 the equality $Z \cap H' = H'$ holds, a case where these methods are redundant. Is it perhaps true that $\gamma_3(G)$ is always locally of finite Prüfer rank for a finitely presented abelian-by-nilpotent group G?

References

[1] L. Auslander and G. Baumslag, Automorphism groups of finitely generated nilpotent groups, *Bull. Amer. Math. Soc.* **73** (1967), 716-717.

[2] S. Bachmuth, G. Baumslag, J. Dyer and H.Y. Mochizuki, Automorphism groups of two generator metabelian groups, preprint.

[3] G.M. Bergman, The logarithmic limit-set of an algebraic variety, *Trans. Amer. Math. Soc.* **157** (1971), 459-469.

[4] R. Bieri and J.R.J. Groves, A rigidity property for the set of all characters induced by valuations, *Trans. Amer. Math. Soc.* **294** (1986), 425-434.

[5] R. Bieri and R. Strebel, Almost finitely presented soluble groups, *Comment. Math. Helv.* **53** (1978), 258-278.

[6] R. Bieri and R. Strebel, Valuations and finitely presented metabelian groups, *Proc. London Math. Soc. (3)* **41** (1980), 439-464.

[7] A. Borel, Arithmetic properties of linear algebraic groups, *Proc. Internat. Congress Math. Stockholm 1962* (1963), 10-22.

[8] C.J.B. Brookes, Modules over polycyclic groups, *Proc. London Math. Soc. (3)* **57** (1988), 88-108.

[9] C.J.B. Brookes and K.A. Brown, Primitive group rings and Noetherian rings of quotients, *Trans. Amer. Math. Soc.* **288** (1985), 605-623.

[10] D.R. Farkas and R.L. Snider, Arithmeticity of stabilizers of ideals in group rings, *Inventiones Math.* **75** (1984), 75-84.

[11] A.W. Goldie, *Rings with maximal condition*, Lecture Notes, Yale Univ. (1964).

[12] P. Hall, Finiteness conditions for soluble groups, *Proc. London Math. Soc. (3)* **4** (1954), 419-436.

[13] P. Hall, On the finiteness of certain soluble groups, *Proc. London Math. Soc. (3)* **9** (1959), 595-622.

[14] P. Hall, *Nilpotent groups*, (Lectures given at the Canadian Math. Congress, Summer Seminar, Univ. of Alberta 1957) Queen Mary College Math. Notes (1969).

[15] A.V. Jategaonkar, *Localization in Noetherian rings*, London Math. Soc. Lecture Notes 98, Cambridge Univ. Press (1986).

[16] G.W. Mackey, On induced representations of groups, *Amer. J. Math.* **73** (1951), 576-592.

[17] Y. Nouazé et P. Gabriel, Ideaux premiers de l'algèbre enveloppante d'une algèbre de Lie nilpotente, *J. Algebra* **6** (1967), 77-99.

[18] R. Rentschler et P. Gabriel, Sur la dimension des anneaux et ensembles ordonnés, *C. R. Acad. Sci. Paris Sér. A* **265** (1967), 712-715.

[19] D.J.S. Robinson, *A course in the theory of groups*, Springer, Berlin (1982).

[20] J.E. Roseblade, The integral group rings of hypercentral groups, *Bull. London Math. Soc.* **3** (1971), 351-355.

[21] J.E. Roseblade, Prime ideals in group rings of polycyclic groups, *Proc. London Math. Soc.* **36** (1978), 385-447.

[22] D. Segal, *Polycyclic groups*, Cambridge Univ. Press (1983).

[23] D.W. Sharpe and P. Vámos, *Injective modules*, Cambridge Univ. Press (1972).

Corpus Christi College
Cambridge CB2 1RH
United Kingdom

Fitting classes after Dark

R.A. Bryce, John Cossey, E.A. Ormerod**

1. Introduction

It is well-known, and easy to prove, that the class of all finite direct powers of a finite, non-abelian simple group, is a Fitting class. That is, it is closed for normal subgroups and for products of normal subgroups. Indeed, if G is a finite, non-abelian simple group, if $\mathcal{X}$ is a Fitting formation and $\mathcal{F}$ a Fitting class, then the class of finite groups

$$\mathcal{D}(G) = \{H : H^{\mathcal{X}}/(H^{\mathcal{X}})_{\mathcal{F}} \in D_0(G)\}$$

is also a Fitting class. The proof of this uses nothing more than the definition of residuals of formations, radicals of Fitting classes, and the fact that $D_0(G)$ is a Fitting class.

Until recently most interest in Fitting classes has lain in the universe of finite soluble groups, and here no simple, general analogue of $\mathcal{D}(G)$ exists: the class of direct powers of an abelian simple group is certainly not a Fitting class. However in Dark [2] we find a construction of Fitting classes with a broad similarity to $\mathcal{D}(G)$. There followed papers by Beidelman and Brewster [1] and Hawkes [3] which elaborated Dark's idea, producing examples of Fitting classes that one has come to term "Dark-type".

More recently, in the as yet unpublished PhD thesis of McCann [5], another Fitting class construction has appeared. It is of special interest because it bears much more directly than the previous examples on the problem of determining the Fitting class generated by a given finite group.

The present report has grown out of an attempt to understand McCann's construction and to relate it to that of Dark. The Fitting classes we construct in §3 have been inspired by McCann's thesis, and are closely based on his, but they are not identical. The changes we have made are motivated by a desire to simplify the definitions; to make the classes as small

* The last two authors acknowledge support of an ARGS grant.

as possible; and to underline their similarity to Dark's classes. The methods used have deep roots in Dr McCann's thesis and we thank him for providing us with a copy of it.

The notation used is generally standard. For $\mathcal{X}$ a Fitting class and G a finite group $G_{\mathcal{X}}$ denotes the $\mathcal{X}$-radical of G. For a formation $\mathcal{F}$, $G^{\mathcal{F}}$ denotes the $\mathcal{F}$-residual. $\mathcal{S}_\pi$ denotes the class of all finite π-groups for a set of primes π and $O_\pi(G)$ and $O^\pi(G)$ denote respectively the $\mathcal{S}_\pi$-radical and $\mathcal{S}_\pi$-residual of the finite group G. When p is a prime number p' denotes the set of primes other than p. $\mathcal{N}$ denotes the class of finite nilpotent groups.

One usage perhaps not standard is this. Suppose that σ is an abstract subgroup function on the class of finite groups. (That is for each finite group G, $\sigma(G)$ is a subgroup of G and, whenever $\alpha : G \to H$ is an isomorphism, $\sigma(H) = \sigma(G)\alpha$.) If N is a normal subgroup of a finite group G then $\sigma(G \div N)$ is that subgroup defined by

$$\sigma(G \div N)/N = \sigma(G/N).$$

Thus for example $\zeta_1(G \div N)$ is the centre of G module N.

2. Dark's construction generalized

The aim of this section is to define Fitting classes which are slight generalizations of those of Dark (Theorem 4 of [2]). It is via these that we explain the close connexion between the Fitting classes of Dark and McCann. The notation of the following definition will be carried through this section.

2.1 Definition. Let Y be a finite soluble group and p a prime. Moreover
(i) $\zeta_1(Y) = 1$;
(ii) Y is monolithic.
Y has an automorphism α of order p with the property that
(iii) $[Y, \alpha] = Y$;
and
(iv) if X is the splitting extension $Y\langle\alpha\rangle$ then, whenever $X_1 \cong X_2 \cong X$ and X_1, X_2 are normal subgroups of a group $H = X_1X_2$ with $\zeta_1(H) = 1$, either $X_1 = X_2$ or $[X_1, X_2] = 1$.

2.2 Theorem (cf. Dark [2]). *Let X, p, be as in* (2.1) *with $p \not| |Y|$, and let $\mathcal{K}$ be a Fitting class satisfying $\mathcal{K}\mathcal{S}_p = \mathcal{K}$. The class of finite soluble groups*

$$\mathcal{D}(\mathcal{K},X) = \{H : O^{p'}(H)/O^{p'}(H)_{\mathcal{K}} \in s_n D_0(X)\}$$

is a Fitting class.

We begin the proof of this theorem now with several lemmas which will be useful later on in §3 as well. This section will conclude with a proof that every Fitting class satisfying Dark's conditions (Theorem 4 of [2]) is of the form $\mathcal{D}(\mathcal{K},X)$ for suitable choice of $\mathcal{K}$ and X.

2.3 Lemma. *$\mathcal{D}(\mathcal{K},X)$ is closed for normal subgroups.*

Proof. Let H_0 be a normal subgroup of a group H in $\mathcal{D}(\mathcal{K},X)$. Then

$$\begin{aligned} O^{p'}(H_0)/O^{p'}(H_0)_{\mathcal{K}} &= O^{p'}(H_0)/O^{p'}(H_0) \cap O^{p'}(H)_{\mathcal{K}} \\ &\cong O^{p'}(H_0)O^{p'}(H)_{\mathcal{K}}/O^{p'}(H)_{\mathcal{K}} \\ &\trianglelefteq O^{p'}(H)/O^{p'}(H)_{\mathcal{K}} \\ &\in s_n D_0(X). \end{aligned}$$

Hence, by definition, $H_0 \in \mathcal{D}(\mathcal{K},X)$, as required.

2.4 Lemma. *Let H be a finite soluble group, N a normal subgroup of H and S a sub-normal subgroup of H of defect at most 2. Moreover suppose that*

(i) *$N = \times_{j=1}^{r} Y_j$ is the unique, unrefinable direct decomposition of N;*

(ii) *$S \leq O^2(H)$;*

and (iii) *for some $i \in \{1,2,\dots,r\}$, $S \cong Y_i$.*

Then S normalizes Y_i.

Proof. For each $h \in H$, conjugation by h induces a permutation $\pi(h)$ on the set $\{1,2,\dots,r\}$ defined by $i\pi(h) = j$ if and only if $Y_i^h = Y_j$. This follows from the hypothesis (i). Indeed $\pi : H \to Sym(r)$ so defined is a homomorphism. The conclusion of the theorem will follow from the following fact.

2.5 *For $x \in S$, $\{i\}$ is an orbit of $\pi(x)$.*

We prove (2.5) in stages, first showing that (cf. Lemma 6 of Dark [2])

2.6 *if $x \in S$ then the orbit of $\pi(x)$ containing i has length less than* 3.

To this end define, for given $x \in S$, the subgroup

$$M = [Y_i, x, x].$$

Suppose that the orbit of $\pi(x)$ containing i has length ≥ 3. Then for $y \in Y_i$,

$$[y, x, x] = yy^{-2x}y^{x^2}$$

so we conclude that M projects fully onto Y_i in the direct decomposition (i). Hence

$$|M| \geq |Y_i| = |S|.$$

However $M \leq S$ since S is 2-step subnormal. Therefore $M = S$. But $M \leq N$, so Y_i is normalized by S. This contradicts the supposition that i is in an orbit of $\pi(x)$ of length ≥ 3.

2.7 *If $x \in S$ and i is in an orbit of length* 2 *for $\pi(x)$ then $O^2(M) = O^2(S)$.*

For, in this case,

$$[y, x, x] = (yy^{x^2})(y^{-x})^2$$

and it follows that $O^2(M)$ projects onto $O^2(Y_{i\pi(x)})$. Hence $|O^2(M)| \geq |O^2(Y_i)|$. However $M \leq S$ so $O^2(M) \leq O^2(S)$. Also, by hypothesis $|O^2(S)| = |O^2(Y_i)|$. It follows that $|O^2(M)| = |O^2(S)|$ and therefore $O^2(S) = O^2(M)$ as required.

The proof of (2.5) is completed by observing that since $S \leq O^2(H)$, and since $O^2(H)$ is generated by elements of odd order, $\pi(S)$ consists of even permutations. Hence if $x \in S$ and if i is in an orbit of length two for $\pi(x)$, for some $j \in \{1, 2, \ldots, r\} \backslash \{i, i\pi(x)\}$, j is in an orbit of length at least two for $\pi(x)$. The arguments in 2.6 and 2.7 show that if

$$M_0 = [Y_j, x, x]$$

then

$$|O^2(Y_j)| \leq |O^2(M_0)| \text{ and } O^2(M_0) \leq O^2(S) = O^2(M).$$

However from the direct decomposition (i) we deduce that $O^2(M_0) \cap O^2(M) = 1$ whence $O^2(M_0) = 1$ and so $O^2(Y_j) = 1$. This is a contradiction to the hypothesis that the direct decomposition in (i) is unique: if Y_j, and therefore $Y_{j\pi(x)}$, were a 2-group the decomposition $Y_j \times Y_{j\pi(x)}$ would not be unique.

This completes the proof of (2.5) and with it that of (2.4).

2.8 Lemma. *Let H be a finite soluble group. Suppose that N, K are normal subgroups of H satisfying the following conditions.*

(i) $N = \times_{j=1}^{r} Y_j$ *where each Y_j is monolithic with non-central monolith.*

(ii) *For some $i \in \{1, \ldots, r\}$,*

$$K = \mathop{\times}_{j=1}^{s} W_j$$

where $W_j \cong Y_i (1 \leq j \leq s)$ and $K \leq O^2(H)$.

If $x \in \mathbf{N}_H(W_j)(1 \leq j \leq s)$ and $[Y_i, x] \leq K$ then $x \in \mathbf{N}_H(Y_i)$.

Proof. Condition (i) ensures that the given direct decomposition is unique. Then (2.4) yields that K normalizes Y_i. Suppose $x \in H$ normalizes all W_j but does not normalize Y_i. Conjugation by x induces a permutation on the set $\{Y_1, \ldots, Y_r\}$, and

$$\begin{aligned} 1 \neq [Y_i, x] &\leq (Y_i \times Y_i^x) \cap K \\ &= K_0, \end{aligned}$$

say, where

$$K_0 \trianglelefteq Y_i \times Y_i^x.$$

The projection of K_0 into Y_i (or into Y_i^x) is a non-trivial normal subgroup and therefore contains the monolith T of Y_i. Since T is non-central,

$$T = [T, Y_i] \leq [K_0, Y_i] \leq K_0 \leq K.$$

Also $T \trianglelefteq K$ since K normalizes Y_i. Then the projection of T onto some W_j is a non-trivial normal subgroup which therefore contains the monolith, U say, of W_j. Hence

$$U = [U, W_j] \leq [T, W_j] \leq T$$

so $U = T$ since they have the same order.

Finally observe that x normalizes W_j, therefore it normalizes T, which contradicts the supposition that $Y_i \cap Y_i^x = \{1\}$. This completes the proof of (2.8).

2.9 Lemma. *Let X be as defined in (2.1). The class of groups*

$$\mathcal{C}(X) = \{R : O^{p'}(R) = R,\ \ O^p(R) = Y_1 \times \ldots \times Y_t,\ \ Y \cong Y_j \trianglelefteq R,$$

$$R/\mathbf{C}_R(Y_j) \cong X,\ \ 1 \le j \le t\}$$

is normal product closed.

Proof. Let $R = R_1 R_2$ where $R_i \trianglelefteq R$ and $R_i \in \mathcal{C}(X)$ $(i = 1, 2)$. Then

$$O^p(R_i) = \mathop{\times}_{j=1}^{t(i)} Y_{ij}$$

where $Y \cong Y_{ij} \trianglelefteq R_i$, and $R_i/\mathbf{C}_{R_i}(Y_{ij}) \cong X$ $(i = 1, 2, 1 \le j \le t(i))$.

First we use (2.8) to show that $Y_{1j} \trianglelefteq R$ $(1 \le j \le t(1))$, and to do this we need only show that Y_{1j} is normalized by R_2. Suppose, on the contrary, that $x \in R_2$ does not normalize Y_{1j}. Then since conjugation by x permutes $\{Y_{11}, Y_{12}, \ldots, Y_{1t(1)}\}$, $[Y_{1j}, x] \cong Y_{1j}$ is generated by elements of order prime to p. Also $[Y_{1j}, x] \le R_2$, so $[Y_{1j}, x] \le O^p(R_2)$. But now by (2.8), $x \in \mathbf{N}_R(Y_{1j})$ a contradiction. Hence $Y_{1j} \trianglelefteq R$ $(1 \le j \le t(1))$ and symmetrically $Y_{2k} \trianglelefteq R$ $(1 \le k \le t(2))$.

Next we show:

2.10 *For $1 \le j \le t(1)$, $1 \le k \le t(2)$, either $[Y_{1j}, Y_{2k}] = 1$ or $Y_{1j} = Y_{2k}$.*

For $1 \le k \le t(2)$ either $R_1 \cap Y_{2k} = 1$ or $R_1 \cap Y_{2k} \ne 1$. In the first case $[Y_{1j}, Y_{2k}] = 1$ for all $j \in \{1, \ldots, t(1)\}$. So consider the second possibility. Let T be the socle of Y_{2k}. Then $T \le R_1$ and, since $Y_{2k} \trianglelefteq R$, $T \trianglelefteq O^p(R_1)$. Now T has non-trivial normal projection in some Y_{1j}, and the argument used in the proof of (2.8) shows that T is the socle of Y_{1j} and so T, and therefore $R_1 \cap Y_{2k}$, projects trivially into all other Y_{1l}.

Let

$$Z = \mathbf{C}_R(Y_{1j}).$$

Since T is not central in Y_{1j} it follows that $Z \cap Y_{1j} = 1$; and from this we deduce that $Z \cap Y_{2k} = 1$. In particular $Z \leq \mathbf{C}_R(Y_{2k})$. A symmetrical argument shows that $\mathbf{C}_R(Y_{2k}) \leq Z$, so $Z = \mathbf{C}_R(Y_{2k})$. It follows that R/Z is a normal product of R_1Z/Z and R_2Z/Z, both of which by hypothesis are isomorphic to X. Moreover $\varsigma_1(R/Z) = 1$. For, if $W = \varsigma_1(R \div Z)$ then $[W, Y_{1j}] \leq Z \cap Y_{1j} = 1$ so $W = Z$. By (iv) of (2.1) therefore, $R_1Z/Z = R_2Z/Z$. In particular

$$Y_{1j}Z = Y_{2k}Z.$$

It follows that

$$Y_{1j}Y_{2k} = Y_{1j} \times (Y_{1j}Y_{2k} \cap Z);$$

and that

$$Y_{1j}Y_{2k} \cap Z \cong Y_{1j}Y_{2k}/Y_{1j} \cong Y_{2k}/Y_{1j} \cap Y_{2k}.$$

However $Y_{1j}Y_{2k} \cap Z$ is abelian, since it is central in $Y_{1j}Y_{2k}$. Also $Y_{2k}/Y_{2k} \cap Y_{1j}$ must be a p'-group, by (iii) of (2.1). Hence

2.11 $$Y_{1j}Y_{2k} \cap Z \leq O_{p'}(O^p(R)).$$

Also since $R_1 \cap Y_{2k}$ projects trivially onto all Y_{1l} except Y_{1j} it follows that

$$\begin{aligned} R_1 \cap Y_{1j}Y_{2k} \cap Z &= Y_{1j}(R_1 \cap Y_{2k}) \cap Z \\ &= Y_{1j} \cap Z \\ &= 1; \end{aligned}$$

and similarly $R_2 \cap Y_{1j}Y_{2k} \cap Z = 1$, so

2.12 $$Y_{1j}Y_{2k} \cap Z \leq \varsigma_1(R).$$

By re-numbering if necessary let $Y_{21}, Y_{22}, \ldots, Y_{2s}$ have the property that $R_1 \cap Y_{2l} = 1$ $(1 \leq l \leq s)$ whilst $R_1 \cap Y_{2l} \neq 1$ $(s+1 \leq l \leq t(2))$. Then for $t = t(1) + s$ define

2.12.1 $$Y_i = \begin{cases} Y_{1i}, & 1 \leq i \leq t(1), \\ Y_{2,i-t(1)}, & t(1)+1 \leq i \leq t. \end{cases}$$

From this definition it follows that

$$[Y_i, Y_l] = 1, \quad 1 \leq i < l \leq t$$

and from (2.11) and (2.12) it follows that for some central, p'-subgroup L of R,

$$O^p(R) = Y_1 \times Y_2 \times \ldots \times Y_t \times L,$$

the product being direct because the Y_i's have trivial centre. But then $R/Y_1Y_2\ldots Y_t$ is nilpotent and $LY_1Y_2\ldots Y_t/Y_1Y_2\ldots Y_t$ is a p'-factor group of it. Since $O^{p'}(R) = R$ we deduce that

$$O^p(R) = Y_1 \times Y_2 \times \ldots Y_t,$$

and, in fact, that $L = 1$.

In order to conclude that $R \in \mathcal{C}(X)$ we need to show that for $i \in \{1,\ldots,t\}$, $R/\mathbf{C}_R(Y_i) \cong X$. This is almost immediate. In the case that $Y_i \cap R_1 \cap R_2 \neq 1$ we have seen above that $R/\mathbf{C}_R(Y_i) \cong X$; and in the other case we have, say, $Y_i \leq R_1$ and $Y_i \cap R_2 = 1$ so that

$$\begin{aligned} R/\mathbf{C}_R(Y_i) &\cong R_1/\mathbf{C}_{R_1}(Y_i) \\ &\cong X. \end{aligned}$$

This concludes the proof of (2.9).

2.13 Lemma. *$\mathcal{D}(\mathcal{K},X)$ is closed for normal products.*

Proof. Let $H = H_1H_2$ where $H_i \trianglelefteq H$ and $H_i \in \mathcal{D}(\mathcal{K},X)$ $(i = 1, 2)$. Write $S = O^{p'}(H)$, $S_i = O^{p'}(H_i)$ $(i = 1,2)$. Then

$$R_i = S_i/(S_i)_{\mathcal{K}} \in s_n D_0(X), \quad i = 1,2.$$

Since $O^{p'}(R_i) = R_i$ it follows that if R_i is expressed as a subnormal subgroup of a direct product $X_1 \times X_2 \times \ldots \times X_r$ of copies of X then no projection of it into any X_j, P_j say, with $P_j \neq 1$ can be into $O^p(X_j)$. Moreover $O^{p'}(P_j) = P_j$ and $P_j \triangleleft\triangleleft X_j$. The next lemma is more than enough to show that, in fact, $P_j = X_j$ $(1 \leq j \leq r)$.

2.14 Lemma. *Let X be as in (2.1), and let P be a subnormal subgroup of X satisfying $O^{p'}(P) = P$, $P \not\leq Y$. Then $P = X$.*

Proof. Let $Y_0 = Y > Y_1 > \ldots > Y_n = 1$ be the descending p-series of Y: $Y_0/Y_1 \in \mathcal{S}_{p'}$; $Y_1/Y_2 \in \mathcal{S}_p$ and so on. We have already seen, it follows from (2.1)(iii), that $O^p(Y) = Y$; and of course $O^{p'}(X) = X$.

Since $P \not\leq Y$ we have

$$PY_0 = X.$$

We show by induction on i that

$$\textbf{(2.15)} \qquad PY_i = X, \quad 0 \leq i \leq n.$$

Suppose $i < n$ and that $Y_i/Y_{i+1} \in \mathcal{S}_{p'}$. Then if (2.15) holds, by (3.3) of the next section we have, modulo Y_{i+1}, $P = O^{p'}(X) = X$. That is $PY_{i+1} = X$. And if $Y_i/Y_{i+1} \in \mathcal{S}_p$ and (2.15) holds we have

$$Y = O^p(X) = O^p(P)Y_i$$

so by (3.3) again, modulo Y_{i+1}, $O^p(P) = O^p(X) = Y$. That is $O^p(P)Y_{i+1} = Y$ which yields

$$X = PY = PY_{i+1}.$$

In any case the inductive step is complete and (2.14) is proved.

We return to the proof of (2.13). It follows now that each $R_i \in \mathcal{C}(X)$ in the notation of (2.9). But $S = S_1S_2$ and $S_i \cap S_{\mathcal{K}} = (S_i)_{\mathcal{K}}$ so $S/S_{\mathcal{K}}$ is a normal product of groups in $\mathcal{C}(X)$. By (2.9), therefore, $S/S_{\mathcal{K}}$ is in $\mathcal{C}(X)$, say $R = S/S_{\mathcal{K}}$ and

$$O^p(R) = Y_1 \times Y_2 \times \ldots \times Y_t$$

with $R/\mathbf{C}_R(Y_j) \cong X$ $(1 \leq j \leq t)$. Let

$$C = \bigcap_{j=1}^{t} \mathbf{C}_R(Y_j).$$

Since $\zeta_1(Y_1 \times \ldots \times Y_t) = 1$ we have $C \cap Y_1Y_2 \ldots Y_t = 1$. This means that $C \in \mathcal{S}_p$ since $R/Y_1Y_2 \ldots Y_t \in \mathcal{S}_p$. However $\mathcal{K} = \mathcal{K}\,\mathcal{S}_p$ so $C = 1$, and it follows that $R \in s_n D_0(X)$ as required.

This completes the proof of (2.13) and therefore that of (2.2).

To conclude this section we prove that the Fitting classes defined by Dark (Theorem 4 in [2]) are all of the form $\mathcal{D}(\mathcal{K}, X)$ for suitable choice of $\mathcal{K}$

and X. All we need to do is to verify that his conditions imply (iv) of (2.1). In fact we do a little better.

2.16 Theorem. *Let Y,α satisfy conditions* (i), (ii) *and* (iii) *of* (2.1). *Suppose moreover that, with $X = Y\langle\alpha\rangle$ and regarding X as a subgroup of* Aut Y, *we also have*

(iv)′ *whenever $Y \cong Y_1 \le \text{Aut } Y$ and $Y \le \mathbf{N}_{\text{Aut } Y}(Y_1)$, $Y_1 = Y$; and whenever $\varphi \in \text{Aut}\, Y \backslash Y$ is such that $\langle\alpha, \alpha^{\varphi}\rangle$ is a p-group, then $\langle\alpha\rangle = \langle\alpha^{\varphi}\rangle$.*

Then (iv) *of* (2.1) *is also satisfied.*

Proof. To this end suppose that X satisfies the hypotheses of the theorem, that $H = X_1X_2$ where $X \cong X_i \trianglelefteq H$ $(i = 1,2)$, and that $\zeta_1(H) = 1$. We need to show that $X_1 = X_2$ or $[X_1, X_2] = 1$.

Let $Y_i = O^p(X_i)$ and let α_i be the automorphism of order p of Y_i required by (2.1). If $X_1 \cap X_2 = 1$ then $[X_1, X_2] = 1$, so we may suppose that $X_1 \cap X_2 \ne 1$. Then $Y_1 \cap Y_2 \ne 1$, and the argument used in (2.10) shows that $\mathbf{C}_H(Y_1) = \mathbf{C}_H(Y_2) = Z$, say. But $Z \cap X_i = 1$ $(i = 1,2)$, so $Z \le \zeta_i(H) = 1$.

Hence we may regard H as a subgroup of Aut Y_1 and as such Y_2 is a subgroup of it isomorphic to Y_1 and normalized by Y_1. Therefore by (iv)′, $Y_1 = Y_2$. Now there is an isomorphism $\psi : X_1 \to X_2$. Since $\alpha_1\psi$ is a p-element of X_2, there is an inner automorphism θ of Y_2 such that $\alpha_1\psi\theta$ and α_1 belong to the same Sylow p-group of H: thus we may suppose ψ is chosen so that $\langle\alpha_1, \alpha_1\psi\rangle$ is a p-group.

Let $\varphi = \psi | Y_1$, an automorphism of Y_1. Then for $y \in Y_1$,

$$\begin{aligned}(\alpha_1^{-1}y\alpha_1)\varphi &= (\alpha_1^{-1}y\alpha_1)\psi \\ &= (\alpha_1\psi)^{-1}y\varphi(\alpha_1\psi)\end{aligned}$$

which is to say that, regarded as elements of Aut Y_1,

$$\alpha_1\varphi = \varphi\alpha_1\psi$$

whence $\alpha_1^{\varphi} = \alpha_1\psi$. By (iv)′ therefore $\langle\alpha_1\rangle = \langle\alpha_1\psi\rangle$, so $X_1 = X_2$, as required.

One final comment: if, in $\mathcal{D}(\mathcal{K}, X)$, the prime dividing the order of the socle of X is in *char* $\mathcal{K}$ then $\mathcal{D}(\mathcal{K}, X) = \mathcal{K}S_{p'}$.

3. McCann's classes

Central to the Dark-type Fitting classes of the last section is the group X. The classes of McCann which we describe in this section also involve such a group playing a similar role.

3.1 Definition. Let p, q, r be primes satisfying $p \neq q \neq r$ (though $p = r$ is allowed), and let $\mathcal{K}$ be a Fitting class contained in $\mathcal{S}_{q'}$.

G is a group with a series of normal subgroups

$$N < U < B < G$$

where:

(i) $N = G_{\mathcal{K}}$;

(ii) U/N is an r-group, B/U is a q-group and $|G/B| = p$;

(iii) $O^{q'}(B) = B$;

(iv) $U/\Phi(U \div N)$ and $B/\Phi(B \div U)$ are faithful and irreducible for the action of G/U and G/B respectively; and

(v) $X = G/\Phi(U \div N)$ satisfies condition (iv) of (2.1).

For convenience we denote $U^* = \Phi(U \div N)$.

In fact it is worth noting that G/U^* is a special case of the group X of (2.1): since U/U^* is complemented in G/U^* (Higman [4]) B/U^* has a complement of order p. Thus if $\alpha \in G/U^* \backslash B/U^*$ has order p it satisfies (iii) of (2.1) with $Y = B/U^*$.

Also note that in another sense G/U^* is more general than the group X that goes into the definition of $\mathcal{D}(\mathcal{K}, X)$: p and r here are allowed to be equal.

Note the following easy fact.

3.2 Lemma *If G satisfies* (3.1) *then* $\zeta_1(G \div N) \leq U^*$.

Proof. Since U/U^* is the non-central unique minimal normal subgroup of G/U^* we could not have $\zeta_1(G \div N) \not\leq U^*$.

Before defining the Fitting classes which are the subject of this section we prove the following useful result of McCann's.

3.3 Lemma ([5, Lemma III.4]). *Let K, L be subgroups of a group H satisfying $H = KL$; K is subnormal, and L normal in H; for some set π of primes $O^\pi(K) = K$ and $L \in S_\pi$. Then $K = O^\pi(H)$.*

Proof. The proof uses induction on $|H|$. When $H = K$ the result is immediate, so suppose that $K < H$. Since K is subnormal in H there is a proper normal subgroup H_0 of H containing K, and with $L_0 = H_0 \cap L$ we have

$$K \lhd\lhd H_0, \quad L_0 \trianglelefteq H_0, \quad H_0 = KL_0;$$

and, of course

$$O^\pi(K) = K, \quad L_0 \in S_\pi.$$

Now $O^\pi(H) \leq H_0$ and therefore $O^\pi(H) = O^\pi(H_0)$. Hence $O^\pi(H_0) = K$ implies $O^\pi(H) = K$, and this inductive step completes the proof.

In the definition to be given now, and in the proofs which follow, we adopt a familiar and useful convention. Isomorphic copies of the group B of (3.1) will be denoted B_i $(1 \leq i \leq t)$ with the understanding that $b \to b_i$ $(b \in B)$ is a given isomorphism $B \to B_i$ $(1 \leq i \leq t)$. We use the logical consequences of this convention: thus U_i, N_i, for example, denote the images of U and N under this isomorphism.

3.4 Definition (cf. McCann [5], Construction IV.2, p.64). Using the notation of (3.1) define, for a group H, the following subgroups:

$$V = O^{p'}(H), \quad W = O^p(V)(= V^{\mathcal{N}}), \quad Z = W_{\mathcal{K}}.$$

The class $\mathcal{M}(\mathcal{K}, G)$ consists of all groups H satisfying conditions (i), (ii), (iii).

(i) $W = B_1 B_2 \ldots B_t Z$, $B \cong B_i \lhd\lhd W (1 \leq i \leq t)$;

(ii) $B_i Z/Z \trianglelefteq V/Z$, $[B_i, B_j] \leq Z$, $1 \leq i < j \leq t$.

From (3.3) it follows that $B_i \trianglelefteq V$ $(1 \leq i \leq t)$: $B_i Z \trianglelefteq V$, $B_i \lhd\lhd V$ and $O^{q'}(B_i) = B_i$ so $B_i = O^{q'}(B_i Z) \trianglelefteq V$. In particular $U_i, U_i^* \trianglelefteq V$ $(1 \leq i \leq t)$.

(iii) Let $D_i = \mathbf{C}_H(B_i/U_i^*)$. Then $V/D_i \cong G/U^*$.

In (i) $t = 0$ is possible: indeed $\mathcal{K} S_p S_{p'} \subseteq \mathcal{M}(\mathcal{K}, G)$ is part of the definition.

3.5 Theorem (cf. McCann [5]). *Let G, $\mathcal{K}$ be as in* (3.4). *Then $\mathcal{M}(\mathcal{K},G)$ is a Fitting class.*

The proof of this theorem falls, of course, into two parts. Normal subgroup closure is proved in §4 and normal product closure in §5.

4. $\mathcal{M}(\mathcal{K},G)$ is normal subgroup closed

We begin with two preliminary lemmas.

4.1 Lemma. *Let $H \in \mathcal{M}(\mathcal{K},G)$. In the notation of* (3.4), $Z \leq D_i$ $(1 \leq i \leq t)$.

Proof. For,

$$\begin{aligned}[Z, B_i] &\leq Z \cap B_i \\ &= N_i \\ &\leq U_i^*\end{aligned}$$

so $Z \leq \mathbf{C}_H(B_i/U_i^*) = D_i$.

4.2 Lemma ([5], Cor. II.15). *Let H be a finite soluble group and suppose that*

$$H^{\mathcal{N}} = \prod_{i=1}^{k} A_i$$

where $A_i \trianglelefteq H$ $(1 \leq i \leq k)$. If $K \trianglelefteq H$ then

$$K^{\mathcal{N}} = \prod_{i=1}^{k} (K^{\mathcal{N}} \cap A_i).$$

Proof. For a positive integer s

$$K^{\mathcal{N}} = [K^{\mathcal{N}}, sK] \leq [H^{\mathcal{N}}, sK] \leq \prod_{i=1}^{k} [A_i, sK].$$

Choose s so that $K^{\mathcal{N}} = \underbrace{[K, K, \ldots, K]}_{s \text{ copies}}$. Then

$$K^{\mathcal{N}} \leq \prod_{i=1}^{k} [A_i, sK] \leq \prod_{i=1}^{k} (K^{\mathcal{N}} \cap A_i) \leq K^{\mathcal{N}}$$

which yields the equality sought.

Now let $H_0 \trianglelefteq H \in \mathcal{M}(\mathcal{K}, G)$. Throughout the remainder of the proof the notation of (3.1) and (3.4) will be used without further comment; and recall that subscripts attached to B indicate a fixed isomorphism and U_i, for example, is the image of U in this isomorphism $B \to B_i$.

We must show that $H_0 \in \mathcal{M}(\mathcal{K}, G)$. We may as well suppose $t \geq 1$ since otherwise $H \in \mathcal{K} S_p S_{p'} \subseteq \mathcal{M}(\mathcal{K}, G)$, and $\mathcal{K} S_p S_{p'}$ is certainly normal subgroup closed.

Since for $1 \leq i \leq t$, with $C_i = \mathbf{C}_H(U_i/U_i^*)$ (and note $D_i U_i = C_i$)

$$V_0 C_i / C_i \trianglelefteq V/C_i \cong G/U$$

we have one of two possibilities. Either

(a) $V C_i / C_i$ is a q-group

or (b) $V_0 C_i / C_i = V/C_i$.

In case (a), $V_0 / V_0 \cap C_i$ is a q-group whence $V_0 = O^{p'}(H_0) \leq V_0 \cap C_i$, so

4.3 $$V_0 \leq C_i.$$

In case (b) we show that

4.4 $$B_i \leq W_0.$$

For,

$$B_i \trianglelefteq V = V_0 C_i,$$

whence

$$B_i = O^p(B_i) \leq O^p(V_0) O^p(C_i) \leq W_0 (C_i \cap W).$$

Using (4.2) we deduce that since $W_0 = V_0^{\mathcal{N}}$, $W = V^{\mathcal{N}}$ and $V_0 \trianglelefteq V$,

$$B_i \le \prod_{j=1}^{t} (B_j \cap W_0)(C_i \cap W).$$

(Recall that $Z \le C_i$ by (4.1), so $Z \cap W \le C_i \cap W$.) By Dedekind's Law we have

$$B_i = (B_i \cap W_0)\Big[B_i \cap \prod_{j \ne i} (B_j \cap W_0)(C_i \cap W)\Big].$$

However for $j \ne i$, B_j centralizes B_i modulo Z by (ii) of (3.4), so

$$B_i \le (B_i \cap W_0)(C_i \cap W).$$

Using Dedekind's Law again we conclude that

$$\begin{aligned} B_i &= (B_i \cap W_0)(B_i \cap C_i \cap W) \\ &= (B_i \cap W_0)U_i. \end{aligned}$$

Finally,

$$\begin{aligned} U_i &= [U_i, B_i] \\ &= [U_i, (B_i \cap W_0)U_i] \\ &= [U_i, B_i \cap W_0]U_i' \\ &\le [U_i, B_i \cap W_0]N_i \end{aligned}$$

since $\quad U_i' \le U_i^* = \Phi(U_i \div N_i)$,
and so

$$U_i \le (B_i \cap W_0)N_i,$$

whence

$$B_i = (B_i \cap W_0)N_i.$$

However $B_i/B_i \cap W_0 \cong N_i/B_i \cap W_0 \cap N_i \in \mathcal{S}_{q'}$, so by (iii) of (3.1)

$$B_i \le B_i \cap W_0,$$

whence

$$B_i \le W_0$$

to confirm (4.4).

Let us suppose, by renumbering if necessary, that 1, 2, ..., s come under case (b) and $s+1, \dots, t$ under case (a). Then

$$(4.5) \qquad W_0 = B_1 B_2 \dots B_s (W_0 \cap B_{s+1} \dots B_t Z).$$

Now

$$\begin{aligned} W_0 \cap B_{s+1} \dots B_t Z &\leq \bigcap_{j=s+1}^{t} C_j \cap B_{s+1} \dots B_t Z \\ &= \bigcap_{j=s+1}^{t-1} C_j \cap B_{s+1} \dots B_{t-1} Z U_t \end{aligned}$$

since

$$\begin{aligned} C_t \cap B_{s+1} \dots B_t Z &= B_{s+1} \dots B_{t-1} Z (C_t \cap B_t) \\ &= B_{s+1} \dots B_{t-1} Z U_t. \end{aligned}$$

Repeated application of this step yields

$$W_0 \cap B_{s+1} \dots B_t Z \leq U_{s+1} \dots U_t Z,$$

and from (4.5) we obtain

$$W_0 = B_1 B_2 \dots B_s S,$$

where $S = W_0 \cap U_{s+1} \dots U_t Z$. Now modulo $B_1 \dots B_s Z$, S is an r-group centralized by V_0. In the case $r \neq p$ this means that V_0 has an r-factor group, a contradiction unless $S \leq B_1 \dots B_s Z$; and when $r = p$, W_0 has a p-factor group unless $S \leq B_1 \dots B_s Z$. In any case

$$W_0 \leq B_1 B_2 \dots B_s Z,$$

from which we deduce that

$$W_0 = B_1 B_2 \dots B_s Z_0$$

where $Z_0 = W_0 \cap Z = (W_0)_K$.

It remains to observe that conditions (i), (ii) of (3.4) are satisfied by H_0 immediately; and condition (iii) is because 1,2, ..., s come under property (b) above.

This concludes the proof that $\mathcal{M}(\mathcal{K},G)$ is normal subgroup closed. It is perhaps worth pointing out that we have not used, so far, property (v) of (3.1). It is crucial to the proof of normal product closure which we take up now.

5. $\mathcal{M}(\mathcal{K},G)$ is normal product closed

Suppose now that $H = H_1H_2$ where $H_i \trianglelefteq H$ and $H_i \in \mathcal{M}(\mathcal{K},G)$ $(i = 1,2)$. We are to show that $H \in \mathcal{M}(\mathcal{K},G)$. We use the notation of Definitions (3.1) and (3.4) with appropriate subscripts. Thus, as usual,

$$V = O^{p'}(H),\ W = O^{p}(V),\ Z = W_{\mathcal{K}},$$

$$V_i = O^{p'}(H_i),\ W_i = O^{p}(V_i),\ Z_i = (W_i)_{\mathcal{K}} \quad (i = 1,2);$$

$$W_i = \prod_{j=1}^{t(i)} B_{ij}Z_i, \qquad i = 1,2,$$

where $B_{ij} \cong B$, and so on. Note that $Z_i = W_i \cap Z$, $V = V_1V_2$, $W = W_1W_2$.

We may as well suppose $t(1) > 0$ since if both $t(1) = 0$ and $t(2) = 0$ then $H = H_1H_2 \in \mathcal{K}\,S_p\,S_{p'} \subseteq \mathcal{M}(\mathcal{K},G)$.

A further piece of notation we use is

$$P_i = \Phi(W_iZ \div Z), \quad i = 1,2.$$

The significance of P_i is summed up in the next lemma.

5.1 Lemma. (a) $$W_iP_i/P_i = \mathop{\times}_{j=1}^{t(i)} B_{ij}P_i/P_i,$$

and this is the unique, unrefinable direct decomposition of W_iP_i/P_i (apart from order of the direct factors).

(b) *There is a homomorphism $\pi_i : H \to Sym(t(i))$ satisfying*

$$j\pi_i(h) = k \quad \textit{if and only if} \quad B_{ij}^hP_i/P_i = B_{ik}P_i/P_i.$$

(c) *For all $h \in H$,*

$$B_{ij}^hP_i/P_i = B_{ik}P_i/P_i \quad \textit{if and only if} \quad B_{ij}^h = B_{ik}.$$

Proof. The proof of (a) is an immediate consequence of the hypothesis (3.1) and the fact that each $B_{ij}P_i/P_i$ has trivial centre by (3.2).

(b) Since each $h \in H$ induces an automorphism on W_iP_i/P_i by conjugation, and since the given decomposition of W_iP_i/P_i is unique we get (b) immediately.

(c) One direction is immediate. For the other direction note that by Lemma 3.3

$$B_{ij}^h = O^{q'}(B_{ij}^h P_i) = O^{q'}(B_{ik}P_i) = B_{ik},$$

as required.

5.2 Lemma. *For $1 \leq j \leq t(i)$, B_{ij} is normalized by $F(H \div P_i)$*

Proof. By (5.1) (c) it suffices to show that $B_{ij}P_i/P_i$ is normalized by $F(H/P_i)$. Suppose, on the contrary, that for some $h \in F(H \div P_i)$, hP_i does not normalize $B_{ij}P_i/P_i$. Then for some $k \neq j$,

$$B_{ij}^h P_i/P_i = B_{ik}P_i/P_i,$$

and the subgroup

$$M/P_i = [B_{ij}P_i/P_i, hP_i]$$

is subdirect in

$$B_{ij}P_i/P_i \times B_{ik}P_i/P_i.$$

However $M/P_i \leq F(H/P_i)$ which is nilpotent, whilst M/P_i is not, a contradiction.

5.3 Lemma. *For $1 \leq i \leq 2$,*

$$\Phi(W \div Z) \cap W_i = P_i.$$

Proof. We may as well suppose $i = 1$. Let

$$Q = \prod_{j=1}^{t(1)} U_{1j}$$

(extending our subscript convention on B to double subscripts). Then $Q \trianglelefteq W$. It suffices to show that

5.4 $$\Phi(QZ \div Z) = W_1 \cap \Phi(W \div Z).$$

For, $\Phi(QZ/Z) \leq \Phi(W/Z)$ since $QY/Y \trianglelefteq W/Y$. Hence

$$\Phi(QZ \div Z) \leq \Phi(W \div Z) \cap W_1.$$

To obtain the other inclusion note that by (5.2), each B_{1j} is normalized by $F(W \div Z)$. Hence $U_{1j}/\Phi(U_{1j} \div N_{1j})$ is an irreducible module for $W_1 F(W \div Z)$. It follows that $O_r(W \div Z)$ centralizes $U_{1j}/\Phi(U_{1j} \div N_{1j})$. Consequently the W/Z-module $Q/\Phi(QZ \div Z)$ is completely reducible. Hence

$$Q \cap \Phi(W \div Z) \leq \Phi(QZ \div Z).$$

Finally, of course,

$$\begin{aligned} W_1 \cap \Phi(W \div Z) &\leq Q \cap \Phi(W \div Z) \\ &\leq \Phi(QZ \div Z) \end{aligned}$$

establishing the reverse inclusion, so (5.4) is verified.

5.5 Lemma. $H \in \mathcal{M}(\mathcal{K}, G)$.

Proof. Let $P = P_1 P_2$. Then H/P is a normal product of H_1P/P and H_2P/P and in particular $O^{p'}(H/P) = V/P$ is a normal product of $O^{p'}(H_1P/P)$ and $O^{p'}(H_2P/P)$. Now by (5.3)

$$W_i \cap P = P_i, \quad i = 1, 2,$$

so

$$\begin{aligned} V_iP/P &= O^{p'}(H_iP/P) \\ &\in \mathcal{C}(G/U^*) \end{aligned}$$

in the notation of (2.9). It follows that $O^{p'}(H/P) \in \mathcal{C}(G/U^*)$.

Then from (2.9), and more particularly (2.12.1), It follows that there is a subset $\{B_1, B_2, \ldots, B_t\}$ of $\{B_{ij} : 1 \leq i \leq 2, \quad 1 \leq j \leq t(i)\}$ such that

5.6 $$W = B_1 B_2 \ldots B_t P, \quad \prod_{i=1}^{t} B_i = \prod_{ij} B_{ij},$$

and

5.7 $$[B_i, B_j] \leq P, \quad 1 \leq i < j \leq t.$$

Finally we prove that $[B_i, B_j] \leq Z$ $(1 \leq i < j \leq t)$. First, (5.7) gives $B_j \leq \mathbf{N}_H(B_i)$ by (5.1)(c). Let S_j be a Sylow q-subgroup of B_j. By (5.7) we have $S_j \leq D_i$ from which we deduce $S_j \leq \mathbf{C}_H(B_i/N_i)$ since U_i/N_i is an r-group and $r \neq q$. However B_j is the normal closure of S_j modulo N_j, so $[B_i, B_j] \leq N_i = B_i \cap Z \leq Z$ as required.

This completes the proof of (5.5) and with it the proof of (3.5).

6. The Dark class associated with $\mathcal{M}(\mathcal{K}, G)$

We begin with a lemma which generalizes the proof of (2.13). The closure operation c_0 in the statement means "central product".

6.1 Lemma. *Let Γ be a group such that $\Gamma/\Phi(\Gamma) = X$ where X is as in (2.1), and suppose that $\Delta/\Phi(\Gamma) = Y$. Moreover we ask that $p \not| |\Delta|$. Let $R \in s_n c_0(\Gamma)$ satisfy $O^{p'}(R) = R$. Then $O^p(R)$ is a central product*

$$O^p(R) = \Delta_1 \curlyvee \Delta_2 \curlyvee \ldots \curlyvee \Delta_m$$

of copies of Δ_i of Δ, with $\Delta_i \trianglelefteq R$ $(1 \leq i \leq m)$ such that $R/\mathbf{C}_R(\Delta_i/\Phi(\Gamma_i)) \cong X$.

Proof. Since $R \in s_n c_0(\Gamma)$ there are copies $\Gamma_1, \ldots, \Gamma_m$ of Γ and a group T such that the following diagram with exact rows and columns commutes.

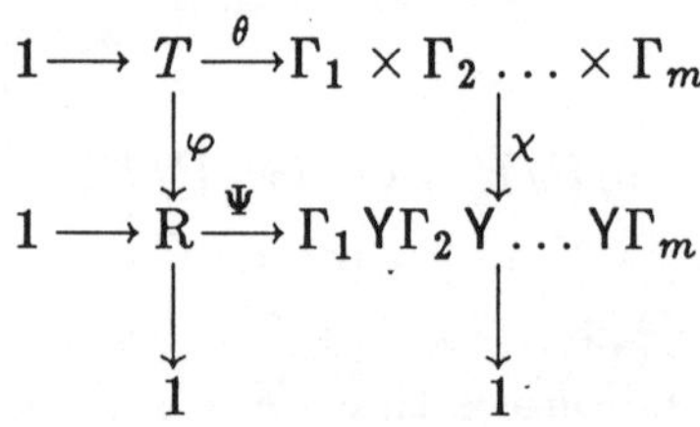

$R\psi$ is subnormal and hence $T\theta$ is subnormal. Moreover since $R = O^{p'}(R)$ we may suppose that $O^{p'}(T) = T$. Also suppose that m is minimal.

Now $O^{p'}(T\theta) = T\theta$, $T\theta$ subnormal, and m minimal mean that the projection of $T\theta$ into each Γ_i is not contained in Δ_i. Hence, by (2.14), $T\theta$ projects onto each Γ_i modulo $\Phi(\Gamma_i)$, and hence onto Γ_i. It follows from subnormality and (iii) of (2.1) that $\Delta_i \leq T\theta$ $(1 \leq i \leq m)$ and $T\theta/\mathbf{C}_{T\theta}(\Delta_i/\Phi(\Gamma_i)) \cong X$.

Finally note that $\Delta_i \cap \ker\chi = 1$, $\ker\chi \leq \zeta_1(\Gamma_1 \times \ldots \times \Gamma_m)$ so the result of the statement of the lemma follows.

6.2 Theorem. *In the notation of* (3.4) *for a group H let $Z_0 = V_{\mathcal{K}}$. Then $H \in \mathcal{M}(\mathcal{K}, G)$ if and only if* (i)′, (ii)′, (iii) *hold, where*

(i)′ $WZ_0 = B_1B_2 \ldots B_tZ_0$ *where* $B_i \cong B$ *and* $B_i \lhd\lhd W$ $(1 \leq i \leq t)$;

(ii)′ $B_iZ_0/Z_0 \trianglelefteq V/Z_0$, $[B_iZ_0/Z_0, B_jZ_0/Z_0] = 1$, $1 \leq i < j \leq t$.

Proof. If $H \in \mathcal{M}(\mathcal{K}, G)$ then we see immediately that it satisfies (i)′, (ii)′. Conversely if H satisfies (i)′ then

$$B_1B_2 \ldots B_t \leq W \leq B_1B_2 \ldots B_tZ_0$$

so

$$W = B_1B_2 \ldots B_tZ$$

because $Z = W \cap Z_0$. And if H satisfies (ii)′ then $B_i \trianglelefteq V$ by (3.3), so $B_iZ/Z \trianglelefteq V/Z$, and

$$\begin{aligned} [B_i, B_j] &\leq Z_0 \cap W \\ &= Z. \end{aligned}$$

Hence if H satisfies (i)′and (ii)′ it satisfies (i) and (ii) as required.

We now consider the case when $p \neq r$. Write

$$C = \bigcap_{i=1}^{t} C_i,$$

using the notation of (4.2), with respect to some group $H \in \mathcal{M}(\mathcal{K}, G)$. Let D be a Sylow p-subgroup of C. Then since $D \leq C_i$, $[D, U_i] \leq U_i^*$ so

$[D, U_i] \leq Z$ $(1 \leq i \leq t)$. Since $C \cap W \leq O_r(W)$ modulo Z it follows that DZ/Z is characteristic in C/Z, so

$$DZ/Z \trianglelefteq V/Z.$$

Hence if we assume that $\mathcal{K}$ satisfies $\mathcal{K} S_p = \mathcal{K}$ then $D \leq Z_0$. Consequently, under these conditions, V/Z_0 is a normal subgroup of a central product of copies of G/N. From this observation we obtain the result

6.3 Theorem. *Suppose that* G *satisfies* (3.1), *that* $p \neq r$, *and* $\mathcal{K} S_p = \mathcal{K}_p$. *Then* $H \in \mathcal{M}(\mathcal{K}, G)$ *if and only if* H *satisfies condition* (i)′ *of* (6.2) *and* (ii)″ $V/Z_0 \in s_n c_0(G/N)$.

Proof. The preamble to the statement of the theorem shows that if $H \in \mathcal{M}(\mathcal{K}, G)$ then H satisfies (ii)″.

Conversely suppose that H satisfies (i)′ of (6.2) and (ii)″. First we note that since V/Z_0 has no p'-factor groups, and since it satisfies (ii)″, V/Z_0 has by (6.1) normal subgroups $\bar{B}_1, \ldots, \bar{B}_s$ satisfying:

$$\bar{B}_i \cong B/N, \ [\bar{B}_i, \bar{B}_j] = 1$$
$$(V/Z_0)/\mathbf{C}_{V/Z_0}(\bar{B}_i/\Phi(\bar{B}_i)) \cong G/U^* \quad 1 \leq i < j \leq s.$$

The theorem will be proved therefore if we can show that $s = t$ and that, renumbering if necessary, $B_i Z_0/Z_0 = \bar{B}_i$ $(1 \leq i \leq t)$. Now $B_i Z_0/Z_0$ is not nilpotent, so for some j, the "projection" of $B_i Z_0/Z_0$ to $\bar{B}_j$ is a non-nilpotent subnormal subgroup of $\bar{B}_j$. Since $\bar{U}_j/\Phi(\bar{U}_j)$ is faithful and irreducible for the action of $\bar{B}_j/\bar{U}_j$, it follows that

$$[\bar{U}_j, B_i Z_0/Z_0] = \bar{U}_j,$$

and consequently

$$[\bar{U}_j, \underbrace{B_i Z_0/Z_0, \ldots, B_i Z_0/Z_0}_{n}] = \bar{U}_j$$

for all n. However $B_i Z_0/Z_0 \trianglelefteq\trianglelefteq V/Z_0$ so, choosing n large enough, we conclude that $\bar{U}_j \leq B_i Z_0/Z_0$. From this we see that

$$\bar{U}_j = U_i Z_0/Z_0,$$

and consequently $B_i Z_0/Z_0$ can project non-trivially to no other $\bar{B}_k$. Hence $B_i Z_0/Z_0 \leq \bar{B}_j$, whence

$$B_i Z_0/Z_0 = \bar{B}_j.$$

Now different B_i' cannot correspond to the same $\bar{B}_j$ in this way. For, if $B_i Z_0/Z_0 = B_j Z_0/Z_0$ then $B_i = B_j$ by (3.3). Hence $t \leq s$ and, renumbering if necessary,

$$B_i Z_0/Z_0 = \bar{B}_i, \quad 1 \leq i \leq t.$$

That $s = t$ is now obvious.

This completes the proof of (6.3).

6.4 Theorem. *In the notation of* (6.3) *let* $\mathcal{D}(\mathcal{K}, G)$ *be the class of groups* H *satisfying* (ii)$''$. *Then* $\mathcal{D}(\mathcal{K}, G)$ *is a Fitting class and, moreover,*

$$\mathcal{D}(\mathcal{K}, G) = \mathcal{K}\mathcal{M}(S_p, G/N).$$

Proof. For an arbitrary group H

$$\begin{aligned} O^{p'}(H/H_{\mathcal{K}}) &= O^{p'}(H)H_{\mathcal{K}}/H_{\mathcal{K}} \\ &\cong O^{p'}(H)/O^{p'}(H) \cap H_{\mathcal{K}} \\ &= O^{p'}(H)/O^{p'}(H)_{\mathcal{K}} \\ &= V/Z_0. \end{aligned}$$

Now if $H \in \mathcal{D}(\mathcal{K}, G)$ then $V/Z_0 \in s_n c_0(G/N)$ and consequently, since $O_p(H/H_{\mathcal{K}}) = 1$, $H/H_{\mathcal{K}} \in \mathcal{M}(S_p, G/N)$. That is

$$\mathcal{D}(\mathcal{K}, G) \subseteq \mathcal{K}\mathcal{M}(S_p, G/N).$$

Conversely if $H \in \mathcal{K}\mathcal{M}(S_p, G/N)$, $H/H_{\mathcal{K}} \in \mathcal{M}(S_p, G/N)$ and $O_p(H/H_{\mathcal{K}}) = 1$ so $O^{p'}(H/H_{\mathcal{K}}) \in s_n c_0(G/N)$. That is, $H \in \mathcal{D}(\mathcal{K}, G)$.

It is this result which establishes the connexion between the Fitting classes of Dark and McCann, at least in cases when both are applicable. The McCann classes for $p = r$ do not feature in (6.4); and of course the only Dark classes which do are those where $X \cong G/U^*$. On the other hand (6.4) provides a small generalization of Dark in that the section

$O^{p'}(H)/\ O^{p'}(H)_{\mathcal{K}}$ can be subnormal in a central, rather than a direct, product; and though this has been proved here only in the case $X \cong G/U^*$ it seems clear that (2.2) can be generalized in the way (6.4) indicates:

6.5 Theorem. *Let Γ be a group satisfying $\Gamma/\Phi(\Gamma) \cong X$ and $p \not| |O^p(\Gamma)|$, where X, p are as in (2.1). Let $\mathcal{K}$ be a Fitting class satisfying $\mathcal{K}\mathcal{S}_p = \mathcal{K}$. The class*

$$\mathcal{D}(\mathcal{K},\Gamma) = \{H : O^{p'}(H)/O^{p'}(H)_{\mathcal{K}} \in s_n c_0(\Gamma)\}$$

is a Fitting class.

7. Realizing the conditions of 3.1

Let G satisfy the conditions (i)-(iv) of (3.1). Write $X = G/U^*$, $Y = B/U^*$. By Higman [4], U/U^* is complemented in X, and therefore X contains an element of order p not in Y. Let α be the automorphism induced in Y by conjugation with this element.

7.1 Lemma. *Let G satisfy the conditions* (i)-(iv) *of* (3.1). *G also satisfies* (v) *of* (3.1) *if either of the two conditions below are satisfied.*

(a) *If $\varphi \in \mathrm{Aut}\, Y \backslash Y$ and $\langle \alpha, \alpha^\varphi \rangle$ is a p-group then $\langle \alpha \rangle = \langle \alpha^\varphi \rangle$.*

(b) *The endomorphism ring of U/U^* regarded as a G-module contains no pth roots of unity other than* 1.

Note that (a) is half of condition (iv)$'$ of (2.16).

We defer the proof of (7.1) to the end of this section.

7.2 Definition. Let L be a normal subgroup of a group K, and r a prime. A normal subgroup H of K is *an r-niche for L in K* if $L \leq H$ and if $O_r(L)/\Phi(O_r(L))$ as $Z_r H$-module is a direct sum of K-conjugate irreducible submodules.

7.3 Lemma. *Suppose that H is an r-niche for L_1, L_2 in K. Suppose further that $O^r(L_i) = L_i$ and $\mathbf{C}_{L_i}(O_r(L_i)) \leq O_r(L_i)$ $(i = 1, 2)$. Then either $O_r(L_1) = O_r(L_2)$ or $[L_1, L_2] = 1$.*

Proof. Suppose that $O_r(L_1) \neq O_r(L_2)$. Then we may suppose without loss of generality, that

$$O_r(L_1) \cap O_r(L_2) < O_r(L_1).$$

Now $[O_r(L_1), L_2] \leq O_r(L_1) \cap L_2$ and therefore, since the latter is a normal r-group of L_2,

$$\begin{aligned}[O_r(L_1), L_2] &\leq O_r(L_1) \cap O_r(L_2) \\ &< O_r(L_1).\end{aligned}$$

Since $[O_r(L_1), L_2]\Phi(O_r(L_1)) < O_r(L_1)$ it follows that $O_r(L_1)/\Phi(O_r(L_1))$ as L_2-module has trivial composition factors. However since $L_2 \trianglelefteq H$ and H is a niche for L_1 in K, $O_r(L_1)/\Phi(O_r(L_1))$ as L_2-module is a direct sum of H-conjugate irreducible submodules, all of which must therefore be trivial. This means that

$$[O_r(L_1), L_2] \leq \Phi(O_r(L_1)).$$

In particular, for every r'-elements x of L_2,

$$[O_r(L_1), x] = 1.$$

Since L_2 is generated by r'-elements we have that

7.4 $$[O_r(L_1), L_2] = 1.$$

From the three subgroup lemma we have

$$\begin{aligned}[L_1, L_2, O_r(L_1)] &\leq [L_2, O_r(L_1), L_1][O_r(L_1), L_1, L_2] \\ &= 1\end{aligned}$$

by (7.4). Hence, by hypothesis,

$$[L_1, L_2] \leq O_r(L_1),$$

from which we deduce, with the help of (7.4), that

$$[L_1, L_2, L_2] = 1.$$

That is, for $x \in L_1$, $y \in L_2$, $[x, y, y] = 1$. It is then a simple induction on n to deduce that

$$[x, y^n] = [x, y]^n.$$

Finally choose n to be the r'-part of the exponent of L_2, and y to be an r'-element of L_2, and we find that

$$[x, y] = 1,$$

for all $x \in L_1$. Since L_2 is generated by r'-elements this means that

$$[L_1, L_2] = 1$$

as required.

Proof of (7.1). Suppose that $H = X_1X_2$ where $X \cong X_1 \cong X_2$, $X_i \trianglelefteq H$ $(i = 1, 2)$ and $\varsigma_1(H) = 1$. We must show that (a), (b) separately imply $[X_1, X_2] = 1$ or $X_1 = X_2$.

For convenience write $\bar{U} = U/U^*$. Now if $\bar{U}_1 \neq \bar{U}_2$ then $X_1 \cap X_2 = 1$ so $[X_1, X_2] = 1$. Hence we may suppose $\bar{U}_1 = \bar{U}_2$. Let

$$C = \mathbf{C}_H(\bar{U}_1).$$

Then $\bar{U}_1 \leq C$ and $X_1 \cap C = X_2 \cap C = \bar{U}_1$. Moreover H/C is a normal product of X_1C/C and X_2C/C each of which is isomorphic to X/U. Hence by (7.3) either $[X_1C/C, X_2C/C] = 1$ or $Y_1C/C = Y_2C/C$. The first alternative is impossible since it would required X_2C/C to be in the centralizer of $\bar{U}_1$ as X_1C/C-module. But this centralizer is commutative whereas X_2C/C is not. Hence

$$Y_1C = Y_2C.$$

It follows from Dedekind's law that

$$Y_1(Y_1Y_2 \cap C) = Y_2(Y_1Y_2 \cap C).$$

Write $C_1 = Y_1Y_2 \cap C$ and let D be a Sylow q-subgroup of C_1. Then $C_1 = \bar{U}_1D$ and $[\bar{U}_1, D] = 1$, so D is normal in H. It follows that for $i = 1, 2$,

$$\begin{aligned}[X_i, D] &\leq D \cap X_i \\ &= C \cap X_i \cap D \\ &= \bar{U}_1 \cap D \\ &= 1.\end{aligned}$$

But this means that $D \leq \varsigma_1(H) = 1$. Consequently $C_1 = \bar{U}_1$ and we conclude from (7.5) that

$$Y_1 = Y_2.$$

That (a) implies $X_1 = X_2$ now follows as in the last three paragraphs of (2.16).

Finally observe that $C/\bar{U}_1$ is a p-group. Also $Y_1/\bar{U}_1 \cap C/\bar{U}_1 = 1$ and, if $X_1 \neq X_2$, $H = X_1C$. However this means that $\bar{U}_1$ is an irreducible module for $X_1/\bar{U}_1 \times C/\bar{U}_1$ or, in other words, End $\bar{U}_1$ contains non-trivial pth roots of 1. Thus (b) also implies that $X_1 = X_2$.

8. The Fitting class generated by S_4 (cf. McCann [5,Chapter 5])

It is clear that S_4, the symmetric group on four symbols, satisfies the hypothesis (i)-(iv) of (3.1) and (b) of (7.1) and therefore it satisfies (3.1). Hence $\mathcal{M}(\{1\}, S_4) = \mathcal{X}$ is a Fitting class by (3.5), and of course, it contains S_4. Let $H \in \mathrm{Fit}(S_4)$. Now H is a normal product

$$H = O^2(H)O^3(H).$$

We examine these normal factors separately. Since $H \in \mathcal{X}$, $V = O^3(H)$ has the following structure: either $V \in \mathcal{S}_2$ or $O^2(V)$ is a direct product of copies $Y_1, Y_2, \ldots, Y_t$ of A_4 normal in V and

$$V/\mathbf{C}_V(Y_i) \cong S_4, \quad 1 \leq i \leq t.$$

Hence if

$$C = \bigcap_{i=1}^{t} \mathbf{C}_V(Y_i)$$

we have that $C \cap O^2(V) = 1$, since $\varsigma_1(O^2(V)) = 1$, so C is a 2-group. Therefore V is a subnormal, subdirect product of copies of S_4 and a 2-group.

On the other hand $O^2(H) \in \mathcal{S}_2\mathcal{S}_3 \cap \mathrm{Fit}(S_4)$ and it is easy to prove that $\mathcal{S}_2\mathcal{S}_3 \cap \mathrm{Fit}(S_4) = \mathrm{Fit}(A_4)$. For, consider the classes $\mathcal{F}_0 = s_n\{S_4\}$, $\mathcal{F}_i = s_0 c_0 \mathcal{F}_{i-1}$, $i \geq 1$. Then $\{\mathcal{F}_i : i \geq 0\}$ forms an ascending chain, and

$$\bigcup_{i=1}^{\infty} \mathcal{F}_i = \mathrm{Fit}(S_4).$$

Now

$$\mathrm{Fit}(A_4) \leq \mathcal{S}_2\mathcal{S}_3 \cap \mathrm{Fit}(S_4) = \mathcal{S}_2\mathcal{S}_3 \cap \bigcup_{i=0}^{\infty} \mathcal{F}_i$$

$$= \bigcup_{i=0}^{\infty} (\mathcal{S}_2\mathcal{S}_3 \cap \mathcal{F}_i)$$

so it suffices to show that $\mathcal{S}_2\mathcal{S}_3 \cap \mathcal{F}_i \subseteq \mathrm{Fit}(A_4)(0 \leq i)$, and this follows by induction on i. For $i = 0$: $\mathcal{S}_2\mathcal{S}_3 \cap \mathcal{F}_0 = \{A_4, C_2 \times C_2, C_2, 1\} \subseteq \mathrm{Fit}A_4$. For $i > 0$: let $H \in \mathcal{S}_2\mathcal{S}_3 \cap \mathcal{F}_i$. Then

$$H \lhd\lhd K = \prod_{j=1}^{n} K_j$$

where $K_j \trianglelefteq K$ and $K_j \in \mathcal{F}_{i-1}(1 \leq j \leq n)$. Therefore

$$O^2(H) \lhd\lhd \prod_{j=1}^{n} O^2(K_j)$$

and

$$O^2(K_j) \in \mathcal{S}_2\mathcal{S}_3 \cap s_n\mathcal{F}_{i-1} = \mathcal{S}_2\mathcal{S}_3 \cap \mathcal{F}_{i-1} \subseteq \mathrm{Fit}(A_4)$$

by induction. Hence $O^2(H) \in \mathrm{Fit}(A_4)$. Since $O^3(H) \in \mathcal{S}_2 \subseteq \mathrm{Fit}(A_4)$, this completes the induction.

To sum up:

8.1 Theorem. *The following conditions on a group H are equivalent.*

(i) $H \in \mathrm{Fit}(S_4)$.

(ii) $O^2(H) \in \mathrm{Fit}(A_4)$, $O^3(H) \in s_n(\mathcal{S}_2 \cup \{A_4\}) \cap R_0(\mathcal{S}_2 \cup \{S_4\})$.

(iii) $H = KL$ *where* $K \trianglelefteq H$, $L \trianglelefteq H$, $K \in \mathcal{C}(S_4)$ *(in the notation of* (2.9)*) and* $L \in \mathrm{Fit}(A_4)$.

As McCann points out ([5], p.80) this description of the groups in $\mathrm{Fit}(S_4)$ is achieved "modulo" the groups in $\mathrm{Fit}(A_4)$. We sidestep a problem which has proved intractable namely that of classifying the groups in $\mathrm{Fit}(H)$ for a metanilpotent group H. Theorem 8.1 may indicate that the real difficulty in classifying the groups in $\mathrm{Fit}(H)$ in general lies in the metanilpotent case.

References

[1] J.C. Beidelman and Ben Brewster, Strict Normality in Fitting classes, II. *J. Algebra* **51** (1978), 218-227.

[2] Rex S. Dark, Some examples in the theory of injectors of finite soluble groups. *Math. Z.* **127** (1972), 145-156.

[3] Trevor Hawkes, On metanilpotent Fitting classes. *J. Algebra* **63** (1980), 459-483.

[4] Graham Higman, Complementation of abelian normal subgroups. *Publ. Math. Debrecen* **4** (1956), 455-458.

[5] Brendan McCann, *On Fitting classes of groups of nilpotent length three.* Ph.D. Thesis, Würzburg, 1985.

Department of Mathematics
Australian National University
GPO Box 4
Canberra ACT 2601
Australia

On groups related to Fibonacci groups

C.M. Campbell, E.F. Robertson and *R.M. Thomas*

1. Introduction

The Fibonacci group $F(r,n)$ is defined by the presentation:

$$\langle x_1, x_2, \ldots, x_n : x_1x_2\ldots x_r = x_{r+1},\ x_2x_3\ldots x_{r+1} = x_{r+2}, \ldots,\\ x_nx_1x_2\ldots x_{r-1} = x_r\rangle,$$

where the subscripts are reduced modulo n. (See [6, 7, 8] for details.) In [9], it is shown that the group $F(r,n)$ is infinite if either

$$d > 3 \qquad (1.1)$$

or

$$d = 3 \text{ and } n \text{ is even} \qquad (1.2)$$

where $d = (r+1,n)$. The proof involves showing that the related group $G(r,n)$ defined by the presentation

$$\langle a,\ b\ :\ a^2 = b^n = abab^rab^{-r}ab^{-1} = 1\rangle$$

is infinite if (1.1) or (1.2) holds. These results are extended to the related groups $F(r,n,k)$ and $H(r,n,s)$ in [2].

In this paper, we shall consider some natural generalizations $G(n;h,i,j,k)$ of the groups $G(r,n)$. We shall show how finite groups of deficiency zero may be obtained from certain classes of these groups. A related problem is considered by Conder in [4], where he determines which of the presentations

$$\langle a,\ b : a^2 = b^3 = w(a,b) = 1\rangle$$

define finite groups when the relator $w(a,b)$ has length at most 24.

The groups $G(n;h,i,j,k)$ are defined by the presentations

$$\langle a,\, b : a^2 = b^n = ab^h ab^i ab^j ab^k = 1\rangle$$

with

$$h+i+j+k=0. \tag{1.3}$$

Thus the group $G(r,n)$ is the group $G(n;1,r,-r,-1)$, which is isomorphic to $G(n;1,-1,-r,r)$ (see Proposition 2.3 of [3]). In [3], the groups $G(n;h,i,j,k)$ satisfying (1.3) and

$$h,i,j,k \in \{\pm 1, \pm 2\} \tag{1.4}$$

are investigated. The related deficiency zero groups defined by the presentations

$$\langle a,b : a^2 = 1,\ ab^2ab^{-2}ab^{-1}ab^{1-n} = 1\rangle$$

are discussed in [1], and the remaining cases of the groups $H(n;h,i,j,k)$ defined by

$$\langle a,b : a^2 = 1,\ ab^h ab^i ab^j ab^{k-n} = 1\rangle$$

and satisfying (1.3) and (1.4) will be discussed in a future paper.

In this paper, we consider some subclasses of the groups $G(n;h,i,j,k)$ and $H(n;h,i,j,k)$ satisfying condition (1.3) but relaxing condition (1.4).

2. The deficiency zero groups $H(-2n;i,-i,n,-n)$

In this section, we investigate the groups $H(-2n;i,-i,n,-n)$, and show that they are finite if $(n,i)=1$. Another subclass of finite deficiency zero groups $H(n;h,i,j,k)$ is given in [1].

Let us denote $H(-2n;i,-i,n,-n)$ by H, so that H has presentation:

$$\langle a,\, b : a^2 = ab^i ab^{-i} ab^n ab^n = 1\rangle.$$

We aim to prove the following theorem:

Theorem 2.1. *The groups* $H = H(-2n;i,-i,n,-n)$ *have:*
(i) *infinite order if* $(i,n) > 1$,
(ii) *order* $4ni(3^n+1)$ *if* $(n,i)=1$ *and* i *is odd,*
(iii) *order* $4ni(3^n-1)$ *if* $(n,i)=1$ *and* i *is even.*

Proof. The proof of (i) is straightforward. If $d = (n, i) > 1$, then we may add relation $b^d = 1$ to those for H to get

$$\langle a, b : a^2 = b^d = 1 \rangle,$$

which is a presentation for the infinite group $C_2 * C_d$. So H is infinite in this case.

We now consider the case $(n, i) = 1$. We introduce new generators $c = b^n$ and $d = b^i$ to get

$$\langle a, b, c, d : a^2 = 1,\ dad^{-1} = ac^{-1}ac^{-1}a,\ [c, d] = 1,\ c = b^n,\ d = b^i \rangle.$$

Let u and v be such that $ui + vn = 1$. Then $b = b^{ui}b^{vn} = d^u c^v$, and we have

$$\langle a, b, c, d : a^2 = 1,\ dad^{-1} = ac^{-1}ac^{-1}a,\ [c, d] = 1,\ c = b^n, d = b^i,\ b = d^u c^v \rangle.$$

We may now delete the generator b via the last relation to get

$$\langle a, c, d : a^2 = 1,\ dad^{-1} = ac^{-1}ac^{-1}a,\ [c, d] = 1,\ c = (d^u c^v)^n,\ d = (d^u c^v)^i \rangle.$$

The relation

$$c = (d^u c^v)^n = d^{un} c^{vn} = d^{un} c^{1-ui}$$

is equivalent to $(d^n c^{-i})^u = 1$, since $[c, d] = 1$. The relation

$$d = (d^u c^v)^i = d^{ui} c^{vi} = d^{1-vn} c^{vi}$$

is equivalent to $(d^{-n} c^i)^v = 1$, and hence, after inversion and conjugation, to $(d^n c^{-i})^v = 1$.

Since $(u, v) = 1$, the relations $(d^n c^{-i})^u = 1$ and $(d^n c^{-i})^v = 1$ are equivalent to $d^n c^{-i} = 1$, and hence to $c^i = d^n$. So we have

$$\langle a, c, d : a^2 = 1,\ dad^{-1} = ac^{-1}ac^{-1}a,\ [c, d] = 1,\ c^i = d^n \rangle.$$

Let N be the normal subgroup $\langle a, c \rangle$ of H. We proceed as in Chapter 7 of [6] to determine a presentation for N. If we choose the coset representatives $1, d, d^2, \ldots, d^{n-1}$ for N, we get generators:

$$a,\ a_1 = dad^{-1},\ a_2 = d^2ad^{-2}, \ldots,\ a_{n-1} = d^{n-1}ad^{-(n-1)},$$
$$c,\ c_1 = dcd^{-1},\ c_2 = d^2cd^{-2},\ \ldots,\ c_{n-1} = d^{n-1}cd^{-(n-1)},\ t = d^n,$$

and relations:

$$\begin{aligned}
&a^2 = a_1^2 = a_2^2 = \ldots = a_{n-1}^2 = 1,\\
&a_1 = ac^{-1}ac^{-1}a,\ a_2 = a_1c_1^{-1}a_1c_1^{-1}a_1, \ldots,\\
&a_{n-1} = a_{n-2}c_{n-2}^{-1}a_{n-2}c_{n-2}^{-1}a_{n-2},\ tat^{-1} = a_{n-1}c_{n-1}^{-1}a_{n-1}c_{n-1}^{-1}a_{n-1},\\
&c = c_1,\ c_1 = c_2, \ldots, c_{n-2} = c_{n-1},\ c_{n-1} = tct^{-1},\\
&c^i = t,\ c_1^i = t,\ \ldots,\ c_{n-1}^i = t
\end{aligned}$$

for N. If we eliminate the generators $c_1, c_2, \ldots, c_{n-1}$ and t, we get the following presentation for N:

$$\begin{aligned}
\langle a, a_1, \ldots, a_{n-1}, c : a^2 &= a_1^2 = \ldots = a_{n-1}^2 = 1,\\
a_1 &= (ac^{-1})^2a,\ a_2 = (a_1c^{-1})^2a_1,\ \ldots,\\
a_{n-1} &= (a_{n-2}c^{-1})^2a_{n-2},\\
tat^{-1} &= (a_{n-1}c^{-1})^2a_{n-1}\rangle.
\end{aligned}$$

The relation $a_1^2 = 1$ is equivalent to

$$(ac^{-1}ac^{-1}a)^2 = 1, \text{ i.e. } ac^{-1}ac^{-2}ac^{-1}a = 1, \text{ i.e. } ac^2a = c^{-2}. \tag{2.1}$$

We may rewrite the relations:

$$a_1 = (ac^{-1})^2a,\ a_2 = (a_1c^{-1})^2a_1,\ \ldots,\ a_{n-1} = (a_{n-2}c^{-1})^2a_{n-2}$$

as:

$$a_1 = (ac^{-1})^2a,\ a_2 = (ac^{-1})^8a,\ \ldots,\ a_{n-1} = (ac^{-1})^ma,$$

where $m = 3^{n-1} - 1$. The relations $a_1^2 = a_2^2 = \ldots = a_{n-1}^2 = 1$ are then all redundant via $ac^2a = c^{-2}$. We may therefore successively delete the generators $a_1, a_2, \ldots, a_{n-1}$ to get

$$\langle a,\ c : a^2 = 1,\ c^iac^{-i} = (ac)^{3m+2}a,\ ac^2a = c^{-2}\rangle.$$

If i is even, then the second relation is equivalent to

$$c^{2i}a = (ac)^{3m+2}a, \text{ i.e. } c^{2i} = (ac)^{3m+2}.$$

If i is odd, then the second relation is equivalent to

$$c^i ac^{-i+1} = (ac)^{3m+3}, \text{ i.e. } c^{2i-1}a = (ac)^{3m+3}, \text{ i.e. } c^{2i} = (ac)^{3m+4}.$$

So we have

$$\langle a, c : a^2 = 1,\ c^{2i} = (ac)^p,\ ac^2a = c^{-2}\rangle,$$

$$\text{where} \qquad p = \begin{cases} 3m+2 = 3^n - 1 & (i \text{ even}), \\ 3m+4 = 3^n + 1 & (i \text{ odd}). \end{cases}$$

Let $e = aca$. Then $M = \langle e, c\rangle$ is a normal subgroup of index 2 in N with presentation:

$$\langle c, e : c^{2i} = (ec)^{p/2},\ e^{2i} = (ce)^{p/2},\ c^2 = e^{-2}\rangle.$$

Given the first and third relations, we have

$$(ce)^{p/2} = (c^{-1}c^2e^2e^{-1})^{p/2} = (c^{-1}e^{-1})^{p/2} = (ec)^{-p/2} = c^{-2i} = e^{2i},$$

so that the second relation is redundant, and we have

$$\langle c, e : c^{2i} = (ec)^{p/2},\ c^2 = e^{-2}\rangle.$$

Introduce $f = ec$ and delete $e = fc^{-1}$:

$$\langle c, f : c^{2i} = f^{p/2},\ c^2 = cf^{-1}cf^{-1}\rangle,$$

i.e.

$$\langle c, f : c^{2i} = f^{p/2},\ c^{-1}fc = f^{-1}\rangle.$$

Since $c^{-1}fc = f^{-1}$, we have $c^{-1}f^{p/2}c = f^{-p/2}$, so that $f^{p/2} = f^{-p/2}$, and hence $f^p = 1$. So we have

$$\langle c, f : c^{2i} = f^{p/2}, f^p = 1,\ c^{-1}fc = f^{-1}\rangle.$$

This is a presentation for a metacyclic group of order $2ip$. So H has order $4inp$ as required.

3. The groups $G(n; i, -i, j, -j)$

In this section, we investigate some classes of groups $G(n; i, -i, j, -j)$. The first result follows from Theorem 2.1:

Theorem 3.1. *The groups $G(2n; i, -i, n, -n)$ have:*
(i) *infinite order if $(n, i) > 1$,*
(ii) *order $2n(3^n + 1)$ if $(n, i) = 1$ and i is odd,*
(iii) *order $2n(3^n - 1)$ if $(n, i) = 1$ and i is even.*

Note. Notice that the order of $H(-2n; i, -i, n, -n)$ contains a factor i, whereas the order of $G(-2n; i, -i, n, -n)$ $(= G(2n; i, -i, n, -n))$ depends on i only to the extent of whether i is odd or even. In fact, if i_1, i_2 are distinct integers of the same parity, then the groups $G(-2n; i_1, -i_1, n, -n)$ and $G(-2n; i_2, -i_2, n, -n)$ are isomorphic whereas $H(-2n; i_1, -i_1, n, -n)$ and $H(-2n; i_2, -i_2, n, -n)$ do not even have the same order.

Proof. Let $G = G(2n; i, -i, n, -n)$, $H = H(-2n; i, -i, n, -n)$, so that G and H have presentations

$$\langle a, b : a^2 = 1,\ b^{2n} = 1,\ ab^i ab^{-i} ab^n ab^{-n} = 1\rangle,$$

and

$$\langle a, b : a^2 = 1,\ ab^i ab^{-i} ab^n ab^n = 1\rangle,$$

respectively. If $d = (n, i) > 1$, we argue as in the proof of Theorem 2.1; add the relation $b^d = 1$ to those for G to get the presentation

$$\langle a, b : a^2 = b^d = 1\rangle,$$

so that G has an infinite homomorphic image $C_2 * C_d$, and is therefore infinite. So henceforth assume that $(n, i) = 1$.

If we let z denote b^{2n}, then, from the proof of Theorem 2.1, we have that $z = c^2$ (where $c = b^n$ as in the proof of Theorem 2.1), and, by (2.1), $ac^2a = c^{-2}$, so that $\langle z\rangle$ is normal in H, and $H/\langle z\rangle$ is isomorphic to G. Since c has order $4i$ in H, z has order $2i$, and

$$|G| = |H|/2i = \begin{cases} 2n(3^n + 1) & (i \text{ odd}) \\ 2n(3^n - 1) & (i \text{ even}) \end{cases}$$

as required.

As an immediate consequence of Theorem 3.1, we have:

Corollary 3.2. *$G(2n;1,-1,n,-n)$ is a finite metabelian group of order $2n(3^n+1)$.*

Notes. (i) The group $G(2n;1,-1,n,-n)$ is the group $G(-n,2n) = G(n,2n)$, so that we have a class of finite groups $G(r,n)$. However, neither (1.1) nor (1.2) hold here, since

$$d = (n+1, 2n) = \begin{cases} 1 & (n \text{ even}), \\ 2 & (n \text{ odd}). \end{cases}$$

(ii) An independent proof of Corollary 3.2 has recently been obtained by Doostie, and will appear in [5].

Before giving the next result, we define the *Lucas sequence* (g_n). The Lucas numbers are defined by

$$g_0 = 2,\ g_1 = 1,\ g_{n+2} = g_n + g_{n+1},$$

where n may assume negative, as well as positive, values. The next result shows that our classes of groups related to the Fibonacci groups retain interesting properties related to the Fibonacci numbers.

Theorem 3.3. *The group $G(2n+1;1,-1,n,-n)$ is a finite metabelian group of order $(4n+2)g_{2n+1}$.*

Proof. We first prove that $G(2n+1;1,-1,n,-n)$ is isomorphic to $G(2n+1;1,-1,-2,2)$. Introducing the extra generator $c = b^n$ into the presentation for $G(2n+1;1,-1,n,-n)$ yields

$$\langle a,b,c : a^2 = b^{2n+1} = c^{2n+1} = 1,\ c = b^n,\ abab^{-1}acac^{-1} = 1\rangle.$$

Since $c^2 = b^{2n} = b^{-1}$, we have that $b = c^{-2}$. The relation $b^{2n+1} = 1$ is now redundant, giving

$$\langle a,b,c : a^2 = c^{2n+1} = 1,\ b = c^{-2},\ c = b^n,\ abab^{-1}acac^{-1} = 1\rangle.$$

The relation $c = b^n$ is redundant, and we may eliminate b to get

$$\langle a, c : a^2 = c^{2n+1} = 1,\ ac^{-2}ac^2acac^{-1} = 1\rangle,$$

which is a presentation for the group $G(2n+1; 1, -1, -2, 2)$. The result now follows from Proposition 2.3 and Theorem 6.1 (iv) of [3].

Note. $G(2n+1; 1, -1, n, -n) = G(-n, 2n+1) = G(n+1, 2n+1)$, and

$$d = (n+2, 2n+1) = \begin{cases} 1 & \text{if } (2n+1, 3) = 1, \\ 3 & \text{if } (2n+1, 3) = 3. \end{cases}$$

This gives a class of finite groups $G(r, n)$ with $d = (r+1, n) = 3$ and n odd, in contrast to the results (1.1) and (1.2).

Acknowledgements

The third author would like to thank Hilary Craig for all her help and encouragement.

References

[1] C.M. Campbell, E.F. Robertson and R.M. Thomas, Finite groups of deficiency zero involving the Lucas numbers, to appear.

[2] C.M. Campbell and R.M. Thomas, On infinite groups of Fibonacci type, *Proc. Edinburgh Math. Soc.* **29** (1986), 225-232.

[3] C.M. Campbell and R.M. Thomas, On $(2, n)$-groups related to Fibonacci groups, *Israel J. Math.*, **58** (1987), 370-380.

[4] M.D.E. Conder, Three-relator quotients of the modular group, University of Auckland preprint, 1986.

[5] H. Doostie, *Fibonacci-type sequences and classes of groups*, Ph.D. Thesis, University of St. Andrews.

[6] D.L. Johnson, *Presentations of groups*, London Math. Soc. Lecture Notes **22** (Cambridge University Press, 1976).

[7] D.L. Johnson, *Topics in the theory of group presentations*, London Math. Soc. Lecture Notes **42** (Cambridge University Press, 1980).

[8] D.L. Johnson, J.W. Wamsley and D. Wright, The Fibonacci groups, *Proc. London Math. Soc.* **29** (1974), 577-592.

[9] R.M. Thomas, Some infinite Fibonacci groups, *Bull. London Math. Soc.* **15** (1983), 383-386.

C.M. Campbell and E. F. Robertson
Mathematical Institute
University of St Andrews
St Andrews KY16 9SS
Scotland

R.M. Thomas
Department of Computing Studies
University of Leicester
Leicester LE1 7RH
England

On the Chern classes of representations of the symmetric groups

Humberto Cárdenas and *Emilio Lluis*

The Chern classes of the natural representation of the symmetric group S_n by permutation matrices have been considered by H. Mùi [10], B.M. Mann and R.J. Milgram [9], L. Evens and D.F. Kahn [6] and others. Evens and Kahn prove, in particular, that the p-primary components c_i of the Chern classes $c_i(\nu) \in H^{2i}(S_{p^n}; Z)$ of the natural representation $\nu : S_{p^n} \to U(p^n)$ for p an odd prime have exactly the orders given as bounds by Grothendieck in [7], namely

$$ord(c_i) = \begin{cases} 1 & \text{if } i \not\equiv 0 \pmod{p-1} \\ p^{v(i)+1} & \text{if } i \equiv 0 \pmod{p-1} \end{cases} \tag{0.1}$$

where $p^{v(i)}$ is the maximal power of p that divides i.

Now, since the restriction map from the p-primary component of the cohomology of S_{p^n} to the cohomology of a p-Sylow subgroup

$$H^*(S_{p^n}; \mathbf{Z})_p \to H^*(S_{p^n,p}; \mathbf{Z})$$

is injective, to study $c_1, \ldots, c_{p^n}$ it is enough to consider the restriction ν_n of the representation ν to $S_{p^n,p}$. This restriction has as a summand the rational representation $\nu_{n-1}g$

$$S_{p^n,p} \xrightarrow{g} S_{p^{n-1},p} \xrightarrow{\nu_{n-1}} U(p^{n-1})$$

where g is the projection. So $\nu_n = \nu_{n-1}g \oplus \mu$ where μ is a rational representation

$$\mu : S_{p^n,p} \longrightarrow U(p^{n-1}(p-1)).$$

But since g^* is a monomorphism and $c_i(\nu_{n-1}g) = g^*(c_i(\nu_{n-1}))$ the Chern classes of $\nu_{n-1}g$ are inductively determined. In this paper we study the Chern classes of the representation μ. The main results are:

The subalgebra of $H^*(S_{p^n,p};\mathbf{Z})$ generated by the Chern classes of μ is isomorphic to a polynomial algebra over $\mathbf{Z}$ in p^{n-1} indeterminates z_i with the only relation given by

$$p^{v(i)+1}z_i \quad (1 \leq i \leq p^{n-1}).$$

Those are the relations given by the orders of the respective Chern classes.

We obtain a similar result about the subalgebra of $H^*(S_{p^n};\mathbf{Z})$ generated by the p-primary components of the Chern classes of degree less than or equal to $2p^{n-1}(p-1)$ of the natural representation of S_{p^n}.

For the proofs we compute, in the first part of the paper, the Chern classes of the regular representation of the cyclic group $\mathbf{Z}/p^n$, p an odd prime. Then we restrict the natural representation ν of S_{p^n} to a family of subgroups

$$\mathbf{Z}/p^n, (\mathbf{Z}/p^{n-1})^p, \ldots, (\mathbf{Z}/p^r)^{p^{n-r}}, \ldots, (\mathbf{Z}/p)^{p^{n-1}}$$

and using Grothendieck bounds (0.1) we prove that those restrictions determine completely the Chern classes of μ and, up to degree $2p^{n-1}(p-1)$, those of ν.

We will assume through the paper that p *is a prime different from* 2.

1. To each complex representation σ of degree n of a finite group G there are associated certain cohomology classes $c_i(\sigma) \in H^{2i}(G;\mathbf{Z})$, the Chern classes of σ. $c_0(\sigma) = 1$ and $c_i(\sigma) = 0$ for $i > n$.

We will use the following properties of these classes ([1], page 285-287).

1. If $f : G \longrightarrow G'$ is a group homomorphism and $\sigma : G' \longrightarrow U(n)$ then
$$c_i(\sigma f) = f^*(c_i(\sigma)). \tag{1.1}$$
2. $c_i(\sigma \oplus \tau) = \sum_{j+k=i} c_j(\sigma)c_k(\tau)$.
3. If $deg\,\sigma = deg\,\tau = 1$ then
$$c_i(\sigma\tau) = c_i(\sigma) + c_i(\tau), \quad (i > 0). \tag{1.2}$$
4. The correspondence $\sigma \longrightarrow c_1(\sigma)$ gives an isomorphism
$$Hom(G, U(1)) \xrightarrow{\cong} H^2(G;\mathbf{Z}). \tag{1.3}$$

Besides, one defines the total Chern class of σ as

$$c(\sigma) = 1 + c_1(\sigma) + ... + c_n(\sigma)$$

and the Chern polynomial $P_\sigma(x) \in H^*(G;\mathbf{Z})[x]$ as

$$P_\sigma(x) = \sum_{i=0}^{n} c_i(\sigma)x^{n-i}$$

5. $c(\sigma \oplus \tau) = c(\sigma)c(\tau)$
6. $P_{\sigma\oplus\tau}(x) = P_\sigma(x)P_\tau(x)$ (1.4)

2. In this section we give some results on the integral cohomology of the group $\mathbf{Z}/p^n$ that will be used later.

The cohomology algebra of the cyclic group $\mathbf{Z}/p^n$ with integral coefficients is a polynomial algebra in one indeterminate of degree 2:

$$H^*(\mathbf{Z}/p^n;\mathbf{Z}) = (\mathbf{Z}/p^n)[y] \tag{2.1}$$

The projection $h : \mathbf{Z}/p^n \longrightarrow \mathbf{Z}/p^{n-1}$ induces

$$h^* : H^*(\mathbf{Z}/p^{n-1};\mathbf{Z}) \longrightarrow H^*(\mathbf{Z}/p^n;\mathbf{Z}) \tag{2.2}$$

and if we write $H^*(\mathbf{Z}/p^{n-1};\mathbf{Z}) = (\mathbf{Z}/p^{n-1})[z]$ we can see that

$$h^*(z^q) = p^q y^q. \tag{2.3}$$

When $p \neq 2$ the group U of units of the ring $\mathbf{Z}/p^{n-1}$ is cyclic of order $p^{n-2}(p-1)$. U operates on the additive group $\mathbf{Z}/p^{n-1}$ by multiplication and it is easy to check that the induced action in cohomology

$$u^* : H^*(\mathbf{Z}/p^{n-1};\mathbf{Z}) \longrightarrow H^*(\mathbf{Z}/p^{n-1};\mathbf{Z}), \quad u \in U$$

is given by

$$u^*(z) = uz \tag{2.4}$$

Proposition 2.5. *The nonzero U-invariant elements of $H^*(\mathbf{Z}/p^{n-1};\mathbf{Z})$ are those of the form az^q with $a \neq 0$, $q = (p-1)s$ and*

$$v(a) \geq n - v(s) - 2$$

where $p^{v(m)}$ denotes the maximal power of p that divides m.

Proof. Let u be a generator of U. If $az^q \neq 0$ is invariant we have from (2.4)

$$p^{v(a)}(u^q - 1) \equiv 0 \pmod{p^{n-1}}.$$

Therefore, since the order of u mod $p^{n-v(a)-1}$ is $p^{n-v(a)-2}(p-1)$ we have $q = (p-1)s$ with $v(s) \geq n - v(a) - 2$.

Corollary 2.6.

$$h^*((H^q(\mathbf{Z}/p^{n-1}; \mathbf{Z})^U) = 0, \quad q > 0.$$

Proof. Let az^q be a non zero invariant. Then

$$q = (p-1)s \geq (p-1)p^{v(s)} \geq (p-1)p^{n-v(a)-2}$$

But since $p > 2$ this implies $q \geq n - v(a)$ and we have

$$h^*(az^q) = ap^q y^q = 0.$$

3. In the next two sections we compute the Chern classes of the regular representation R of $\mathbf{Z}/p^n$.

Let ω be a primitive p^n-th root of the unity and ρ, σ the representations

$$\rho : \mathbf{Z}/p^n \longrightarrow U(1), \quad \rho(1) = \omega \tag{3.1}$$

$$\sigma : \mathbf{Z}/p^{n-1} \longrightarrow U(1), \quad \sigma(1) = \omega^p$$

Then $R = 1 \oplus \rho \oplus \ldots \oplus \rho^{p^n - 1}$ and $R' = 1 \oplus \sigma \oplus \ldots \oplus \sigma^{p^{n-1}-1}$ are the regular representations of $\mathbf{Z}/p^n$ and $\mathbf{Z}/p^{n-1}$.

It is easy to see that:

Lemma 3.2. *The Chern classes $c_i(R')$ are U-invariant.*

Proposition 3.3. *The Chern polynomial of the representation $R'h$*

$$\mathbf{Z}/p^n \xrightarrow{h} \mathbf{Z}/p^{n-1} \xrightarrow{R'} U(p^{n-1})$$

is

$$P_{R'h}(x) = x^{p^{n-1}},$$

i.e. the Chern classes $c_i(R'h)$ are zero for $i \neq 0$.

Proof. From (1.1), (3.2) and (2.6) we have $c_i(R'h) = h^*(c_i(R')) = 0$, $(i \neq 0)$.

This result is equivalent to

Corollary 3.4. *The elementary symmetric functions in the elements of the radical of the ring $\mathbf{Z}/p^n$ are zero.*

In other words, $\bmod p^n$

$$x(x+py)(x+2py)\dots(x+(p^{n-1}-1)py) = x^{p^{n-1}}. \tag{3.5}$$

4.

Theorem 4.1. *The Chern polynomial of the regular representation R of $\mathbf{Z}/p^n$ is*

$$P_R(x) = x^{p^{n-1}}(x^{p-1} - y^{p-1})^{p^{n-1}} \in H^*(\mathbf{Z}/p^n; \mathbf{Z})[x].$$

Proof. Since $R = 1 \oplus \rho \oplus \dots \oplus \rho^{p^n-1}$ we have to prove that

$$\prod_{k=0}^{p^n-1}(x+ky) = (x^p - xy^{p-1})^{p^{n-1}}.$$

The left hand side product is

$$\prod_{r=0}^{p-1}\prod_{s=0}^{p^{n-1}-1}(x+ry+spy).$$

Then using (3.5) we have

$$\prod_{s=0}^{p^{n-1}-1} (x + ry + spy) = (x + ry)^{p^{n-1}}$$

and therefore our product is equal to

$$\begin{aligned}(\prod_{r=0}^{p-1} (x + ry))^{p^{n-1}} &= (x^p - xy^{p-1} + pv)^{p^{n-1}} \\ &= (x^p - xy^{p-1})^{p^{n-1}}.\end{aligned}$$

Corollary 4.2. *The non zero Chern classes of the regular representation R of $\mathbf{Z}/p^n$ are*

$$c_{(p-1)s}(R) = (-1)^s \binom{p^{n-1}}{s} y^{(p-1)s}, \quad s = 0, 1, \ldots, p^{n-1}.$$

This is equivalent to:

Corollary 4.3. *The elementary symmetric functions s_1, $s_2, \ldots, s_{p^n}$ in the numbers, 1, 2,...,p^n are as follows:*

$$\begin{cases} s_i \equiv 0 \pmod{p^n} \quad \text{for } p-1 \nmid i \\ s_{i(p-1)} \equiv (-1)^i \binom{p^{n-1}}{i} \pmod{p^n}. \end{cases}$$

Corollary 4.4. *The orders of the Chern classes of the regular representation R of $\mathbf{Z}/p^n$ are*

$$ord(c_q(R)) = \begin{cases} 1 & \text{if } p-1 \nmid q \\ p^{\bar{v}(q)+1} & \text{if } p-1 \mid q. \end{cases}$$

Proof. Follows from (4.2) and the fact that $v(\binom{p^{n-1}}{s}) = n - v(s) - 1$.

Theorem 4.5. *The Chern classes of the regular representation of the group $\mathbf{Z}/p^n$ generate the U-invariant subalgebra of $H^*(\mathbf{Z}/p^n;\mathbf{Z})$, where U is the group of units of $\mathbf{Z}/p^n$.*

Proof. This follows from (2.5) since, as we just have seen, $c_{(p-1)s}(R) = ay^{(p-1)s}$ with $v(a) = n - v(s) - 1$.

5. Let $\nu : S_{p^n} \longrightarrow U(p^n)$ be the natural representation of the symmetric group S_{p^n} by permutation matrices and consider the p-primary components c_i of the non zero Chern classes $c_{i(p-1)}(\nu)$, $0 \leq i \leq (p^n - 1)/(p-1)$.

Since the restriction map

$$H^*(S_{p^n};\mathbf{Z})_p \to H^*(S_{p^n,p};\mathbf{Z})$$

from the p-primary component of the cohomology of S_{p^n} to the cohomology of a p-Sylow subgroup $S_{p^n,p}$ is a monomorphism, to study $c_1, \ldots, c_{p^{n-1}}$ it is enough to consider the restriction ν_n of ν to $S_{p^n,p}$.

Let S_{p^n} be the group of permutations of the set $\{1,2,\ldots,p^n\}$ and $S_{p^{n-1}}$ the subgroup of permutations of the p^{n-1} rows of

$$\begin{array}{cccc} 1 & 2 & \ldots & p \\ p+1 & p+2 & & 2p \\ \multicolumn{4}{c}{\cdots\cdots\cdots\cdots\cdots} \\ p^n-p+1 & p^n-p+2 & \ldots & p^n \end{array}$$

Let $S_{p^{n-1},p}$ be a p-Sylow subgroup of $S_{p^{n-1}}$ and $(\mathbf{Z}/p)^{p^{n-1}}$ the subgroup generated by the permutations

$$(1\,2\ldots p),\ (p+1\ldots 2p),\ \ldots,\ (p^n-p+1\ldots p^n)$$

Then these two subgroups generate $S_{p^n,p}$ and

$$S_{p^n,p} \cong S_{p^{n-1},p} \ltimes (\mathbf{Z}/p)^{p^{n-1}}$$

We will interpret now our representation ν_n as a module. Let M be the $QS_{p^n,p}$-module with Q-basis $u_1, \ldots, u_{p^n}$ and the operation given by $\sigma(u_i) = u_{\sigma(i)}$, $\sigma \in S_{p^n,p}$. Let $\tau \in S_{p^n,p}$ be the cyclic permutation of rows and consider the following elements of M:

$$\begin{array}{ccc} v_1 = u_1 + \ldots + u_p & \omega_{11} = u_1 - u_2 & \ldots \\ v_2 = \tau v_1 & \omega_{12} = \tau\omega_{11} & \ldots \\ \ldots & \ldots & \\ v_{p^{n-1}} = \tau^{p^{n-1}-1} v_1 & \omega_{1,p^{n-1}} = \tau^{p^{n-1}-1}\omega_{11} & \ldots \end{array}$$

$$\begin{array}{c} \omega_{p-1,1} = u_{p-1} - u_p \\ \omega_{p-1,2} = \tau\omega_{p-1,1} \\ \ldots \\ \omega_{p-1,p^{n-1}} = \tau^{p^{n-1}-1}\omega_{p-1,1}. \end{array}$$

Let M_1 be the Q-subspace of M generated by all v_j and M_2 generated by all ω_{ij}.

Lemma 5.1. *M_1 and M_2 are $S_{p^n,p}$ -submodules of M and $M = M_1 \oplus M_2$.*

Proof. Straightforward.

It is also easy to verify that the representation corresponding to the submodule M_1 is equivalent to the representation $\nu_{n-1}g$

$$S_{p^n,p} \xrightarrow{g} S_{p^{n-1},p} \xrightarrow{\nu_{n-1}} U(p^{n-1})$$

where g is the projection. So we have

Corollary 5.2. $\nu_n = \nu_{n-1}g \oplus \mu$.

Notice that the degree of the representation μ is $p^{n-1}(p-1)$.

Consider for $1 \le r \le n$ the subgroup $(\mathbf{Z}/p^r)^{p^{n-r}}$ of S_{p^n} generated by the cycles $(12\ldots p^r)$, $(p^r+1\ldots 2p^r)$, ..., $(p^n - p^r + 1 \ldots p^n)$. In what follows we will consider the restrictions of the representations ν_n, $\nu_{n-1}g$ and μ to this family of subgroups.

Consider the diagrams (for $1 \le r \le n$)

$$\begin{array}{ccc} S_{p^n,p} & \xrightarrow{\nu_n} & U(p^n) \\ \uparrow f & \nearrow \nu_n f & \\ (\mathbf{Z}/p^r)^{p^{n-r}} & & \\ \downarrow q_i & & \\ \mathbf{Z}/p^r & \xrightarrow{R} & U(p^r) \end{array} \tag{5.3}$$

where f is the inclusion, q_i the projections and R the regular representation.

It is easy to see that

Lemma 5.4. The representation $\nu_n f$ is the sum of the representations Rq_i:

$$\nu_n f = \oplus_{i=1}^{p^{n-r}} Rq_i.$$

Let $H^*(\mathbf{Z}/p^r;\mathbf{Z}) = (\mathbf{Z}/p^r)[y]$ and f^*, q_i^* the induced homomorphisms

$$H^*(S_{p^n,p};\mathbf{Z}) \xrightarrow{f^*} H^*((\mathbf{Z}/p^r)^{p^{n-r}};\mathbf{Z}) \xleftarrow{q_i{}^*} H^*(\mathbf{Z}/p^r;\mathbf{Z}).$$

Proposition 5.5. *The Chern polynomial of* $\nu_n f$ *is*

$$P_{\nu_n f}(x) = x^{p^{n-1}} \prod_{i=1}^{p^{n-r}} (x^{p-1} - y_i^{p-1})^{p^{r-1}}$$

where $y_i = q_i^*(y)$.

Proof. From (5.4), (4.1), (1.4) and (1.1) we have

$$\begin{aligned} P_{\nu_n f}(x) &= \prod P_{Rq_i}(x) = \prod q_i^*(x^{p^{r-1}}(x^{p-1} - y^{p-1})^{p^{r-1}} \\ &= \prod x^{p^{r-1}}(x^{p-1} - y_i^{p-1})^{p^{r-1}}. \end{aligned}$$

When $r = n$, $\nu_n f = R$. Then from (4.3), (5.5) and the Grothendieck bounds (0.1) we have for $1 \leq i \leq p^{n-1}$:

Proposition 5.6. *The order of the* $i(p-1)$*-Chern class of* ν_n *is* $p^{v(i)+1}$.

In [6], Theorem p. 319, Evens and Kahn prove this result for all i $(1 \leq i \leq 1 + p + \ldots + p^{n-1})$.

Consider now the commutative diagrams

$$\begin{array}{ccccc}
S_{p^n,p} & \xrightarrow{g} & S_{p^{n-1},p} & \xrightarrow{\nu_{n-1}} & U(p^{n-1}) \\
\uparrow f & & \uparrow f' & & \\
(\mathbf{Z}/p^r)^{p^{n-r}} & \xrightarrow{g'} & (\mathbf{Z}/p^{r-1})^{p^{n-r}} & & \\
\downarrow q_i & & \downarrow q_i' & & \\
\mathbf{Z}/p^r & \xrightarrow{h} & \mathbf{Z}/p^{r-1} & \xrightarrow{R'} & U(p^{r-1})
\end{array}$$

We have

Lemma 5.8.

$$\nu_{n-1} g f = \oplus_{i=1}^{p^{n-r}} R' h q_i$$

Proposition 5.9. *The Chern polynomial of $\nu_{n-1} g f$ is*

$$P_{\nu_{n-1} g f}(x) = x^{p^{n-1}}.$$

Proof. Follows from (5.8) and (3.3).

Proposition 5.10. *The restrictions of the representations ν_n and μ of $S_{p^n,p}$ to the subgroups*

$$(\mathbf{Z}/p^r)^{p^{n-r}}, \quad 1 \leq r \leq n$$

have the same Chern classes.

Proof. Follows from (5.2), (5.5) and (5.9).

Notice that since ν and ν_n have the same restrictions to these groups, this result can also be stated for ν.

In the next section we will consider the restriction of ν to this family of subgroups and it will follow that the results there obtained will hold also for the representation μ.

6. Let $s_1, \ldots, s_{p^{n-r}} \in H^*((\mathbf{Z}/p^r)^{p^{n-r}};\mathbf{Z})$ be the elementary symmetric functions in

$$y_1^{(p-1)p^{r-1}}, \ldots, y_{p^{n-r}}^{(p-1)p^{r-1}}.$$

Proposition 6.1. $f^*(p^{r-1}c_{i_1p^{r-1}}^{n_1} \ldots c_{i_tp^{r-1}}^{n_t}) = p^{r-1}s_{i_1}^{n_1} \ldots s_{i_t}^{n_t}$.

Proof. From (5.5) we have

$$f^*(p^{r-1}P_{\nu_n}(x)) = x^{p^{n-1}}p^{r-1}\prod_{i=1}^{p^{n-r}}(x^{(p-1)p^{r-1}} - y_i^{(p-1)p^{r-1}}).$$

The coefficient of $x^{p^n - i(p-1)p^r}$ in this polynomial is $p^{r-1}s_i$. Therefore

$$f^*(p^{r-1}c_{ip^{r-1}}) = p^{r-1}s_i$$

and from here follows the proposition.

We will need the two following known results:

(6.2) If as in (5.5) $y_i = q_i^*(y)$ with $y = c_1(\rho)$ then the subalgebra of $H^*((\mathbf{Z}/p^r)^{p^{n-r}};\mathbf{Z})$ generated by $y_1, \ldots, y_{p^{n-r}}$ is a polynomial algebra in $y_1, \ldots, y_{p^{n-r}}$ over $\mathbf{Z}/p^r$. (We have not found explicitly this statement in the standard texts on homology of groups but it is not very difficult to prove.)

(6.3) If $A[y_1, \ldots, y_m]$ is the polynomial ring in m indeterminates over a commutative ring A then the set of monomials in the elementary symmetric functions $s_1, \ldots, s_m$ in $y_1, \ldots, y_m$ is A-free. (This is a classical result.)

Let C be the subalgebra of $H^*(S_{p^n,p};\mathbf{Z})$ generated by the Chern classes $c_1, \ldots, c_{p^n-1}$ of ν. Notice that the possible orders of the monomials m in these classes are $p, p^2, \ldots, p^n$.

Proposition 6.4. *An element $u = \sum a_m m$ ($a_m \in \mathbf{Z}$, m monomial) of C is zero if and only if $a_m \equiv 0 \pmod{p^r}$ for all monomials m of order p^r, i.e. each summand is zero.*

Proof. Assume inductively that $a_m \equiv 0 \pmod{p^t}$ for all m of order $\geq p^t$ with $t < r$. Then by induction we have $0 = u = \sum a_m m = \sum p^{r-1}a'_m m$ (m of order $\geq p^r$). From (6.1) we have

$$f^*(p^{r-1}a'_m m) = a'_m p^{r-1}s_{i_1}{}^{n_1} \ldots s_{i_t}^{n^t}$$

and from (6.2) and (6.3) it follows that $a'_m p^{r-1} \equiv 0 \pmod{p^r}$ and therefore $a_m \equiv 0 \pmod{p^r}$ for all m of order $\geq p^r$.

So we have just proved:

Theorem 6.5. *If $c_i = c_{i(p-1)}(\mu)$ or $c_i = c_{i(p-1)}(\nu_n)$ then the subalgebra of $H^*(S_{p^n,p};\mathbf{Z})$ generated by $c_1, \ldots, c_{p^{n-1}}$ is isomorphic to the polynomial algebra $\mathbf{Z}[z_1,\ldots,z_{p^{n-1}}]$ in p^{n-1} indeterminates modulo the ideal generated by*

$$p^{v(i)+1}z_i, \quad (1 \leq i \leq p^{n-1}).$$

Corollary 6.6. *The subalgebra of $H^*(S_{p^n};\mathbf{Z})$ generated by the p-primary components of the Chern classes of degree smaller than or equal to $2p^{n-1}(p-1)$ of the natural representation of S_{p^n} is isomorphic to the algebra described in* (6.5).

References

[1] M.F. Atiyah, *Characters and cohomology of finite groups.* Inst. Hautes Ét. Sci. Publ. Math. No. 9 (1961).

[2] N. Bourbaki, *Éléments de mathématique.*Livre II, Chap. 7. Hermann (1964).

[3] H. Cárdenas and E. Lluis, El álgebra de cohomología del grupo simétrico S_4 con coeficientes enteros. *An. Inst. Mat.* **22** (1982), 137-188.

[4] H. Cárdenas and E. Lluis, On the integral cohomology of the symmetric group of degree p^2. (To appear)

[5] H. Cartan and S. Eilenberg, *Homological algebra,* Princeton U. P. (1956).

[6] L. Evens and D. S. Kahn, Chern classes of certain representations of symmetric groups, *Trans. Am. Math. Soc.* **245** (1978), 309-330.

[7] A. Grothendieck, Classes de Chern et représentations linéaires des groupes discrets. *Dix exposés sur la cohomologie des schémas.* Adv. Studies

in Pure Math. Vol. 3. North Holland, Amsterdam, (1968).

[8] B. M. Mann, The cohomology of the symmetric groups. *Trans. Am. Math. Soc.* **242** (1978), 157-184.

[9] B. M. Mann. and R. J. Milgram, On the Chern classes of the regular representations of some finite groups. *Proc. Edinburgh Math. Soc.* **25** (1982), 319-369.

[10] H. Múi, Modular invariant theory and cohomology algebras of symmetric groups. *J. Fac. Sc. U. of Tokyo* **22** (1975), 319-369.

[11] B. B. Venkov, On Characteristic classes for finite groups. *Dokl. Ak. Nauk SSSR* **137** (1961), 1274-1277.

Instituto de Matemáticas
Universidad Nacional Autónoma de México
Ciudad Universitaria
04510 México D. F.
Mexico

Solutions of equations in free groups

Leo P. Comerford, Jr. and *Charles C. Edmunds*

1. Introduction.

Let F be the free group on $X = \{x_1, x_2, \ldots\}$ and H the free group on $A = \{a_1, a_2, \ldots\}$. We call the elements of $X^{\pm 1}$ *variables* and the elements of $A^{\pm 1}$ *constants*. An *H-map* is an endomorphism of the free group $F * H$ which fixes H elementwise. An *equation over H* is an expression $W = 1$ with $W \in F * H$, and a *solution* of $W = 1$ is an H-map ϕ such that $W\phi = 1$. We sometimes speak of $U = V$ with $U, V \in F * H$ as an equation; by this we mean the equation $UV^{-1} = 1$.

Clearly if ϕ is a solution of $W = 1$, so is $\alpha\phi\psi$ for any H-map α in the stabilizer in Aut $(F * H)$ of W and any H-map ψ. The principal result of this paper, Theorem 3.2, is that if W is *quadratic* in terms of its variables, that is, if each variable which occurs in W occurs exactly twice, then there is a finite set of "basic" soltuions of $W = 1$ such that every solution is obtained from one of these finitely many in the way described above. This was established for quadratic equations without constants by D. Piollet [10].

The first result of this sort was proved by Ju. I. Hmelevskiĭ [3] for equations

$$[x_1, x_2] = U(a_1, a_2, \ldots) \tag{1.1}$$

where U is a word on constants only. Hmelevskiĭ's theorem was sharpened slightly and given a short, combinatorial proof by R. G. Burns, C. C. Edmunds, and I. H. Farouqi in [1]. Their conjecture that one "basic" solution of (1.1) would suffice was shown to be false by R. C. Lyndon and M. J. Wicks in [6].

M. Culler [2] proved a result similar to Hmelevskiĭ's for the equations

---

Research support from the Natural Sciences and Engineering Research Council of Canada is gratefully acknowledged.

$$[x_1, x_2] \ldots [x_{2n-1}, x_{2n}] = U(a_1, a_2, \ldots) \tag{1.2}$$

and

$$x_1^2 \ldots x_n^2 = U(a_1, a_2, \ldots) \tag{1.3}$$

where, as before, U is a word on constants only and in (1.2) U is a product of n but not fewer commutators and in (1.3) U is a product of n but not fewer squares.

The solutions ϕ of the equations considered by Hmelevskii, by Burns, Edmunds, and Farouqi, and by Culler, have the common feature that $(F * H)\phi \subseteq H$, so that following ϕ by an H-map would have no additional effect. Their results say that the given equations have only finitely many orbits of solutions under the action of the stabilizer in Aut (F) of the left-hand side of the equation.

A characterization of solutions of two-variable equations over a free group was given by Yu. I. Ozhigov [8]. For arbitrary systems of equations over a free group, A. A. Razborov [11], using methods developed by G. S. Makanin [7], describes solutions of given bounded exponent of periodicity, that is, solution for which images of variables have no subwords containing proper powers beyong a fixed bound.

Our method of proof is based on Nielsen transformations, and was first used to describe solutions of equations in free groups by Lyndon [4] in his solution of the Vaught conjecture. This approach was later used by H. Zieschang [12] and Piollet [9] to calculate ranks of solutions of quadratic equations without constants, and by Piollet [10] in his characterisation of solutions of quadratic equations without constants. A synopsis of the method and some of its applications may be found in §I.6 and §I.7 of the book by Lyndon and P. E. Schupp [5].

In section 2, we adapt previous treatments of Nielsen's method to the case of words which contain constants. Section 3 contains the statement and proof of the main theorem and an example. Our development will largely parallel that of Lyndon [4] and Piollet [10]. Unfortunately, there are some errors in [10]; we shall point them out at the corresponding stages of our work.

2. The Nielsen method

Let $Y = A \cup X$, a free generating set of $F * H$. We call elements of $Y^{\pm 1}$ *letters* and, as in the introduction, distinguish between constants (from $A^{\pm 1}$) and variables (from $X^{\pm 1}$). When we speak of elements of $F * H$, we shall assume that they are represented by reduced words on Y. An H-map which is an automorphism of $F * H$ is an *H-automorphism.* In defining H-maps, we mean them to fix all variables for which an image is not specified.

Following Lyndon, we call an H-map ρ defined by

$$x\rho = xy^{-1}$$

where x is a variable and y is a letter, $y \neq x^{\pm 1}$, an *elementary regular H-map.* The map ρ is *attached* to a word $W \in F * H$ (Lyndon uses "belongs to W") if $(xy)^{\pm 1}$ is a subword of W. We also include the identity map as an elementary regular H-map attached to every word. An H-map σ defined by

$$x\sigma = 1$$

where x is a variable is called an *elementary singular H-map.* The map σ is *attached* to $W \in F * H$ if $x^{\pm 1}$ occurs in W. An H-map τ is an H-map *attached* to W if $\tau = \tau_1 \dots \tau_k$ with each τ_i an elementary regular or singular H-map attached to $W\tau_1 \dots \tau_{i-1}$. We call τ a *regular* or *singular* H-map if the τ_i are all regular or all singular.

An H-map ϕ is called a *specialization* of an H-map ξ if there is an H-map ψ with $\phi = \xi\psi$. An analogue of Lemma 5 of [4] which applies to the present situation is :

Theorem 2.1. *If W is in $F * H$ and ϕ is a solution of $W = 1$, then there is an H-map τ attached to W such that $W\tau = 1$ and ϕ is a specialization of τ.*

As preliminaries to the proof of this theorem, we state mild generalizations of Lemmas 2 and 3 of [4]. Here K is a function from $F * H$ to the positive integers which Lyndon defines in [4] to establish an ordering of elements of $F * H$. This function has the property that $|U| < |V|$ implies $K(U) < K(V)$, where $|U|$ denotes the length of U.

Lemma 2.2. *If $u_1, \ldots, u_t$ $(t \geq 1)$ are in $F * H$ with no $u_i = 1$ and with $|u_1 \ldots u_t| < t$, then for some i, $1 \leq i \leq t-1$, $K(u_i u_{i+1}) < K(u_i)$ or $K(u_i u_{i+1}) < K(u_{i+1})$.*

Lyndon's statement has the hypothesis that $u_1 \ldots u_t = 1$ instead of $|u_1 \ldots u_t| < t$, but his proof uses only the latter condition.

We let gen(W) denote the set of elements x of X such that $x^{\pm 1}$ occurs in W. If $V \in F * H$ and θ is an H-map, we define $K(V, \theta)$ to be $\sum_{x \in \mathrm{gen}(V)} K(x\theta)$.

Lemma 2.3. *If $W \in F * H$ and ϕ is an H-map such that $|W\phi| < |W|$, then either $x\phi = 1$ for some $x \in$ gen(W) or there is an elementary regular H-map ρ attached to W such that $K(W\rho, \rho^{-1}\phi) < K(W, \phi)$.*

Lyndon's hypothesis in Lemma 3 of [4] is that $W\phi = 1$, rather than just $|W\phi| < |W|$. Our set of elementary regular H-maps is smaller than Lyndon's set of elementary transformations, so we must extend Lyndon's proof.

Let $W = y_1 \ldots y_t$, where $y_1, \ldots, y_t \in Y^{\pm 1}$, and suppose that $x\phi \neq 1$ for all $x \in$ gen(W). Since $|W\phi| < |W|$, it follows that $t \geq 2$. By Lemma 2.2, there is a j, $1 \leq j \leq t-1$, such that either $K((y_j\phi)(y_{j+1}\phi)) < K(y_j\phi)$ or $K((y_j\phi)(y_{j+1}\phi)) < K(y_{j+1}\phi)$. We consider the first case, the second being similar. Since W is reduced, $y_{j+1} \neq y_j^{-1}$, and since $K((y_j\phi)(y_{j+1}\phi)) < K(y_j\phi)$, $y_{j+1} \neq y_j$. If y_j is a variable, we may let ρ be defined by $y_j\rho = y_j y_{j+1}^{-1}$, and the conclusion follows. If y_j is a constant, then $y_j\phi = y_j$ and, since $K((y_j\phi)(y_{j+1}\phi)) < K(y_j\phi)$, y_{j+1} is a variable and $y_{j+1}\phi$ begins with y_j^{-1}. Thus $K((y_j\phi)(y_{j+1}\phi)) < K(y_{j+1}\phi)$ and we choose ρ so that $y_{j+1}\rho = y_j^{-1} y_{j+1}$. This concludes the proof of Lemma 2.3.

Proof of Theorem 2.1. The rest of the proof proceeds as in [4]. Lyndon shows in Lemma 4 of [4] that if $W \neq 1$ and ϕ is a solution of $W = 1$, there is a regular H-map ρ attached to W such that $y\rho^{-1}\phi = 1$ for some y occuring in W. This follows by Lemma 2.3 and induction on $K(W, \phi)$. The theorem is then established by induction on the number of elements of gen(W). If gen(W) is empty, $W = 1$ and we may let τ be the identity map. Otherwise, by Lemma 4 of [4], there is a regular H-map ρ attached to W such that $x\rho^{-1}\phi = 1$ for some variable x occuring in W. If x occurs in $W\rho$, the elementary singular map σ which sends x to 1 is admissible to $W\rho$ and $\rho^{-1}\phi$ is a solution of $W\rho\sigma = 1$. Since $W\rho\sigma$ has fewer variables than W, there is an H-map τ_1 attached to $W\rho\sigma$ such that $W\rho\sigma\tau_1 = 1$ and $\rho^{-1}\phi = \tau_1\psi$ for some H-map ψ. Now $x\rho^{-1}\phi = 1$, so $\rho^{-1}\phi = \sigma\rho^{-1}\phi = \sigma\tau_1\psi$, and so $\rho\sigma\tau_1\psi = \phi$. We let $\tau = \rho\sigma\tau_1$. If x does not occur in $W\rho^{-1}$, we find that $W\rho$ has fewer variables than W and so again there is a H-map τ_1 attached to $W\rho$ with $W\rho\tau_1 = 1$ and $\tau_1\psi = \rho^{-1}\phi$ for some H-map ψ. We then let $\tau = \rho\tau_1$ and the proof of the theorem is complete.

We record another consequence of Lemmas 2.2 and 2.3. An element W of $F * H$ is called *minimal* if $|W\alpha| \geq |W|$ for all H-automorphisms α.

Proposition 2.4. *If $W \in F * H$, there is an H-map ρ attached to W such that $W\rho$ is minimal.*

Proof. If W is minimal, we may let ρ be the identity map. If not, let α be an H-automorphism such that $W\alpha$ is minimal. Now $x\alpha \neq 1$ for all x in gen(W) since α is an H-automorphism, and $|W\alpha| < |W|$. By Lemma 2.3, there is a regular H-map ρ_1 attached to W such that $K(W\rho_1, \rho_1^{-1}\alpha) < K(W, \alpha)$. If $W\rho_1$ is minimal, we are finished. If not, $|W\alpha| = |(W\rho_1)(\rho_1^{-1}\alpha)| < |W\rho_1|$ and as before there is a regular H-map ρ_2 attached to $W\rho_1$ such that $K(W\rho_1\rho_2, \rho_2^{-1}\rho_1^{-1}\alpha) < K(W\rho_1, \rho_1^{-1}\alpha)$. Since the function K has values in the positive integers, we arrive within finitely many steps at a minimal word $W\rho_1 \ldots \rho_k$ with $\rho_1 \ldots \rho_k$ a regular H-map attached to W.

3. Quadratic equations

In this section, we restrict our attention to equations $W = 1$, where $W \in F * H$ is *quadratic*, that is, each element of X which occurs in W occurs exactly twice, each time with exponent 1 or -1.

We begin with an extension of a result of Piollet [10] :

Theorem 3.1. *If W is a quadratic element of $F * H$ and $\tau = \rho_0\sigma_1\rho_1 \ldots \sigma_k\rho_k$ is attached to W with $\rho_0, \ldots, \rho_k$ regular and $\sigma_1, \ldots, \sigma_k$ singular, then there is a regular H-map ρ attached to W such that $\tau = \rho\sigma_1 \ldots \sigma_k$.*

Proof. We follow the argument given by Piollet in Remarque 4 of section 4 of [10]. First note that it suffices to prove that if σ_1 is an elementary singular H-map and ρ_1 is an elementary regular H-map with $\sigma_1\rho_1$ attached to W, there is a regular H-map ρ attached to W such that $\rho\sigma_1 = \sigma_1\rho_1$. If $x\rho_1 = xy^{-1}$ and $z\sigma_1 = 1$, then W has a subword $(xVy)^{\pm 1}$ with $V\sigma_1 = 1$. Since W is quadratic, x does not occur in V and so the regular H-map $x\rho = xy^{-1}V^{-1}$ is attached to W. It is easy to see that $\rho\sigma_1 = \sigma_1\rho_1$.

[In Lemme 1 of [10], Piollet asserts that this result is true for arbitary $W \in F$. The proof is faulty, however, in that the map he produces is not in general attached to W. If we take $W = x^2zuzu^{-1}y^2$ with x, y, z, u variables and follow Piollet's construction (taking U_i to be $Q_iV_iQ_i^{-1}$, which seems to be intended instead of $Q_iU_iQ_i^{-1}$), we get a map ρ, the map Piollet calls α, such that $x\rho = xy^{-1}$ and $z\rho = yx^{-1}zxy^{-1}$. We find that $|W| = 8$ and $|W\rho| = 14$. Since an attached H-map never increases the length of a quadratic word, this ρ is not attached to W.]

If we drop the condition that ρ is attached to W, the statement of Theorem 3.1 becomes true for arbitrary $W \in F * H$. To see this, observe that if $1 \le i \le j \le k$, the varible killed by σ_i does not occur in $W\rho_0\sigma_1\rho_1 \dots \sigma_j$, to which ρ_j is attached. Thus ρ_j fixes the variable killed by σ_i, σ_i and ρ_j commute, and so $\tau = \rho_0\rho_1 \dots \rho_k\sigma_1 \dots \sigma_k$.

To establish our main result, Theorem 3.2, we need another tool. We associate with a minimal quadratic word W in $F * H$ a graph $R[W]$ of *regular transforms* of W. The set $R^0[W]$ of vertices of $R[W]$ is the set of $W\rho$ where ρ is a regular H-map attached to W, and the set $R^1[W]$ of (directed) edges of $R[W]$ is the set of pairs (U, ρ) where U is in $R^0[W]$ and ρ is a nonidentity elementary regular H-map attached to U. Edge (U, ρ) has initial vertex U and terminal vertex $U\rho$.

Since W is minimal and quadratic, so are all elements of $R^0[W]$. If (U, ρ) is in $R^1[W]$ and $x\rho = xy^{-1}$, then one occurrence of $x^{\pm 1}$ in U is in a subword $(xy)^{\pm 1}$, but the other is not since $|U\rho| = |U|$. Therefore $(xy^{-1})^{\pm 1}$ is a subword of $U\rho$, and so $(U, \rho)^{-1} = (U\rho, \rho^{-1})$ is in $R^1[W]$ as well.

[Piollet [10] associates a graph $C[W]$ with a not-necessarily-quadratic word W of F which is minimal in the sense that $|W\alpha| \ge |W|$ for all $\alpha \in \mathrm{Aut}(F)$. The vertices of $C(W)$ are words $W\tau$ where τ is an H-map attached to W and the edges of $C[W]$ are pairs (U, τ) where U is a vertex of $C[W]$ and τ is an elementary H-map attached to U. The inital vertex of (U,τ) is U and its terminal vertex is $U\tau$. For U a vertex of $C[W]$, $C_r[U]$ is the connected component containing U of $C'[W]$, the graph obtained from $C[W]$ by deleting edges associated with elementary singular H-maps. Piollet claims that $C_r[U]$ does not depend on W. This is false, even for quadratic words. For example, if $W = xyzxzy$, $z\sigma = 1$, $x\rho = xy^{-1}$, and $U = W\sigma\rho = x^2$, we see that W is minimal and that $W\sigma = xyxy$ is in $C_r[U]$ in $C[W]$ but not in $C_r[U]$ in $C[U]$.]

Since the number of images of a quadratic word W in $F * H$ under attached regular H-maps is finite, we can effectively construct a regular H-map β such that $W\beta$ is minimal. An H-map ϕ is a solution of $W = 1$ if and only if $\beta^{-1}\phi$ is a solution of $W\beta = 1$. Thus in discussing solutions of $W = 1$, there is no loss of generality in assuming that W is minimal.

For $W \in F * H$, we let RegStab(W) be the group of regular H-maps α attached to W such that $W\alpha = W$.

Theorem 3.2. *Let W be a minimal quadratic word in $F * H$. There is a finite set Ξ of solutions of $W = 1$ such that every solution of $W = 1$ is a specialization of some $\alpha\xi$ where α is in* RegStab(W) *and ξ is in Ξ.*

Proof. Choose a maximal tree T in $R[W]$. For $U \in R^0[W]$, let $\gamma_U = \rho_1 \ldots \rho_k$ where $(W,\rho_1)\ldots(U\rho_k^{-1},\rho_k)$ is a reduced path in T from W to U, that is, a path in which no edge is followed immediately by its inverse. By Theorem 2.1, every solution of $W = 1$ is a specialization of some τ attached to W. By Theorem 3.1, $\tau = \rho\sigma$ with ρ a regular H-map attached to W and σ a singular H-map. Now $W\rho$ is in $R^0[W]$, $\alpha = \rho\gamma_{W_\rho}^{-1}$ is in RegStab(W), and $W\gamma_{W_\rho}\sigma = W\rho\sigma = 1$. We may therefore let Ξ be the set of $\gamma_U\sigma$ where U is in $R^0[W]$ and σ is a singular H-map such that $U\sigma = 1$. This concludes the proof of Theorem 3.2.

Observe that a finite generating set for RegStab(W) may be calculated by the technique used to get generators for the fundamental group of a graph. If $e = (U,\rho)$ is an edge of $R[W]$ which is not in the maximal tree T, we let $\alpha_e = \gamma_U\rho\gamma_{U_\rho}^{-1}$; RegStab($W$) is generated by $\{\alpha_e \mid e \in R^1[W] - T\}$.

We illustrate Theorem 3.2 in a rather simple case. The graph $R[W_1]$ for the minimal quadratic word $W_1 = x^{-1}abxa^{-1}b^{-1}$ is shown in Figure 1.

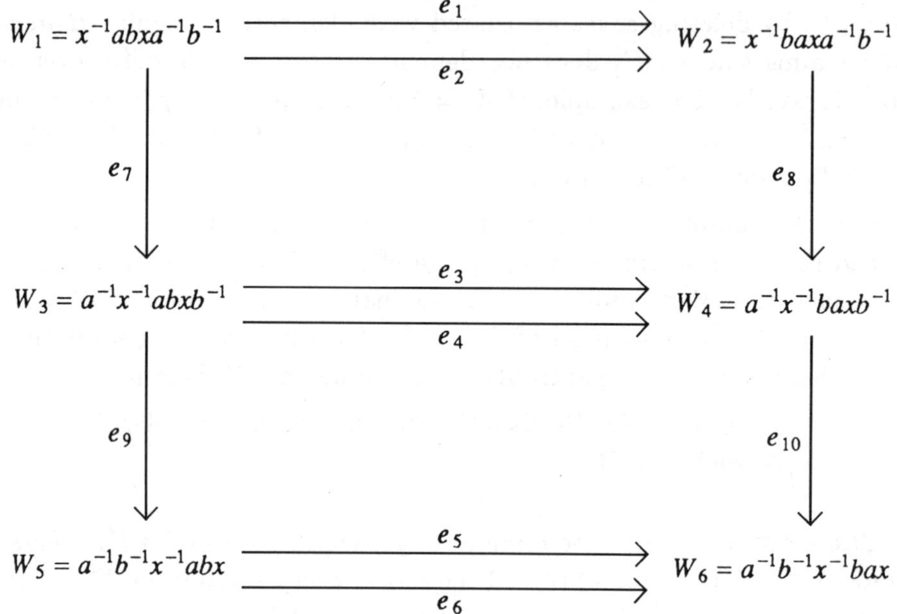

Figure 1.

The map $x^{-1} \to x^{-1}a^{-1}$ is associated with edges e_1, e_3 and e_5, $x \to b^{-1}x$ with e_2, e_4 and e_6, $x \to xa$ with e_7 and e_8, and $x \to xb$ with e_9 and e_{10}. Each of these edges has an inverse, to which the inverse map is associated. If we let our maximal tree T contain edges e_1, e_3, e_5, e_7, and e_9, we find that $\alpha_{e_2}, \alpha_{e_4}$ and α_{e_6} send x to $b^{-1}a^{-1}x$, and that α_{e_8} and $\alpha_{e_{10}}$ are the identity. Since only $W_2 = 1$, $W_3 = 1$, and $W_6 = 1$ have singular solutions, Ξ contains only the maps $x\xi_1 = a$ and $x\xi_2 = aba$. We conclude that every solution of $W_1 = 1$ sends x to $(ab)^n a$ for some integer n.

In closing, we remark that since $W = 1$ and $U^{-1}WU = 1$ have the same solutions for any $U \in F * H$, we could have chosen to work with cyclic words. This would have reduced the number of vertices in our example above, identifying W_5 with W_1 and W_6 with W_2. The number of edges of the graph, however, would have remained the same.

References

[1] R. G. Burns, C. C. Edmunds and I. H. Farouqi, On commutator equalities and stabilizers in free groups, *Canad. Math. Bull.* **19** (1976), 263-267.

[2] M. Culler, Using sufaces to solve equations in free groups, *Topology* **20** (1981), 133-145.

[3] Ju. I. Hmelevskii, Systems of equations in a free group, I, *Math. USSR Izvestiya* **5** (1971), 1245-1276.

[4] R. C. Lyndon, The equation $a^2b^2 = c^2$ in free groups, *Michigan Math. J.* **6** (1959), 89-95.

[5] R. C. Lyndon and P. E. Schupp, *Combinatorial Group Theory*, Springer Verlag, Berlin/Heidelberg/New York, (1977).

[6] R. C. Lyndon and M. J. Wicks, Commutators in free groups, *Canad. Math. Bull.* **24** (1981), 101-106.

[7] G. S. Makanin, Equations in a free group, *Math. USSR Izvestiya* **21** (1983), 483-546.

[8] Yu. I. Ozhigov, Equations in a free group, *Math. USSR Doklady* **27** (1983), 177-181.

[9] D. Piollet, Equations quadratiques dans le groupe libre, *J. Algebra* **33** (1975), 395-404.

[10] D. Piollet, Solutions d'une equation quadratique dans le groupe libre, *Discrete Math.* **59** (1986), 115-123.

[11] A. A. Razborov, On systems of equations in a free group, *Math. USSR Izvestiya* **25** (1985), 115-162.

[12] H. Zieschang, Alternierende Produkte in freien Gruppen, *Abh. Math. Sem. Univ. Hamburg* **27** (1964), 13-31.

Department of Mathematics
University of Wisconsin - Parkside
Kenosha, Wisconsin 53141-2000
U. S. A.

Department of Mathematics
Mount Saint Vincent University
Halifax, Nova Scotia B3M 2J6
Canada

Present address of L. Comerford :
Department of Mathematics
Eastern Illinois University
Charleston, Illinois 61920
U. S. A.

Block covers and module covers of finite groups

Fan Yun

§1. Introduction

Terminologies and notations follow [F] usually.

G denotes a finite group, R a complete local principal ideal domain and $p \in J(R)$ with p being a fixed prime (see [F] Ch. 1, §17). Let $H \leq G$, $\widetilde{B}$ an RH-block with idempotent $\tilde{e}$ and B an RG-block with idempotent e. These assumptions are preserved throughout, except for other illumination.

B is regarded as a two-sided G-module, i.e. an $R(G \times G)$-module. The restriction of B to an $H \times H$-module is written as $B{\downarrow}_{H\times H}$. Similarly, $B{\uparrow}^{G\times G}$ denotes the induced module.

We say B covers $\widetilde{B}$ if $\widetilde{B}|B{\downarrow}_{H\times H}$, and $\widetilde{B}$ covers B if $B|\widetilde{B}{\uparrow}^{G\times G}$. We define the module cover in the same way. Let U be an RH-module and V an RG-module. V covers U if and only if $U|V{\downarrow}_H$. U covers V if and only if $V|U{\uparrow}^G$.

A module belonging to B is also called a B-module.

We are concerned with the relations between block cover and module cover. Remark that it is easy to see that the results and ideas of this paper are all valid in the case of $\bar{R}G$ where $\bar{R}$ is the residue field of R.

Under the hypothesis that R is sufficiently large, [W] proves:

(1) B covers $\widetilde{B} \Rightarrow B$-modules cover all $\widetilde{B}$-modules;

(2) $\widetilde{B}$ covers $B \Rightarrow \widetilde{B}$-modules cover all B-modules.

Firstly, we show these two assertions are true in a very general ring-theoretic sense. Some other results of [W] and a result of [JT] are also extended. Then we prove the following theorem.

Theorem A. *If $P \cdot C_G(P) \leq H \leq G$ and P is a defect group of both B and $\widetilde{B}$, where P is a p-subgroup of G, then the following six statements are equivalent:*

(i) $\widetilde{B}|B\downarrow_{H\times H}$.

(ii) $B|\widetilde{B}\uparrow^{G\times G}$.

(iii) *Every $\widetilde{B}$-module is covered by a B-module.*

(iv) *Every B-module is covered by a $\widetilde{B}$-module.*

(v) *One of indecomposable $\widetilde{B}$-modules with vertex P is covered by a B-module.*

(vi) *One of indecomposable B-modules with vertex P is covered by a $\widetilde{B}$-module.*

Our arguments are based on the module theory and ring theory, and does not involve characters at all. Hence the hypothesis of splitting fields is not necessary to our results. The concepts such as defect groups, Brauer correspondences, etc. are also understood in the ways of module and ring theory. However, all different definitions of a certain concept coincide, provided they make senses respectively, see [AB].

We restate the module form of the Brauer's First Main Theorem and its proof for the latter quotations. Recall $\delta : \begin{matrix} G \to G\times G \\ x\mapsto (x,x)\end{matrix}$ is the diagonal map, and so, P^δ is the diagonal subgroup of $G\times G$.

First Main Theorem. *Let P be a p-subgroup of G, $\mathbf{N}_G(P)\le H\le G$. Block$(RG,P)$ denotes the set of blocks of G with defect group P. The Green correspondence with respect to $(G\times G,\ H\times H,\ P^\delta)$ is an one-to-one correspondence between Block(RG,P) and Block(RH,P), which is coincident with the Brauer correspondence, and consequently, the Brauer correspondence is in one-to-one correspondence between Block(RH,P) and Block(RG,P).*

Proof. $\forall\widetilde{B}\in Block(RH,P)$, set $B=\widetilde{B}^G$, i.e. $\widetilde{B}|B\downarrow_{H\times H}$, then P^δ is a vertex of B by [BC] Thm 5, so B is Green correspondent to $\widetilde{B}$ and $B\in Block(RG,P)$. Conversely, $\forall B\in Block(RG,P)$, $f(B)$, the image of B under Green correspondence, has P^δ as its vertex, and hence, $f(B)\in Block(RH,P)$ because any component with vertex P^δ of $(RG)\downarrow_{H\times H}$ must be a block of RH by [AB] Lemma 3.3.

§2. Some general results

Let $\tilde{S}$ be a subring of a ring S, B a block of S. The restriction of B as a two-sided S-module to a two-sided $\tilde{S}$-module is denoted by $B{\downarrow}_{\tilde{S}^o\otimes\tilde{S}}$, where $\tilde{S}^o$ is the opposite ring of $\tilde{S}$.

Proposition 1. *Let $\tilde{S}$ be a subring of a ring S. Assume the block decompositions of S and $\tilde{S}$ both exist, B is an S-Block with idempotent e and $\widetilde{B}$ is an $\tilde{S}$-block with idempotent $\tilde{e}$.*

(1) *If $\widetilde{B}|B{\downarrow}_{\tilde{S}^o\otimes\tilde{S}}$, then $U|((U{\uparrow}^S)e){\downarrow}_{\tilde{S}}$ for any $\widetilde{B}$-module U.*

(2) *If $B|\widetilde{B}{\uparrow}^{S^o\otimes S}$, then $V|((V{\downarrow}_{\tilde{S}})\tilde{e}){\uparrow}^S$ for any B-module V.*

Proof.

(1) Let $B{\downarrow}_{\tilde{S}^o\otimes\tilde{S}} = M \oplus N$, where $M \cong \widetilde{B}$, then $((U{\uparrow}^S)e){\downarrow}_{\tilde{S}} = (U \otimes_{\tilde{S}} S)e = U \otimes_{\tilde{S}} B_{\tilde{S}} = U \otimes_{\widetilde{B}} B = (U \otimes_{\widetilde{B}} M) \oplus (U \otimes_{\widetilde{B}} N)$ and $U \otimes_{\widetilde{B}} M \cong U \otimes_{\widetilde{B}} \widetilde{B} \cong U$.

(2) Clearly, $\widetilde{B}{\uparrow}^{S^o\otimes S} = S \otimes_{\tilde{S}} \widetilde{B} \otimes_{\tilde{S}} S$. Then $V = V \otimes_S S = V \otimes_S B | V \otimes_S \widetilde{B}{\uparrow}^{S^o\otimes S}$, $V \otimes_S \widetilde{B}{\uparrow}^{S^o\otimes S} = V \otimes_S (S \otimes_{\tilde{S}} \widetilde{B} \otimes_{\tilde{S}} S) = (V \otimes_S S) \otimes_{\tilde{S}} \widetilde{B} \otimes_{\tilde{S}} S = ((V{\downarrow}_{\tilde{S}}) \otimes_{\tilde{S}} \widetilde{B}) \otimes_{\tilde{S}} S$ $= ((V{\downarrow}_{\tilde{S}}))\tilde{e}){\uparrow}^S$.

This proposition shows that the Thm 3 of [W] is, in fact, a general ring-theoretic result and is easy to prove.

Now we turn to observe the group algebras.

Let η be the $H \times H$-homomorphism from RG to RH defined by

$$\eta x = \begin{cases} x & x \in H \\ o & x \in G \backslash H\,. \end{cases}$$

Another natural $H \times H$-homomorphism from RH to $\widetilde{B}$ is the multiplication by $\tilde{e}$. So $B \xrightarrow{\tilde{e}\eta} \widetilde{B}$ makes a plain sense.

The dual homomorphism is as follows.

There are two descriptions for $(RH)\uparrow^{G\times G}$: $RH \underset{R(H\times H)}{\otimes} R(G\times G) = (RH)\uparrow^{G\times G} = RG \underset{RH}{\otimes} RG$. The identification of the two descriptions is as follows:

$$RH \underset{R(H\times H)}{\otimes} R(G\times G) \longrightarrow RG \underset{RH}{\otimes} RG$$

$$h\otimes(x,y)\mapsto x^{-1}\otimes hy$$

and

$$RG \underset{RH}{\otimes} RG \longrightarrow RH \underset{R(H\times H)}{\otimes} R(G\times G)$$

$$x\otimes y\mapsto 1\otimes(x^{-1},y)\,.$$

A $G\times G$-homomorphism μ is defined as

$$\mu:\ RG \underset{RH}{\otimes} RG \longrightarrow RG$$

$$x\otimes y\mapsto xy\,.$$

It is easy to see that μ is an $H\times H$-split surjection. From this view of point the Corollary 6 of [W] is trivial.

Let $Tr_H^G(z)$ denote the trace map as usual.

The next two propositions are, in essential, restatements of Thm 1 and Thm 2 of [W] because the hypothesis of splitting field does not play a crucial role in the two cases.

Proposition 2. *The following three statements are equivalent:*

(i) $\widetilde{B}|B\downarrow_{H\times H}$.

(ii) $B\xrightarrow{e\eta}\widetilde{B}$ *is an* $H\times H$*-split surjection.*

(iii) *There is a* $z\in B$ *such that* $hz=zh\ \ \forall h\in H$ *and* $\eta z=\tilde{e}$.

Proof. (i) $\Rightarrow$ (iii). By [W] there is $z'\in B$, $hz'=z'h$, $\forall h\in H$ and $\eta z'\in Z(\widetilde{B})\backslash J(Z(\widetilde{B}))$ (in [W], the argument for this assertion is independent of the hypothesis of splitting field). But $Z(\widetilde{B})$ is a local ring because it has an unique idempotent (see [F] Ch.1, Cor. 12.7), hence $\eta z'$ is invertible in $Z(\widetilde{B})$. So there is $y\in Z(\widetilde{B})$ such that $y(\eta z')=\tilde{e}$. Put $z=yz'$, then it is all right.

(iii) $\Rightarrow$ (ii). Set $\lambda:\underset{\tilde{e}\mapsto z}{\widetilde{B}\longrightarrow B}$, which is just an $H\times H$-lifting of $B\xrightarrow{\tilde{e}\eta}\widetilde{B}$.

(ii) $\Rightarrow$ (i). Trivial.

Proposition 2*. *The following three statements are equivalent:*

(i) $B|\widetilde{B}\uparrow^{G\times G}$.

(ii) $\widetilde{B}\uparrow^{G\times G} \xrightarrow{e\mu} B$ *is a* $G\times G$*-split surjection.*

(iii) *There is a* $z \in \tilde{e}\cdot RG$ *such that* $hz = zh, \forall h \in H$, *and* $Tr_H^G(z) = e$.

Proof. (i) $\Rightarrow$ (iii). See [W], whose proof for this is valid in any case.

(iii) $\Rightarrow$ (ii). Let $x_1, \ldots, x_n$ be a cross section of H in G. Set

$$\tau : B \longrightarrow RG \underset{RH}{\otimes} (\tilde{e}\cdot RG) \quad (= \widetilde{B}\uparrow^{G\times G})$$

$$e \mapsto \sum_{i=1}^{n} x_i^{-1} \otimes (zx_i).$$

Since $\forall x \in G$, $x\cdot(\sum_{i=1}^n x_i^{-1}\otimes(zx_i)) = (\sum_{i-1}^n x_i^{-1}\otimes(zx_i))\cdot x$, the above τ is $G\times G$-homomorphism, consequently, it is a $G\times G$-lifting of $\widetilde{B}\uparrow^{G\times G} \xrightarrow{e\mu} B$.

(ii) $\Rightarrow$ (i). Trivial.

The next corollary includes the inverse proposition of Lemma 4 of [JT] and also extends Lemma 4 of [JT].

Corollary 3. $\widetilde{B}|B\downarrow_{H\times H}$ *if and only if there is a* $\hat{e} \in B$ *such that* $h\hat{e} = \hat{e}h$, $\forall h \in H$ *and* $\hat{e} = \tilde{e} + d$, $d \in R(G\backslash H)$, $\tilde{e}d = d = d\tilde{e}$, $\widetilde{B} = \hat{e}\cdot RH$. (Remark: it is implied that $hd = dh$, $\forall h \in H$.)

Proof. Take z in (iii) of Proposition 2 for $\hat{e}$ here.

Proposition 4. *If* $\widetilde{B}^G = B$ *in the sense of Alperin, then* $\widetilde{B} \xrightarrow{e} B$ *is a split injection.*

Proof. Refine the decomposition $RG = B \oplus (1-e)RG$ into

$$RG = X_1 \oplus \ldots \oplus X_n \oplus Y_1 \oplus \ldots \oplus Y_m \tag{1}$$

where $X_1 \oplus \ldots \oplus X_n = B$, $Y_1 \oplus \ldots \oplus Y_m = (1-e)RG$. Then refine the decomposition $RG = RH \oplus R(G\backslash H)$ into

$$RG = B_1 \oplus \ldots \oplus B_t \oplus D_1 \oplus \ldots \oplus D_s \tag{2}$$

where $B_1 = \widetilde{B}$. Of course, (1) and (2) are both indecomposable. Using Krull -Schmidt-Azumaya Theorem, one can replace $X_1, \ldots, X_n$ in the formula (1) by certain terms in the (2) one by one, and then obtain

$$RG = U_1 \oplus \ldots \oplus U_n \oplus Y_1 \oplus \ldots \oplus Y_m$$

where each U_i is either a some B_j or a some D_j. Since $\widetilde{B} \nmid (1-e)RG$, one of U_i is just the $\widetilde{B}$. Hence we have

$$RG = \widetilde{B} \oplus U \oplus (1-e)RG \tag{3}$$

We denote the projection from the decomposition (3) onto its summand $\widetilde{B}$ by ρ. From $e\tilde{e} = \tilde{e} - (1-e)\tilde{e}$, we have $\rho(e\tilde{e}) = \tilde{e}$. So $B \xrightarrow{\rho} \widetilde{B}$ is a contraction of $\widetilde{B} \xrightarrow{e} B$.

Proposition 5. *Let $H \trianglelefteq G$, and H contain a defect group of B. Then $\widetilde{B}|B{\downarrow_{H\times H}}$ if and only if $B|\widetilde{B}{\uparrow^{G\times G}}$.*

Proof. Firstly we note that $\widetilde{B}{\uparrow^{G\times G}}{\downarrow_{H\times H}}$ is a direct sum of conjugate modules of $\widetilde{B}$ by the Mackey decomposition.

If $B|\widetilde{B}{\uparrow^{G\times G}}$, then $B{\downarrow_{H\times H}} \,|\widetilde{B}{\uparrow^{G\times G}}{\downarrow_{H\times H}}$, hence $\widetilde{B}^{(x,y)}|B$, in another word, $\widetilde{B}|(B{\downarrow_{H\times H}})^{(x^{-1},y^{-1})} = B{\downarrow_{H\times H}}$.

Conversely, suppose $\widetilde{B}|B{\downarrow_{H\times H}}$. Then $B{\downarrow_{H\times H}}$ is a sum of some $G\times G$-conjugate modules of $\widetilde{B}$ (see [AB] §4). By Going-down Theorem, there is a component U of $B{\downarrow_{H\times H}}$ such that $B|U{\uparrow^{G\times G}} = \widetilde{B}^{(x,y)}{\uparrow^{G\times G}} = \widetilde{B}{\uparrow^{G\times G}}$.

§3. The proof of Theorem A

We begin the proof with a lemma.

Lemma 6. *Assume P is a normal p-subgroup of G, $M := PC_G(P) \leq H \leq G$, e is a block idempotent of RG. Then*

(1) $e = e_1 + \ldots + e_m$, each e_i is a block idempotent of RH; and $e_i = e_{i1} + \ldots + e_{in}$, each e_{ij} is a block idempotent of RM.

(2) For any i, $e_{i1}, \ldots, e_{in}$ is just a H-conjugate class, while all e_{ij} is a G-conjugate class.

(3) *If e' is another block idempotent of RG and $e' \neq e$, $e' = e'_1 + \ldots + e'_{m'}$, then $e_1, \ldots, e_m$, $e_1{}' \ldots, e'_{m'}$ is a system of orthogonal central idempotents.*

Proof. We prove the lemma in the case of $\bar{R}G$ because of the one-to-one correspondence between RG-block idempotents and $\bar{R}G$-block idempotents.

Write e as $e = \sum_{x \in G} e_x \cdot x$ with $e_x \in \bar{R}$. Using Brauer homomorphism, $e = e^\beta + e'$, where $e^\beta = \sum_{x \in C_G(P)} e_x \cdot x$, $e' = \sum_{x \in G \setminus C_G(P)} e_x \cdot x$. Since $C_G(P) \trianglelefteq G$, e^β can be expanded in G-class sums, so e^β is a central idempotent. e' can also be expanded in G-class sums, and the defect groups of these classes all do not contain P. Hence $e' \in J(RG)$, see [F] Ch. III, Cor. 6.11 and Ch. I, Lemma 8.15. So the difference of two central idempotents $e - e^\beta \in J(RG)$, which holds only in the case $e = e^\beta$. (1) follows.

For any $i, \{e_{i1}, \cdots, e_{in_i}\}$ is invariant under H-conjugate obviously; on the other hand, the sum of RM-block idempotents of an H-conjugate class is clearly a central idempotent of RH. From this observation (2) follows.

(3) is evident.

As first step of the proof of Theorem A, we prove the following proposition.

Proposition 7. *Let $P \trianglelefteq G$, $M := PC_G(P) \leq H \leq G$, then the six statements of Theorem A are equivalent.*

Proof. (i) $\Rightarrow$ (ii). Suppose $\widetilde{B} | B{\downarrow}_{H \times H}$. Using the notations in Lemma 6, taking e as the block idempotent of B, we can assume $\tilde{e} = e_1$ (recall that the Brauer correspondence in the sense of Brauer and in the sense of Alperin are the same one). Put $\approx{B} = e_{11} \cdot RM$, then $\approx{B} | B{\downarrow}_{M \times M}$, hence $B | \approx{B}{\uparrow}^{G \times G}$. Now Proposition 2* shows that there is a $z_{11} \in e_{11} \cdot RG$ such that $yz_{11} = z_{11} y$, $\forall y \in M$, and $Tr_M^G(z_{11}) = e$. Set $z_1 = Tr_M^H(z_{11})$. Clearly $hz_1 = z_1 h$, $\forall h \in H$. Using $e_{11} z_{11} = z_{11}$ and $e_1 = e_{11} + \cdots + e_{1n_1}$, it follows at once that $e_1 z_1 = z_1$, i.e. $z_1 \in e_1 \cdot RG$. $Tr_H^G(z_1) = Tr_M^G(z_{11})$. We have $B | \widetilde{B}{\uparrow}^{G \times G}$ by Proposition 2*.

(ii) $\Rightarrow$ (i). Suppose $B | \widetilde{B}{\uparrow}^{G \times G}$. From Lemma 6 we have an RM-block $\approx{B}$ such that $\approx{B} | \widetilde{B}{\downarrow}_{M \times M}$ and $\widetilde{B} | \approx{B}{\uparrow}^{H \times H}$ (see Proposition 5 also). So $B | \approx{B}{\uparrow}^{G \times G}$, hence $\approx{B}|B\downarrow_{M\times M}$. Using Lemma 6 again, e is the block idempotent of B, one of the e_{ij}, say e_{11}, must be the block idempotent of $\approx{B}$. Then $\approx{B}|\widetilde{B}\downarrow_{M\times M}$ implies $e_1 = \tilde{e}$. Hence $\widetilde{B}|B\downarrow_{H\times H}$.

(i) $\Rightarrow$ (iii). Proposition 1.(1).

(iii) $\Rightarrow$ (v). Trivial (see [AB] Cor. 3.8.).

(v) $\Rightarrow$ (i). By Lemma 6 and notations there, the restriction of any module V belonging to a block $e' \cdot RG$ of G to an RH-module has the form $V\downarrow_H = Ve'_1 \oplus \cdots \oplus Ve'_{m'}$. That is to say the modules belonging to $e'_i RH$ ($i = 1, \ldots, m'$) and only these RH-modules can be covered by the modules belonging to $e'RG$.

(ii) $\Rightarrow$ (iv) $\Rightarrow$ (vi). Similar to the above.

(vi) $\Rightarrow$ (ii). Use Lemma 6 and notations there again. let e and e' be distinct idempotents, $e = e_1 + \ldots + e_m$. Then $e_1 e' = 0$. So $U\uparrow^G \cdot e' = 0$ for any module U belonging to $e_1 \cdot RH$.

The following is also a general proposition, which reduces the proof of Theorem A to Proposition 7.

Proposition 8. *Let P be a p-subgroup of G, $PC_G(P) \leq H \leq G$, $N := \mathbf{N}_G(P)$, $L := H \cap N = \mathbf{N}_H(P)$, $M := PC_G(P)$. Assume B is an RG-block with defect group P and $\widetilde{B}$ is an RH-block with defect group P; B^* and B^{**} are an RN-block and an RL-block respectively, $(B^*)^G = B$, $(B^{**})^H = \widetilde{B}$. Then*

$$(1) \qquad \widetilde{B}|B\downarrow_{H\times H} \quad \text{if and only if} \quad B^{**}|B^*\downarrow_{L\times L};$$

$$(2) \qquad B|\widetilde{B}\uparrow^{G\times G} \quad \text{if and only if} \quad B^*|B^{**}\uparrow^{N\times N}.$$

Proof. From the First Main Theorem (see the Introduction of this paper), the Brauer correspondence on $Block(RH, P)$ and $Block(RG, P)$ coincides with the Green correspondence. So we have

$$B^*|B\downarrow_{N\times N} \quad \text{and} \quad B|B^*\uparrow^{G\times G}$$

$$B^{**}|\widetilde{B}\downarrow_{L\times L} \quad \text{and} \quad \widetilde{B}|B^{**}\uparrow^{H\times H}$$

consequently, B^* and B^{**} both have vertex P.

(1) Suppose $\widetilde{B}|B\downarrow_{H\times H}$. Then $B^{**}|B\downarrow_{L\times L}$. $B\downarrow_{L\times L}=(B\downarrow_{N\times N})\downarrow_{L\times L}$, and $B\downarrow_{N\times N}$ has an unique component, that is, B^*, with vertex P. Hence $B^{**}|B^*\downarrow_{L\times L}$.

Conversely, suppose $B^{**}|B^*\downarrow_{L\times L}$. Then $B^{**}|B\downarrow_{L\times L}$. $B\downarrow_{L\times L}=(B\downarrow_{H\times H})\downarrow_{L\times L}$, this is a component X of $B\downarrow_{H\times H}$ such that $B^{**}|X\downarrow_{L\times L}$, X has the vertex P because B and B^{**} both have the vertex P. Therefore X is correspondent to B^{**} under the Green correspondence with respect to $(H\times H, L\times L, P^{\delta})$. However, the $\widetilde{B}$ is also Green correspondent to B^{**}. So $\widetilde{B}\cong X|B\downarrow_{H\times H}$.

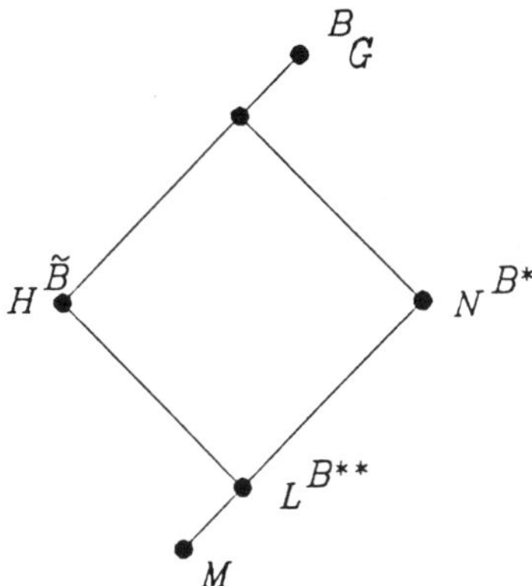

(2) Suppose $B|\widetilde{B}\uparrow^{G\times G}$. Then $B|B^{**}\uparrow^{G\times G}$. $B^{**}\uparrow^{G\times G}=(B^{**}\uparrow^{N\times N})\uparrow^{G\times G}$, hence $B^{**}\uparrow^{N\times N}$ has a component Y such that $B|Y\uparrow^{G\times G}$. The P is a vertex of Y, so B is correspondent to Y under Green correspondence with respect to $(G\times G,\ N\times N,\ P^{\delta})$. But B is also Green correspondent to B^*, this shows $B^*\cong Y|B^{**}\uparrow^{N\times N}$.

Conversely, if $B^*|B^{**}\uparrow^{N\times N}$, then $B|B^{**}\uparrow^{G\times G}$. From the facts that $B^{**}\uparrow^{G\times G}=(B^{**}\uparrow^{H\times H})\uparrow^{G\times G}$ and $B^{**}\uparrow^{H\times H}$ has an unique component with vertex P, that is $\widetilde{B}$, we have $B|\widetilde{B}\uparrow^{G\times G}$.

The above Proposition is, in fact, a general result on modules over group rings, i.e. we have the following

Proposition 9. *Let P, H, G, N, L, M be the same as those in*

Proposition 8. Assume U and V are an RH-module and an RG-module respectively, and U, V both have the vertex P. Let f_G and f_H be the Green correspondence with respect to (G, N, P) and the Green correspondence with respect to (H, L, P) respectively. Then

(1) $$U|V{\downarrow_H} \quad \text{if and only if} \quad f_H(U)|f_G(V){\downarrow_L};$$

(2) $$V|U{\uparrow^G} \quad \text{if and only if} \quad f_G(V)|f_H(U){\uparrow^N}.$$

The proof of Theorem A.

(i) $\Leftrightarrow$ (ii). Proposition 8 + Proposition 7.

(i) $\Rightarrow$ (iii) $\Rightarrow$ (v) $\Rightarrow$ (i). Only (v) $\Rightarrow$ (i) remains to be proved. Suppose a $\widetilde{B}$-module U with vertex P is covered by a B-module V.$f_G(V)$ and $f_H(U)$ (notations of Proposition 9) are B^*-module and B^{**}-module respectively by Nagao-Green Theorem ([F] Ch. III, Thm 7.8). Now the conclusion follows from Proposition 9, Proposition 7 and Proposition 8.

(ii) $\Rightarrow$ (iv) $\Rightarrow$ (vi) $\Rightarrow$ (ii). Similar to the above.

References

[AB] Alperin, J.L., Burry, D.W., Block theory with modules, *J. Algebra* **65** (1980), 225-233.

[BC] Burry, D.W., Carlson, J.F., Representations of modules to local subgroups, *Proc. Amer. Math. Soc.*, **84** (1982), 181-184.

[F] Feit, W., *The Representation Theory of Finite Groups*, North-Holland Publishing Company, 1982.

[JT] Juhasz, A., Tsushima, Y., A proof of the Brauer's second main theorem and related results, *Hokkaido Math. J.*, **14** (1985), 33-37.

[W] Watanabe, A., Relations between blocks of a finite group and its subgroup. *J. Algebra*, **78** (1982), 282-291.

Department of Mathematics
Wuhan University, Wuhan
People's Republic of China

A theorem in the commutator calculus

*Anthony M. Gaglione** and *Hermann V. Waldinger*

Background and statement of result

The purpose of this paper is to establish a group theoretical identity, Theorem 1.1, which involves basic commutators. The basic commutators derived from the generators of a free group of finite rank, $F = < c_1, c_2, \ldots, c_r >$, are implicit in the work of Philip Hall as they are the commutators which arise in his "collection process". The basic commutators, c_i, of a given weight n, $W(c_i) = n$, do form a basis for the free abelian group F_n/F_{n+1} where F_n denotes the nth term of the lower central series of F. As such, these basic commutators have been studied extensively by many others.

The commutator identity established here generalizes a result of [A] (see Lemma 3.6). It arose as the result of research on the lower central series of a free product of cyclic groups, at least one of which is of finite order. In [A], a procedure for finding presentations of the factors of the lower central series of such free products was established. Many of the results proved in [A] require special cases of our identity (1.1). Moreover this identity should be of interest in itself as it may be of use in solving other group theoretic problems.

We shall use the terminology and notation of [A]. Furthermore, the numbering $a.b.$ of any equation, theorem, definition, or lemma in [A] will correspond to $a.b'$ here. For instance, Lemma 3.6′ shall mean Lemma 3.6 of [A]. Moreover, the references in [A] are also relevant here.

Our identity is as follows:

Theorem 1.1 *Let* $c = (c_{j_1}, c_{j_2}, \ldots, c_{j_n})$ *be a* F*-simple basic commutator and let* d *be a generator of* F *such that* (i) $n > 1$ *and* (ii) $d < c_{j_n}$.

* Research of this author was supported by a grant from the Naval Academy Research Council.

Let $v = (c_{j_2}, d, e_1, e_2, \ldots, e_{n-1})$ *be such that* $e_1 = c_{j_1}$ *for* $n = 2$, *but* $e_1 \leq e_2 \leq \ldots \leq e_{n-1}$ *is a rearrangement of* $c_{j_1}, c_{j_3}, \ldots, c_{j_n}$ *for* $n > 2$. *(Note that* v *is a* F*-simple basic commutator with* $W(v) = n+1$ *if* $d < c_{j_2}$.*)* *The identity*

$$(c,d) = \begin{cases} \Pi_1 M(c,d)\Pi_2 & \text{if } d \geq c_{j_2} \\ \Pi_1 v^{-1}\Pi_2 M(c,d)\Pi_3 & \text{if } d < c_{j_2} \end{cases} \tag{1.1}$$

holds in F *where the* $\Pi_j (j = 1,2,3)$ *are either* $= 1$ *or are words in finitely many* F*-simple basic commutators* v_i *such that*

(i) $1 < W(v_i) \leq n$;

(ii) *if* $v_i = (w_1, w_2, \ldots, w_t)$, *then* $w_1, w_2, \ldots, w_t$ *is a rearrangement of a subsequence of* $c_{j_1}, c_{j_2}, \ldots, c_{j_n}, d$;

(iii) *if the* F*-simple basic commutator* v_i *occurs on the right hand side of* (1.1) *and* $1 < W(v_i) \leq n$, *then it occurs with exponent sum* 0.

For our definitions of basic commutators, maximal component of a, $M(a)$, and F-simple basic commutator see Definitions 2.1′, 3.1′, and 3.6′, respectively. Also "exponent sum" is defined as follows: if $w \in F$, $w = \prod_{i=1}^{I} c_{j_i}{}^{\varepsilon_i}$ where $I > 0$, the c_{j_i} are F-simple basic commutators with $W(c_{j_i}) > 1$, and the $\varepsilon_i = \pm 1$, then let $\delta_{hi} = \varepsilon_i$ if $c_{j_i} = c_h$ but $\delta_{hi} = 0$ if $c_{j_i} \neq c_h$ and let $n_h = \sum_{i=1}^{I} \delta_{hi}$, where c_h is an F-simple basic commutator with $W(c_h) > 1$. The F-simple basic commutator c_l is said to occur in w with exponent sum n_l if $c_l \in \{c_{j_1}, \ldots, c_{j_I}\}$.

Finally for convenience, we recall two identities. First, the well known identity (3.20)′: for a, b, c, $\in F$,

$$(a,b,c) = (b,a)(c,a)(c,b)(c,b,a)(a,b)(a,c)(a,c,b)(b,c). \tag{1.2}$$

Clearly (1.2) has the alternate version

$$(a,b,c) = (b,a)(c,a)a^{-1}(c,b)a(a,b)(a,c)(a,c,b)(b,c). \tag{1.2a}$$

Second, the trivial identity (3.19)′: for a_i, $b \in F$, $\varepsilon_i = \pm 1$,

$$(\prod_{i=1}^{I} a_i^{\varepsilon_i}, b) = (\prod_{i=1}^{I} a_i^{\varepsilon_i})^{-1} \prod_{i=1}^{I} (a_i(a_i, b))^{\varepsilon_i}. \tag{1.3}$$

Proof of Theorem 1.1 The proof proceeds by induction on the place of c in the ordering of basic commutators in Definition 2.1′. Let us start with $W(c) = n = 2$; then $d < c_{j_2}$ by hypothesis. We note that

$$(d, c_{j_2})(d, c_{j_2}, c_{j_1}) = [c_{j_1}^{-1}(c_{j_2}, d)^{-1}c_{j_1}(c_{j_2}, d)]\,(c_{j_2}, d)^{-1}$$
$$= (c_{j_2}, d, c_{j_1})^{-1}(c_{j_2}, d)^{-1}. \tag{2.1}$$

Making use of (1.2) and (2.1), we immediately obtain

$$(c, d) = (c_{j_1}, c_{j_2})^{-1}(c_{j_1}, d)^{-1}(c_{j_2}, d, c_{j_1})^{-1}(c_{j_2}, d)^{-1}\cdot$$
$$(c_{j_1}, c_{j_2})(c_{j_1}, d)(c_{j_1}, d, c_{j_2})(c_{j_2}, d). \tag{2.2}$$

We note that $v = (c_{j_2}, d, c_{j_1})$ and that $M(c, d) = (c_{j_1}, d, c_{j_2})$ according to Lemma 3.5′; moreover, (c_{j_1}, c_{j_2}), (c_{j_1}, d), and (c_{j_2}, d) are F-simple basic commutators of weight 2 and each occurs in (c, d) with exponent sum 0 from (2.2). Thus the identity (1.1) is established for $n = 2$.

Now let $W(c) = n > 2$ and suppose that our proof has been completed for all F-simple basic commutators c_i such that $W(c_i) > 1$ but $c_i < c$ in the ordering of Definition 2.1′. To find our conclusion for c, we apply (1.2a) to yield

$$(c, d) = (c^L, c_{j_n}, d) = u(c_{j_n}, d)^{-1}(c^L, d, c_{j_n})(c_{j_n}, d) \tag{2.3}$$

where

$$u = c^{-1}(c^L, d)^{-1}(c^L)^{-1}(c_{j_n}, d)^{-1}c^L c(c^L, d)(c_{j_n}, d). \tag{2.4}$$

By the definition of F-simple basic commutators (Definition 3.6′) and the induction hypothesis, u is a word in finitely many F-simple basic commutators v_i such that each v_i which occurs in u does so with exponent sum 0. (The v_i here are as in the conclusion of our theorem.) Let

$$w = (c_{j_n}, d)^{-1}(c^L, d, c_{j_n})(c_{j_n}, d). \tag{2.5}$$

It remains to show that w has all the required properties of our identity (1.1). For this purpose, we consider three mutually exclusive cases:

(I) $c_{j_{n-1}} \leq d < c_{j_n}$; (II) $c_{j_2} \leq d < c_{j_{n-1}}$; (III) $d < c_{j_2} \leq c_{j_{n-1}}$.

In case (I), our conclusion is immediate since $M(c,d) = (c^L, d, c_{j_n})$ in this case by Lemma 3.5′.

From the induction hypothesis, we have

$$(c^L, d) = \begin{cases} \Pi_{11} M(c^L, d) \Pi_{21} & \text{in case (II)} \\ \Pi_{11}(v_-)^{-1} \Pi_{21} M(c^L, d) \Pi_{31} & \text{in case (III)} \end{cases} \tag{2.6}$$

where

(a) $v_- = (c_{j_2}, d, e_{11}, e_{21}, \ldots, e_{n-2,1})$ such that $e_{11} = c_{j_1}$ if $n = 3$ but $e_{11} \leq e_{21} \leq \ldots \leq e_{n-2,1}$ is a rearrangement of $c_{j_1}, c_{j_3}, c_{j_4}, \ldots, c_{j_{n-1}}$ if $n > 3$;
(b) the Π_{j1} ($j = 1, 2, 3$) are either $= 1$ or are words in finitely many F- simple basic commutators v_{i1} such that (i) $1 < W(v_{i1}) \leq n - 1$, and (ii) if $v_{i1} = (w_{11}, w_{21}, \ldots, w_{t1})$, then $w_{11}, \ldots, w_{t1}$ is a rearrangement of a subsequence of $c_{j_1}, c_{j_2}, \ldots, c_{j_{n-1}}, d$.
(c) if the F-simple basic commutator v_{i1} occurs in (c^L, d) and $1 < W(v_{i1}) < n$, then it occurs with exponent sum 0.

Applying (1.3) to compute (c^L, d, c_{j_n}) from (2.6), we find

$$(c^L, d, c_{j_n}) = \begin{cases} (c^L, d)^{-1} \Pi_{12} M(c^L, d)(M(c^L, d), c_{j_n}) \Pi_{22} & \text{in case (II)} \\ (c^L, d)^{-1} \Pi_{12} (v_-, c_{j_n})^{-1} (v_-)^{-1} \Pi_{22}. & \\ M(c^L, d)(M(c^L, d), c_{j_n}) \Pi_{32} & \text{in case (III)} \end{cases} \tag{2.7}$$

where the Π_{j2} ($j = 1, 2, 3$) are words in the v_{i1} occurring in (2.6) and the commutators (v_{i1}, c_{j_n}). If the commutator (v_{i1}, c_{j_n}) is not basic, then we may apply the induction hypothesis to it because $v_{i1} < c$ (from (b)(i) above $W(v_{il}) < W(c)$). Applying the induction hypothesis to those (v_{i1}, c_{j_n}) which are not basic, (2.7) allows us to write (2.5) as

$$w = \begin{cases} \Pi_{13}(M(c^L, d), c_{j_n}) \Pi_{23} & \text{in case (II)} \\ \Pi_{13}(v_-, c_{j_n})^{-1} \Pi_{23}(M(c^L, d), c_{j_n}) \Pi_{33} & \text{in case (III)} \end{cases} \tag{2.8}$$

where the Π_{j3} ($j = 1, 2, 3$) are either $= 1$ or are words in finitely many F-simple basic commutators v_{i2} which have properties (i) and (ii) of our conclusion.

To see that property (iii) of our conclusion also holds, note that if a commutator has exponent sum 0 in (2.6), then it is clear that it will also have exponent sum 0 in (2.7). The only commutators which do not have exponent sum 0 in (2.6) are $M(c^L, d)$ and v_-. But these commutators have exponent sum 0 in (2.7). Since the v_{i1} have exponent sums 0 in (2.7), the v_{i1} and the (v_{i1}, c_{j_n}) if basic or the basic commutators occurring in the expression for (v_{i1}, c_{j_n}) will have exponent sum 0 in (2.8). Thus (iii) holds for (2.8).

We note that since $d < c_{j_{n-1}}$ in cases (ii) and (iii), $(M(c^L, d), c_{j_n}) = M(c, d)$ by Lemma 3.5′. Thus from (2.8), to complete our induction proof it remains to examine (v_-, c_{j_n}) in case (III). Either (v_-, c_{j_n}) is basic in which case, we are done, or if (v_-, c_{j_n}) is not basic then we note that $v_- < c$ by our lexicographic ordering of basic commutators of the same weight (see Definition 2.1′). Thus if (v_-, c_{j_n}) is not basic, we may apply our induction hypothesis to write it in the form (1.1). Since $v_- = (c_{j2}, d, e_{11}, \ldots, e_{n-2,1})$ and $d < c_{j_n}$, the induction hypothesis gives

$$(v_-, c_{j_n}) = \Pi_{14} M(v_-, c_{j_n}) \Pi_{24} \qquad (2.9)$$

where Π_{j4} ($j = 1, 2$) have the properties of our conclusions. Finally, substitution of (2.9) into (2.8) in case (III) completes our proof, by virtue of Lemma 3.5′.

Reference

[A] A. M. Gaglione and H. V. Waldinger, Factor Groups of the Lower Central Series of Free Products of Finitely Generated Abelian Groups, *Proceedings of Groups - St. Andrews 1985*, London Math. Soc. Lecture Notes, Ser 2 (1987), 164-203.

Department of Mathematics
U.S. Naval Academy
Annapolis, Maryland 21012, U. S. A.

Department of Mathematics
Polytechnic University
Brooklyn, New York 11201, U. S. A.

Topological methods in equivariant cohomology

J.P.C. Greenlees

0. Ideal

My purpose here is to advertise the fruits of applying topological methods to the study of group cohomology and equivariant cohomology. The topological methods I mean are those used to study theories themselves, and not simply the use of topological constructions on spaces. These methods are usually known as stable equivariant topology [8], but we will avoid using them here, and only explain their consequences.

For example it is well known that the ordinary equivariant cohomology or Tate cohomology of a group or a space is a module over the group cohomology ring; we will explain various methods for exploiting the homological properties of the group cohomology ring in the study of these cohomology theories.

The present account follows the talk I gave at the conference, but provides a little more body by giving some sample calculations. These calculations are in Section 3 and may be read independently of the intervening sections. In Section 1 we clear up several possible sources of misunderstanding, introduce a homology theory and a cohomology theory, and summarise various descriptions of well known theories and their relationships. In Section 2 we explain and then state the main Universal Coefficient Theorem of [11].

1. Three equivariant theories

Before we begin we must be quite clear about certain conventions. In fact G will be a finite group throughout, all homology and cohomology theories will be unreduced unless provided with a tilde and will have coefficients in a ring k on which G acts trivially. It is possible to introduce coefficients on which G acts, but it seems hard to analyse the result. Spaces will be well behaved; for example simplicial G- complexes or G CW-complexes. An asterisk always refers to grading over the integers.

In fact, there are three well known theories.

(1) Ordinary equivariant cohomology $H_G^*(X)$

For this we have the chain level description

$$H_G^*(X) = H^*(\text{tot Hom}_{\mathbf{Z}G}(P_\bullet, C^*(X)))$$

where $P_\bullet \to \mathbf{Z} \to 0$ is a projective resolution of $\mathbf{Z}$ by $\mathbf{Z}G$-modules, and $C^*(X)$ is the cochain complex of X. We also have the topological description

$$H_G^*(X) = H^*(EG \times_G X)$$

where EG is a nonequivariantly contractible G-space on which G acts freely, and $EG \times_G X$ is the quotient of $EG \times X$ by the diagonal action of G.

(2) Ordinary equivariant homology $H_*^G(X)$

For this we have the chain level description

$$H_*^G(X) = H_*(\text{tot } P_\bullet \otimes_{\mathbf{Z}G} C_*(X))$$

where $C_*(X)$ is the chain complex of X. We also have the topological description

$$H_*^G(X) = H_*(EG \times_G X).$$

Finally, we have

(3) Tate homology and cohomology $\hat{H}_*^G(SX)$ and $\hat{H}_G^*(X)$

Here we have for finite dimensional X the chain definitions

$$\hat{H}_G^*(X) = H^*(\text{tot Hom}_{\mathbf{Z}G}(T_\bullet, C^*(X)))$$

and

$$\hat{H}_*^G(X) = H_*(\text{tot } T_\bullet \otimes_{\mathbf{Z}G} C_*(X))$$

where $T_\bullet$ is a Tate resolution of $\mathbf{Z}$.

In this case the topological definition is not so well known, but has been independently discovered by several people. It involves a little further explanation, and we give it below. This approach has the advantage of leading to the correct definition when X is not finite dimensional, as well as revealing useful formal properties [9].

The reader may have noticed that ordinary equivariant cohomology and homology appeared as different theories, and that an extra suspension was inserted in Tate homology. This is my first important point. To explain it we must understand when a homology and cohomology theory "correspond". The answer is that they should obey duality. Here we have a choice of various forms of duality, but the resulting correspondences are the same : we concentrate on Alexander duality (see for example [13]).

Suppose then that we have a finite (nonequivariant) simplicial complex K; we may embed K in a simplicial fashion in a suitably triangulated sphere S^{n+1} :

$$K \hookrightarrow S^{n+1}.$$

Alexander duality then gives the first isomorphism below; the second follows since $S^{n+1}\backslash pt$ is contractible.

$$H^q(K, pt) \cong H_{(n+1)-q}(S^{n+1}\backslash pt, S^{n+1}\backslash K) \cong H_{n-q}(S^{n+1}\backslash K, pt).$$

Futhermore this isomorphism is, in a suitable sense, independent of n and the embedding. To emphasise this we use the formal n-fold desuspension

$$DK = S^{-n}(S^{n+1}\backslash K)$$

which is known as the Spanier-Whitehead dual of X [12,14]. This is independent of all choices up to a suitable notion of equivalence. It is of course not a space, but it allows us to write [12]

$$\tilde{H}^q(K) = \tilde{H}_{-q}(DK).$$

Futhermore, since K is also homotopy equivalent to the complement of its complement it follows that the homology of K is determined by the cohomology of DK. In short the homology and cohomology of finite complexes determine each other.

In the equivariant case we must allow S^{n+1} to be replaced by the one point compactification $S^{V\oplus 1}$ of the representation $V \oplus 1$, so that we have enough spheres to allow us to embed an arbitrary finite complex. We then say that the equivariant homology theory $E_*(\bullet)$ corresponds to the cohomology theory $E^*(\bullet)$ when we have a natural correspondence as before

$$E^q(K, pt) \cong E_{V-q}(S^{V\oplus 1}\backslash K, pt)$$

i.e.

$$\tilde{E}^q(K) \cong \tilde{E}_{-q}(DK)$$

where $DK = S^{-V}(S^{V\oplus 1}\backslash K)$ by analogy with the nonequivariant case [2]. The appearance of a subscript $V - q$ shows that we require our equivariant cohomology theory to admit suspension isomorphisms for all real representations. Such theories may be graded over the real representation ring $RO(G)$, but it is clearer to restrict attention to integer gradings by the above device of suitable suspension or desuspension. Note in particular that for the coefficient groups we have

$$E^q = E^q(pt) = E^q(S^0, pt) = E_{-q}(S^1\backslash S^0, pt) = E_{-q}(pt) = E_{-q}.$$

We have now seen two examples where superscripted and subscripted objects correspond under the rule $M^q = N_{-q}$. It is useful to be systematic about this, and reverse the grading whenever we raise or lower an index. If the grading appears as a subscript we refer to it as a degree, and if as a superscript as a codegree. Henceforth we will refer to an object M^* as cograded; the object M_* obtained by taking $M_q = M^{-q}$ is referred to as the corresponding graded object.

Thus we may rewrite the above equalities as

$$E^* = E_*$$

and, more generally,

$$\tilde{E}^*(K) = \tilde{E}_*(DK).$$

We may now return to the cases (1) and (2) above. In fact the coefficient ring $H_G^* = H_G^*(pt) = H^*(BG)$ is nonzero in positive codegrees, which is to say *negative* degrees. On the other hand $H_*^G = H_*^G(pt) = H_*BG$

is nonzero in *positive* degrees. Consequently, $H^*_G(X)$ and $H^G_*(X)$ are *not* corresponding homology and cohomology theories.

Similarly, we have the well known isomorphism [5]

$$\hat{H}^i_G = \hat{H}^i(G) = \hat{H}_{i-1}(G) = \hat{H}^G_{i-1}$$

and so we must insert a suspension to hope for a correspondence between Tate homology and cohomology.

It is therefore useful to have available a notation allowing us to refer to corresponding homology and cohomology theories by the same letter. For Tate cohomology we use the letter t, so that

$$t^*(X) = \hat{H}^*_G(X)$$

and

$$t_*(X) = \hat{H}^G_*(SX).$$

For equivariant cohomology the term Borel cohomology is also current, (since Borel introduced the topological form [4]) and so we use the letter b and we have

$$b^*(X) = H^*(EG \times_G X).$$

The corresponding homology theory is called Borel homology and written $b_*(X)$. Finally, since equivariant homology is clearly closely related to Borel cohomology we use the letter c and we have

$$c_*(X) = H_*(EG \times_G X).$$

We refer to this as coBorel homology, and to the corresponding cohomology theory $c^*(X)$ as coBorel cohomology.

We summarise these points in Table 1 below. The descriptions of $b^*(X)$ and $c_*(X)$ are well known (see for instance [5]). The original definition of Tate cohomology of spaces is due to Swan [15], and a discussion of $\hat{H}^*_G(X)$ and $\hat{H}^G_*(X)$ may be found in [5]. The correspondence of these theories with the topologists' definition of the Tate theory has been discovered several times, and appears in [9]. Finally, the chain level descriptions of $c^*(X)$ and $b_*(X)$ follow by the methods of [9] together with the description of the representing objects [10]. Alternatively, by adopting Dold-Puppe's treatment

TABLE 1

	(1)
Theory :	Borel
Representing object :	b
Cohomology :	
Name and notation	Borel cohomology $b^*(X)$
Alternatives	Ordinary equivariant cohomology $H^*_G(X)$
Chain description	$b^*(X) = H^*(\text{tot Hom}_{\mathbf{Z}G}(P_\bullet, C^*(X)))$
Topological description	$= H^*(EG \times_G X)$
Homology :	
Name and notation	Borel homology $b_*(X)$
Alternatives	——
Chain description	$b_*(X) = H_*(\text{tot } Q_\bullet \otimes_{\mathbf{Z}G} C_*(X))$
Topological description	——

Comments : Here $\cdots \longrightarrow P_2 \longrightarrow P_1 \longrightarrow P_0 \longrightarrow \mathbf{Z} \longrightarrow 0$ is a projective $\mathbf{Z}G$-resolution and the Tate resolution $T_\bullet$ may be taken to be $\cdots \longrightarrow P_1$

$0 \longrightarrow S_{-1}Q_\bullet \longrightarrow T_\bullet \longrightarrow P_\bullet \longrightarrow 0.$

Continued on next page

Table 1. *continued*

(2) coBorel $c = b \wedge EG_+$	(3) Tate $t = b \wedge S^0 * EG$
coBorel cohomology $c^*(X)$ —— $c^*(X) = H^*(\text{tot Hom}_{\mathbf{Z}G}(Q_\bullet, C^*(X))$ $= \varinjlim_n b^*(D(EG_+^{(n)}) \wedge X)$ (X finite dimensional)	Tate cohomology $t^*(X)$ $\hat{H}_G^*(X)$ $t^*(X) = H^*(\text{tot Hom}_{\mathbf{Z}G}(T_\bullet, C^*(X)))$ $= \varinjlim_n b^*(D(S^0 * EG^{(n)}) \wedge X)$ (X finite dimensional)
coBorel homology $c_*(X)$ ordinary equivariant homology $H_*^G(X)$ $c_*(X) = H_*(\text{tot } P_\bullet \otimes_{\mathbf{Z}G} C_*(X))$ $= H_*(EG \times_G X)$	Tate homology $t_*(X)$ $\hat{H}_*^G(SX)$ $t_*(X) = H_*(\text{tot } S_1 T_\bullet \otimes_{\mathbf{Z}G} C_*(X))$ $= b_*(X \wedge S^0 * EG)$

ZG-resolution, $0 \longrightarrow \mathbf{Z} \longrightarrow Q_0 \longrightarrow Q_{-1} \longrightarrow Q_{-2} \longrightarrow \cdots$ is an injective $\longrightarrow P_0 \longrightarrow Q_0 \longrightarrow Q_{-1} \longrightarrow \cdots$ so that we have an exact sequence (with $P_0 \searrow \mathbf{Z} \nearrow Q_0$)

of duality [7] the chain level descriptions follow from those of $c_*(X)$ and $b^*(X)$ respectively. Similarly, these methods also allow an "algebraic" proof that $t^*(X)$ and $t_*(X)$ correspond.

It is clear from the chain descriptions in Table 1 that there are natural long exact sequences

$$\cdots \longrightarrow t^*(SX) \longrightarrow c^*(X) \longrightarrow b^*(X) \longrightarrow t^*(X) \longrightarrow \cdots$$

and

$$\cdots \longrightarrow c_*(X) \longrightarrow b_*(X) \longrightarrow t_*(X) \longrightarrow c_*(SX) \longrightarrow \cdots$$

(the existence of the first even when X is not finite dimensional follows easily once one has the relevant definitions).

It is well known that $b^* = H^*(BG)$ is a ring and that $b^*(X)$ is a module over it via cup product and the projection map $EG \times X \longrightarrow EG$; it is also well known that $c_*(X) = H_*(EG \times_G X)$ is a module over b^* via cap product and $t^*(X)$ is a module over b^* in a way combining these actions. In fact the exact sequences above are sequences of b^*-modules, and indeed b is a ring object in a suitable sense with t and c as module objects over it.

2. Universal Coefficient and Künneth Theorems

In the case of ordinary singular homology and cohomology Universal Coefficient Theorems (UCT's) give us information about $H_*(X; M)$ when we provide information about $H_*(X; R)$ and M is a module over R. The best known case is when $R = \mathbf{Z}$. In this classical case the UCT takes the form of a short exact sequence involving derived functors of $\otimes_{\mathbf{Z}}$. We are especially fortunate that $\mathbf{Z}$ is a pid and so $\otimes_{\mathbf{Z}}$ only has one derived functor.

In the general case we expect a UCT to give us information about $N_*(X)$ when provided with information about $S_*(X)$, at least when N is a module over S so that it is reasonable to hope for it to do so. (Of course it is not reasonable to expect to learn all about $H_*(X; \mathbf{Z})$ from $H_*(X; \mathbf{Q})$.) Now R is replaced by S_* , and so we expect it to have global dimension greater than one in general. Therefore we can only hope for a UCT to give a spectral sequence. Similar comments apply to cohomology, and to Künneth Theorems (KT's). The correct formulation of UCT's and KT's was provided by Adams [1], and he also elaborated a general strategy for proving such theorems. This involves geometrically resolving spaces by sums of finite spaces P with $S_*(P)$ S_*-projective, by analogy with homological algebra. In fact he proved that in the nonequivariant case, provided there are "enough well behaved S-projectives" we have UCT's and KT's for S-module theories.

In the equivariant case one must again prove there are enough well behaved S_*-projectives, but there is now an additional problem : in general the spectral sequences will not converge to the groups of interest. In the case when S is the Borel theory b we have shown elsewhere that there are enough well behaved b_*-projectives for any elementary abelian group G and any finite field k; we have also shown that such spectral sequences converge when G is a p-group and $k = \mathbf{F}_p$ (and do not converge in general otherwise) [11]. The resulting theorem runs as follows.

Theorem [11] : *If G is an elementary abelian p-group and $k = \mathbf{F}_p$, then if X and Y are any G-spaces and F is a b-module we have the following spectral sequences*

(UCT1) $$E^2_{s,t} = \mathrm{Tor}^{b_*}_{s,t}(b_*X, F_*) \Rightarrow F_*X$$

(UCT2) $$E_2^{s,t} = \mathrm{Ext}^{s,t}_{b_*}(b_*X, F^*) \Rightarrow F^*X$$

(KT1) $$E^2_{s,t} = \mathrm{Tor}^{b_*}_{s,t}(b_*X, b_*Y) \Rightarrow b_*(X \times Y)$$

(KT2) $$E_2^{s,t} = \mathrm{Ext}^{s,t}_{b_*}(b_*X, b^*Y) \Rightarrow b^*(X \times Y)$$

For finite G-spaces A we also have

(UCT3) $$E^2_{s,t} = \mathrm{Tor}^{b^*}_{s,t}(b^*A, F^*) \Rightarrow F^*A$$

(UCT4) $$E_2^{s,t} = \mathrm{Ext}^{s,t}_{b^*}(b^*A, F_*) \Rightarrow F_*A$$

(KT3) $$E^2_{s,t} = \mathrm{Tor}^{b^*}_{s,t}(b^*A, b^*Y) \Rightarrow b^*(A \times Y)$$

(KT4) $$E_2^{s,t} = \mathrm{Ext}_{b^*}^{s,t}(b^*A, b_*Y) \Rightarrow b_*(A \times Y).$$

The convergence is strong in all cases with Tor E_2-term and conditional in the sense of Boardman [3] in cases with Ext E_2 term (see Note (4) below). Furthermore the edge homomorphisms are of the usual form [1], and the differentials d_r are of bidgree $(-r, r-1)$ or bicodegree $(r, 1-r)$ in each case.

Notes :

(1) Recall that $b^* = H^*(BG)$, so that all the spectral sequences result from doing homological algebra over the group cohomology ring.

(2) Amongst the modules F of interest, for any subgroup H of G we have
   (a) b_H representing H-equivariant Borel theory
   (b) t_H representing H-equivariant Tate theory
   (c) c_H representing H-equivariant coBorel theory
   (d) Suitably localised ordinary, Borel, coBorel or Tate theory of H-fixed points.

(3) The grading conventions remain in force. The total degree in the E_∞ or E^∞ term corresponds to degree in the target. For example with (UCT3) the subquotients of F^nA are the $E^\infty_{s,t}$ with $s+t=-n$. In this connection we use the convention that Hom^t refers to morphisms decreasing degree by t (i.e. increasing codegree by t). This convention is not universal.

(4) Since Boardman's marvellously clear account [3] of convergence has not yet been published we briefly explain his notion of conditional convergence. It is appropriate to a spectral sequence when the filtration of the target, A, is infinite and decreasing. Suppose that the filtration $\{F^sA\}_{s\geq 0}$ arises from an inverse system

$$\cdots \longrightarrow A^s \longrightarrow \cdots \longrightarrow A^2 \longrightarrow A^1 \longrightarrow A^0 = A$$

by taking $F^sA = \mathrm{im}(A^s \longrightarrow A)$, as happens in our case. The best convergence is strong convergence, when the E_∞ term gives us the subquotiently $F^sA/F^{s+1}A$ and the filtration $\{F^sA\}_{s\geq 0}$ defines a complete Hausdorff topology (i.e. $\lim\limits_{\leftarrow s} F^sA$ and $R\lim\limits_{\leftarrow s} F^sA$ are zero). Conditional convergence occurs when $\lim\limits_{\leftarrow s} A^s$ and $R\lim\limits_{\leftarrow s} A^s$ are zero. The point is that even a basically well behaved spectral sequence can converge very poorly when applied to an unpleasant or infinite space. Conditional convergence records the basic good behaviour of the type of spectral sequence, and Boardman gives necessary and sufficient conditions (the vanishing of a certain $R\ \underleftarrow{\lim}$) for a conditionally convergent spectral sequence to be strongly convergent. For example in our case the spectral sequences are certainly strongly convergent if $E_2^{s,t}$ is finite for all s and t. Furthermore even if the spectral sequence converges conditionally but not strongly, various comparison theorems may be used.

We comment briefly that since $c^* = \mathrm{Hom}_k^*(b_*, k)$, c^* is b_*-injective and so (UCT2) and (UCT4) with $F = c$ re-express the obvious fact that Borel and coBorel theories are especially closely related.

In the next section we will perform specific calculations with (UCT3) and (KT3) involving only ordinary equivariant cohomology.

3. Specific calculations

Here we will concentrate on two very special cases in which the spectral sequences involve only familiar theories. We recall that G is a p-group and homology and cohomology have mod p coefficients. We revert to classical notation to avoid alienating those who omitted Sections 1 and 2.

Firstly, (UCT3) with F representing K-equivariant Borel cohomology for some subgroup K of G provides a spectral sequence

$$(3.1) \qquad E^2_{s,t} = \mathrm{Tor}^{H^*BG}_{s,t}(H^*_G(A),\ H^*(BK)) \Rightarrow H^*_K(A)$$

for any finite G-space A. Secondly (KT3) provides a spectral sequence

$$(3.2) \qquad E^2_{s,t} = \mathrm{Tor}^{H^*BG}_{s,t}(H^*_G(A), H^*_G(Y)) \Rightarrow H_G(A \times Y)$$

for any finite G-space A and any G-complex Y at all. Note here the special case that $G = \tilde{G}/N$ and $Y = (E\tilde{G})/N$ then we have

$$\begin{aligned} H^*_G(A \times Y) &= H^*_G(A \times (E\tilde{G})/N) \\ &= H^*((A \times (E\tilde{G})/N)/G) \\ &= H^*_{\tilde{G}}(A) \end{aligned}$$

thus the spectral sequence reads

$$(3.3) \qquad E^2_{s,t} = \mathrm{Tor}^{H^*BG}_{s,t}(H^*_G(A), H^*(B\tilde{G})) \Rightarrow H^*_{\tilde{G}}(A)$$

for any finite G-space A.

We recall that these spectral sequences are only proved to exist when G is elementary abelian. Some of the algebra in Case 2 is relevant to $G = C_p$ for odd primes p.

Let us consider various abelian 2-groups G.

Case 1 $G = C_2$.

Of course $H^*BG = \mathbf{F}_2[t]$ is a pid and therefore of global dimension 1. In particular (3.1) and (3.2) become the short exact sequences

$$0 \to H^*_{C_2}(A)/tH^*_{C_2}(A) \to H^*(A) \to {}_tH^*_{C_2}(A) \to 0$$

where ${}_tN = \{n \in N \mid tn = 0\}$ is the t-torsion of N, and

$$0 \to H^*_{C_2}(A) \otimes_{\mathbf{F}_2[t]} H^*_{C_2}(Y) \to H^*_{C_2}(A \times Y) \to S_1 \mathrm{Tor}^{\mathbf{F}_2[t]}_{1,*}(H^*_{C_2}(A), H^*_{C_2}(Y)) \to 0.$$

Here we have used the homological suspension functor defined by $(S_1 M_*)_i = M_{i-1}$. This should not be confused with the cohomological suspension functor S^1 defined by $(S^1 M^*)^i = M^{i-1}$.

Case 2 $G = C_{2^k}$ for $k \geq 2$.

In this case we have $H^*BG = \mathbf{F}_2[x] \otimes \Lambda[\tau]$ where $\tau \in H^1$, $x \in H^2$; we write b^* for this ring. From the relevant Lyndon-Hothschild-Serre Spectral Sequences we deduce the following.

Lemma (3.4) :
(a) $\check{C}_{2^\ell} \rightarrowtail C_{2^k}$ *(for* $1 \leq \ell \leq k-1$*) induces the map*

$$\begin{array}{ccccccc} \mathbf{F}_2[x] & \otimes & \Lambda[\tau] & \to & \mathbf{F}_2[\check{x}] & \otimes & \Lambda[\check{\tau}] \\ x & \longmapsto & & & \check{x} & & \\ & & \tau & \longmapsto & & & 0 \end{array}$$

(b) $\tilde{C}_{2^m} \twoheadrightarrow C_{2^k}$ *(for* $m \geq k+1$*) induces the map*

$$\begin{array}{ccccccc} \mathbf{F}_2[x] & \otimes & \Lambda[\tau] & \to & \mathbf{F}_2[\tilde{x}] & \otimes & \Lambda[\tilde{\tau}] \\ x & \longmapsto & & & 0 & & \\ & & \tau & \longmapsto & & & \tilde{\tau} \end{array}$$

This allows us to deduce the b^*-module structures. In fact if

$$L := b^*/(\tau)$$
$$M := b^*/(x)$$

we have the following identifications where $\check{G}$ is the subgroup $\check{C}_{2^\ell}$ of G and $\tilde{G}$ is the covering group $\tilde{C}_{2^m}$ of G.

Corollary (3.5) : *We have the following isomorphisms of b^*-modules*

(a) $H^*(B\check{G}) \cong L \otimes \Lambda[\check{\tau}]$

and (b) $H^*(B\tilde{G}) = M \otimes \mathbf{F}_2[\tilde{x}]$.

It is now easy to write down free resolutions of L and M.

(a) $$\cdots \longrightarrow S^3 b^* \xrightarrow{\tau} S^2 b^* \xrightarrow{\tau} S^1 b^* \xrightarrow{\tau} b^* \longrightarrow L \longrightarrow 0$$

and

(b) $$0 \longrightarrow S^2 b^* \xrightarrow{x} b^* \longrightarrow M \longrightarrow 0.$$

Hence we find

(a) (i) $N \otimes_{b^*} L \cong N/\tau N$

(ii) $\mathrm{Tor}_s^{b^*}(N, L) \cong S^s H(N;\tau)$ for $s \geq 1$

and

(b) (i) $N \otimes_{b^*} M = N/xN$

(ii) $$\mathrm{Tor}_s^{b^*}(N, M) = \begin{cases} \ker(S^2 N \xrightarrow{x} N) & s = 1 \\ 0 & s \geq 2 \end{cases}$$

Using (a) we can at least find from (3.1) a useful condition for a very simple expression for $H^*_{\check{G}}(A)$.

Corollary (3.6) : *If $H(H^*_G(A);\tau)$ is zero in codegrees $\leq c$ then the edge map $\{H^*_G(A)/\tau H^*_G(A)\} \otimes \Lambda[\check{\tau}] \longrightarrow H^*_{\check{G}}(A)$ is iso in codegrees $\leq c$ and mono in codegree $c+1$.*

Using (b) we have from (3.3)

Corollary (3.7) : *There is a short exact sequence*

$$0 \longrightarrow H^*_G(A)/x \otimes \mathbf{F}_2[\tilde{x}] \longrightarrow H^*_{\tilde{G}}(A) \longrightarrow S^1[{}_x\{H^*_G(A)\}] \otimes \mathbf{F}_2[\tilde{x}] \longrightarrow 0.$$

Remark: Notice that τ is zero on L and x is zero on M. More generally if τ is zero on the b^*-module λ we may use the fact that $b^*//\Lambda[\tau] \cong \mathbf{F}_2[x]$ and the Cartan-Eilenberg Spectral Sequence ([6]p.349) to obtain the short exact sequence

$$0 \rightarrow \mathrm{Tor}_s^{\Lambda[\tau]}(N, \mathbf{F}_2) \otimes_{\mathbf{F}_2[x]} \lambda \rightarrow \mathrm{Tor}_s^{b_*}(N, \lambda) \rightarrow \mathrm{Tor}_1^{\mathbf{F}_2[x]}(\mathrm{Tor}_{s-1}^{\Lambda[\tau]}(N, \mathbf{F}_2), \lambda) \rightarrow 0$$

On the other hand $\mathrm{Tor}_*^{\Lambda[\tau]}(N, \mathbf{F}_2)$ is periodic due to the exact sequence $0 \longrightarrow S^1\mathbf{F}_2 \longrightarrow \Lambda[\tau] \longrightarrow \mathbf{F}_2 \longrightarrow 0$. This means one may obtain from (UCT3) analogues of Corollary (3.6) for theories F for which τ acts trivially on the coefficients.

Corollary (3.6)$^+$: *Suppose that F is a theory with the property that τ is zero on F^* ; then if $H(H_G^*(A); \tau) = 0$ we have a short exact sequence*

$$0 \longrightarrow H_G^*(A)/\tau \otimes_{\mathbf{F}_2[x]} F^* \longrightarrow F^*(A) \longrightarrow S_1 \mathrm{Tor}_1^{\mathbf{F}_2[x]}(H_G^*(A)/\tau, F^*) \longrightarrow 0.$$

If in addition F^* is zero in negative codegrees and $H(H_G^*(A); \tau) = 0$ only in codegrees $\leq c$ then the short exact sequence is still valid in codegrees $\leq c$.

Similarly, if μ is a b^*-module on which x is zero we may use the fact that $b^*//\mathbf{F}_2[x] \cong \Lambda[\tau]$ and the Cartan-Eilenberg Spectral Sequence gives a Gysin long exact sequence

$$\cdots \rightarrow \mathrm{Tor}_{s+1}^{\Lambda[\tau]}(N/x, \mu) \rightarrow$$

$$\mathrm{Tor}_{s-1}^{\Lambda[\tau]}(S^2[{}_xN], \mu) \rightarrow \mathrm{Tor}_s^{b^*}(N, \mu) \rightarrow \mathrm{Tor}_s^{\Lambda[\tau]}(N/x, \mu) \rightarrow$$

$$\mathrm{Tor}_{s-1}^{\Lambda[\tau]}(S^2[{}_xN], \mu) \rightarrow \cdots \quad .$$

Therefore if x is zero on F^* and F^* is flat over $\Lambda[\tau]$ we obtain from (UCT3) a precise analogue of (3.7). Otherwise we must impose further conditions on $H_G^*(A)$ to obtain a theorem worth stating.

Analogous theorems can be proved using the spectral sequences with Ext E_2 term; for this purpose it is useful to record an elementary proposition.

Propsoition (3.8) : *The b^*-module I is injective iff*

(a) $H(I;\tau)=0$

and

(b) *For any pair of elements i,j with $xi=\tau j$ there is an element k such that $xk=j$ and $\tau k=i$.*

Remarks :

(i) It follows from (a) and (b) that I is x-divisible

(ii) It is easy to see there are *not* enough injectives.

Indeed if we have an element $m\in M$ and $x\tau m=0$, $\tau m\neq 0$ M cannot be embedded in an injective.

In the present case where $G=C_{2^k}$, we can make explicit all cyclic b^*-modules N, and for any such N it is easy to construct a finite G-space X so that N is a summand of $H^*_{C_{2^k}}(X)$. Indeed we may use quotients S^V/S^0 and smash products of them, since every element of b^* is the Euler class of some representation. This shows the richness of the topological setup is quite sufficient to require the homological complication.

Case 3 G is an elementary abelian 2-group

In this case let us write V for G and suppose $\dim_{\mathbf{F}_2}(V)=n$. Then of course H^*BV is the symmetric algebra $S(V^*)$.

Now suppose $\check{V}$ is a subgroup of V and $i:\check{V}\to V$ is the inclusion; the induced map of cohomology is

$$S(i^*):S(V^*)\longrightarrow S(\check{V}^*)$$

and clearly surjective. If U is a hyperplane it is quite easy to see that $\ker i^*$ is the cyclic module generated by the quotient map $V\longrightarrow V/U\cong\mathbf{F}_2$. Hence $S(U^*)$ is of projective dimension 1 in this case. Similar considerations show that if $\check{V}$ is of codimension r then $S(\check{V}^*)$ has projective dimension r; of course we can prove this by induction on r.

In any case if A is a V-space and we wish to deduce $H^*_{\check{V}}(A)$ from $H^*_V(A)$ we can do so using the r short exact sequences arising from restriction to a hyperplane of which the first is

$$0\to H^*_V(A)\otimes_{S(V^*)}S(U^*)\to H^*_U(A)\to S_1\operatorname{Tor}^{S(V^*)}_{1,*}(H^*_V(A),S(U^*))\to 0.$$

Note that since i^* is surjective we do know $H^*_U(A)$ as an $S(U^*)$-module up to extension. In particular if $H^*_V(A)$ is $S(V^*)$ projective, $H^*_{\check{V}}(A)$ is $S(\check{V}^*)$-projective.

On the other hand if we have a quotient map

$$\pi : \tilde{V} \longrightarrow V$$

this induces

$$S(\pi^*) : S(V^*) \longrightarrow S(\tilde{V}^*)$$

and $S(\tilde{V}^*)$ is a free $S(V^*)$-module, indeed

$$S(\tilde{V}^*) \cong S(V^*) \otimes S((\ker \pi)^*).$$

Hence (3.3) always collapses to show

$$H^*_{\tilde{V}}(A) \cong H^*_V(A) \otimes S((\ker \pi)^*).$$

In fact it follows in this case that if $H^*_V(A)$ is $S(V^*)$ projective then $H^*_{\tilde{V}}(A)$ is $S(\tilde{V}^*)$ projective.

It is certainly possible to do calculations with other small 2-groups, but perhaps we have provided sufficient illustration with these abelian examples.

References

[1] J.F. Adams, Lectures on generalised cohomology, Lecture notes in maths. **99**, Springer-Verlag (1969), 1-138.

[2] J. F. Adams, Prerequisites (on equivariant stable homotopy) for Carlsson's Lecture, Lecture notes in maths. **1051**, Springer-Verlag (1982), 483-532.

[3] J. M. Boardman, Conditionally convergent spectral sequences, The John Hopkins University, Preprint, August 1981.

[4] A. Borel, *Seminar on transformation groups*, Princeton University Press (1960), (Chapter IV).

[5] K. S. Brown, *Cohomology of groups*, Springer-Verlag (1982).

[6] H. Cartan and S. Eilenberg, *Homological algebra*, Princeton University Press (1956).

[7] A. Dold and D. Puppe, Duality, trace and transfer, *Proc. Steklov Inst. (1984) No. 4., AMS Translations*, 85-103.

[8] L. G. Lewis, J. P. May, M. Steinberger, with contributions by J. E. McClure, *Equivariant Stable Homotopy Theory*, Lecture notes in maths. **1213**, Springer-Verlag (1986).

[9] J. P. C. Greenlees, Representing Tate cohomology of G-spaces, *Proceedings of the Edinburgh Math. Soc.* **30** (1987), 435-443.

[10] J. P. C. Greenlees, Stable maps into free G-spaces, *Trans. American Math. Soc.* **309** (1988) (to appear).

[11] J. P. C. Greenlees, Borel homology I, Under revision.

[12] E. H. Spanier, Function spaces and duality, *Annals of Maths.* **70** (1959), 338-378.

[13] E. H. Spanier, *Algebraic Topology*, McGraw-Hill (1966).

[14] E. H. Spanier and J. H. C. Whitehead, A first approximation to homotopy theory, *Proc. Nat. Acad Sci.*, USA, **39**, (1953) 258-294.

[15] R. G. Swan, A new method in fixed point theory, *Comm. Math. Helv.* **34** (1960), 1-16.

Department of Mathematics
National University of Singapore
Kent Ridge
Singapore 0511

Solvable subgroups of free products of profinite groups

Wolfgang Herfort and *Luis Ribes*

Introduction

Let A and B be profinite groups, and let $G = A \amalg B$ be their free profinite product, i.e. their coproduct in the category of profinite groups. This paper is a contribution to the study of the structure of the closed subgroups of G. The main thrust of previous work on this topic, has been directed towards finding subgroups H that can again be expressed as free profinite products of certain subgroups of H related to the free factors A and B. The model for this approach is the classical Kurosh subgroup theorem for free products of abstract groups (cf. [1], [4], [5], [7], [10]).

In this paper we consider closed solvable subgroups of free profinite products. It is quite clear a priori, that a Kurosh subgroup theorem type of approach could not work for such groups. The main reason for this is that a non-trivial profinite free product contains a free profinite subgroup; and a free profinite group of rank greater or equal to 2, is not solvable.

Let H be a solvable subgroup of $G = A \amalg B$. One obvious possibility for H is that it is a conjugate of a subgroup of A or B, and in that sense it is arbitrary. But otherwise, it has a very rigid structure; it must be of one of the following types. If H is torsion-free, it is a semi-direct product of two infinite procyclic groups of relatively prime orders. If H contains non-trivial elements of finite orders, then there are two possibilities:

(a) H is of dihedral type, i.e., a semidirect product $\widehat{\mathbf{Z}}_\sigma \rtimes C_2$, where $\widehat{\mathbf{Z}}_\sigma$ is an infinite torsion-free procyclic group of order divisible by 2, C_2 is the group of order two, and C_2 acts on $\widehat{\mathbf{Z}}_\sigma$ by inversion; or

(b) H is a profinite Frobenius group of the form $\widehat{\mathbf{Z}}_\sigma \rtimes C_n$ where C_n is a finite cyclic group of order n relatively prime to the order of $\widehat{\mathbf{Z}}_\sigma$, and the action of every non-trivial element of C_n on $\widehat{\mathbf{Z}}_\sigma$ is fixed point free. (See Theorem 2.6 and 3.2 for a more general and detailed statement of these results in the context of pro-$\mathcal{C}$-groups).

1. Notation and preliminaries

Throughout this paper $\mathcal{C}$ denotes a non-empty class of finite groups satisfying the following conditions

(i) every subgroup and every homomorphic image of a group in $\mathcal{C}$, is also in $\mathcal{C}$; and

(ii) $\mathcal{C}$ is closed under extensions, i.e., if $1 \to L \to G \to H \to 1$ is an exact sequence of groups and L and H are in $\mathcal{C}$, then G is also in $\mathcal{C}$.

Following Haran and Jarden, we will refer to such a class as a *full class* of finite groups. A *pro-$\mathcal{C}$-group* is a projective limit of groups in $\mathcal{C}$, i.e., a compact, totally disconnected topological group, all whose finite continuous homomorphic images are in $\mathcal{C}$. Important examples, in this paper, of full classes are 1) all finite groups; 2) all finite solvable groups; 3) for a fixed prime number p, the class of all finite p-groups. We refer to the corresponding pro-$\mathcal{C}$-groups as profinite, prosolvable, and pro-p, respectively. For general properties of these groups, [13] and [16] may be consulted.

In this paper, all homomorphisms among profinite groups are assumed to be continuous, and all subgroups of profinite groups are assumed to be closed. If H is a subgroup of a group G, we write $H \leq G$; and $H < G$ will indicate that H is a proper subgroup of G.

Let $A_1, \cdots, A_n$ be pro-$\mathcal{C}$-groups. Their *free pro-$\mathcal{C}$-product* G is, by definition, their coproduct in the category of pro-$\mathcal{C}$-groups, i.e., G is a pro-$\mathcal{C}$-group together with homomorphisms $\rho_i : A_i \to G$, such that if $\psi_i : A_i \to H$ are homomorphisms $(i = 1, \cdots, n)$ into a pro-$\mathcal{C}$-group H, then there exists a unique homomorphism $\psi : G \to H$ with $\psi\rho_i = \psi_i$ $(i = 1, \cdots, n)$. See [1], [4] for an explicit construction of G. We say that each A_i is a *free factor* of the free product G. The free pro-$\mathcal{C}$-product of groups $A_1, \cdots, A_n$ will be denoted by $A_1 \amalg_{\mathcal{C}} \cdots \amalg_{\mathcal{C}} A_n$ or $\coprod_{\mathcal{C}\, i=1}^{n} A_i$, or as we will do more often, by $A_1 \amalg \cdots \amalg A_n$ or $\amalg_{i=1}^n A_i$ if there is no danger of confusion. If p is a prime number, and A and B are pro-p-groups, their pro-p-product will be denoted by $A \amalg_p B$. Finally, we will have occasion to use free pro-$\mathcal{C}$-products of groups A_x indexed by a topological space X in the sense of [1], [4] or [7]; then we will use the notation $\coprod_X A_x$.

If $G = A_1 \amalg \cdots \amalg A_n$ is a free pro-$\mathcal{C}$-product of the pro-$\mathcal{C}$-groups A_i, let $\rho : G \to A_1 \times \cdots \times A_n$ be the homomorphism which sends A_i to A_i

identically. The kernel K of this homomorphism is called the *cartesian subgroup* of G. It is well known that K is a free pro-$\mathcal{C}$-group (cf. [4], Th. 5.5, for the case when each A_i is finite).

If σ is a set of prime numbers, we denote by $\widehat{\mathbf{Z}}_\sigma$ the direct product $\prod_{p\in\sigma}\widehat{\mathbf{Z}}_p$, where $\widehat{Z}_p$ denotes the additive group of p-adic numbers. Similarly, if $\mathcal{C}$ is a full class of finite groups, $\widehat{\mathbf{Z}}_c$ has the same meaning as $\widehat{\mathbf{Z}}\sigma$, where σ is the set of all primes dividing the order of some group in $\mathcal{C}$. For a prosolvable group H and set of prime numbers σ, H_σ will denote one of its σ-Hall subgroups. For a natural number n, C_n denotes the cyclic group of order n. A final piece of notation: $L \rtimes H$ denote the semiproduct of the groups L and H with respect to a certain given action of H on L.

To end this section we state a result to which we will refer often. Its proof can be deduced easily from Theorems A' and B' in [9].

Lemma 1.1. *Let $\mathcal{C}$ be a full class of finite groups, and let $G = \amalg_{i=1}^n A_i$ be the free pro-$\mathcal{C}$-product of (finite) groups A_i in $\mathcal{C}$. Let $1 \neq x \in G$, x of finite order. Then the centralizer $\mathrm{C}_G(x)$ of x in G, is finite.*

2. First results

In this section we will describe the structure of a solvable subgroup H of a free pro-$\mathcal{C}$-product $G = A_1 \amalg \cdots \amalg A_n$ when each of the groups A_i is finite (i.e., $A_i \in \mathcal{C}$). We start with a special case.

Proposition 2.1. *Let $G = \amalg_{i=1}^n A_i$ be a free pro-$\mathcal{C}$-product of the finite groups $A_i \in \mathcal{C}$, and let p be a fixed prime number. Assume H is a solvable pro-p-subgroup of G. Then one of the following holds:*
(1) H is finite;
(2) $H \approx \hat{\mathbf{Z}}_p$;
(3) $p = 2$ and $H \approx C_2 \amalg_2 C_2$ (the pro-2-product of two groups of order two).

Proof. Assume that H is infinite.

Case 1: H is torsion-free. Consider the cartesian subgroup K of G (cf. §1). Then K is a free pro-$\mathcal{C}$-subgroup of G of finite index. Hence $K \cap H$ is free pro-p (cf. [16], p. 218), and of finite index in H. Since H is torsion-free, it follows that H is also free pro-p (cf. [14], p. 413). But being solvable, $H \approx \widehat{\mathbf{Z}}_p$.

Case 2: H contains some torsion elements. By Case 1, $K \cap H \approx \widehat{\mathbf{Z}}_p$. Recall that $\mathrm{Aut}(\widehat{\mathbf{Z}}_p) \approx C_{p-1} \times \widehat{\mathbf{Z}}_p$ if $p \neq 2$, and $\mathrm{Aut}(\widehat{\mathbf{Z}}_2) \approx C_2 \times \widehat{\mathbf{Z}}_2$ (cf.[15], p.17). Consider the homomorphism $\eta : H \to \mathrm{Aut}(H \cap K)$ given by $\eta(x) =$ restriction to $H \cap K$ of the inner automorphism of H determined by x. Since $\eta(K \cap H) = 1$ and $K \cap H$ has finite index in H, it follows that $\eta(H)$ is finite. Now, if $p \neq 2$, this means that $\eta(H) \leq C_{p-1}$, and since $\eta(H)$ is a p-group, $\eta(H) = 1$; i.e., each element of H commutes with every element of $K \cap H$. Let $1 \neq h \in H$ be an element of finite order. By Lemma 1.1, the centralizer $\mathbf{C}_G(h)$ of h in G, is finite. However, $\widehat{\mathbf{Z}}_p \approx K \cap H \leq \mathbf{C}_G(h)$, a contradiction. Thus $p = 2$. Then H contains an element of order 2, say h. Then, by the above argument $\eta(h) \neq 1$, and hence $\eta(H) = C_2$. Moreover, the map $\eta : H \to C_2$ splits. Finally since the elements in $\ker\eta$ centralize $K \cap H$, $\ker\eta$ must be torsion-free; so by case 1, $\ker\eta \approx \widehat{\mathbf{Z}}_2$. Therefore we have a split exact sequence. $1 \to \widehat{\mathbf{Z}}_2 \to H \to C_2 \to 1$, i.e., H is the dihedral pro-2-group, or equivalently $H \approx C_2 \amalg_2 C_2$. □

The following Corollary is an immediate consequence of the above proposition (alternative 2).

Corollary 2.2. *Let G be as in Proposition 2.1, and let H be a torsion-free solvable subgroup. Then every non-trivial p-Sylow subgroup of H is isomorphic to $\widehat{\mathbf{Z}}_p$.*

A finite group H all whose p-Sylow subgroups are cyclic (a so-called Zassenhaus group) has a well-known structure: it is a semidirect product $H = C \rtimes C'$ of two cyclic groups of relatively prime orders (cf. [17], p.174). From this and a standard inverse limit argument one obtains the following result.

Lemma 2.3. *Let H be a profinite group all whose p-Sylow subgroups are procyclic. Then H is the semidirect product of two procyclic groups of relative prime orders. If in addition H is torsion-free, then $H \approx \widehat{\mathbf{Z}}_\sigma \rtimes \widehat{\mathbf{Z}}_{\sigma'}$, where σ and σ' are disjoint sets of prime numbers.*

Lemma 2.4. *Let $G = \amalg_{i-1}^n A_i$ be a free pro-$\mathcal{C}$-product of finite groups*

$A_i \in \mathcal{C}$, and let H be an infinite solvable subgroup of G. Then every finite subgroup Γ of H is cyclic.

Proof. Let K be the cartesian subgroup of G. Since $K \cap H$ has finite index in H, $K \cap H$ is infinite. By Cor. 2.2 and Lemma 2.3, $H \cap K = L \rtimes L'$ where $L \approx \widehat{\mathbf{Z}}_\sigma$, $L' \approx \widehat{\mathbf{Z}}_{\sigma'}$ and σ and σ' are disjoint sets of prime numbers. We may assume $\sigma \neq \emptyset$. Clearly $L \triangleleft H$, and so if $p \in \sigma$ and L_p denotes the p-Sylow subgroups of L, it follows that $L_p \triangleleft H$. Thus H acts on L_p by conjugation, and so we have a natural homomorphism

$$\eta : H \to \mathrm{Aut}(L_p) \approx \mathrm{Aut}(\widehat{\mathbf{Z}}_p).$$

As we have seen before (see the proof of Prop. 2.1), $\eta_{|\Gamma}$ is $1-1$. Therefore $\Gamma \approx \eta(\Gamma)$ is contained in the torsion part of $\mathrm{Aut}(\widehat{\mathbf{Z}}_p)$, which is cyclic. □

Before we state the next result, we establish some notation. Let H be a solvable profinite group. Let π denote the set of prime numbers p for which a p-Sylow subgroup H_p of H is finite, but not trivial. Then let $\pi' = \{p \mid H_p \text{ is infinite}\}$. Then $H = H_\pi H_{\pi'}$ where H_π and $H_{\pi'}$ are π- and π'-Hall subgroups of H respectively (cf. [2]).

Lemma 2.5. *Let H be an infinite solvable subgroup of the free pro-$\mathcal{C}$-product $G = \amalg_{i=1}^n A_i$ with each $A_i \in \mathcal{C}$. With the notation above, assume $H_\pi \neq 1$. Then*
(1) H_π is a finite cyclic group;
(2) $H_{\pi'} \approx \widehat{\mathbf{Z}}_{\pi'}$; and
(3) $H = H_{\pi'} \rtimes H_\pi$ is a Frobenius profinite group with kernel $H_{\pi'}$, i.e., each non-trivial element of H_π acts on $H_{\pi'}$ by conjugation without fixed points.

Proof. Let σ be a finite subset of π. Then H_σ is a finite subgroup, and therefore H_σ is conjugate in G to a subgroup of some A_i, and in fact i is independent of σ (cf. [9], Th. 2). Since A_i is finite, it follows that H_π (and so also π) is finite. Therefore statement (1) is a consequence of Lemma 2.4. Next we will see that $H_{\pi'}$ is torsion-free. If this were not the case, then $2 \in \pi'$, by Proposition 2.1, and the 2-Sylow subgroups of H would be isomorphic to $C_2 \amalg_2 C_2$. Let L be a $\pi \cup \{2\}$-Hall subgroup of H. Then $L = H_\pi H_2$. Since H_π is finite and H_2 contains a subgroup of finite index isomorphic to $\widehat{\mathbf{Z}}_2$, we must have that H_π normalizes a subgroup T of H_2 and $T \approx \widehat{\mathbf{Z}}_2$. Let $\eta : H_\pi \to \mathrm{Aut}(T) \approx C_2 \times \widehat{\mathbf{Z}}_2$ be the action of H_π on T by conjugation. Since H_π is finite $\ker\eta = 1$, by Lemma 1.1. But this is not possible since $2 \notin \pi$. It follows that $2 \notin \pi'$, and hence $H_{\pi'}$ is torsion-free. Now, to prove that H is a Frobenius group in the profinite sense (cf. [3]) with kernel $H_{\pi'}$, it suffices to prove that H_π is an isolated subgroup of H. To see this, let $h \in H$, and assume $H_\pi^h \cap H_\pi \neq 1$. Write $h = rs$ with $r \in H_\pi$ and $s \in H_{\pi'}$. Then $H_\pi^s \cap H_\pi \neq 1$. Since H_π is finite, there is some $g \in G$ and some $i \in \{1, 2, \cdots, n\}$ such that $H_\pi \leq A_i^g$ (cf. [9], Th. 2). It follows that $A_i^{gsg^{-1}} \cap A_i \neq 1$, and therefore $gsg^{-1} \in A_i$ (cf. [9], Th. 2). Thus $s \in g^{-1}A_ig \cap H_{\pi'} = 1$ since A_i is finite and $H_{\pi'}$ torsion-free. Hence we have $h = r \in H_\pi$, and so H_π is isolated in H. This proves part (3). To prove part (2), note that the kernel of a Frobenius profinite group is nilpotent (cf. [3], Cor. 3.7). So $H_{\pi'} = \prod_{p\in\pi'} H_p$. But by Prop. 2.1, $H_p \approx \widehat{\mathbf{Z}}_p$, $\forall p \in \pi'$. Therefore $H_{\pi'} \approx \widehat{\mathbf{Z}}_{\pi'}$. □

Theorem 2.6. *Let $G = \amalg_{i=1}^n A_i$ be a free pro-$\mathcal{C}$-product of finite groups $A_i \in \mathcal{C}$. Let H be a solvable subgroup of G. Then one of the following statements holds:*

(a) *H is finite. Then H is conjugate to a subgroup of one of the free factors A_i;*

(b) *H is torsion-free, non-trivial. Then $H \approx \widehat{\mathbf{Z}}_\sigma \rtimes \widehat{\mathbf{Z}}_{\sigma'}$, where σ and σ' are disjoint sets of prime numbers;*

(c) *H is a profinite Frobenius group of the form $\widehat{\mathbf{Z}}_{\pi'} \rtimes C$ with Frobenius kernel $\widehat{\mathbf{Z}}_{\pi'}$, and C is a non-trivial finite cyclic group;*

(d) *$H \approx C_2 \amalg_{\mathcal{C}'} C_2 \approx \widehat{\mathbf{Z}}_{\mathcal{C}'} \rtimes C_2$ the dihedral pro-$\mathcal{C}'$-group, where $\mathcal{C}'$ is a full subclass of $\mathcal{C}$, and $C_2 \in \mathcal{C}'$.*

Proof. Using the notation of Lemma 2.5, we have $H = H_\pi H_{\pi'}$. If H is finite, then it is conjugate to a subgroup of some A_i (cf. [9], Th. 2) and (a) holds. If H is infinite and $H_\pi \neq 1$, then according to Lemma 2.5, H is Frobenius and (c) holds. If H is infinite and torsion-free, then (b) holds by Cor. 2.2 and Lemma 2.3. Therefore the only remaining case is that $H_\pi = 1$ but $H_{\pi'}$ has non-trivial elements of finite order. Then $H = H_{\pi'} = H_2 H_{p_1} H_{p_2} \cdots$, where $\pi' = \{2, p_1, p_2, \cdots\}$. (Note $2 \in \pi'$, for otherwise $H = H_{\pi'}$ would be torsion-free, according to Prop. 2.1). We will see that in this case, (d) holds. First we assume that π' is finite, say $\pi' = \{2\} \cup \lambda$ with $1 < |\lambda| < \infty$. Let K be the cartesian subgroup of G. Then by Cor. 2.2 and Lemma 2.3, $L := K \cap H = L_\sigma \rtimes L_{\sigma'}$ where $L_\sigma \approx \widehat{\mathbf{Z}}_\sigma$, $L_{\sigma'} \approx \widehat{\mathbf{Z}}_{\sigma'}$, and $\sigma \cap \sigma' = \emptyset$. We may assume $2 \in \sigma'$, for otherwise if $p \in \sigma'$, the action $L_p \to \mathrm{Aut}(L_2) \approx C_2 \times \widehat{\mathbf{Z}}_2$ of L_p on L_2 given by conjugation must be trivial. Consider the action $\eta : K \cap H \to \mathrm{Aut}(L_\sigma)$ given by conjugation. Since $\sigma \subseteq \pi'$, σ is finite, and so the torsion subgroup of $\mathrm{Aut}(L_\sigma)$ is finite. Hence $\eta(K \cap H) = \eta(L_{\sigma'})$ is finite. It follows that $\ker\eta \approx \widehat{\mathbf{Z}}_\sigma \times \widehat{\mathbf{Z}}_{\sigma'}$, and so H has a normal subgroup S of finite index isomorphic to $\widehat{\mathbf{Z}}_\sigma \times \widehat{\mathbf{Z}}_{\sigma'}$. Thus $S_2 \approx \widehat{\mathbf{Z}}_2$ is normal in H. Consider the action by conjugation of H on S_2

$$\psi : H \to \mathrm{Aut}(S_2) \approx C_2 \times \widehat{\mathbf{Z}}_2.$$

Clearly $\psi(H)$ is finite, for $\psi(H_\lambda) = 1$, and S_2 has finite index in H_2. Since H contains an element of order 2, and this element cannot centralize S_2 (Lemma 1.1), we have that $\psi(H) = C_2$ and ψ splits. Say $H = (\ker\psi) \rtimes \langle\alpha\rangle$ with order $(\alpha) = 2$. We need to prove that $\ker\psi \approx \widehat{\mathbf{Z}}_{\pi'} = \widehat{\mathbf{Z}}_\lambda \times \widehat{\mathbf{Z}}_2$. Since $\ker\psi$ is torsion-free by Lemma 1.1, we have $T := \ker\psi = T_\tau \rtimes T_{\tau'}$ where $T_\tau \approx \widehat{\mathbf{Z}}_\tau$, $T_{\tau'} \approx \widehat{\mathbf{Z}}_{\tau'}$ and $\tau \cap \tau' = \emptyset$ (Cor. 2.2 and Lemma 2.3). Again (see a similar argument above), we may assume $2 \in \tau'$. Note $\tau \cup \tau' = \lambda \cup \{2\}$. It is clear that T_λ is characteristic in T, and so normal in H. Hence α acts on T_λ by conjugation, and this action is fixed point free by Lemma 1.1. Since order $(\alpha) = 2$ and $2 \in \lambda$, T_λ is abelian (to see this, use Cor. 1.4 in [3] to express T_λ as a projective limit of finite quotients of T_λ on which α acts fixed point free, and therefore are abelian). Thus $T_\lambda \approx \widehat{\mathbf{Z}}_\lambda$,and $T = T_\lambda \rtimes T_2$ with $T_2 \approx \widehat{\mathbf{Z}}_2$. Let $\langle t\rangle = T_\lambda$, $\langle z\rangle = T_2$, and $p \in \lambda$. Since $\mathrm{Aut}(T_p) \approx C_{p-1} \times \widehat{\mathbf{Z}}_p$, z acts on T_p as an automorphism of order a power of 2. Say $t^z = t^\rho$ with $\rho \in \widehat{\mathbf{Z}}_p$ a unit of order 2^r in the ring $\widehat{\mathbf{Z}}_p$, where r is a natural number. Now $(z\alpha)^2 = z\alpha z\alpha = zz^{-1} = 1$. So $t = t^{z\alpha z\alpha} = t^{\rho\alpha z\alpha} = t^{-\rho z\alpha} = t^{\rho^2}$. Therefore $\rho^2 = 1$. It follows that $\rho = 1$ or $\rho = -1$, then $t^z = t^\rho = t^{-1} = t^\alpha$. So $t^{z\alpha} = t$. But then $z\alpha$, which is an element of order 2 in G, would centralize an infinite subgroup, contradicting Lemma 1.1. thus $\rho = 1$, i.e., $\ker\psi = T = T_\lambda \times T_2 \approx \widehat{\mathbf{Z}}_\lambda \times \widehat{\mathbf{Z}}_2 = \widehat{\mathbf{Z}}_{\pi'}$ as desired.

To finish the proof we need to consider the case where π' is not necessarily finite. In general we have $H = H_2 H_{p_1} H_{p_2} \cdots = \langle\alpha\rangle \bar{H}_2 H_{p_1} H_{p_2} \cdots$, where order $(\alpha) = 2$, $\bar{H} \approx \widehat{\mathbf{Z}}_2$, $H_{p_i} \approx \widehat{\mathbf{Z}}_{p_i}$. By the case considered above, $H_2 H_{p_1} \cdots H_{p_t} = \langle\alpha\rangle \ltimes (\bar{H}_2 \times H_{p_1} \times \cdots \times H_{p_t})$, for every natural number

t. Therefore the elements of $\bar{H}_2$, H_{p_1}, H_{p_2}, $\cdots$ commute with each other. Thus $H = \langle\alpha\rangle \ltimes \widehat{\mathbf{Z}}_{\pi'} \approx C_2 \amalg_{\pi'} C_2 = C_2 \amalg_{\mathcal{C}'} C_2$, where $\mathcal{C}'$ is the class of all finite groups whose order involve primes in π'. $\square$

3. Solvable subgroups of a free product

In this section we extend Theorem 2.6 to the case when the free product is completely general, i.e., the free factors are not necessarily finite and there may be infinitely many of them. Throughout this section we consider a free pro-$\mathcal{C}$-product $G = \amalg_X A_x$ of pro-$\mathcal{C}$-groups A_x indexed by a topological space X, in the sense of [7], [12], [4], [5]. For our purpose, the only fact we need to recall about such free product is the following, which indeed can be taken as its definition, as pointed out by Haran (cf. [7], Lemma 2.2; [4], Prop. 2.1; and more explicitly in [10], Lemma 4.2).

Fact 3.1. *Let $G = \amalg_X A_x$ be a free pro-$\mathcal{C}$-product of pro-$\mathcal{C}$-groups A_x in the sense of* [7], [12] *or* [4]. *Then G is a projective limit of pro-$\mathcal{C}$-groups $G = \varprojlim_I G_i$ over a directed set I with canonical epimorphisms $\psi_i : G \to G_i$ for $i \in I$, and $\psi_{ij} : G_i \to G_j$ for $i \geq j$, such that*

(1) *Each G_i is a free pro-$\mathcal{C}$-product $G_i = \amalg_{k=1}^{n_i} G_{ik}$ of a finite number of finite groups $G_{ik} \in \mathcal{C}$;*
(2) *For every $i \in I$ and every $x \in X$, $\psi_i(A_x) = G_{ik}$ for some $k \in \{1, \cdots, n_i\}$;*
(3) *If $i \geq j$ (in I) then ψ_{ij} maps every G_{ik} onto some G_{jl}.*

Theorem 3.2. *Let $G = \amalg_X A_x$ be a free pro-$\mathcal{C}$-product of pro-$\mathcal{C}$-groups A_x in the sense of* [7], [12] *or* [4]. *Let H be a solvable subgroup of G. Then one of the following holds:*

(1) *H is conjugate to a subgroup of one of the free factors A_x of G;*
(2) *$H \approx \widehat{\mathbf{Z}}_\sigma \rtimes \widehat{\mathbf{Z}}_{\sigma'}$, where σ and σ' are disjoint sets of prime numbers;*
(3) *H is a profinite Frobenius group of the form $\widehat{\mathbf{Z}}_\sigma \rtimes C$ with Frobenius kernel $\widehat{\mathbf{Z}}_\sigma$, where C is a finite cyclic group;*
(4) *$H \approx C_2 \amalg_{\mathcal{C}'} C_2$ for some full subclass $\mathcal{C}'$ of $\mathcal{C}$.*

Proof. We will assume that $H \neq 1$. We shall use the notation established in Fact 3.1. Set $H_i = \psi_i(H)$ for $i \in I$. By Th. 2.6, H_i must satisfy one of the alternatives (a), (b), (c), (d) in it. Consider the following cases.

Case I: H_i is finite for every $i \in I$. Since $H \neq 1$, we may assume, by taking a cofinal subset of I if necessary, that $H_i \neq 1$ for every $i \in I$. Then H_i is conjugate to a subgroup of a (unique) free factor $G_{ik(i)}$ of G_i (cf. Th. 2,[9]). Consider the set $X_i = \{x \in G_i | H_i^x \leq G_{ik(i)}\}$. Then X_i is a non-empty compact set (for, let $1 \neq h \in H_i$; then also $X_i = \{x \in G_i | h^k \in G_{ik(i)}\}$, and clearly $X_i = \rho^{-1}(G_{ik(i)})$, where $\rho : G_i \to G_i$ is the continuous map given by $x \to h^x$). Now (X_i, ψ_{ij}) is a projective system, for $\psi_{ij}(X_i) \subseteq X_j$. Since each X_i is compact and non-empty, we have $\varprojlim X_i \neq \emptyset$. Let $y \in \varprojlim X_i$. Clearly $(G_{ik(i)}, \psi_{ij})$ form a projective system, and $H^y = \varprojlim H_i^{\psi_i(y)} \leq \varprojlim G_{ik(i)} = A_x$, where A_x is some free factor of G.

Case II: For some $i \in I$, H_i has the form $C_2 \amalg_{\mathcal{C}_i} C_2 \approx \widehat{\mathbf{Z}}_{\mathcal{C}_i} \rtimes C_2$, where $\mathcal{C}_i$ is a full subclass of $\mathcal{C}$. Let $j \geq i$. Then, since $\psi_{ji}(H_j) = H_i$, H_j cannot have the form indicated in the alternatives (a), (b), or (c) of Th.2.6, since $C_2 \in \mathcal{C}_i$ (recall that in alternative (b), $\sigma \cap \sigma' = \emptyset$ and in alternative (c), $(|\widehat{\mathbf{Z}}_\sigma|, |C|) = 1$). Thus, according to Th. 2.6, H_j must be of the form $C_2 \amalg_{\mathcal{C}_j} C_2 \approx \widehat{\mathbf{Z}}_{\mathcal{C}_j} \rtimes C_2$, $\mathcal{C}_j \supseteq \mathcal{C}_i$. Consider the cofinal subset $I' = \{j \in I | j \geq i\}$. Then it is easy to see that $H = \varprojlim_{I'} H_j \approx \widehat{\mathbf{Z}}_{\mathcal{C}'} \rtimes C_2$, for some class $\mathcal{C}'$.

Case III: For some i, H_i is of the form $\widehat{\mathbf{Z}}_\sigma \rtimes \widehat{\mathbf{Z}}_{\sigma'}$ ($\sigma \neq \emptyset \neq \sigma'$, and the action of $\widehat{\mathbf{Z}}_{\sigma'}$ on $\widehat{\mathbf{Z}}_\sigma$ is not trivial). If $j \geq i$, $\psi_{ji}(H_j) = H_i$, and so H_j cannot be finite or of the forms $C_2 \amalg_{\mathcal{C}'} C_2$ or $\widehat{\mathbf{Z}}_\pi \rtimes C$ (C finite). Therefore H_j must be of the form $\widehat{\mathbf{Z}}_\sigma \rtimes \widehat{\mathbf{Z}}_{\sigma'}$, (not necessarily the same σ, σ' as above). I.e., for each $j \geq i$, H_j is torsion-free and all its p-Sylow subgroups are cyclic. It follows that $H = \varprojlim_{j \geq i} H_j$ is also torsion-free and its p-Sylow subgroups are cyclic. Therefore, by Lemma 2.3, H is of the form $\widehat{\mathbf{Z}}_\sigma \rtimes \widehat{\mathbf{Z}}_{\sigma'}$.

Case IV: Assume that one H_i is Frobenius of the form $\widehat{\mathbf{Z}}_{\sigma_j} \rtimes C_i$ ($\sigma_i \neq \emptyset$, C_i cyclic non-trivial, $\widehat{\mathbf{Z}}_{\sigma_i}$ the Frobenius kernel), and that for every $j \in I$, H_j is not of the form $\widehat{\mathbf{Z}}_{\sigma_i} \rtimes \widehat{\mathbf{Z}}_{\sigma'_j}$ ($\sigma_j \neq \emptyset \neq \sigma'_j$, $\widehat{\mathbf{Z}}_{\sigma_j} \rtimes \widehat{\mathbf{Z}}_{\sigma'_j} \not\approx \widehat{\mathbf{Z}}_{\sigma_j} \times \widehat{\mathbf{Z}}_{\sigma'_j}$), and H_j is not of the form $C_2 \amalg_{\mathcal{C}'} C_2$. Then according to Th. 2.6, for every $j \geq i$, we have that H_j is a profinite Frobenius group of the form $\widehat{\mathbf{Z}}_{\sigma_j} \rtimes C_j$, with $\widehat{\mathbf{Z}}_{\sigma_j}$ its Frobenius kernel. Since the action of C_j on $\widehat{\mathbf{Z}}_{\sigma_j}$ is fixed point free, it is clear that $\psi_{jj'}$ $(j \geq j' \geq i)$ sends the Frobenius kernel of H_j onto the Frobenius kernel of $H_{j'}$. It then follows that $H = \varprojlim_{j \geq i} H_j \approx \widehat{\mathbf{Z}}_\sigma \rtimes C$ where C is a procyclic group acting fixed point free on $\widehat{\mathbf{Z}}_\sigma$ and whose order is relatively prime to that of $\widehat{\mathbf{Z}}_\sigma$. Therefore C is finite (cf. [3], Th. 3.6).

Case V: H_i is procyclic for each $i \in I$. Then $H = \varprojlim H_i$ is procyclic, and so H satisfies either (1) or (2) in the statement of the theorem.

These cases exhaust all possibilities. □

Corollary 3.3. *Let F be a free pro-$\mathcal{C}$-group, and let H be a solvable subgroup of F. Then H is of the form $\widehat{\mathbf{Z}}_\sigma \rtimes \widehat{\mathbf{Z}}_{\sigma'}$, where σ and σ' are disjoint, possibly empty, sets of prime numbers. (This corollary is, of course, also a direct consequence of Lemma 2.3.)*

4. Realization of solvable groups

The purpose of this section is to point out that none of the alternatives of Theorem 2.6 can be excluded. More explicitly, if H is a profinite solvable group of one of the forms $\widehat{\mathbf{Z}}_\mathcal{C} \rtimes C_2 \approx C_2 \amalg_\mathcal{C} C_2$, $\widehat{\mathbf{Z}}_\sigma \rtimes \widehat{\mathbf{Z}}_{\sigma'}$, or $\widehat{\mathbf{Z}}_\sigma \rtimes C$ (Frobenius with kernel $\widehat{\mathbf{Z}}_\sigma$, and C cyclic), then H can be realized as a subgroup of a free pro-$\mathcal{C}$-product $A \amalg_\mathcal{C} B$ of two finite groups in $\mathcal{C}$, for an appropriate full class $\mathcal{C}$. This is obviously the case if $H = C_2 \amalg_\mathcal{C} C_2$. In fact one has the following result.

Proposition 4.1. *Let $H = C_2 \amalg_\mathcal{C} C_2$ $(C_2 \in \mathcal{C})$. Let $\mathcal{C}'$ be a full class of finite groups such that $\mathcal{C}' \supseteq \mathcal{C}$. Let A, B be finite groups in $\mathcal{C}'$ of even order. Then H is isomorphic to a subgroup of $A \amalg_{\mathcal{C}'} B$.*

Proof. Since $C_2 \leq A, B$, there is a natural embedding

$$C_2 \amalg_{\mathcal{C}'} C_2 \hookrightarrow A \amalg_{\mathcal{C}'} B$$

(cf. [11], Lemma 5.3). But on the other hand we have a natural embedding

$$H = C_2 \amalg_\mathcal{C} C_2 \approx \widehat{\mathbf{Z}}_\mathcal{C} \rtimes C_2 \hookrightarrow \widehat{\mathbf{Z}}_{\mathcal{C}'} \rtimes C_2 \approx C_2 \amalg_{\mathcal{C}'} C_2.$$

For groups of the form $\widehat{\mathbf{Z}}_\sigma \rtimes \widehat{\mathbf{Z}}_{\sigma'}$, we have the following.

Proposition 4.2. *Let σ and σ' be disjoint, possibly empty, sets of prime numbers and let $H = \widehat{\mathbf{Z}}_\sigma \rtimes \widehat{\mathbf{Z}}_{\sigma'}$ be a semidirect product of the procyclic groups $\widehat{\mathbf{Z}}_\sigma$ and $\widehat{\mathbf{Z}}_{\sigma'}$. Let $\mathcal{C}$ be any full class of finite groups containing those finite groups whose orders involve only primes in $\sigma \cup \sigma'$. Let A and B be non-trivial groups in $\mathcal{C}$ with $|A| + |B| \geq 5$. Then $G = A \amalg_{\mathcal{C}} B$ contains a subgroup isomorphic to H.*

Proof. Note that the cartesian subgroups of G is a free pro-$\mathcal{C}$-group of rank at least 2. In fact we will show that a free pro-$\mathcal{C}$-group of rank 2 contains a subgroup isomorphic to H. Indeed, let $\psi : F \to H$ be any epimorphism. Note that H is a projective profinite group, since every p-Sylow subgroup of H is free pro-p-group (cf. [6], Th. 4). Therefore, there exists $\rho : H \to F$ with $\psi\rho = id_H$. Thus ρ is a monomorphism. □

Finally we consider the Frobenius case.

Proposition 4.3. *Let A and B be non-trivial finite solvable groups, and let C be a cyclic subgroup of A. Let σ be a non-empty set of primes such that for each prime number p dividing $|C|$, and each $q \in \sigma$, one has $p|(q-1)$. Consider any full class $\mathcal{C}$ of finite solvable groups containing A, B, and all finite solvable groups whose orders involve only primes in σ. Then, the group $G = A \amalg_{\mathcal{C}} B$ contains a subgroup isomorphic to the (unique) profinite Frobenius group $H = \widehat{\mathbf{Z}}_\sigma \rtimes C$ with Frobenius kernel $\widehat{\mathbf{Z}}_\sigma$.*

The proof of this proposition is a consequence of some properties of automorphisms of free prosolvable groups, and it is contained in [11].

References

[1] E. Binz, J. Neukirch and G.H. Wenzel, A subgroup theorem for free products of profinite groups, *J. Algebra*, **19** (1971), 104-109.

[2] E.D. Bolker, Inverse limits of solvable groups, *Proc. Amer. Math. Soc.*, **14** (1963), 147-152.

[3] D. Gildenhuys, W. Herfort and L. Ribes, Profinite Frobenius groups, *Archiv Math.*, **33** (1979), 518-528.

[4] D. Gildenhuys and L. Ribes, A Kurosh subgroup theorem for free pro-$\mathcal{C}$-products of pro-$\mathcal{C}$-groups, *Trans. Amer. Math. Soc.*, **186** (1973), 309-329.

[5] D. Gildenhuys and L. Ribes, Profinite groups and Boolean graphs, *J. Pure Appl. Algebra*, **12** (1978), 21-47.

[6] K.W. Gruenberg, Projective profinite groups, *J. London Math. Soc.*, (3) **42** (1967), 155-165.

[7] D. Haran, On closed subgroups of free products of profinite groups, *J. London Math. Soc.*, (3) **55** (1987), 266-298.

[8] D. Haran and A. Lubotzky, Maximum abelian subgroups of free profinite groups, *Math. Proc. Camb. Phil. Soc.*, **97** (1985), 51-55.

[9] W. Herfort and L. Ribes, Torsion elements and centralizers in free products of profinite groups, *J. Reine Angew. Math.*, **358** (1985), 155-161.

[10] W. Herfort and L. Ribes, Subgroups of free pro-p-products, *Math. Proc. Camb. Phil. Soc.*, **101** (1987), 197-206.

[11] W. Herfort and L. Ribes, Frobenius subgroups of free products of prosolvable groups, to appear.

[12] J. Neukirch, Frei Produkte pro-endlicher Gruppen und ihre Kohomologie, *Archiv Math.*, **22** (1971), 337-357.

[13] J-P. Serre, *Cohomologie Galoisienne*, Lect. Notes Math. 5, Springer-Verlag, Berlin-Heidelberg - New York, 1965.

[14] J-P. Serre, Sur la dimension cohomologique des groupes profinis, *Topology*, **3** (1965), 413-420.

[15] J-P. Serre, *A course in Arithmetic*, Springer-Verlag, New York - Heildelberg - Berlin, 1973.

[16] L. Ribes, Introduction to profinite groups and Galois cohomology, *Queen's Papers Pure Appl. Math.*, Kingston, Canada 1970.

[17] H. Zassenhaus, *The theory of groups*, Chelsea, New York, 1956.

Institut für Numerische und Angewandte Mathematik
Technische Universität
Vienna
Austria - 1040

Department of Mathematics
Carleton University
Ottawa, Ontario K1S 5B6
Canada

Non-cancellation and non-abelian tensor squares

D. L. Johnson

Let G, H and K be groups, with G finitely-generated, such that

$$G \times H \cong G \times K. \tag{1}$$

Then what can be said about the relationship between H and K? It is known, for example, that H and K are necessarily isomorphic in each of the following cases:

G and H both finite [10],
G finite [8], [11], [6],
$G = Z$, the infinite cyclic group, and H abelian [11], [4],
$G = Z$ and $\mathbf{Z}(H)$ torsion [11],
$G = Z$ and H perfect [11].

Thus, Walker's paper [11] gives an affirmative answer to Test Problem III of Kaplansky [9], of which the following special case is already a consequence of [10] and the Basis Theorem for finitely-generated abelian groups.

Lemma 1. *If G, H and K are finitely-generated abelian groups such that* (1) *holds, then $H \cong K$.*

It is natural to ask whether (1) forces H and K to be isomorphic in all cases. When $G = Z$, a negative answer is provided by the groups

$$\begin{aligned} H &= \langle x, y \mid y^n = 1, xyx^{-1} = y^l \rangle, \\ K &= \langle x, y \mid y^n = 1, xyx^{-1} = y^{l^m} \rangle, \end{aligned} \tag{2}$$

for which (1) holds but $H \not\cong K$ whenever

$$(l, n) = 1, \ (m, t) = 1, \ \text{ and } \ m \not\equiv \pm 1 \ (\text{mod } t),$$

where t is the order of l mod n.

This is proved in general in [1], [5], and stated in [6] (in slightly different form), [11] (in a special case attributed to W.R. Scott).

On the other hand, (1) ensures that H and K have many invariants in common. When H and K (as well as G) are finitely generated, for example, it is known that

H and K have the same finite images [1],

if $G = Z$ and H' is is finite, then $H^{\times r} \cong K^{\times r}$ for some $r \in \mathbf{N}$ [12],

if $G = Z$ and H is nilpotent, then $H_i(H) = H_i(K)$ for all $i \geq 0$ [5],

if $(G \times H)'$ is cyclic, then $H \otimes H \cong K \otimes K$.

In the last result (which is proved below), $\otimes$ stands for the non-abelian tensor product of [3]. The properties of $\otimes$ needed here are as follows (see [2] for details).

For any group G, $G \otimes G$ is an isomorphism invariant of G which may be defined by a presentation with generators $\{g \otimes h \mid g, h \in G\}$. The defining relations ensure that there are homomorphisms

$$\left.\begin{array}{rl}\psi: \Gamma G_{ab} \to G \otimes G \\ \gamma(gG') \mapsto g \otimes g\end{array}\right\}, \qquad \left.\begin{array}{rl}\kappa: G \otimes G \to G' \\ g \otimes h \mapsto ghg^{-1}h^{-1}\end{array}\right\},$$

where Γ is Whitehead's universal quadratic functor. It is clear that κ is onto and that $I = \operatorname{Im} \psi \subseteq K = \operatorname{Ker} \kappa$.

Lemma 2. *There is an epimorphism from $G \otimes G$ onto G' whose kernel K contains a homomorphic image I of ΓG_{ab}. Moreover, $K \subseteq \mathbf{Z}(G \otimes G)$ and $K/I \cong H_2(G)$.*

Lemma 3. *For any groups G, H,*

$$(G \times H) \otimes (G \times H) \cong G \otimes G \times G \otimes H \times H \otimes G \times H \otimes H,$$

where the cross terms are ordinary (abelian) tensor products of groups.

Proposition 1. *Let G, H and K be finitely generated groups such that $G \times H \cong G \times K$ and $(G \times H)'$ is cyclic. Then $H \otimes H \cong K \otimes K$.*

Proof. Let $T = (G \times H) \otimes (G \times H)$ and apply Lemma 2 to $G \times H$. Since $K \subseteq \mathbf{Z}(T)$ and $T/K \cong (G \times H)'$ is cyclic, it follows that T is abelian. To show that T is finitely generated, note first that T/K is cyclic. Next, $K/I \cong H_2(G)$ is finitely generated as our conditions imply that $G \times H$ is finitely presented. Finally, the application of Γ to a finitely generated abelian group yields another, so that I is finitely generated.

Now observe that

$$G_{ab} \times H_{ab} \cong (G \times H)_{ab} \cong (G \times K)_{ab} \cong G_{ab} \times K_{ab},$$

whence it follows from Lemma 1 that $H_{ab} \cong K_{ab}$. Hence,

$$G \otimes H \cong G \otimes K \quad \text{and} \quad H \otimes G \cong K \otimes G.$$

Now apply Lemma 3 to G, H and G, K in turn. The resulting decompositions are isomorphic finitely-generated abelian groups. By cancelling isomorphic terms in accordance with Lemma 1, it follows that $H \otimes H \cong K \otimes K$.

Taking $G = Z$, this result shows that the groups H and K in (2) have the same tensor square. This can be shown by direct computation, as carried out in [2], [7] for finite split metacyclic groups. The strategy is to reduce the presentation defining $H \otimes H$ to recognizable form by judicious use of Tietze transformations. The cases n odd, even have to be treated separately, and the result is as follows.

Proposition 2. *Let H be the group presented by (2). Then $H \otimes H$ is the direct product of the four cyclic groups generated by*

$$x \otimes x,\ y \otimes y,\ (x \otimes y)(y \otimes x),\ x \otimes y,$$

of orders

$$\infty,\ (n, \varepsilon(l-1)),\ (n, l-1),\ n,$$

respectively, where ε is 1 or 2 according as n is odd or even. Moreover, $H_2(H)$ is cyclic of order $(n, l-1)$.

References

[1] G. Baumslag, Residually finite groups with the same finite images, *Compositio Math.* **29** (1974), 249-252.

[2] R. Brown, D.L. Johnson, E.F. Robertson, Some computations of non-abelian tensor products of groups, *J. Algebra,* **111**:1 (1987), 177-202.

[3] R. Brown, J.-L. Loday, Excision homotopique en basse dimension, *C.R. Acad. Sci. Paris Ser. I Math.* **298**: 15 (1984), 353-356.

[4] L. Fuchs, *Abelian groups*, Pergamon Press, New York 1967.

[5] P.J. Hilton, Non-cancellation properties for certain finitely-presented groups, *Quaestiones Math.* **9** (1986), 281-292.

[6] R. Hirshon, On cancellation in groups, *Amer. Math. Monthly* **76** (1969), 1037-1039.

[7] D.L. Johnson, The non-abelian tensor square of a finite split metacyclic group, *Proc. Edinburgh Math. Soc.* **30** (1987), 91-95.

[8] B. Jónsson, A. Tarski, *Direct decompositions of finite algebraic systems,* Notre Dame Math. Lectures, Notre Dame, Indiana 1947.

[9] I. Kaplansky, *Infinite abelian groups,* Univ. of Michigan Press, Ann Arbor 1954.

[10] R. Remak, Über die Zerlegung der endlichen Gruppen in direkte unzerlegbare Faktoren, *J. reine angew. Math.* **139** (1911), 293-308.

[11] E.A. Walker, Cancellation in direct sums of groups, *Proc. Amer. Math. Soc.* **7** (1956), 898-902.

[12] R. Warfield, Genus and cancellation for a group with finite commutator subgroup, *J. Pure and Applied Algebra* **6** (1975), 132-152.

Department of Mathematics
University of Nottingham
Nottingham NG7 2RD
United Kingdom

A class of SQ-Universal groups

Frank Levin and *Gerhard Rosenberger*

1. A group is *SQ-Universal (SQU)* if the factor groups of G embed all countable groups. The earliest published example of a finitely generated SQU group was the free group of rank 2 (Higman, Neumann, Neumann [2]). This example was later generalized to free products of two nontrivial cyclic groups, one of which has order ≥ 3 (Levin [3]). The latter result follows also from the fact that a finite extension of an SQU group is SQU (P.M. Neumann [5]). (For a more comprehensive account of SQU-group see Lyndon-Schupp [4]).

The present note was prompted by a problem posed to the second author by P.M. Neumann of determining $k \geq 2$ such that a group with presentation $\langle x, y \mid x^m = y^n = [x,y]^k = 1\rangle$ will be SQU. In this paper we determine for each pair $m \geq 2, n \geq 3, m \leq n$, the values of k for which the group is SQU. (The condition on m, n is necessary since for $m = n = 2$ the resulting group is dihedral.) Our methods are combinatorial, unlike those of Fine, Howie, Rosenberger [1], whose results anticipate some of the present results, especially for m, n, k not too small. However, their methods do not apply in general and especially not for the cases $m = n = 3$ and $m = 2, n = 4$, which are included here.

For the following proofs we shall need, in addition to the P.M. Neumann result [5] cited above, the obvious fact that pre-images of SQU groups are SQU and the generalization of the results in [3] to free products of groups, at least one of which has order greater than 2. (Cf., Lyndon-Schupp [4]).

2. Let $G = G(m,n,k) = \langle x, y \mid x^m = y^n = [x,y]^k = 1\rangle$, $m, k \geq 2$, $n \geq 3$, $m \leq n$. The commutator subgroup G' has finite index mn, and the set

$$T = \{x^i y^j : 0 \leq i < m,\ 0 \leq j < n\}$$

forms a transversal for G' in G. A straightforward application of the Reidemeister-Schreier process gives the following presentation for G': it has a generating set

$$\{x^i y^j x y^{-j} x^{-i-1},\ x^i y^j y y^{-j-1} x^{-i}\},$$

$0 \le i < m$, $0 \le j < n$. The generators on the right are trivial in G, and if we rewrite the left ones as $[x^{-i}, y^{-j}][y^{-j}, x^{-i-1}]$ it follows that

$$a_{ij} = [x^{-i}, y^{-j}] \tag{2.1}$$

will also form a generating set for G'. In terms of the latter generators G' has the following three sets of defining relators corresponding to the three relators of G:

(a) $a_{0,j} = a_{m,j} = 1,\ j = 0, \ldots, n,$ (2.2)
(b) $a_{i,0} = a_{i,n} = 1,\ i = 0, \ldots, m,$ (2.3)
(c) $(a_{i,j} a_{i+1,j}^{-1} a_{i+1,j+1} a_{i,j+1}^{-1})^k = 1,$ (2.4)
$i = 0, \ldots, m-1,\ j = 0, \ldots, n-1.$

Case 1. $G = G(2, 3, k)$.

For $k \ge 4$, G is SQU. This follows from results in [6], but for completeness we include a short proof here.

Using (2.1) - (2.4), $G' = \langle a, b \mid a^k = b^k = (ab^{-1})^k = 1\rangle$, $a = a_{1,1}, b = a_{1,2}$. The normal closure K of a in G' is presented by $K = \langle b_0, \ldots, b_{k-1} \mid b_i^k = 1, i < k, b_0 \cdots b_{k-1} = 1\rangle$. For $k < 4$, K is solvable so G is not SQU. For $k \ge 4$ adding the relations $b_1 b_2 = 1, b_i = 1, i > 3$, gives a free product of two cyclic groups of order k, which is SQU. Thus G is SQU for $k \ge 4$.

Case 2. $G = G(2, 4, k)$.

In this case G' has the presentation

$$G' = \langle a_{1,1}, a_{1,2}, a_{1,3} \mid a_{1,1}^k = a_{1,3}^k = (a_{1,1} a_{1,2}^{-1})^k = (a_{1,2} a_{1,3}^{-1})^k = 1\rangle.$$

As above the group defined by G' with the additional relator $a_{1,2} = 1$ is SQU for $k > 2$, as is G.

For $k = 2$, $G(2, 4, k)$ is a solvable group and, hence, not SQU (see [1]). In fact, the third term of the lower central series of $G(2, 4, 2)$ is free abelian of rank 2 (cf. next section).

Case 3. $G = G(2, n, k), n \ge 5$.

In this case adding the relations $a_{i,j} = 1,\ 2 < j < (n-1)$ yields the free product of $\langle a_{11}, a_{12} \mid a_{11}^k = a_{12}^k = (a_{11}a_{12}^{-1})^k = 1\rangle$ and $\langle a_{1,n-1} \mid a_{1,n-1}^k = 1\rangle$. Hence, by the above remark, G' is SQU for $k \geq 2$, as is G.

Case 4. $G = G(3,3,k)$, $k \geq 4$.

In this case, (2.1) - (2.4) gives the following presentation for G':

$$G' = \langle a,b,c,d \mid a^k = b^k = c^k = d^k = (ca)^k = (a^{-1}b)^k = (acd^{-1}b^{-1})^k = (c^{-1}d)^k = (bd)^k = 1\rangle.$$

Adding the relations $c = d = 1$ to this presentation yields the group

$$H = \langle a,b \mid a^k = b^k = (a^{-1}b)^k = 1\rangle \tag{3.1}$$

which is SQU for $k \geq 4$. It follows, as above, that $G(3,3,k)$ is SQU for $k \geq 4$.

Case 5. $G = G(3,n,k)$, $n \geq 4$, $k \geq 2$.

If we add the relations $a_{ij} = 1, 1 \leq j \leq n-2, i \leq m$, to the presentation (2.1) - (2.4) of G' we obtain the free product $H = H_1 * H_2$, where

$$H_1 = \langle a_{1,1}, a_{2,1} \mid a_{1,1}^k = a_{2,1}^k = (a_{1,1}a_{2,1}^{-1})^k = 1\rangle$$

and

$$H_2 = \langle a_{1,n-1}, a_{2,n-1} \mid a_{1,n-1}^k = a_{2,n-1}^k = (a_{1,n-1}a_{2,n-1}^{-1})^k = 1\rangle.$$

Since both H_1 and H_2 have orders ≥ 3, H is SQU as are G' and G.

Case 6. $G = G(m,n,k)$, $m,n \geq 4$.

Map G' to H whose generators and relations are those of G' plus the relations $a_{ij} = 1$ if $2 \leq i \leq m-2$ or $2 \leq j \leq n-2$. This group H has a presentation

$$\langle a,b,c,d \mid a^k = b^k = c^k = d^k = 1\rangle,$$

where $a = a_{1,1}, b = a_{1,n-1}, c = a_{m-1,1}, d = a_{m-1,n-1}$.

For $k \geq 2$, H is a free product of 4 nontrivial cyclic groups and, hence, by the above remarks, H is SQU, as is G'. Finally, G is SQU since G' has finite index in G.

The following lemma summarizes the results of this section.

Lemma 1. *Let* $G(m,n,k) = \langle x, y \mid x^m = y^n = [x,y]^k = 1\rangle$.

(i) *For* $m = 2$, $n = 3$, $G(m,n,k)$ *is* SQU *if* $k \geq 4$ *only.*
(ii) *For* $m = 2$, $n = 4$, $G(m,n,k)$ *is* SQU *for* $k \geq 3$ *only.*
(iii) *For* $m = 2$, $n \geq 5$, $G(m,n,k)$ *is* SQU *for all* $k \geq 2$.
(iv) *For* $m = 3$, $n = 3$, $G(m,n,k)$ *is* SQU *for* $k \geq 4$.
(v) *For* $m = 2$, $n \geq 4$, $G(m,n,k)$ *is* SQU *for all* $k \geq 2$.
(vi) *For* $m, n \geq 4$, $G(m,n,k)$ *is* SQU *for all* $k \geq 2$.

3. Lemma 1 leaves open the cases $G(3,3,k)$, $k = 2,3$. We first consider the case $k = 3$. The proof that $G(3,3,3)$ is SQU does not follow immediately from the abelianizing methods of the last section. However, the desired result can be obtained essentially by showing that the third term G_3 of the lower central series of $G = G(3,3,3)$ is SQU. We will give two proofs that $G(3,3,3)$ is SQU. The first proof is relatively short. The second is a bit longer but gives a better description of G_3 and is more easily generalizable in some cases.

First proof: The above results give the following presentation for G':

$$G' = \langle a, b, c, d \mid a^3 = b^3 = c^3 = d^3 = (ca)^3 = (a^{-1}b)^3 = (acd^{-1}b^{-1})^3 = (c^{-1}d)^3 = (bd)^3 = 1\rangle$$

where $a = a_{11}$, $b = a_{12}$, $c = a_{21}^{-1}$, $d = a_{22}^{-1}$.

Since $[G : G'] = 9$, to complete the proof it will be sufficient to show that G' has a homomorphic image H which is SQU. In fact, we contend that the group H constructed from G' by adding the relation $a = b$ is SQU. This group H has the presentation

$$H = \langle a, c, d \mid a^3 = c^3 = d^3 = (ad)^3 = (ac)^3 = (cd^{-1})^3 = 1\rangle.$$

Next, let K be the normal closure of the elements ac^{-1}, ad^{-1}, cd^{-1} in H. The group K has index 3 in H. With respect to the transversal $1, d, d^2$, the Reidemeister-Schreier process gives the following presentation for K:

$$K = \langle a_i, c_i, i = 0,1,2 \mid c_0^3 = c_1^3 = c_2^3 = a_0 a_1 a_2 = a_0 a_2 a_1 = c_0 c_1 c_2 = a_0 c_1 a_2 c_0 a_1 c_2 = 1 \rangle,$$

where $a_i = d^i a d^{-(i+1)}$, $c_i = d^i c d^{-(i+1)}$. The final step is to add the relations $a_2 = c_2 = 1$ to K to obtain $\bar{K}$, which has the presentation

$$\bar{K} = \langle a_1, c_1 \mid c_1^3 = 1 \rangle = Z * Z_3.$$

Hence $\bar{K}$ is SQU, and it follows from the construction that $G = G(3,3,3)$ is SQU.

Second proof: If we present G in the form $G = \langle x, y, z \mid x^3 = y^3 = z^3 = z[x,y] = 1 \rangle$, then G_3 has the transversal $\{x^i y^j z^k, 0 \le i, j, k < 3\}$. The corresponding generators for G_3 are

$$x_{ijk} = x^i y^j z^k x z^{-k-j} y^{-j} x^{-i-1}, \; y_{ijk} = x^i y^j z^k y z^{-k} y^{-j-1} x^{-i}, \; 0 \le i, j, k < 3.$$

(Observe that $y_{ij0} = 1$ for all i, j.) The corresponding relations for G_3 are

(a) $x_{ijk} x_{i+1,j,k+j} x_{i+2,j,k+2j} = 1 \, (= x^i y^j z^k x^3 z^{-k} y^{-j} x^{-i})$

(b) $y_{ijk} y_{i,j+1,k} y_{i,j+2,k} = 1$

(c) $x^{-1}_{i-1,j,k+1-j} y^{-1}_{i-1,j-1,k+1-j} x_{i-1,j-1,k+1-j} y_{i,j-1,k+1} = 1$
$0 \le i, j, \le 2$.

Since the relations (b) are independent of cyclic permutations of the index j, we may replace these by the reduced set

(b′) $y_{i0k} y_{i1k} y_{i2k} = 1$, $i = 0,1,2$; $k = 1,2$.

The next step is to add relations equating all the x_{ijk} to form H. The group H has a presentation

$$\langle t, y_{ijk} \mid t^3 = 1, \, t^{-1} y^{-1}_{i-1,j-1,k+1-j} t y_{i,j-1,k+1} = 1, \, y_{i0k} y_{i1k} y_{i2k} = 1 \rangle,$$

$0 \le i, j \le 2$, $k = 1,2$, and it is an easy exercise to show that H is a split extension of $F = \langle y_{ijk}, 0 \le i, j \le 2, k = 1,2 \mid y_{i0k} y_{i1k} y_{i2k} = 1 \rangle$

by $\langle t \mid t^3 = 1\rangle$. It follows that F is freely generated by the set $\{y_{ijk}\}$, $0 \le i \le 2$, $1 \le j,k \le 2$, and, being a free group of rank 12, is SQU, as is H since $[H:F] = 3$. It follows that G_3 is SQU, which by the above remarks, shows that $G(3,3,3)$ is SQ-universal.

As was observed in [1], $G(3,3,2)$ has an abelian normal subgroup of finite index and, hence, is not SQU. In fact, $G = G(3,3,2)$ is finite of order 288. This can be seen by observing that the map $f : x \to (1,2,3)$, $y \to (1,2,4)$ defines an epimorphism of G to A_4. It follows that $K = \ker(f)$ has a transversal

$$\{1, x, x^2, y, y^2, xy, xy^2, x^2y, x^2y^2, y^2x, yx^2, xy^2x\},$$

and a routine application of the Reidemeister-Schreier method yields the presentation

$$K = \langle a, b \mid a^3b^3 = ab^{-1}ab^2 = 1\rangle.$$

If we add the relation $a = b$ to those of K we obtain a group H of order 3. Using the transversal $1, a, a^2$, it is again routinely established that the kernel K_1 of the map from K to H has a presentation

$$K_1 = \langle c, d \mid c^2d^{-2} = c^2(cd)^{-2} = 1\rangle$$

and, hence, $|K_1| = 8$. It follows that $|K| = 24$ and $|G| = 12 \cdot 24 = 288$.

The following lemma summarizes the above results.

Lemma 2. (i) *$G(3,3,2)$ has order 288.*
(ii) *$G(3,3,3)$ is SQU.*

Finally, we observe that an analogous argument to that used in the second proof for $G(3,3,3)$ shows that for $G = G(2,4,2)$, G_3, which has index 16 in G, is free abelian of rank 2.

4. The following theorem summarizes the above results.

Theorem. *Let $G = G(m,n,k) = \langle x, y \mid x^m = y^n = [x,y]^k = 1\rangle$, $2 \le m \le n$, $k \ge 2$. The group G is SQU in the following cases and not SQU, otherwise.*

(i) $m = 2,\ n = 3,\ k \geq 4.$

(ii) $m = 2,\ n = 4,\ k \geq 3.$

(iii) $m = 2,\ n \geq 5,\ k \geq 2.$

(iv) $m = 3,\ n = 3,\ k \geq 3.$

(v) $m \geq 3,\ n \geq 4,\ k \geq 2.$

References

[1] B. Fine, J. Howie and G. Rosenberger, Ree-Mendelsohn pairs in generalized triangle groups (Preprint, Dortmund 1986).

[2] G. Higman, B. H. Neumann and H. Neumann, Embedding theorems for groups. *J. London Math. Soc.* **24**, 247-254 (1949).

[3] F. Levin, Factor groups of the modular group. *J. London Math. Soc.* **43**, 195-203 (1968).

[4] R. C. Lyndon and P. E. Schupp, *Combinatorial Group theory*. Springer-Verlag, Berlin, Heidelberg, New York (1977).

[5] P. M. Neumann, The *SQ*-Universality of some finitely presented groups. *Proc. Australian Math. Soc.* **16**, 1-6 (1973).

[6] H. Zieschang, E. Vogt and H. -D. Coldewey, *Surfaces and Planar Discontinuous Groups*. Springer-Verlag, Berlin, Heidelberg, New York (1980).

Fakultät für Mathematik
Ruhr-Universität Bochum
4630 Bochum 1
Federal Republic of Germany

Fachbereich Mathematik
Universität Dortmund
4600 Dortmund 50
Federal Republic of Germany

Action tables for the Fischer group $\bar{F}_{22}$

Jamshid Moori

Abstract

Let $\bar{F}_{22}$ denote the full automorphism group of the smallest Fischer group F_{22}. In $\bar{F}_{22}$ there are 3 classes of involutory outer automorphisms of F_{22} denoted by e, f and θ. The purpose of this paper is to compute the actions of e, f and θ on the F_{22}-conjugacy class D of 3-transpositions. We study the graphs related to $C_D(e)$, $C_D(f)$ and $C_D(\theta)$. Generators for the groups $C_{F_{22}}(e)$, $C_{F_{22}}(f)$ and $C_{F_{22}}(\theta)$ are also obtained.

1. Introduction

The sporadic simple group F_{22} is described by Fischer in [5]. It is generated by a conjugacy class D of involutions, called 3-transpositions. By using a subgroup $S \cong S_{10}$, Enright in [3] gives a complete list for the transform table of F_{22}, that is a table of a^b for all $a, b \in D$. The full automorhism gorup of F_{22} is denoted by $\bar{F}_{22}$ and it is well known that (see Moori [7,8]) $\bar{F}_{22} = F_{22} \cdot \langle e \rangle$, where e is an involutory outer automorphism of F_{22}. In [4], Enright computed the full action of e on the class D of F_{22}. In $\bar{F}_{22}$ there are 3 classes of involutory outer sutomorphisms of F_{22}, denoted by e, f and θ, represented in the ATLAS [1] by $2D$, $2F$ and $2E$ respectively.

The purpose of this paper is to compute the action of f and θ on D. By using the action tables of e, f and θ on D we study the graphs $\Gamma_e, \Gamma_f, \Gamma_\theta$, related to $C_D(e)$, $C_D(f)$ and $C_D(\theta)$ respectively, and compute their parameters. Generators for the groups $C_{F_{22}}(e)$, $C_{F_{22}}(f)$ and $C_{F_{22}}(\theta)$ are also obtianed.

1980 *Mathematics Subject Classification*, Primary 20D08; Secondary 20D45

2. Preliminaries

If we denote the character of the permutation representation of $\bar{F}_{22}$ on the conjugacy class D by $\chi(\bar{F}_{22}|D)$, then by referring to the character table of $\bar{F}_{22}$ (ATLAS [1]) we see that

$$\chi(\bar{F}_{22}|D) = \underline{1}a + \underline{429}a + \underline{3080}a,$$

where $\underline{m}a$ denotes the first irreducible character of degree m in the ATLAS character table of $\bar{F}_{22}$. For any x in $\bar{F}_{22}$ we have

$$|C_D(x)| = \chi(\bar{F}_{22}|D)(x),$$

and it is not difficult to see that

$$|C_D(e)| = 360, \quad |C_D(f)| = 64, \quad |C_D(\theta)| = 72.$$

In F_{22} [4] we have two conjugacy classes of D-subgroups isomorphic to S_{10}. Under the action of S_{10}, D splits into three orbits and we present the following information about the action of S_{10} on D:

Orbit Name	Representative	Size	Stablizer in S_{10}
A	(12)	45	$C_2 \times S_8$
B	$1234/5678/9x$	1575	$(S_4 \times S_4 \times S_2).2$
C	$12/34/56/78/9x$	1890	$2^5.A_5$

2.1 Action of e on D

The outer automorphism e interchanges the two conjugacy classes of S_{10} in F_{22}. Referring to Enright [4, Table 2] we may assume that e centralizes each of the graphs

(12) (23) (34) (45)   (67) (78) (89) $(9x)$

isomorphic to S_5, and interchanges the transposition (56) and $1234/789x/56$. Now we are able to write $e = 12345/6789x$. Then with the aid of the tranform table, Enright [4] computed the full action of e on the class D of F_{22}. We will display this in the following table:

TABLE I
Action of e

Orbit Type	Representative	Size	Image
Non-Split	(12)	20	Fixed
Split	(16)	25	$2345/789x/16$
40/04/11	$1234/789x/56$	25	(56)
40/13/02	$1234/5678/9x$	100	Fixed
31/13/11	$1236/4789/5x$	400	$123x/5789/46$
31/22/02	$1236/4578/9x$	600	$1236/459x/78$
22/22/11	$1267/3489/5x$	450	$12/34/67/89/5x$
5-Split	$16/27/38/49/5x$	240	Fixed
3-Split	$16/27/38/45/9x$	1200	$17/36/28/45/9x$
1-Split	$16/23/45/78/9x$	450	$2378/459x/16$

3. The graph Γ_e

3.1 Parameters of Γ_e

Let A_e, B_e and C_e denote the centralizers of e in A, B and C respectively. By using the information in TABLE I we have

$$|A_e| = 20, \quad |B_e| = 100, \quad |C_e| = 240$$

and

$$A_e = [(12)] = \{(a,b) | (a,b) \text{is non-split}\}$$
$$B_e = [1234/5678/9x] = \text{ orbit of type } 40/13/02$$
$$C_e = [16/27/38/49/5x] = \text{ orbit of type 5-split.}$$

Taking the elements of $C_D(e)$ as vertices and joining distinct commuting pairs, we get the regular graph Γ_e (Fig.1), whose parameters can be deduced by using TABLE I and the Transform Table (Enright [3]). In Γ_e the sets A_e, B_e and C_e are denoted by **20, 100** and **240** reapectively.

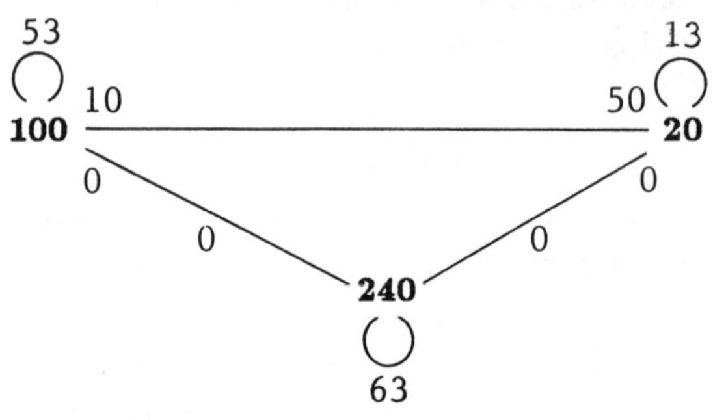

FIGURE 1

The graph Γ_e is not strongly regular, but it is the union of the three strongly regular graphs Γ'_e, Γ''_e, Γ''_e. Each of these graphs is isomorphic to the following strongly regular graph (Fig.2)

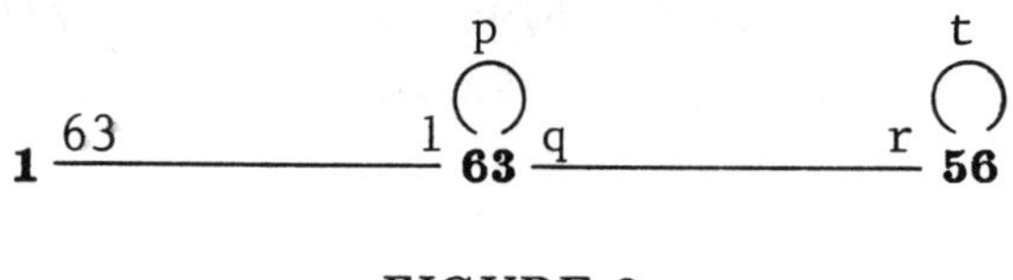

FIGURE 2

The graph Γ'_e is produced by $A_e \cup B_e$, and the graph related to C_e is the union of Γ''_e and Γ''_e.

We will determine the parameters p, q, r and t by the following computations.

It is obvious that $p+q+1 = r+t = 63$. Choose the element (12) in **20**. Then there are 63 elements in $C_D(e)$, distinct from (12), commuting with (12). Assume $\Delta \in \{\mathbf{20}, \mathbf{100}\}$ and let $\Delta_{(12)}$ denote the set of all elements in Δ commuting with (12) but distinct from (12). Then by using TABLE I and the Transform Table for the action of (12) on D we obtain the following information.

$\Delta_{(12)}$	$\mathbf{20}_{(12)}$		$\mathbf{100}_{(12)}$	
$\lvert\Delta_{(12)}\rvert$	13		50	
Elements of	Type	Number	Type	Number
$\Delta_{(12)}$	(ab); $a,b \in \{3,4,5\}$	3	$12ab/c\alpha\beta\gamma/\delta\theta$	30
	$(\alpha\beta)$; $\alpha,\beta \in \{6,7,8,9,x\}$	10	$abc\alpha/\beta\delta\gamma\theta/12$	5
			$12a\alpha/\beta\gamma\delta\theta/bc$	15
			$a,b,c \in \{3,4,5\}$	
			$\alpha,\beta,\gamma,\delta,\theta \in \{6,7,8,9,x\}$	

Now choose $(34) \in \mathbf{20}_{(12)}$. If $\Delta_{(12)(34)}$ denotes the set of elements in $\Delta_{(12)}$ fixed by (34), then, using the same method employed above, we obtain the following information.

$\Delta_{(12)(34)}$	$\mathbf{20}_{(12)(30)}$		$\mathbf{100}_{(12)(34)}$	
$\lvert\Delta_{(12)(34)}\rvert$	10		20	
Elements of	Type	Number	Type	Number
$\Delta_{(12)(34)}$	(α,β)	10	$1234/5\alpha\beta\gamma/\theta\delta$	10
	$\alpha,\beta \in \{6,7,8,9,x\}$		$345\alpha/\beta\delta\theta\gamma/12$	5
			$125\alpha/\beta\gamma\theta\delta/bc$	15
			$\alpha,\beta,\gamma,\delta,\theta \in \{6,7,8,9,x\}$	

Now since

$$p = \lvert\mathbf{20}_{(12)(34)}\rvert + \lvert\mathbf{100}_{(12)(34)}\rvert = 10 + 20 = 30$$

and $p + q = 62$, we have $q = 32$.

Since the number of edges from **63** to **56** is $63q = 56r$, we have $r = 36$. Now from $r + t = 63$, we get $t = 27$.

3.2. The group $C_{F_{22}}(e)$

The group $C_{F_{22}}(e)$ is generated by $C_D(e)$, and it is a D-subgroup of F_{22} (Enright [4]). This group is isomorphic to the group $O_8^+(2).S_3$, the full automorphism group of $O_8^+(2)$, and it is also a maximal subgroup of F_{22} (Wilson [9]). In [7], the author studied this group and calculated it character table.

By using TABLE I and information produced in (3.1), it is not difficult to see that the group generated by **63** is isomorphic to the group $Sp_6(2)$ of all 6×6 matrices over $GF(2)$ preserving a non-singular symplectic form. This group has the following diagram (Fig. 3) for its generators and relations.

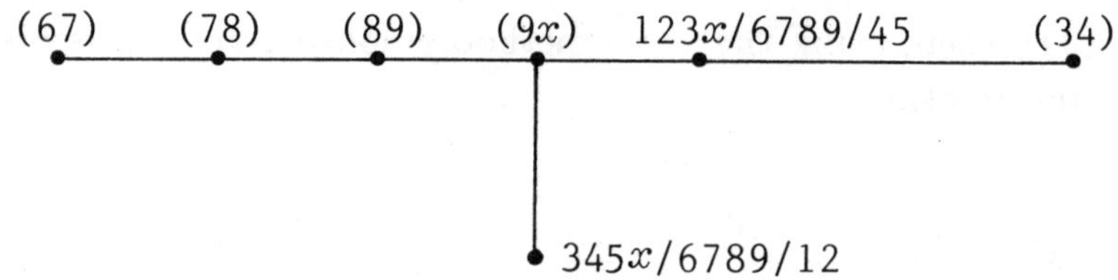

FIGURE 3

The group $S_8 = \langle (67), (78), (79), (9x), 123x/6789/45,\ (34), (45) \rangle$ is a maximal subgroup of $\langle \mathbf{63} \rangle$.

If we choose the element $(14) \in 20^*$, then we are able to show that the group $\langle \mathbf{63}, (14) \rangle$ contains $\langle \mathbf{63} \rangle \times \langle (12) \rangle$ as a subgroup. In fact $\langle \mathbf{63}, (14) \rangle$ contains $20^* \cup 100^*$ and is isomorphic to the group $O_8^+(2).2$. The group $\langle \mathbf{63} \rangle \times \langle (12) \rangle$ will be a maximal subgroup of this group. The group $\langle \mathbf{63}, (14) \rangle \cong O_8^+(2).2$ has the following diagram (Fig. 4) for its generators and relations.

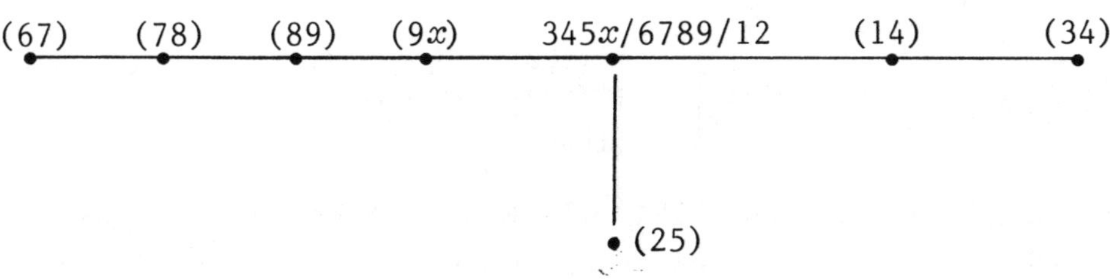

FIGURE 4

Finally choose the element 16/27/38/49/5x in 240. Then since $O_8^+(2).2$ is a maximal subgroup of $O_8^+(2).S_3$, the group $C_{F_{22}}(e)$ is generated by $\mathbf{63} \cup \{(14), 16/27/38/49/5x\}$ and it has the following diagram (Fig.5) for its generators and relations.

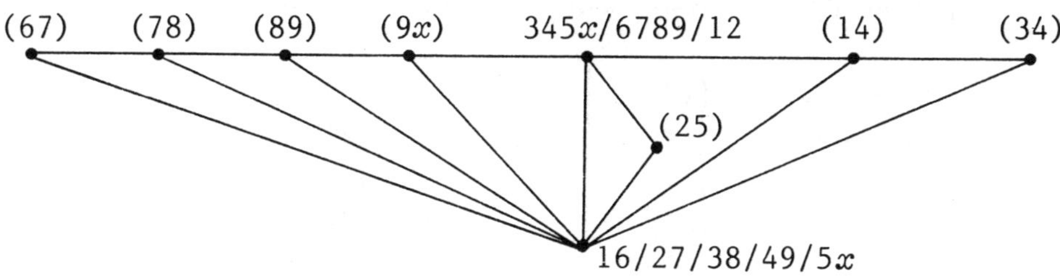

FIGURE 5

4. The graph Γ_f

4.1 The action of f on D

Since $(12) \in C_D(e)$, the element $e.(12)$ is an involutory outer automorphism for the group F_{22}. Using the action of (12) on D (Enright [3,TABLE I]) and the action table of e on D (TABLE I) we are able to compute the action of $e.(12)$ on D. Using this action, we can easily see that $|C_D(e.(12))|$ is **64**. Hence $\bar{F}_{22}$-conjugacy classes of f and $e.(12)$ are identical.

The complete list of action of f on D is given in TABLE II. In this table a B-element is represented by $[B_1/B_2/B_3]_{a_1a_2a_3}$, where

$$a_i = \begin{cases} 2 & \text{if } \{1,2\} \subseteq B_i \\ 0 & \text{if } \{1,2\} \cap B_i = \phi \\ 1 & \text{otherwise.} \end{cases}$$

For example, the element $1234/5678/9x$ is of type $[40/13/02]_{200}$ and $1345/789x/26$ is of type $[40/04/11]_{101}$. It is easy to see that $[22/22/11]_{101}$ $= [22/22/11]_{011}$ and $[22/22/11/]_{200} = [22/22/11]_{020}$. A t-Split element in C, where $t \in \{1,3\}$ is called of type $[t - \text{Split}]_1$ if (12) is one of the five 2-blocks, otherwise, it is of type $[t - \text{Split}]_0$.

TABLE II
Action of $f = e.(12)$

Orbit Type	Representative	Size	Image
Non-Split	(12)	1	Fixed
Non-Split	(13)	6	(23)
Non-Split	(34)	3	Fixed
Non-Split	(67)	10	Fixed
Split	(16)	10	$1345/789x/26$
Split	(36)	15	$1245/789x/36$
$[40/04/11]_{101}$	$1345/789x/26$	10	(16)
$[40/04/11]_{200}$	$1234/789x/56$	15	(56)
$[40/13/02]_{110}$	$1345/2678/9x$	20	$2345/1678/9x$
$[40/13/02]_{200}$	$1234/5678/9x$	30	Fixed
$[31/04/20]_{200}$	$1236/789x/45$	15	Fixed
$[31/04/20]_{101}$	$1346/789x/25$	30	$2346/789x/15$
$[31/04/20]_{002}$	$3456/789x/12$	5	Fixed
$[31/13/11]_{200}$	$1236/4789/5x$	120	$123x/5789/46$
$[31/13/11]_{110}$	$1346/2789/5x$	120	$234x/5789/16$
$[31/13/11]_{101}$	$1346/5789/2x$	120	$134x/2789/56$
$[31/13/11]_{011}$	$3456/1789/2x$	40	$345x/1789/26$
$[31/22/02]_{200}$	$1236/4578/9x$	90	$2346/159x/78$

Continued on next page

Table II. - *Continued*

Orbit Type	Representative	Size	Image
$[31/22/02]_{110}$	$1346/2578/9x$	180	$2346/159x/78$
$[31/22/02]_{020}$	$3456/1278/9x$	30	$3456/129x/78$
$[13/22/20]_{110}$	$1678/239x/45$	60	$2678/459x/13$
$[13/22/20]_{101}$	$1678/349x/25$	60	$2678/159x/34$
$[13/22/20]_{011}$	$3679/149x/25$	120	$3678/159x/24$
$[13/22/20]_{020}$	$3678/129x/45$	30	$3678/459x/12$
$[13/22/20]_{002}$	$3678/459x/12$	30	$3678/129x/45$
$[22/22/11]_{110}$	$1367/2489/5x$	180	$23/14/67/89/5x$
$[22/22/11]_{101}$	$1367/4589/2x$	180	$23/45/67/89/5x$
$[22/22/11]_{200}$	$1267/3489/5x$	90	$12/34/67/89/5x$
5–Split	$16/27/38/49/5x$	240	$26/17/38/49/5x$
$[\text{3-Split}]_1$	$12/36/47/58/9x$	120	$12/37/56/48/9x$
$[\text{3-Split}]_0$	$16/27/38/45/9x$	1080	$27/36/18/45/9x$
$[\text{1-Split}]_1$	$36/12/45/78/9x$	90	$1278/459x/36$
$[\text{1-Split}]_0$	$16/23/45/78/9x$	360	$1378/459x/26$

4.2 Parameters of Γ_f

Let A_f, B_f and C_f denote the centralizers of f in A, B and C respectively. Then by using the information in TABLE II we have

$$|C_D(f)| = 64,\ |A_f| = 14,\ |B_f| = 50,\ |C_f| = 0.$$

Taking the elements of $C_D(f)$ as vertices and joining distinct commuting pairs, we get the graph Γ_f (Fig. 6), whose parameters can be deduced by using TABLE II and the Transform Table (Enright [3]).

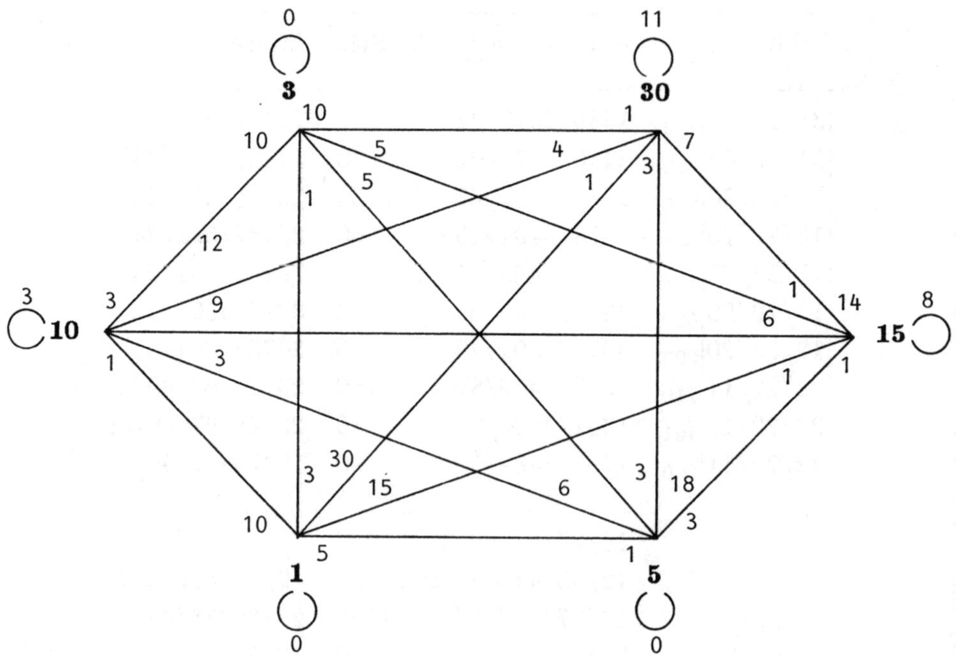

FIGURE 6

In the Graph Γ_f (Fig. 6) we have

$$A_e = \mathbf{10} \cup \mathbf{3} \cup \mathbf{1}; \; B_e = \mathbf{30} \cup \mathbf{15} \cup \mathbf{5}$$

and it is clear that the element $(12) \in \mathbf{1}$ commutes with all elements of $C_D(f)$. Hence the group generated by $C_D(f)$ is the direct product of the groups $\langle(12)\rangle$ and $\langle C_D^*(f)\rangle$, where $C_D^*(f) = C_D(f)\backslash\{(12)\}$.

Now let Γ_f^* denote the graph $\Gamma_f\backslash\{(12)\}$. Then Γ_f^* can be represented by the followng graph (Fig. 7). Then by

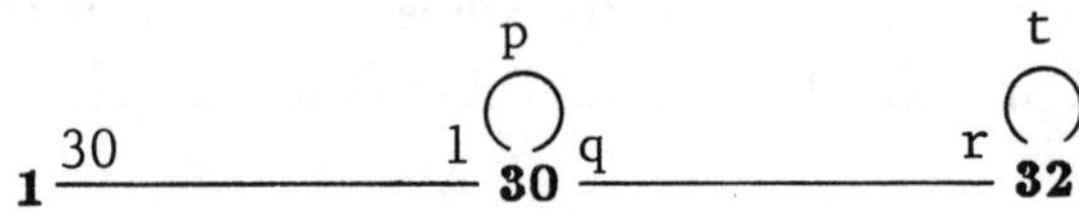

FIGURE 7

using the same method and notation employed in (3.1) together with the action table of f on D (TABLE II) we were able to obtain the following information for Δ and Δ^*, where $\Delta \in \{\mathbf{3}, \mathbf{10}, \mathbf{5}, \mathbf{10}, \mathbf{30}\}$.

$\Delta_{(67)}$	$\mathbf{3}_{(67)}$	$\mathbf{10}_{(67)}$	$\mathbf{5}_{(67)}$	$\mathbf{15}_{(67)}$	$\mathbf{30}_{(67)}$	
$\lvert\Delta_{(67)}\rvert$	3	3	3	9	12	
Elements	(34)	(89)	$3458/9x67/12$	$12a\alpha/\beta\gamma67/bc$	Type	Number
of	(35)	$(9x)$	$3459/8x67/12$	$a, b, c \in \{3,4,5\}$	$12ab/c\alpha67/\beta\gamma$	9
$\Delta_{(67)}$	(45)	$(8x)$	$345x/9867/12$	$\alpha, \beta, \gamma \in \{8,9,x\}$	$12ab/c89x/67$	3
					$a, b, c \in \{3,4,5\}$	
					$\alpha, \beta, \gamma \in \{8,9,x\}$	

$\Delta_{(67)(34)}$	$\mathbf{3}_{(67)(34)}$	$\mathbf{10}_{(67)(34)}$	$\mathbf{5}_{(67)(34)}$	$\mathbf{15}_{(67)(34)}$	$\mathbf{30}_{(67)(34)}$
$\lvert\Delta_{(67)(34)}\rvert$	0	3	3	3	4
Elements		(89)	$3458/9x67/12$	$1258/9x67/34$	$1234/5867/9x$
of		$(9x)$	$3459/8x67/12$	$1259/8x67/34$	$1234/5967/8x$
$\Delta_{(67)(34)}$		$(8x)$	$345x/9867/12$	$125x/8967/34$	$1234/5x67/89$
					$1234/589x/67$

Now since $p = \Sigma|\Delta_{(67)(34)}| = 0+3+3+3+4 = 13$ and $p+q = 29$, we have $q = 16$.

Since the number of edges from **30** to **32** is $30q = 32r$, we have $r = 15$. Now from $r+t = 30$, we get $t = 15$.

4.3 The group $C_{F_{22}}(f)$

The group $C_{F_{22}}(f)$ is generated by $C_D(f)$ and it is a D-subgroup of F_{22} (Enright [4]). This group is equal to $\langle C_D^*(f)\rangle \times \langle (12)\rangle$ (see 4.2) which is isomorphic to the group $Sp_6(2)\times 2$. Since (12) commutes with all the elements of $C_D(f)$, we have $C_D(f) \subseteq C_D(e)$ and $C_{F_{22}}(f) \subseteq C_{F_{22}}(e)$. Therefore the group $\langle C_D^*(f)\rangle$, which is represented by the graph Γ_f^* (Fig. 7), can be generated by the generators and relations given in Fig. 3 in section 3.2.

5. The graph Γ_θ

5.1 The action of θ on D

The element $e.(12)(34)$ is an involutary outer automorphism for the group F_{22}. Using the actions of $e.(12)$ (TABLE II) and of (34) (Enright [3], TABLE I) on D we computed the action of $e.(12)(34)$ on D. By using this action, we are able to see that $|C_D(e.(12)(34))| = 72$. Hence $\bar{F}_{22}$-conjugacy classes of Γ and $e.(12)(34)$ are identical.

The complete list of this action is given in TABLE III. Extending the notation used in 4.1, a B-element in TABLE III is represented by $[B_1/B_2/B_3]_{a_1a_2a_3/b_1b_2b_3}$. In this representation, a_i's are defined as in 4.1, and b_i's are defined similarly by changing (12) to (34). A t-split element in C where $t \in \{1,3\}$, represented by $[t\text{-split}]_{\alpha/\beta}$. The parameter α takes 1 as its value if (12) is one of the five 2-blocks, otherwise its value will be 0. Similarly β takes 1 or 0 for its value according to whether (34) is one of the five 2-block or not.

TABLE III
Action $\theta = e.(12)(34)$

Orbit Type	Representative	Size	Image
Non-Split	(12)	2	Fixed
Non-Split	(13)	4	(24)
Non-Split	(15)	4	(25)
Non-Split	(67)	10	Fixed
Split	(16)	10	1345/789x/26
Split	(36)	10	1235/789x/46
Split	(56)	5	1234/789x/56
$[40/04/11]_{101/200}$	1345/789x/26	10	(16)
$[40/04/11]_{200/200}$	1234/789x/56	5	(56)
$[40/04/11]_{200/101}$	1235/789x/46	10	(36)
$[40/13/02]_{110/200}$	1345/2678/9x	20	2345/1678/9x
$[40/13/02]_{200/200}$	1234/5678/9x	10	Fixed
$[40/13/02]_{200/010}$	1235/4678/9x	20	1245/3678/9x
$[40/13/02]_{200/101}$	1236/789x/45	10	1246/789x/35
$[31/40/20]_{200/002}$	1256/789x/34	5	Fixed
$[31/40/20]_{101/200}$	1346/789x/25	10	2346/789x/15
$[31/40/20]_{101/101}$	1356/789x/24	20	2456/789x/13
$[31/40/20]_{002/200}$	3456/789x/12	5	Fixed
$[31/13/11]_{200/110}$	1236/4789/5x	40	124x/5789/36
$[31/13/11]_{200/011}$	1256/3789/4x	40	125x/3789/4x
$[31/13/11]_{200/101}$	1236/5789/4x	40	123x/4789/56
$[31/13/11]_{110/200}$	1346/2789/5x	40	234x/5789/16
$[31/13/11]_{110/101}$	1356/2789/4x	80	245x/3789/16
$[31/13/11]_{101/200}$	1346/5789/2x	40	134x/2789/56
$[31/13/11]_{101/110}$	1356/4789/2x	80	145x/2789/36
$[31/13/11]_{011/200}$	3456/1789/2x	40	345x/1789/26
$[31/22/02]_{200/110}$	1236/4578/9x	60	1246/359x/78

Continued on next page

Table III. - *continued*

Orbit Type	Representative	Size	Image
$[31/22/02]_{200/020}$	$1256/3478/9x$	30	$1256/349x/78$
$[31/22/02]_{110/200}$	$1346/2578/9x$	60	$2346/159x/78$
$[31/22/02]_{110/110}$	$1356/2478/9x$	120	$2456/139x/78$
$[31/22/02]_{020/200}$	$3456/1278/9x$	30	$3456/129x/78$
$[31/22/20]_{110/011}$	$1678/239x/45$	40	$2678/359x/14$
$[31/22/20]_{110/002}$	$1678/259x/34$	20	$2678/349x/15$
$[13/22/20]_{101/020}$	$1678/349x/25$	20	$2678/159x/34$
$[13/22/20]_{101/011}$	$1678/359x/24$	40	$2678/139x/45$
$[13/22/20]_{011/110}$	$3678/149x/25$	40	$4678/159x/23$
$[13/22/20]_{011/101}$	$3678/159x/24$	40	$4678/139x/24$
$[13/22/20]_{011/011}$	$5678/139x/24$	40	Fixed
$[13/22/20]_{020/101}$	$3678/129x/45$	20	$4678/359x/12$
$[13/22/20]_{020/002}$	$5678/129x/34$	10	$5678/349x/12$
$[13/22/20]_{002/110}$	$3678/459x/12$	20	$4678/129x/38$
$[13/22/20]_{002/020}$	$5678/349x/12$	10	$5678/129x/34$
$[22/22/11]_{110/110}$	$1367/2489/5x$	60	$24/13/67/89/5x$
$[22/22/11]_{110/101}$	$1367/2589/4x$	120	$24/15/67/89/3x$
$[22/22/11]_{101/110}$	$1367/4589/2x$	120	$24/35/67/89/1x$
$[22/22/11]_{101/020}$	$1567/3489/2x$	60	$25/34/67/89/1x$
$[22/22/11]_{200/020}$	$1267/3489/5x$	30	$12/34/67/89/5x$
$[22/22/11]_{200/010}$	$1267/3589/4x$	60	$12/45/67/89/3x$
5-Split	$16/27/38/49/5x$	240	$26/16/48/39/5x$
$[\text{3-Split}]_{1/0}$	$12/36/47/58/9x$	120	$12/47/56/38/9x$
$[\text{3-Split}]_{0/1}$	$16/27/58/34/9x$	360	$27/56/18/34/9x$
$[\text{3-Split}]_{0/0}$	$16/27/38/45/9x$	720	$27/46/18/35/9x$
$[\text{1-Split}]_{1/0}$	$36/12/45/78/9x$	60	$1278/359x/46$
$[\text{1-Split}]_{0/1}$	$16/25/34/78/9x$	120	$1578/349x/26$
$[\text{1-Split}]_{1/1}$	$56/12/34/78/9x$	30	$1278/349x/56$
$[\text{1-Split}]_{0/0}$	$16/23/45/78/9x$	240	$1478/359x/26$

5.2 Parameters of Γ_θ

Let A_θ, B_θ and C_θ denote the centralizers of θ in A, B and C respectively. Then by using the information produced in TABLE III we have

$$|C_D(\theta)| = 72, \ |A_\theta| = 12, \ |B_\theta| = 20, \ |C_\theta| = 40.$$

The following graph (Fig. 8) will represent the graph Γ_θ. The parameters of Γ_θ can be deduced by using TABLE III and the Transform Table (Enright [3]).

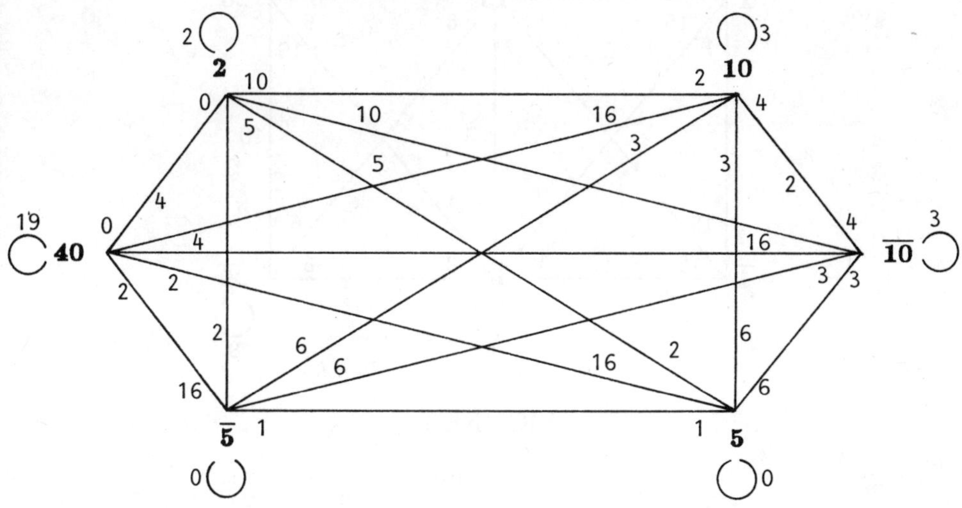

FIGURE 8

In the graph Γ_θ (Fig. 8) we have

$$A_\theta = \mathbf{10} \cup \mathbf{2}, \ B_\theta = \overline{\mathbf{10}} \cup \mathbf{5} \cup \bar{\mathbf{5}}, \ C_\theta = \mathbf{40}$$

Let

$$\mathbf{1} = \{(12)\}, \ \bar{\mathbf{1}} = \{(34)\}, \ \mathbf{15} = \mathbf{10} \cup \mathbf{5},$$

$$\overline{\mathbf{15}} = \overline{\mathbf{10}} \cup \bar{\mathbf{5}}, \ \mathbf{40} = \mathbf{20} \cup \overline{\mathbf{20}}$$

when

$$\mathbf{20} = \{a3\alpha\beta/5\gamma\delta\mu/b4|\alpha,\beta,\gamma,\delta,\mu, \in \{6,7,8,9,x\}, a,b \in \{1,2\}\}$$
$$\overline{\mathbf{20}} = \{a4\alpha\beta/5\gamma\delta\mu/b3|\alpha,\beta,\gamma,\delta,\mu, \in \{6,7,8,9,x\}, a,b \in \{1,2\}\}$$

Then, by using the same methods employed in 3.1 and 4.2 together with the information produced in TABLE III, we are able to show that Γ_θ can be represented by the following diagram (Fig. 9).

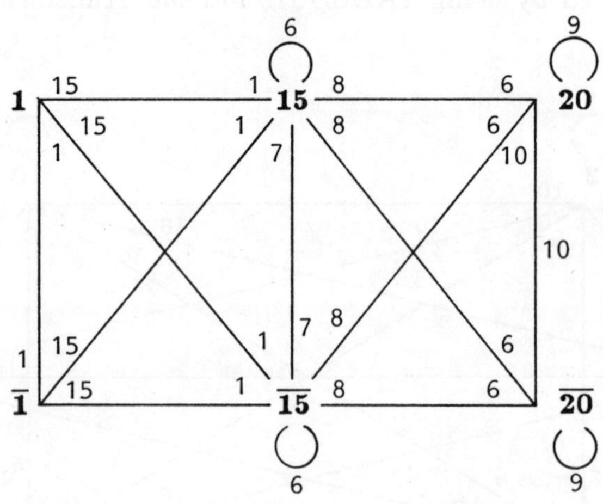

FIGURE 9

5.3 The group $C_{F_{22}}(\theta)$

The group $C_{F_{22}}(\theta)$ is generated by $C_D(\theta)$. It is a D-subgroup of F_{22} isomorphic to the group $2^6.0_6^-(2)$ (Enright [4]). The group $2^6./0_6^-(2)$ is a subgroup of the maximal subgroup $2^6.Sp_6(2)$ of F_{22} ([6,9], ATLAS [1]).

It is not difficult to see that the group $\langle 15 \rangle$ is isomorhic to S_6 and it is generated by $\{(67),(78),(89),(9x),125x/6789/34\}$. Choosing $139x/6785/24 \in 20$, we have

$$\langle 15, 139x/6785/24 \rangle = \langle \mathbf{1}, \mathbf{15}, \mathbf{20} \rangle.$$

The group $\langle \mathbf{1}, \mathbf{15}, \mathbf{20} \rangle$ is isomophic to the group $0_6^-(2)$ and it has the following diagram (Fig. 10) for its generators and relations

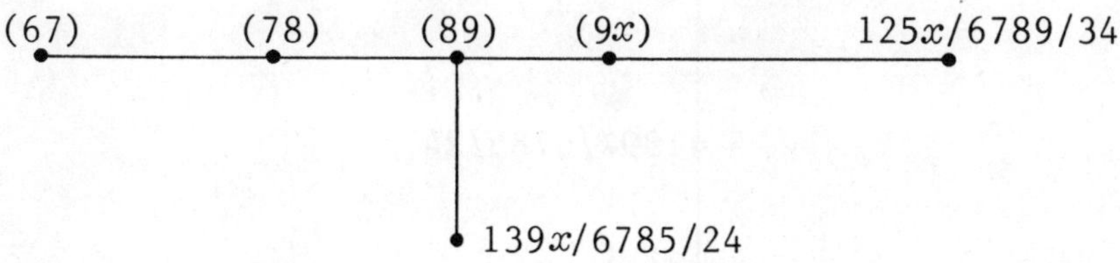

FIGURE 10

Note that the group $\langle 1, 15\rangle$, which is isomprhic to $S_6 \times 2$, is a maximal subgroup of the group described by the diagram in Fig. 10.

Now since (34) transforms **20** to $\overline{\mathbf{20}}$, we have

$$\overline{\mathbf{20}} \subseteq \langle \mathbf{1}, \bar{\mathbf{1}}, \mathbf{15}, \mathbf{20}\rangle.$$

Since also

$$139x/6785/24: \ \ 1478/5x9x/23 \to 3456/789x/12,$$

we have

$$\bar{\mathbf{5}} \subseteq \langle \mathbf{1}, \bar{\mathbf{1}}, \mathbf{15}, \mathbf{20}\rangle.$$

Finally we note that

$$1256/789x/34: \ \ 345x/7896/12 \to 1234/5789/67,$$

and hence

$$\overline{\mathbf{10}} \subseteq \langle \mathbf{1}, \bar{\mathbf{1}}, \mathbf{15}, \mathbf{20}\rangle.$$

Therefore

$$C_{F_{22}}(f) = \langle C_D(f)\rangle = \langle \mathbf{1}, \bar{\mathbf{1}}, \mathbf{15}, \mathbf{20}\rangle.$$

It follows that the group $C_{F_{22}}(f)$ has the following diagram (Fig. 11) for its generators and relations

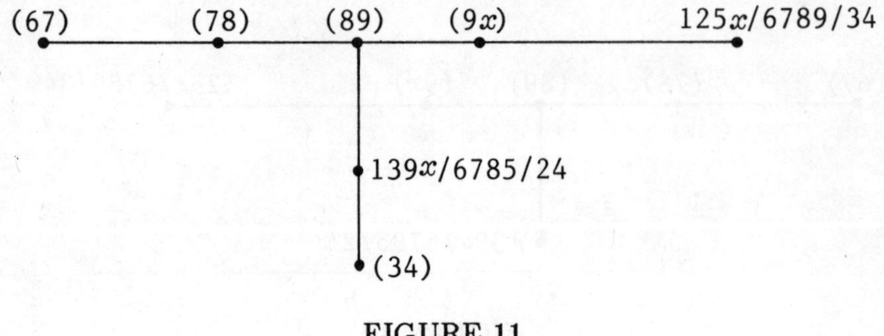

FIGURE 11

References

[1] J.H. Conway, R.T. Curtis, S.P. Norton, R.A. Parker and R.A. Wilson, *An ATLAS of Finite Groups*, Oxford University Press, 1985.

[2] G.M. Enright, *The Structure and Subgroups of the Fischer Groups F_{22} and F_{23}*, Ph.D. Thesis, University of Cambridge, 1976.

[3] G.M. Enright, A description of the Fischer group F_{22}, *J. Algebra* 46 (1977), 334-343.

[4] G.M. Enright, Subgroups generated by transpositions in F_{22} and F_{23}, *Comm. Algebra* 6 (1978), 823-837.

[5] B. Fischer, *Finite Groups generated by 3-transpositions*, Notes, Mathematics Institute, University of Warwick, 1970.

[6] P.B. Kleidman and R.A. Wilson, The maximal subgroups of Fl_{22}, *Math. Proc. Cambridge Philos. Soc.* **102** (1987), 17-23.

[7] J. Moori, On certain groups associated with the smallest Fischer group, *J. London Math. Soc.* (2), **23** (1981), 61-67.

[8] J. Moori, On the automorphism group of the group $D_4(2)$, *J. Algebra* **80** (1983), 216-225.

[9] R.A. Wilson, On maximal subgroups of the Fischer group Fi_{22}, *Math. Proc. Cambridge Philos. Soc.* **95** (1984), 197-222.

Department of Mathematics
University of Bophuthatswana
Post. Bag X2046
Mafikeng 8670
Republic of Bophuthatswana

A CAYLEY library for the groups of order dividing 128

M.F. Newman and *E.A. O'Brien*

> ***... progress in Group Theory depends primarily on an intimate knowledge of a large number of special groups.***
>
> – G. Higman (1958) in a review of Coxeter-Moser (1957).

To acquire an intimate knowledge of groups it is desirable to have good access to collections of information about groups. The monograph of Coxeter and Moser is a useful source of information; so are the tables of Hall and Senior (1964) and the Atlas (Conway *et al.* 1985). Information is starting to become available in electronic form. For example the system CAYLEY (Cannon 1984) has libraries of group descriptions attached to it, and the system CAS (Neubüser *et al.* 1984; sect. 6) has character tables available. Recently Campbell and Robertson (1985) have produced information about simple groups of order less than a million in the form of CAYLEY libraries. In this note we announce the existence of a CAYLEY library containing descriptions of groups with order dividing 128. This library was demonstrated at the Singapore Group Theory Conference. It will be available to all CAYLEY users as it will be supplied with all future releases of the system. This makes available for the first time the 2328 (isomorphism types of) groups of order 128. The method used to generate the descriptions of these groups and some details about the groups are given in James, Newman and O'Brien (in preparation). Our library also gives electronic access to the groups in the Hall-Senior tables.

The most convenient way of describing groups of prime-power order uniformly is by using power-commutator presentations, that is, describing a

1980 *Mathematics Subject Classification* (1985 *Revision*) Primary 20-04, 20D15.

group of order p^n by giving a generating set $a_1, \ldots, a_n$ with defining relations:

$$a_i^p = \prod_{k=i+1}^{n} a_k^{\alpha(i,k)}, \quad 0 \leq \alpha(i,k) < p,$$

$$[a_j, a_i] = \prod_{k=j+1}^{n} a_k^{\alpha(i,j,k)}, \quad 0 \leq \alpha(i,j,k) < p, \ i < j\,.$$

This is, apart from a different naming scheme for the elements of the generating set, the basic way used by Hall and Senior to describe groups of order dividing 64 and it is the way groups of prime-power order are described in CAYLEY. For a discussion of the computational value of such presentations see, for example, Havas and Newman (1980).

The basic data in our library is a sequence of power-commutator presentations, one for each group of order dividing 128. With this it is possible, using the facilities available in CAYLEY, to study these groups in considerable detail. Thus we have, for example, calculated the conjugacy classes for each of these groups and the terms of their lower central series. Moreover for each group of order dividing 64 enough information has been calculated to recognise it in the Hall-Senior tables and hence to access the information about it in those tables. This recognition has been incorporated into the library so that if a group of order dividing 64 comes up in a CAYLEY session its Hall-Senior number can be recovered easily and quickly; its number in our sequence, which is usually different, can also be recovered. It would not be too difficult to recompute all the information in the Hall-Senior tables (but this has not yet been done). It would also be possible to calculate other information such as the character tables (duplicating the work of McKay (1970) in a publicly accessible form).

The power-commutator presentations are actually stored in a much more compact form. We will not go into the technicalities. The library contains a program (a CAYLEY procedure) which produces a power-commutator presentation from the compact description so that a user can then study the group with all the power of CAYLEY. The data is actually stored in a number of sublibraries: for each non-trivial divisor of 64 there is a sublibrary with the compact forms for the power-commutator presentations for the groups of that order and for each d in $\{1, \ldots, 7\}$ there is a sublibrary with the compact forms for the power-commutator presentations for the d-generator groups of order 128. The average time taken to generate a power-commutator presentation for a group of order 128 from the stored compact form is about 0.4 seconds on a **VAX 8700**. A more precise and detailed description of the information is given with the library.

The power-commutator presentations were obtained using a p-group generation program based on the p-group generation algorithm described in Newman (1977). The program provides information about the automorphism group of each group. This information allows one (in general) to get the automorphism group more quickly than with the CAYLEY function *automorphism group*. We have therefore included in the library the order of each automorphism group and a set of generators for a supplement to the subgroup of inner automorphisms; and a program which converts this generating set into a set of permutations on the elements of the group.

Before giving a couple of examples of how the library can be used, we comment briefly on the steps we have taken to ensure the accuracy of the information we provide. As explained in the paper of James *et al.* considerable care has been taken to ensure that the p-group generation program works as claimed and that it produces a complete irredundant set of power-commutator presentations and corresponding automorphism information. To avoid transcription errors this information is massaged by machine into the form in which it appears in the library. The massaging includes reducing, using CAYLEY, the set of automorphisms to an irredundant set. As we mentioned above the information has been used to recognise all the groups of order dividing 64. In the process we got complete agreement with all information we checked in the Hall-Senior tables. This gives us added confidence in our data and added admiration for the work of Hall and Senior. (There are some errors in the generating permutations given there which were pointed out by McKay (1969) and a few obvious misprints in the lattice diagrams.) Calculating enough invariants from the descriptions of the groups of order 128 will eventually give the information needed to provide independent confirmation that the groups described are pairwise non-isomorphic.

The first serious use of the data was made in connection with a question of Cossey (see Ormerod (1987)). The Wielandt subgroup $\omega(G)$ of a group G is the intersection of the normalisers of all the subnormal subgroups of G. The upper Wielandt chain of a group G is defined to be the sequence of subgroups

$$\omega_0(G) \leq \omega_1(G) \leq \ldots$$

where $\omega_0(G)$ is the identity subgroup and $\omega_{i+1}(G)/\omega_i(G)$ is the Wielandt subgroup of $G/\omega_i(G)$. The Wielandt length $wl(G)$ of G is the least non-negative integer k such that $\omega_k(G) = \omega_{k+1}(G)$. From now on we restrict G to being a group of prime-power order. Clearly $\omega(G)$ contains the centre of G, so $wl(G)$ is bounded above by the nilpotency class, $ncl(G)$, of G. Since every subgroup of G is subnormal, a result of Schenkman (1960) gives that $\omega(G)$ lies in the second centre of G and so $2.wl(G) \geq ncl(G)$. Cossey noticed in all groups he examined that $wl(G) \geq ncl(G)-1$ and asked whether this is always the case. We used the library and a short CAYLEY program to calculate the Wielandt length for all groups of order dividing 128. This showed that the answer is yes for these groups. It also showed that the set of Wielandt quotients $G/\omega(G)$ of a given order is a significantly restricted subset of the set of groups of that order. It should be noted that some care needs to be taken with the design of the CAYLEY program if the cost of the calculation is to be kept reasonable.

Another use we have made of the data takes advantage of our having available the order of the automorphism group of each group. There is a long-standing question (see Schenkman (1955) and Gorenstein (1970; sect. 126)) whether for a group G of prime-power order the order of G divides the order of its automorphism group, Aut(G) - apart from some obvious exceptions: all cyclic groups and elementary abelian groups of order p^2. The answer is yes for groups of order dividing 128. We also looked for all groups of order dividing 128 with "relatively small" automorphism groups. It is well-known that the automorphism group of a cyclic group of order 2^n has order 2^{n-1}. A check of the data shows that the order of Aut(G) is at least the order of G for non-cyclic groups G of order dividing 128. There are 14 groups for which equality occurs. These include the first few terms of two infinite series of groups for which equality occurs: for $n \geq 3$

1. the direct products of cyclic groups of order 2^{n-1} with a cyclic group of order 2;
2. the groups $\langle\, a, b \,:\, a^{2^{n-1}} = b^2 = \emptyset,\ a^b = a^{1+2^{n-2}} \,\rangle$.

Looking at the remaining 4 groups, one each of order 16,. . .,128, reveals a similarity between the one of order 64 and the one of order 128 which points to another infinite series of groups with the above property, namely, for $n \geq 6$ the groups:

$$\langle\, a, b, c \;:\; a^{2^{n-2}} = b^2 = c^2 = [b, a] = \emptyset, a^c = a^{1+2^{n-4}} b,\ b^c = ba^{2^{n-3}} \,\rangle.$$

These groups are extensions of groups of type 1 by a cyclic group of order 2. The other two groups are the semidihedral group of order 16 and the group of order 32 $\langle\, a, b \;:\; a^4 = [b, a, a] = [b, a, b],\ b^2 = \emptyset \,\rangle$. Further checking of the data shows that in addition to the above groups there are 6 groups G of order dividing 128 with the order of G equal to the order of the Sylow 2-subgroups of $\mathrm{Aut}(G)$. These observations suggest many questions which we refrain from formulating explicitly.

In conclusion it should be noted that the basic data can be used in contexts other than CAYLEY by people willing to do their own programming; we are prepared to make the data available for such use.

References

Michael D. Atkinson (1984), *Computational Group Theory, (Durham 1982)*, Academic Press, London, New York.

C.M. Campbell and E.F. Robertson (1985), A CAYLEY file of finite simple groups, *Lecture Notes in Comput. Sci.* **204**, pp.243-244, Springer-Verlag, Berlin, Heidelberg, New York.

John J. Cannon (1984), An Introduction to the Group Theory Language, Cayley, see Atkinson pp.145-183.

J.H. Conway, R.T. Curtis, S.P. Norton, R.A. Parker and R.A. Wilson (1985), *Atlas of finite groups*, Clarendon Press, Oxford.

H.S.M. Coxeter and W.O.J. Moser (1957), *Generators and Relations for Discrete Groups*, Ergebnisse der Mathematik und ihrer Grenzgebiete 14, Springer-Verlag, Berlin, Göttingen, Heidelberg.

Daniel Gorenstein (1974), *Reviews on FINITE GROUPS*, Amer. Math. Soc., Providence, Rhode Island.

Marshall Hall, Jr. and James K. Senior (1964), *The Groups of Order* 2^n $(n \leq 6)$, Macmillan, New York.

George Havas and M.F. Newman (1980), Application of computers to questions like those of Burnside, *Burnside Groups, (Bielefeld 1977)*, Lecture Notes in Math. **806**, pp.211-230, Springer-Verlag, Berlin, Heidelberg, New York.

Graham Higman (1958), *Mathematical Reviews* **19**, 527.

R.K. James, M.F. Newman, E.A. O'Brien, The groups of order 128 (in preparation).

John McKay (1969), Table errata : *The Groups of Order* 2^n $(n \leq 6)$, (Macmillan, New York, 1964) by M. Hall and J.K. Senior, *Math. Comp.* **23**, 691-692.

John McKay (1970), The construction of the character table of a finite group from generators and relations, *Computational problems in abstract algebra, (Oxford 1967)*, pp.89-100, Pergamon Press, London, New York.

J. Neubüser, H. Pahlings and W. Plesken (1984), CAS; Design and Use of a System for the Handling of Characters of Finite Groups, see Atkinson pp.195-247.

M.F. Newman (1977), Determination of groups of prime-power order, *Group Theory, (Canberra 1975)*, Lecture Notes in Math. **573**, pp.73-84. Springer-Verlag, Berlin, Heidelberg, New York.

E.A. O'Brien (1988), A computer based description of 2-groups, *Gazette Austral. Math. Soc.* **15**, 1-5.

Elizabeth A. Ormerod (1987), The Wielandt subgroup of metacyclic *p*-groups, Research Report No. 19, Department of Mathematics, Faculty of Science, Australian National University, Canberra, Australia.

Eugene Schenkman (1955), The existence of outer automorphisms of some nilpotent groups of class 2, *Proc. Amer. Math. Soc.* **6**, 6-11.

Eugene Schenkman (1960), On the norm of a group, *Illinois J. Math.* **4**, 150-152.

Department of Mathematics, Institute of Advanced Studies
Australian National University, Canberra, Australia

On octic field extensions and a problem in group theory

Cheryl E. Praeger

Abstract

Suppose that L and K are finite extensions of a field k such that L contains K, $K : k$ is Galois, and L is contained in a Galois extension M of k. Suppose further that A, H, U are the Galois groups of the extensions $M : k$, $M : K$ and $M : L$ respectively so that $U \leq H \trianglelefteq A$. It has been shown that K and L are Kronecker equivalent with respect to k if and only if $\bigcup_{a\in A} U^a = H$. If $|A : H| \leq 2$, that is $[K : k] \leq 2$, it has been shown that this can happen only if $U = H$, that is $K = L$, and if $|A : H| = 4$ there are essentially two possible groups A if $U \neq H$. This paper considers the general problem and classifies all the groups A and subgroups $H > U$ in the next case of interest $|A : H| = 8$, that is for octic Galois extensions $K : k$.

Keywords : Algebraic number fields, Kronecker classes of fields, Galois theory, finite permutation groups.

Acknowledgement : The author acknowledges support of the Australian Research Grants Scheme grant number A685161.

1. Introduction

In [1,2] Jehne and Klingen showed that the Kronecker equivalence of two field extensions is determined by a certain group theoretic property: Suppose that L and K are finite extensions of a given field k which are both contained in a Galois extension M of k and that A, U, H are the Galois groups of the extensions $M : k$, $M : K$ and $M : L$ respectively. Then $K : k$ and $L : k$ are Kronecker equivalent if and only if $\bigcup_{a\in A} U^a = \bigcup_{a\in A} H^a$. For a subgroup U of A the A-*cocore* of U is defined as $U^A = \bigcup_{a\in A} U^a$, so the group theoretic condition can be restated as : the A-cocores of U and H are equal.

Saxl [6] showed that in the special case when $|A : H| = 2$ and $U^A = H^A (= H)$ we must have $U = H$. Thus if $K : k$ is (Galois) of degree 2 then $K : k$ is Kronecker equivalent only to itself. Similarly in [5] it was shown that if H is a normal subgroup of A of index 4 and $U^A = H^A = H$ with $U \neq H$ and core $U_A = \bigcap_{a \in A} U^a = \{1\}$ then A is $Z_3^2 Z_8$ or $Z_3^2 Q_8$, $H = Z_3^2 Z_2$ and $U \simeq S_3$. This classification led to results about Galois quartic field extensions: On the one hand it could be shown that for any k there are Galois quartic extensions $K : k$ which are Kronecker equivalent over k to some cubic extension $L : K$. On the other hand it led to a characterization of those Galois quartic field extensions $K : k$ which are Kronecker equivalent only to themselves (Klingen and Jehne, private communication). The next case of interest for Jehne and Klingen is that of Galois octic (that is degree 8) extensions. The equivalent group theoretic question here is to determine groups A with a normal subgroup H of index 8 and subgroup $U < H$ with $U^A = H$ and trivial core $U_A = \{1\}$. We obtain this classification here.

Theorem. *Suppose that A is a group with a normal subgroup H of index 8, and suppose that H has a maximal subgroup U such that its A-cocore $U^A = \bigcup_{a \in A} U^a = H$ and its A-core $U_A = \bigcap_{a \in A} U^a = \{1\}$. Then one of the following holds.*

(a) *$A = Z_3^2 . Y$, where Y is a Sylow 2-subgroup (of order 16) of $GL(2,3)$, acting naturally on Z_3^2, $H = Z_3^2 . Z_2$ and $U = S_3$.*

(b) *$A = Z_7^2 . Y$, where $Y \leq GL(2,7)$ is Q or $Q \times Z_3$, and Q is Z_{16} or a generalized quaternion group Q_{16} or order 16. Also H is $Z_7^2 . Z_2$ or $Z_7^2 \cdot Z_6$, and U is D_{14} or a Frobenius group F_{42} of order 42, according as $|Y|$ is 16 or 48 respectively.*

(c) *$A = Z_3^4 . Y$ where Y has order 8, 16 or 32.*

The examples in cases (a) and (b) are described explicitly in the statement of the Theorem while the description of the examples arising in case (c) has not been included in the statement. In fact there are more than 20 examples of groups arising in case (c) and these are elaborated in Section 4. In Section 2 we consider the general problem when H has nonabelian socle and show that in the case $|A : H| = 8$ the socle of H must be abelian. The abelian socle case is dealt with in Section 3.

2. Nonabelian case

One general problem associated with the problem of determining Kronecker equivalence classes of field extensions is the following (see [1,2]) :

Problem. *Given a finite group A with a normal subgroup H determine whether H has a proper subgroup U such that $U^A = \bigcup_{a \in A} U^a = H$.*

Clearly we may assume that U is a maximal subgroup of H and that the A-core $U_A = \bigcap_{a \in A} U^a = \{1\}$. Consider the faithful transitive action of A on the set $\Omega = [A : U]$ of right cosets of U in A. Then H has $r = |A : H|$ orbits in Ω, and if $\{a_1 = 1, a_2, \ldots, a_r\}$ is a transversal for H in A the H-orbits are the sets $\Omega_i = \{U a_i h | h \in H\}$ for $i = 1, \ldots, r$. The action of H on Ω_i is primitive since U is maximal in H, and is equivalent to the action of H on the set of cosets of $U_i = U^{a_i}$ in H, so we may identify Ω_i with $[H : U_i]$ for $i = 1, \ldots, r$.

Let S be the socle of H, that is S is the product of the minimal normal subgroups of H. Then as $H \lesssim \prod_{1 \leq 1 \leq r} H^{\Omega_i}$ and H is primitive on Ω_i we have $S \cong T^b$ for some simple group T and integer $b \geq 1$ (see for example [4]). Set $S_i = (U_i \cap S)_H$ the kernel of the action of S on Ω_i for $i = 1, \ldots, r$, so the S_i are normal subgroups of H conjugate in A. Then we have that S is A-invariant, $S_1 = (U \cap S)_H$ and $S = (U \cap S)^A$, and so by [5, Lemma 3.1] it follows that S_1 is not A-invariant. Thus the S_i are nontrivial (that is S is unfaithful on Ω_i).

In the rest of this section we shall consider the case where S is nonabelian. First we obtain a general result and then we look at the special case where $|A : H| = 8$. We note that when S is nonabelian we have $S = N_1 \times \ldots \times N_s$ where the N_i are minimal normal subgroups of H, and $s \geq 2$ since the S_i are nontrivial.

Proposition 2.1. *Suppose that H is a normal subgroup of index r in a finite group A and that U is a maximal subgroup of H such that $U^A = \bigcup_{a\in A} U^a = H$ and $U_A = \bigcap_{a\in A} U^a = \{1\}$. Suppose further that the socle S of H is the product $N_1 \times \ldots \times N_s$ of nonabelian minimal normal subgroups N_i of H for some $s \geq 2$. Then $s \geq 3$, A acts transitively on $\{N_1, \ldots, N_s\}$ by conjugation and, renumbering the N_i if necessary, $U \cap S = D \times N_3 \times \ldots \times N_s$ with D a diagonal subgroup of $N_1 \times N_2$. Further $s/(s,2)$ divides r and $r \geq s$.*

Proof. With the notation above S/S_i is isomorphic to the socle of the primitive group H^{Ω_i} and hence is isomorphic to a minimal normal subgroup of H or the product of two such subgroups (see [4]). Suppose first that H^{Ω_i} has a unique minimal normal subgroup; then we may renumber the N_i so that $S_1 = N_2 \times \ldots \times N_s$ (since a normal subgroup of S is a product of some of the N_i). In this case, as each N_i acts nontrivially on some Ω_j and since A is transitive on the Ω_j, the subgroups $N_1, \ldots, N_s$ form a complete A-conjugacy class. It follows from [5, Lemma 3.2] that $(U \cap N_1)_H$ is not invariant in $N_A(N_1)$; this is a contradiction since $(U \cap N_1)_H = S_1 \cap N_1 = \{1\}$.

Thus H^{Ω_i} has two nonabelian minimal normal subgroups so that we can renumber the N_i such that $S/S_1 = S^{\Omega_1} \cong N_1 \times N_2$. Since $S_1 \neq 1$ we must have $s \geq 3$, $S_1 = N_3 \times \ldots \times N_s$, and $U \cap S = D \times N_3 \times \ldots \times N_s$ where D is a diagonal subgroup of $N_1 \times N_2$ (see [4]). Since each N_i acts nontrivially on some Ω_j it follows that each orbit of A on $\{N_1, \ldots, N_s\}$, acting by conjugation, has length at least $s/2$. So if A is not transitive on the N_i, it has two orbits of length $s/2$, say $\{N_i \mid i \in I\}$ and $\{N_i \mid i \notin I\}$ where I is a subset of $\{1, \ldots, s\}$ of size $s/2$ containing 1. Then $R = \prod_{i\in I} N_i$ is normal in A, $U \geq \prod_{i\in I\setminus\{1\}} N_i$ but $U \not\geq R$, and $R = (U \cap R)^A$. By [5, Lemma 3.2] it follows that $(U \cap N_1)_H$ is not $N_A(N_1)$-invariant. This again is a contradiction since $(U \cap N_1)_H = S_1 \cap N_1 = \{1\}$. Thus A is transitive on the N_i.

Consider A acting on the set $\underline{s} = \{1, 2, \ldots, s\}$ in the same way as it acts on the set $\{N_1, \ldots, N_s\}$ by conjugation. Then the pairs $\{i, j\}$ such that $S^{\Omega_k} \cong (N_i \times N_j)^{\Omega_k}$ (that is $S_k = \prod_{\ell \neq i,j} N_\ell$) for some k are precisely those pairs in the orbit $J = \{1, 2\}^A$ of A on pairs. Clearly $N_i \times N_j$ acts faithfully on $r/|J|$ of the Ω_k. Furthermore $|J|\delta = |A : A_{\{1,2\}}| . |A_{\{1,2\}} : A_{12}| = |A : A_{12}| = |A : A_1| . |A_1 : A_{12}| = s|A_1 : A_{12}|$ where δ is 1 or 2. It follows that $s/(s,2)$ divides $|J|$ which divides r and $r \geq |J| \geq s/(s,2)$. If $|J| = s/2$ then each positive integer $i \leq s$ is in a unique pair $\{i,j\}$ in J and we have $S = M_1 \times \ldots \times M_{s/2}$ with each M_k being the product $N_i \times N_j$ for some $\{i,j\} \in J$, and $M_1 = N_1 \times N_2$. Then [5, Lemma 3.2] implies that $(U \cap M_1)_H$, which is $S_1 \cap (N_1 \times N_2) = \{1\}$, is not A-invariant. This contradiction shows that $|J| \neq s/2$ and hence that $r \geq |J| \geq s$. Thus Proposition 2.1 is proved.

There are of course examples of such groups, two of which were given in [5] in Example 2.4 and one of the examples in 4.4 : the first was $A = S_5 \text{ wr } S_5, H = S = A_5^5$ so that $r = 2^8.3.5$ and $s = 5$. The second was $A = A_5 \text{ wr } L_2(5)$ with $H = S = A_5^6$ so that $r = 60$ and $s = 6$. It is fairly easy to argue that s must be at least 5.

Proposition 2.2. *The integer s in Proposition 2.1 must be at least* 5 *and if $s = 5$ then the $N_i \simeq A_5$.*

Proof. Suppose that $s \leq 5$. Since $U \cap S = D \times N_3 \times \ldots \times N_s$ it follows that each element of $(U \cap S)^A$, with respect to the direct decomposition $S = N_1 \times \ldots \times N_s$, must have at least two entries of equal order, in fact at least two entries in the same conjugacy class of Aut N, where $N \simeq N_i$. Since $(U \cap S)^A = S$ it follows that N must be the union of at most $s-1 \leq 4$ Aut N-conjugacy classes. Now $N \simeq T^x$ where T is a nonabelian simple group and $x \geq 1$, and $|T|$ is divisible by at least three distinct primes. It follows that $s = 5, N = T$ and T is the union of 4 Aut T-conjugacy classes. In particular $|T| = 2^a p^b q^c$ for p and q distinct odd primes and T contains elements of orders only $1, 2, p$ and q. In particular T has elementary abelian Sylow 2-subgroups. By a theorem of Walter [8], T is $L_2(q)$ ($q \equiv 3$ or $5 \pmod 8$, $q > 3$, or $q = 2^n$), $J(11)$ or T is of Ree type : of these groups the only one satisfying the conditions above is $T = L_2(4) \cong A_5$. Thus Proposition 2.2 is proved.

In the rest of this section we consider the special case $|A : H| = 8$ and S nonabelian, and show that this cannot arise. From Propositions 2.1 and 2.2 it follows that $S = N_1 \times \ldots \times N_8$ is the product of $s = 8$ minimal normal subgroups of H all conjugate in A. We note that all elements in $(U \cap S)^A$ have entries in positions i and j of equal order for some $\{i,j\} \in J$ where $\{i,j\} \in J$ if and only if $N_i \times N_j$ acts faithfully on some Ω_k : by the proof of

Proposition 2.1, $|J| = 8$. Let t, u, v be nontrivial elements of $N \simeq N_i$ with distinct orders.

We may assume that $N_1 \times N_2$ acts faithfully on Ω_1, so $J = \{1,2\}^A$. Let $H < A_2 < A_1 < A$ be such that $|A : A_1| = 2$ and $|A : A_2| = 4$ with A_2 normal in A. Then A_1 has orbits $\Omega_1 \cup \ldots \cup \Omega_4$ and $\Omega_5 \cup \ldots \cup \Omega_8$ say and A_2 has orbits $\Omega_{2j-1} \cup \Omega_{2j}$ for $j = 1, \ldots, 4$. Also A_1, A_2 have orbits of length 4 and 2 respectively on $\{N_1, \ldots, N_8\}$. Since $|J| = 8$ it follows that $\{N_1, N_2\}$ is not an A_2-orbit and we may assume that $\{N_1, N_3\}$ and $\{N_2, N_4\}$ are A_2-orbits so that $N_3 \times N_4$ acts faithfully on Ω_2 (that is $\{3,4\} \in J$). If $\{N_i \mid 1 \leq i \leq 4\}$ is an A_1-orbit then $N_5 \times N_6 \times N_7 \times N_8$ fixes $\Omega_1 \cup \Omega_2 \cup \Omega_3 \cup \Omega_4$ pointwise, all pairs $\{i, j\}$ in J are contained in $\{1,2,3,4\}$ or $\{5,6,7,8\}$, and the element $\underline{x} = (1, t, u, v, 1, t, u, v) \in S \setminus (U \cap S)^A$ (since the i, j entries have different orders for each $\{i, j\} \in J$).

Thus we may assume that $\{N_1, N_3, N_5, N_7\}$ and $\{N_2, N_4, N_6, N_8\}$ are A_1-orbits and that $\{5,6\}$ and $\{7,8\}$ are in J with $N_5 \times N_6$ and $N_7 \times N_8$ faithful on Ω_3 and Ω_4 respectively. Further we may suppose that $N_1 \times N_i$ acts faithfully on Ω_5 for some $1 < i \leq 8$. We showed above that N_i must lie in a different A_1-orbit from N_1 so i is even, and since $|J| = 8, i \neq 2$. By renumbering the $N_j, j \geq 5$, we can assume that i is 4 or 6. Then $J = J_1 \cup J_2$ where $J_1 = \{\{2j-1, 2j\} \mid j = 1, \ldots, 4\}$ and either $J_2 = \{\{1,4\}, \{3,2\}, \{5,x\}, \{7,y\}\}$ with $\{x, y\} = \{6,8\}$ or $J_2 = \{\{1,6\}, \{3,8\}, \{5,8\}, \{7,y\}\}$ with $\{x,y\} = \{2,4\}$. In the first case $\underline{x} = (1, t, u, v, 1, t, u, v) \notin (U \cap S)^A$ and in the second case $\underline{x} = (1, t, u, v, 1, v, u, t) \notin (U \cap S)^A$. This contradiction shows:

Lemma 2.3. *If A, H, U are as in the Theorem then the socle S of H is elementary abelian.*

3. Proof of the Theorem : the abelian case

We suppose that A, H, U are as in the Theorem and use the notation introduced in Section 2. By Lemma 2.3 the socle S of H is elementary abelian. Also we showed in Section 2 that $S_i = (U_i \cap S)_H$, the kernel of S on Ω_i, is nontrivial for $i = 1, \ldots, 8$. In this case we note that $S = (U \cap S)^A = S_1 \cup S_2 \cup \ldots \cup S_8$ (since $S_i = U_i \cap S$ when S is abelian).

Set $X = \{1, 2, \ldots, 8\}$ and consider the natural action of A on X which is equivalent to its action on the set $\{\Omega_1, \Omega_2, \ldots, \Omega_8\}$ of H-orbits in Ω. We may assume that A preserves the partitions $\{12|34|56|78\}$ and $\{1234|5678\}$ of X. Let A_1, A_2 denote the setwise stabilizers of $\{1,2,3,4\}$ and $\{1,2\}$ in A respectively. For each nonempty subset I of X define $S(I) = \bigcap_{i \in I} S_i$; we shall also write $S(i) = S_i$ and $S(ij) = S(i) \cap S(j)$ etc.

Suppose first that for $I = \{2, 3, \ldots, 8\}$, $S(I) \neq \{1\}$. Then setting $I(i) = X \setminus \{i\}$ for $i \leq 8$ we obtain a contradiction on applying [5, Lemma 3.2] to $M = S(I(1)) \times \ldots \times S(I(8))$. Thus $S(I) = \{1\}$. Set $I(i,j) = X \setminus \{i,j\}$ for $i \neq j$. Suppose next that $S(I(12)) \neq \{1\}$. Then applying [5, Lemma 3.2] to $M = S(I(12)) \times S(I(34)) \times S(I(56)) \times S(I(78))$ leads to a contradiction. It follows that if B is any block of imprimitivity of size 2 for A in X then $S(X \setminus B) = \{1\}$.

Next we consider the case where $S(1234) \neq \{1\}$.

Lemma 3.1. *Suppose that* $S(1234) \neq \{1\}$. *Then*

(i) $S = Z_3^4 = R_1 \times R_2$ *with* $|R_i| = 9$.

(ii) $A = SY$ *where* $Y \leq W \text{ wr } Z_2 = (W_1 \times W_2) . Z_2$; *the group* W *is* Z_8 *or* Q_8 *and* W_i *is transitive on the four subgroups of* R_i *of order* 3 *and centralizes* R_j, $j \neq i$. *Further* Y *interchanges* W_1 *and* W_2 *and* $Y \cap (W_1 \times W_2)$ *is either a diagonal subgroup* D *or* $D . \Omega_1(W_1 \times W_2)$. *Thus* $|Y|$ *is* 16 *or* 32.

(iii) $H = S . (Y \cap \Omega_1(W_1 \times W_2))$ *and* U *contains* R_2 *and has index* 3 *in* H.

(Note : For a 2-group G, $\Omega_1(G)$ denotes the subgroup generated by all elements of order 2).

Proof. Suppose that $R_2 = S(1234) \neq \{1\}$; then $R_1 = S(5678) \neq \{1\}$. By [5, Lemma 3.2] applied to $M = R_1 \times R_2$ we have $V_i = S(i5678) = S_i \cap R_1 \neq 1$ for $i = 1, \ldots, 4$. Now $M = (U \cap M)^A = ((U \cap R_1)^{A1} \times R_2) \cup (R_1 \times (U^a \cap R_2)^{A1})$ where $a \in A \setminus A_1$. If there is an element $r_1 \in R_1 \setminus (U \cap R_1)^{A1}$ then its conjugate $r_2 = r_1^a \in R_2 \setminus (U^a \cap R_2)^{A1}$ and then $r_1 r_2 \notin (U \cap M)^A$. Thus $R_1 = (U \cap R_1)^{A1}$. Now as $V_1 \cap V_2 = S(125678) = \{1\}$ we have $R_1 = V_1 \times V_2$ of order $|V_1|^2 = m^2$, and as $R_1 = (U \cap R_1)^{A1} = V_1 \cup V_2 \cup V_3 \cup V_4$

we have $|R_1| = 4(m-1)+1$. It follows that $m = 3$ and $Z_3^4 = R_1 \times R_2 \le S$ with $|S| \le 3^6$. We shall show that $S = R_1 \times R_2$. The subgroups $V_1, \ldots, V_4$ must be the 4 distinct subgroups of R_1 of order 3 since their union is R_1. It follows that each nontrivial element of R_i moves exactly 9 points of Ω for $i = 1, 2$. If $|S| = 3^6$ then for each subset I of X of size $b \le 5$ we have $|S(I)| = 3^{6-b}$ and so by the inclusion-exclusion principle, $729 = |S| \le 8.3^5 - \binom{8}{2}.3^4 + \binom{8}{3}.3^3 - \binom{8}{4}.3^2 + \binom{8}{5}.3 = 726$ which is a contradiction. Suppose then that $|S| = 3^5$. Then the group induced on $\Omega' = \Omega_1 \cup \ldots \cup \Omega_4$ has order 3^3 and hence if $x \in S$ acts nontrivially on Ω' then x moves at least 6 points of Ω'. Further, if x moves exactly 6 points of Ω' then $x \notin R_1$ so x acts nontrivially on $\Omega'' = \Omega \setminus \Omega'$ and hence x moves at least 6 points of Ω'' also. Let I be a subset of X. If I has size $b \le 3$ then $|S(I)| = 3^{5-b}$, while if I has size $b \ge 6$ then $S(I)$ is trivial. If I has size 5 then $S(I)$ fixes at least 9 points in one of Ω', Ω'' and so $S(I) \le R_i$ for $i = 1$ or 2; in fact $|S(I)|$ is 3 if I contains $X' = \{1,2,3,4\}$ or $X'' = \{5,6,7,8\}$ and $S(I) = \{1\}$ for the other 48 subsets I of size 5. If I has size 4 then $|S(I)| = 9$ if I is X' or X'', and $|S(I)| = 3$ for the other 68 subsets I of size 4. By the inclusion-exclusion principle then $243 = |S| = 8.3^4 - \binom{8}{2}.3^3 + \binom{8}{3}.3^2 - (2.9+68.3) + (8.3+48.1) - \binom{8}{6} + \binom{8}{7} - 1 = 225$. From this contradiction we conclude that $S = R_1 \times R_2$.

Then as S is self-centralizing we have $A \lesssim AGL(2,3)$ wr $Z_2 < AGL(4,3)$. As H normalizes every subgroup of R_1 and R_2 it follows that $H \le S.(C_1 \times C_2)$ where $C_i \simeq Z_2$ inverts R_i and centralizes R_j for $i \ne j$. Let $\bar{A}_1$ be the group induced by A_1 on Ω' and let $\bar{H}, \bar{U}$ be the groups induced by H, U respectively on Ω'. Then $\bar{A}_1 \le AGL(2,3)$ and $\bar{U}$ is a maximal subgroup of $\bar{H}$ such that $\bar{H} = (\bar{U} \cap \bar{H})^{\bar{A}_1}$ and $|\bar{A}_1 : \bar{H}| = 4$. By [5, Theorem 4.3], $\bar{A}_1 = Z_3^2.Z_8$ or $Z_3^2.Q_8$ and $\bar{H} = Z_3^2.Z_2$. It follows that $A = SY$ where Y is any subgroup of W wr Z_2 (where W is Z_8 or Q_8) such that Y interchanges the two copies of W, and $Y \cap W^2$ is a diagonal subgroup D of W^2 or is $D\Omega_1(W^2)$. This completes the proof of Lemma 3.1.

Thus we may assume that $S(B) = 1$ for any block in X of size 4. Next we deal with the case $S(123) \ne \{1\}$.

Lemma 3.2. *Suppose that $S(123) \ne \{1\}$ (but $S(1234) = \{1\}$). Then*

(i) $S = Z_3^4 = S(1256) \times S(1278) \times S(3456) \times S(3478)$ *and A preserves this direct decomposition.*

(ii) $A = S..Y \le S_3$ *wr* D_8 *with* $Y \le Z_2$ *wr* D_8; $Y/(Y \cap H)$ *has order* 8 *and* $Y \cap H \le Z(Z_2$ *wr* $D_8) \cong Z_2$. *Also* Y *projects onto a subgroup of* D_8 *of order* 4 *acting regularly on the four direct factors* S_3 *of the base group. Thus* $Y \cap Z_2^4$ *has order* 2 *or* 4. (*Here* S_3 *denotes the symmetric group of degree* 3.)

(iii) $H = S(Y \cap H)$ *is* S *or* $S\,.\,Z_2$, *and* U *is* $Z_3^3\,.\,(Y \cap H)$.

Proof. Suppose that $S(123) \ne \{1\}$. Then for $i = 1, \ldots, 4$, the subgroup $V_i = S(\{1,2,3,4\} \setminus \{i\}) \ne \{1\}$ and the V_i are four distinct normal subgroups of H of order $|\Omega_i| = m$ and we have $S = V_1 \times V_2 \times V_3 \times V_4$ (since $S(1234) = 1$). Now $S = (U \cap S)^A = \bigcup_{i \le 8} S_i$ and $|\bigcup_{i \le 4} S_i| = 4m^3 - 6m^2 + 4m - 1 = m^4 - (m-1)^4 < m^4 = |S|$ (using the inclusion-exclusion principle); hence the S_i, $1 \le i \le 8$, are all distinct. Now $|\bigcup_{i \le 4} S_i| = |\bigcup_{i \ge 5} S_i|$ so $m^4 < 2|\bigcup_{i \le 4} S_i| = 2(m^4 - (m-1)^4)$ and it follows that $m < 7$.

Suppose that $S(12)$ fixes Ω_j pointwise for some $j > 2$; then $j > 4$, say $j = 5$. Then applying an element of A_2 which interchanges 5 and 6 we see that $S(12) = S(56)$ fixes precisely $\bigcup_{j \in B} \Omega_j$ where $B = \{1,2,5,6\}$. Now B is a block of imprimitivity for A for if $a \in A$ is such that $B \cap B^a$ contains say 1^a then $\{1,2\}^a$ is $\{1,2\}$ or $\{5,6\}$ and so $S(12)^a = S(12)$; since the fixed point set of $S(12)$ is $\bigcup_{i \in B} \Omega_i$ it follows that $B^a = B$. this contradicts our assumption that $S(B) = \{1\}$ for all blocks B of size 4 (for here $S(B) = S(12) \ne \{1\}$). It follows that, for each block C of size 2, $S(C)$ fixes exactly $2m$ points.

Next suppose that $S(ij)$ has more than $2m$ fixed points for some $i \ne j$. Then $S(ij)$ fixes $2m$ points in either $\bigcup_{i \le 4} \Omega_i$ or $\bigcup_{i \ge 5} \Omega_i$ so we can assume that i and j lie in the same block of size 4, say $i = 1$, $j = 3$ and $S(13)$ fixes Ω_5 pointwise. Since $\{1,3\}$ cannot be a block of size 2 we may assume that $A_1^X = \langle (1324)(5768) \rangle$. Considering the action of A_1 we have $S(32) = S(327)$, $S(24) = S(246)$, and $S(41) = S(418)$. Now (conjugating by some element of $A \setminus A_1$) $S(57)$ also fixes Ω_i for some $i \le 4$ and we must have $i = 4$ (for if $i = 3$ say then $S(57) = S(35) \cap S(37) = S(135) \cap S(327) = S(12357) = S(12)$ fixes only $2m$ points which is not the case and if $i = 1$ then $S(57) = S(15) = (135)$ fixes Ω_3, and if $i = 2$ then $S(57) = S(27) = S(327)$ fixes Ω_3). Thus $S(57) = S(457)$, $S(76) = S(176)$, $S(68) = S(368)$

and $S(85) = S(268)$. Then $S(123) = S(13) \cap S(23) = S(135) \cap S(237) \leq S(57) = S(574) < S(4)$. This is a contradiction since $S(1234) = \{1\}$. Thus all the $S(ij)$, $i \neq j$ are distinct.

Therefore, $m^4 = |S| \leq 8|S_1| - \binom{8}{2}|S(12)| + \binom{8}{3}|S(123)| = 8m^3 - 28m^2 + 56m$, by the inclusion-exclusion principle, and hence $m \leq 4$. Now a Sylow 2-subgroup L of A is transitive on $\{S_1, \ldots, S_8\}$ and if m is even then L centralizes an element $x \neq 1$ of S. Now as $S = \cup S_i$, $x \in S_i$ for some i, and as L is transitive on the S_j, $x \in \bigcap_j S_j = \{1\}$, a contradiction. Thus $m = 3$.

Now applying elements of $A_2 \setminus H$ to $S(125)$ we find that $S(125) = S(1256)$ and similarly $S(127) = S(1278)$ so that $V_3 \times V_4 = S(12) \cong S(1256) \times S(1278)$. Further $V_3 = S(124) \neq S(1256) \neq S(123) = V_4$ so $S(1256)$, and similarly $S(1278)$, are diagonal subgroups of $V_3 \times V_4$. Similarly we obtain $V_1 \times V_2 = S(34) \cong (3456) \times S(3478)$, with $S(3456)$ and $S(3478)$ diagonal subgroups of $V_1 \times V_2$. This gives the direct decomposition of S in part (i), a decomposition preserved by A.

We claim that $H = S$ or $H = S \,.\, Z_2$ with Z_2 inverting each element of H : Now S is self centralizing in A so $H \lesssim AGL(4,3)$ and H normalizes each S_i, $i = 1, \ldots, 8$. We saw in the previous paragraph that H normalizes the four subgroups of $S(12)$ of order 3, namely V_3, V_4, $S(1256)$ and $S(1278)$, and so H centralizes or inverts $S(12)$; similarly H centralizes or inverts $S(34)$. Since $S \cong S(12) \times S(34)$ we have $H/S \leq Z_2 \times Z_2$: further H normalizes $S(567)$ which is contained in a diagonal subgroup of $S(12) \times S(34)$. Thus $S \leq H \leq S.Z_2$ with the Z_2 inverting H.

Now the subgroup of $GL(4,3)$ preserving the direct decomposition in (i) is $Z_2 \text{ wr } S_4$ and hence $A = S\,.\,Y$ with $Y \leq Z_2 \text{ wr } D_8$, $Y \cap H \leq Z(Z_2 \text{ wr } D_8) \cong Z_2$, and $Y/(Y \cap H) \cong A/H$ of order 8; also Y projects onto a regular subgroup of order 4 of D_8.

So now we can assume that $S(C) = \{1\}$ for each subset of size 3 of a block in X of size 4.

Lemma 3.3. *If $S_i = S_j$ for some $i \neq j$ then $S = Z_3^2$ and $A = Z_3^2\,.\,Y$, where Y is a Sylow 2-subgroup of $GL(2,3)$ acting naturally on S; also $H = Z_3^2\,.\,Z_2$ and $U = S_3$, the symmetric group of degree 3.*

Proof. Suppose that $S_i = S_j$ for some $i \neq j$. Then it is easy to show that $B = \{j \mid S_i = S_j\}$ is a block of imprimitivity for A in X, and as $S_i \neq \{1\}$ we may assume that $i = 1$ and $\{1,2\} \subseteq B \subseteq \{1,2,3,4\}$. Since $S(123) = S(124) = \{1\}$ it follows that $B = \{1,2\}$, and hence that there are four distinct S_i, namely S_1, S_3, S_5, S_7. Thus S has order m^2 and we have $m^2 = |S| = |\cup S_i| = 4(m-1)+1$; so $m = 3$. Now $A \leq AGL(2,3)$ and as H normalizes all subgroups of S, $H \leq S.Z_2$. It follows that $A = SY$ where Y is a Sylow 2-subgroup of $GL(2,3)$, $H = S\,.\,Z_2$, and $U = S_3$.

Thus we may assume that the S_i are all distinct.

Lemma 3.4. *The subgroup $S(12) = \{1\}$ so that $S \cong S_1 \times S_2$ has order m^2.*

Proof. Suppose that $S(12) \neq \{1\}$. Then S has order m^3 and since $m^3 = |S| = |\cup S_i| < 8m^2$ we have $m \leq 7$. Suppose first that $S(12)$ fixes Ω_j pointwise for some $j > 2$. Then $\{1,2,j\}$ does not lie in any block of size 4 of A in X and we may take $j = 5$. Then applying an element of $A_2 \setminus H$ we see that $S(12) = S(1256)$; similarly $S(34) = S(3478)$ and $\hat{S} = S(12) \times S(34)$ is an A-invariant subgroup of S of order m^2. We then have $m^2 = |\hat{S}| = |(U \cap \hat{S})^A| = |S(12) \cup S(34)| = 2m-1$ which is not true. Thus $S(12)$ fixes only $\Omega_1 \cup \Omega_2$.

If $S(ij)$ fixes only $\Omega_i \cup \Omega_j$ for all $i \neq j$ then for each i, $|S_i \setminus \bigcup_{j\neq i} S(ij)| = m^2 - 7(m-1) - 1$ and so $m^3 = |S| = |\cup S_i| = 8(m^2-7m+6)+28(m-1)+1 = 8m^2-28m+21$ which is a contradiction. Thus some $S(ij)$, $i \neq j$, fixes some Ω_k pointwise, $k \notin \{i,j\}$. We may assume that $\{i,j\}$ lies in a block of A in X of size 4 (but is not itself a block) and k is not in this block of size 4; say $(i,j,k) = (1,3,5)$. Since $\{1,3\}$ is not a block we may assume that $A_1^X = \langle(1324)(5768)\rangle$. If $S(13)$ fixes only $\Omega_1 \cup \Omega_3 \cup \Omega_5$ pointwise then (considering the action of A) we have exactly 8 distinct subgroups $S(ij)$ of order m fixing more than $2m$ points, namely $S(135)$, $S(327)$, $S(246)$, $S(418)$, and $S(574)$, $S(761)$, $S(683)$, $S(852)$ (for $S(57)$ does not fix Ω_1 or Ω_3 as $S(51) = S(135)$ and it does not fix Ω_2 as $S(27) = S(327)$, and so $S(57) = S(457)$). Also we have 4 subgroups $S(ij)$ fixing 2m points, namely those with $\{ij\}$ a block of the partition $\{12|34|56|78\}$. Each $S(i)$ contains exactly 4 of these $S(ij)$ and so $|S_i \setminus \cup S(jk)| = m^2 - 4(m-1) - 1$, and $m^3 = |S| = |\cup S_i| = 8(m^2-4m+3) + 12(m-1) + 1 = 8m^2 - 20m + 13$ which is a contradiction. Thus $S(13)$ fixes Ω_k for some $k > 5$, and as $\{5,k\}$ is not a block, k is 7 or 8. In either case we get 4 distinct $S(ij)$ fixing $4m$ points (the A-conjugates of $S(13) = S(135k)$) and 4 subgroups $S(ij)$ fixing $2m$ points (the A-conjugates of $S(12)$), and each S_i contains 3 distinct $S(ij)$. Thus $m^3 = |\cup S_i| = 8(m^2 - 3(m-1) - 1) + 8(m-1) + 1 = 8m^2 - 16m + 9$ which is again a contradiction. This completes the proof of Lemma 3.4.

We can now complete the proof of the theorem :

Lemma 3.5. *In the case $S(12) = \{1\}$ we have $S = Z_7^2$ and $A = S\,.Y$, where Y is Q or $Q \times Z_3$ and Q is Z_{16} or a generalized quaternion group Q_{16} (acting transitively on the eight subgroups of S of order 7). Further $H = S\,.(Y \cap H)$ with $Y \cap H$ a group of scalar transformations of order 2 or 6 and U is D_{14} or a Frobenius group F_{42} of order 42 respectively.*

Proof. Here S has order m^2 and S is the union of the eight distinct subgroups S_i, so $m^2 = 8(m-1)+1$, that is $m = 7$. We have $A \leq AGL(2,7)$ since S is self-centralizing in A, and so $A = SY$ with $Y \leq GL(2,7)$. Now $Y \cap H \lesssim Z_6$ is a group of scalar transformations. Also Y acts regularly on the eight S_i with kernel $Y \cap H$ and it follows that a Sylow 2-subgroup Q of Y is Z_{16} or Q_{16} and either $Y = Q$ or $Y = Q \times Z_3$. We have $H = S\,.Z_2$ or $H = S\,.Z_6$ and $U = D_{14}$ or $U = F_{42}$ respectively.

This completes the proof of the Theorem.

4. Discussion of the examples

The examples in parts (a) and (b) of the Theorem are described explicitly in the statement of the Theorem and we note here simply that A/H is D_8 in case (a) and is Z_8 or D_8 in case (b) according as Q is Z_{16} or Q_{16}. The purpose of this section is to elaborate on the examples arising in case (c). These examples arise at two separate stages in the proof and some extra information is given about them in Lemmas 3.1 and 3.2. We examine these cases separately :

Examples from Lemma 3.1. Here $S = Z_3^4 = R_1 \times R_2$ with $|R_i| = 9$, $A = SY$ where $Y \leq V = W\ wr\ Z_2 = (W_1 \times W_2)\,.Z_2$ is an imprimitive linear group acting on S. The group W is Z_8 or Q_8 and W_i is transitive on the four subgroups of R_i of order 3 and W_i centralizes R_j, $j \neq i$. Further Y interchanges W_1 and W_2 and $Y \cap (W_1 \times W_2)$ is either a diagonal subgroup D of $W_1 \times W_2$ or is DM where $M = \Omega_1(W_1 \times W_2) \simeq Z_2 \times Z_2$ is the subgroup of $W_1 \times W_2$ generated by the elements of order 2. We will determine here the possible subgroups Y up to conjugacy in Aut V. We note that $A/H \simeq YM/M \leq V/M$ and V/M is either $Z_4 \text{ wr } Z_2$ or $Z_2^2 \text{ wr } Z_2$; YM/M has order 8, interchanges the two direct factors of the base group V/M and its intersection with the base group is a diagonal subgroup.

Lemma 4.1. (a) *Suppose that* $W = Z_8$ *so that* $YM/M \lesssim \bar{V} = \mathbf{Z}_4 \text{ wr } Z_2 = \{(i,j)\sigma^\delta \mid i,j \in \mathbf{Z}_4,\ \delta = 0 \text{ or } 1\}$ *where* $\sigma^2 = 1$ *and* $(i,j)^\sigma = (j,i)$. *Then up to conjugacy in* $\bar{V}$, YM/M *is one of the following groups.*

(i) $\langle (1,1), \sigma \rangle = \langle (1,1) \rangle \times \langle \sigma \rangle \cong Z_4 \times Z_2$,

(ii) $\langle (0,1)\sigma \rangle \cong Z_8$,

(iii) $\langle (1,3), \sigma \rangle \cong D_8$,

(iv) $\langle (1,3), (0,2)\sigma \rangle \cong Q_8$.

(b) *Suppose that* $W = Q_8$ *so that* $YM/M \leq \bar{V} = \mathbf{Z}_2^2 \text{ wr } Z_2 = \{(u,v)\sigma^\delta \mid u,v \in \mathbf{Z}_2^2,\ \delta = 0 \text{ or } 1\}$ *where* $\sigma^2 = 1$ *and* $(u,v)^\sigma = (v,u)$. *Then, up to conjugacy in Aut* $\bar{V}$, YM/M *is one of the following groups.*

(i) $\langle (10,10), (01,01), \sigma \rangle \cong Z_2^3$,

(ii) $\langle (01,01), (00,10)\sigma \rangle \cong Z_4 \times Z_2$,

(iii) $\langle (10,01), \sigma \rangle \cong D_8$,

(iv) $\langle (10,01), (00,11)\sigma \rangle \cong D_8$.

Proof. (a) Let $\bar{Y} = YM/M$. Since $\bar{Y} \cap \mathbf{Z}_4^2$ is a diagonal subgroup it is generated by x where x is (1,1) or (1,3). Also $\bar{Y}$ contains an element $y = y'\sigma$ for some $y' \in \mathbf{Z}_4^2$; by multiplying by an element of $\bar{Y} \cap \mathbf{Z}_4^2$ we may take $y' = (0,i)$ for some i. We must have $y^2 = (i,i) \in \bar{Y} \cap \mathbf{Z}_4^2$ so if $x = (1,3)$ then i is 0 or 2. In all cases $\bar{Y} = \langle x, y \rangle$.

Consider first $x = (1,1)$. Here $x^y = x$ so $\bar{Y}$ is abelian. If $i = 0$ then $\bar{Y} = \langle (1,1), \sigma \rangle \cong Z_4 \times Z_2$; if $i = 1$ then $\bar{Y} = \langle (1,1), (0,1)\sigma \rangle = \langle (0,1)\sigma \rangle \cong Z_8$ and similarly if $i = 3$ then $\bar{Y} = \langle (1,1), (0,3)\sigma \rangle = \langle (0,3)\sigma \rangle = \langle (0,1)\sigma \rangle^{(0,3)}$ is conjugate in $\bar{V}$ to $\langle(0,1)\sigma\rangle$. If $i = 2$ then $\bar{Y} = \langle(1,1),(0,2)\sigma\rangle \cong \langle(1,1)\rangle \times \langle(1,3)\sigma\rangle$, but the conjugate of this group by (0,3) is $\langle(1,1),\sigma\rangle$.

Now consider $x = (1,3)$. If $i = 0$ then $\bar{Y} = \langle(1,3),\sigma\rangle \simeq D_8$, while if $i = 2$ then $\bar{Y} = \langle(1,3),(0,2)\sigma\rangle \cong Q_8$.

(b) Now we have $\bar{Y} = YM/M \leq \bar{V} = \mathbf{Z}_2^2 \text{ wr } Z_2$. Since $\bar{Y} \cap (\mathbf{Z}_2^2 \times \mathbf{Z}_2^2)$ is a diagonal subgroup, $\bar{Y} \cap (\mathbf{Z}_2^2 \times \mathbf{Z}_2^2) = \{(u,u^\tau) \mid u \in \mathbf{Z}_2^2\}$ for some $\tau \in \text{Aut}\mathbf{Z}_2^2 \simeq S_3$. Also $\bar{Y}$ contains $y = (0,v)\sigma$ for some $v \in \mathbf{Z}_2^2$. Now $\bar{Y}$ must contain each $(u,u^\tau)^y = (u^\tau,u)$ and hence $u^{\tau^2} = u$ for each u, that is $\tau^2 = 1$. Thus we may assume that $\tau = 1$ or τ is the automorphism which interchanges (10) and (01). Also $y^2 = (v,v) \in \bar{Y}$ so if $\tau \neq 1$ then v is (00) or (11). We have $\bar{Y} = \langle(10,10^\tau),(01,01^\tau),y\rangle$.

Suppose that $\tau = 1$. Then $\bar{Y}$ is abelian. If $v = (00)$ then $\bar{Y} = \langle(10,10),(01,01),\sigma\rangle \cong Z_2^3$. If $v \neq (00)$ then we may assume that $v = (10)$ and $\bar{Y} = \langle(10,10),(01,01),(00,10)\sigma\rangle = \langle(01,01)\rangle \times \langle(00,10)\sigma\rangle \cong Z_2 \times Z_4$. Finally suppose that τ interchanges (10) and (01). If $v = (00)$ then $\bar{Y} = \langle(10,01),\sigma\rangle \cong D_8$ and if $v = (1,1)$ then $\bar{Y} = \langle(10,01),(00,11)\sigma\rangle \cong D_8$.

We note that each group of order 8 occurred as A/H in Lemma 4.1. Now we have to identify all possibilities for Y from the possible factor groups given in Lemma 4.1 for the two cases $M \leq Y$ and $Y \cap M = Z(V)$. We look at the case $V = \mathbf{Z}_8 \text{ wr } Z_2$ first.

Lemma 4.2. *Suppose that* $Y \leq V = \mathbf{Z}_8 \text{ wr } Z_2 = \{(i,j)\sigma^\delta \mid i,j \in \mathbf{Z}_8,\ \delta = 0 \text{ or } 1\}$ *where* $\sigma^2 = 1$ *and* $(i,j)^\sigma = (j,i)$.

(a) *If* Y *contains* $M = \langle(0,4),(4,0)\rangle$ *we have the following possibilities for* Y *:*

(i) $\langle(1,1),\sigma,(0,4)\rangle = \langle x,y,z \mid x^8 = y^2 = z^2 = 1,\ [x,y] = [x,z] = 1,\ z^y = x^4 z\rangle = (Z_8 \times Z_2) \rtimes Z_2$, *group* 32/17 *of* [7].

(ii) $\langle(0,1)\sigma,(0,4)\rangle = \langle x,y \mid x^{16} = y^2 = 1,\ x^y = x^9\rangle = Z_{16} \rtimes Z_2$, *group* 32/22 *of* [7].

(iii) $\langle(1,7),\sigma,(0,4)\rangle = \langle x,y,z \mid x^8 = y^2 = 1,\ [x,z] = 1,\ x^y = x^{-1},\ z^y = x^4 y\rangle = (Z_8 \times Z_2) \rtimes Z_2$, *group* 32/26 *of* [7].

(iv) $\langle(1,7),(0,2)\sigma,(0,4)\rangle = \langle(1,7),(0,2)\sigma\rangle = \langle x,y \mid x^8 = 1,\ y^4 = x^4,\ x^y = y^{-1}\rangle$, *group* 32/32 *in* [7].

(b) *If* $Y \cap M = Z(V)$ *we have the following possibilities for* Y :

(i) $\langle(1,1),\sigma\rangle = Z_8 \times Z_2$.

(i)′ $\langle(1,5),\sigma = \langle x, y \mid x^8 = y^2 = 1,\ x^y = x^5\rangle = Z_8 \rtimes Z_2$, *group* 16/11 *of* [7].

(ii) $\langle(0,1)\sigma\rangle = Z_{16}$.

(iii) $\langle(1,7),\sigma,\rangle = \langle x, y, \mid x^8 = y^2 = 1,\ x^y = x^{-1}\rangle = D_{16}$, *group* 16/12 *of* [7].

(iii)′ $\langle(1,7),(0,4)\sigma\rangle = \langle x, y \mid x^8 = 1,\ y^2 = x^4,\ x^y = y^{-1}\rangle = Q_{16}$, *group* 16/14 *of* [7].

(iii)″ $\langle(1,3),\sigma\rangle = \langle x, y \mid x^8 = y^2 = 1,\ x^y = x^3\rangle$, *group* 16/13 *of* [7].

We note that the representation in terms of the elements of V shows how Y acts on S while the generator and relations presentation (and the reference to [7]) identifies Y as an abstract group. The numbers (i), (i)′ denote that for these groups $Y/(Y \cap M)$ satisfies (a) (i) of Lemma 4.1 etc; there are no groups in case (iv) with $|Y \cap M| = 2$.

Proof. If $Y \geq M$ it is straightforward to check that we obtain the four possibilities listed and no others. So we assume that $Y \cap M = Z(V) = \langle(4,4)\rangle$.

For case (a) (i) of Lemma 4.1, $Y = \langle x = m_1(1,1),\ y = m_2\sigma, (4,4)\rangle$ for some $m_1 \in M$ and as $x^4 = (4,4)\varepsilon Y$ we may assume that each m_i is (0,0) or (0,4). Suppose first that $m_1 = (0,0)$. Then either $Y = \langle x = (1,1),\ y = \sigma\rangle \cong Z_8 \times Z_2$ or $Y = \langle x = (1,1),\ y = (0,4)\sigma\rangle = \langle x, x^2y = (2,6)\sigma\rangle$ and these two groups are conjugate in V. If $m_1 = (0,4)$ then $Y = \langle x = (1,5),\ y = \sigma\rangle = \langle x, y \mid x^8 = y^2 = 1,\ x^y = x^5\rangle$ or $Y = \langle x = (1,5),\ y = (0,4)\sigma\rangle = \langle x = (1,5),\ y = (1,5)^2(0,4)\,\sigma = (2,6)\sigma\rangle = \langle x, y \mid x^8 = y^2 = 1,\ x^y = x^5\rangle$ and again these two groups are conjugate in V.

For case (a) (ii) of Lemma 4.1 we have $x = m(0,1)\,\sigma \in Y$ for some $m \in M$ and as $x^8 = (4,4) \in Y$ we may assume that m is (0,0) or (0,4), that is x is $(0,1)\sigma$ or $(0,5)\sigma$ and $Y = \langle x, (4,4)\rangle = \langle x\rangle$. Clearly the two groups obtained are conjugate in Aut V.

For case (a) (iii) of Lemma 4.1, $Y = \langle x = m_1(1,7),\ y = m_2\sigma\rangle$ with $m_i \in M$ and as before we may assume that the m_i are (0,0) or (0,4). If $m_1 = (0,0)$ then $Y = \langle x = (1,7),\ y = \sigma\rangle = \langle x, y \mid x^8 = y^2 = 1,\ x^y = x^{-1}\rangle = D_{16}$ or

$Y = \langle x = (1,7),\ y = (0,4)\sigma\rangle = \langle x, y \mid x^8 = 1,\ y^2 = x^4,\ x^y = x^{-1}\rangle = Q_{16}$. If $m_1 = (0,4)$ then $Y = \langle x = (1,3),\ y = \sigma\rangle = \langle x, y \mid x^8 = y^2 = 1,\ x^y = x^3\rangle$ or $Y = \langle x = (1,3),\ y = (0,4)\sigma\rangle = \langle x = (1,3),\ y = (1,3)(0,4)\sigma = (1,7)\sigma\rangle = \langle x, y \mid x^8 = y^2 = 1,\ x^y = x^3\rangle$ and these two groups are conjugate in V.

For case (a) (iv) of Lemma 4.1, $Y = \langle x = m_1(1,7),\ y = m_2(0,2)\sigma\rangle$ with the m_i equal to (0,0) or (0,4). Here $y^2 = m_2(0,2)m_2^\sigma(0,2)^\sigma = m_2 m_2^\sigma(2,2)$ is either (2,2) or (6,6) and must lie in $\langle x\rangle$ since $|Y| = 16$; this is never satisfied so no examples arise in this case.

Now we consider the case where $V = Q_8$ wr Z_2.

Lemma 4.3. *Suppose that* $Y \le V = Q_8$ *wr* Z_2. *We represent* Q_8 *as* $\langle a, b \mid a^4 = 1,\ b^2 = a^2,\ a^b = a^{-1}\rangle$ *so that* $V = \{(u,v)\sigma^\delta \mid u, v \in Q_8,\ \delta = 0 \text{ or } 1\}$ *where* $\sigma^2 = 1$ *and* $(u,v)^\sigma = (v,u)$.

(a) *If* Y *contains* $M = \langle (1,a^2),\ (a^2,1)\rangle$ *we have the following possibilities for* Y :

(i) $\langle (a,a),(b,b),\sigma,(1,a^2)\rangle = \langle (a,a),(b,b)\rangle * \langle \sigma,(1,a^2)\rangle = Q_8 * D_8$, *group* 32/43 *of* [7].

(ii) $\langle (a,a),(1,b)\sigma,(1,a^2)\rangle = \langle x,y,z \mid y^8 = z^2 = 1,\ x^2 = y^4,\ y^x = zy,\ y^z = y^5,\ [x,z] = 1\rangle$, *group* 32/48 *of* [7].

(iii) $\langle (a,b),\sigma,(1,a^2) = (1,b^2)\rangle = \langle (a,b^3),(a,b)\sigma,(1,a^2)\rangle = \langle x,y,z \mid y^8 = z^2 = 1,\ x^2 = y^4,\ y^x = y^{-1},\ y^z = y^5,\ [x,z] = 1\rangle$, *group* 32/45 *of* [7].

(iv) $\langle (a,b),(1,ab)\sigma,(1,a^2)\rangle = \langle x,y,z \mid y^8 = z^2 = 1,\ x^2 = y^4,\ y^z = y^5,\ y^x = y^{-1},\ [x,z] = 1\rangle$, *group* 32/45 *of* [7].

(b) *If* $Y \cap M = Z(V) = \langle (a^2,a^2)\rangle$ *we have the following possibilities for* Y :

(i) $\langle (a,a),(b,b)\rangle \times \langle\sigma\rangle = Q_8 \times Z_2$, *group* 16/7 *of* [7].

(i)′ $\langle (a,a),(b,b),(1,a^2)\sigma\rangle = \langle x = (a,a^3)\sigma,\ y = (b,b^3)\sigma, z = (1,a^2)\sigma\rangle = \langle x,y,z \mid x^2 = y^2 = z^4 = 1,\ [x,z] = [y,z] = 1,\ xy = yxz^2\rangle$, *group* 16/8 *of* [7].

(iii) $\langle (a,b),\sigma\rangle = \langle x = (a,b)\sigma,\ y = \sigma\rangle = \langle x,y \mid x^8 = y^2 = 1\ x^y = x^3\rangle$, *group* 16/13 *of* [7].

(iii)′ $\langle (a,b),(1,a^2)\sigma\rangle = \langle x = (a,b),\ y = (a,b^3)\sigma\rangle = \langle x,y \mid y^8 = 1,\ x^2 = y^4,\ y^x = y^{-1}\rangle = Q_{16}$, *group* 16/14 *of* [7].

(iv) $\langle (a,b),(1,ab)\sigma\rangle = \langle x,y \mid y^8 = 1,\ x^2 = y^4,\ y^x = y^{-1}\rangle = Q_{16}$, *group* 16/14 *of* [7].

(iv)′ $\langle (a,b),(1,a^3b)\sigma\rangle = \langle x = (a,a^3)\sigma,\ y = (1,a^3b)\sigma\rangle = \langle x,y \mid y^8 = x^2 = 1,\ y^x = y^3\rangle$, *group* 16/13 *of* [7].

As for the previous Lemma the numbers (i), (i)′ denote that these groups correspond to Lemma 4.1 (b) (i) etc. In case (b) (ii) of Lemma 4.1 there are no examples with $|Y \cap M| = 2$.

Proof. Again case (a) is straightforward so we assume that $Y \cap M = \langle (a^2,a^2)\rangle$.

In case (b) (i) of Lemma 4.1, $Y = \langle x = m_1(a,a),\ y = m_2(b,b),\ z = m_3\sigma\rangle$ with the $m_i \in M$ and as $x^2 = (a^2,a^2) \in Y$ we may take the m_i to be (1,1) or $(1,a^2)$. If all $m_i = 1$ then $Y = Y_1 = \langle (a,a),\ (b,b)\rangle \times \langle\sigma\rangle = Q_8 \times Z_2$. If $m_1 = m_2 = 1$ and $m_3 = (1,a^2)$ then $Y = Y_2 = \langle (a,a),(b,b),(1,a^2)\sigma\rangle = \langle x = (a,a^3)\sigma,\ y = (b,b^3)\sigma,\ z = (1,a^2)\sigma\rangle = \langle x,y,z \mid x^2 = y^2 = z^4 = 1,\ [x,z] = [y,z] = 1,\ xy = y\ x\ z^2\rangle$. Next suppose that one of m_1, m_2 is non-trivial, say $m_1 = 1,\ m_2 = (1,a^2)$. If $m_3 = 1$ then $Y = \langle (a,a),(b,b^3),\sigma\rangle = Y_2^{(1,a)}$ while if $m_3 \neq 1$ then $Y = \langle (a,a),(b,b^3),(1,a^2)\sigma\rangle = Y_1^{(1,a)}$. Finally suppose $m_1 = m_2 = (1,a^2)$. If $m_3 = 1$ then $Y = \langle (a,a^3),(b,b^3),\sigma\rangle = Y_1^{(1,a^3b)}$, while if $m_3 \neq 1$ then $Y = \langle (a,a^3),(b,b^3),(1,a^2)\sigma\rangle = Y_2^{(1,ab^3)}$.

In case (b) (ii) of Lemma 4.1, $Y = \langle x = m_1(a,a),\ y = m_2(1,b)\sigma\rangle$ with the m_i equal to (1,1) or $(1,a^2)$. Suppose $m_1 = 1$. Then $x^y = (a^3,a) \in Y$ so $x^y x^{-1} = (a^2,1) \in Y$ which is a contradiction. Thus $m_1 = (1,a^2)$ and $x^y = (a,a)$ so $x^y x^{-1} = (1,a^2) \in Y$, again a contradiction. Thus there are no examples in this case.

In case (b) (iii) of Lemma 4.1, $Y = \langle x = m_1(a,b),\ y = m_2\sigma\rangle$ with the m_i equal to (1,1) or $(1,a^2)$. Suppose $m_2 = 1$. Then we may assume that $m_1 = 1$ (by replacing b by b^3 if necessary), and then $Y = \langle (a,b),\sigma\rangle = \langle x = (a,b)\sigma,\ y = \sigma\rangle = \langle x,y \mid x^8 = y^2 = 1,\ x^y = x^3\rangle$. Suppose then that $m_2 = (1,a^2)$; we may assume that $m_1 = 1$ (by replacing a^2b by b if necessary). Then $Y = \langle (a,b),(1,a^2)\sigma\rangle = \langle x = (a,b),\ y = (a,b)(1,a^2)\sigma = (a,b^3)\sigma\rangle = \langle x,y \mid y^8 = 1,\ x^2 = y^4,\ y^x = y^{-1}\rangle = Q_{16}$.

In case (b) (iv) of Lemma 4.1, $Y = \langle x = m_1(a,b),\ y = m_2(1,ab)\sigma\rangle$ with the m_i equal to (1,1) or $(1,a^2)$. If $m_1 = m_2$ we may assume that $m_1 = m_2 = 1$ (by replacing a by a^3 if necessary) and then $Y = \langle (a,b),(1,ab)\sigma\rangle =$

$\langle x, y \mid y^8 = 1,\ x^2 = y^4,\ y^x = y^{-1}\rangle = Q_{16}$. If $m_1 \neq m_2$ we may assume that $m_1 = 1$, $m_2 = (1, a^2)$ (again by replacing a by a^3 if necessary). Then $Y = \langle (a,b), (1, a^3 b)\sigma\rangle = \langle x = (a,b)(1,a^3b)\sigma = (a,a^3)\sigma,\ y = (1,a^3b)\sigma\rangle = \langle x, y \mid y^8 = x^2 = 1,\ y^x = y^3\rangle$.

Examples from Lemma 3.2. Here $S = Z_3^4$ and $A = SY \leq S_3\ wr\ D_8$ with $Y \leq W = Z_2\ wr\ D_8$; $Y \cap H \leq Z(W) \cong Z_2$, and $Y/(Y \cap H)$ has order 8. Also Y projects onto a subgroup of D_8 of order 4 which acts regularly on the four direct factors of the base group Z_2^4. So $Y \cap Z_2^4$ has order 2 or 4.

With the notation of Lemma 3.2,

$$\begin{aligned} S &= S(1256) \times S(1278) \times S(3456) \times S(3478) \\ &= R_1 \times R_2 \times R_3 \times R_4, \end{aligned}$$

and we shall represent S as $\mathbf{Z}_3^4$ with respect to this direct decomposition. Now $S(123)$ was shown in Lemma 3.2 to be a diagonal subgroup of $S(12) = S(1256) \times S(1278)$ and we may assume that $S(123) = \langle (1,1,0,0)\rangle$ so that $S(124) = \langle (1,2,0,0)\rangle$. Similarly $S(567)$ is a diagonal subgroup of $S(56) = S(1256) \times S(3456)$ so we may assume that $S(567) = \langle (1,0,1,0)\rangle$ and hence that $S(568) = \langle (1,0,2,0)\rangle$. Again $S(234)$ is a diagonal subgroup of $S(34) = S(3456) \times S(3478)$ and we may assume that $S(234) = \langle (0,0,1,1)\rangle$ and hence that $S(134) = \langle (0,0,1,2)\rangle$. Finally $S(578) = \langle (0,1,0,k)\rangle$ and $S(678) = \langle (0,1,0,2k)\rangle$ where k is 1 or 2. Now $Y \cap A_2$, in its action on $X = \{1, 2, \ldots, 8\}$, interchanges 1 and 2, 3 and 4, 5 and 6, 7 and 8; hence $Y \cap A_2$ interchanges $S(123)$ and $S(124)$, $S(567)$ and $S(568)$, $S(134)$ and $S(234)$, $S(578)$ and $S(678)$. It follows that elements in $(Y \cap A_2) \setminus (Y \cap H)$ either invert R_1 and R_4 and centralize R_2 and R_3, or invert R_2 and R_3 and centralize R_1 and R_4. Thus if we represent $W = Z_2 \,\mathrm{wr}\, D_8$ by $\{x\sigma \mid x \in \mathbf{Z}_2^4,\ \sigma \in D\}$ where $D_8 \cong D \leq S_4$ permutes the entries naturally, then $Y \cap A_2$ is contained in $\langle (1,0,0,1), (0,1,1,0)\rangle$ and $Y \cap A_2$ contains at least one of (1,0,0,1) and (0,1,1,0). Further under the projection onto D the image of Y is either $\langle (1243)\rangle \cong Z_4$ or $\langle (14)(23), (12)(34)\rangle \cong Z_2 \times Z_2$. Thus Y contains an element $x(1243)$ or $x(12)(43)$ for some $x \in \mathbf{Z}_2^4$ and since both of these elements interchange (1,0,0,1) and (0,1,1,0) it follows that $|Y \cap A_2| = 4$.

Consider first $Y/(Y \cap H) \cong Z_4$ so that Y is generated by $Y \cap A_2$ and an element $x(1243)$. Now the elements $x(1243)$ for $x \in \mathbf{Z}_2^4$ lie in two conjugacy classes of W; those for which x has an even number of nonzero entries lie in one class and those where x has an odd number of nonzero entries lie in the other. Thus we may assume that x is 0 or (1,0,0,0). Then either $Y = \langle(1,0,0,1),(0,1,1,0),(1243)\rangle = \langle x = (1243),\ y = (1,1,1,1)(14)(23),\ z = (1,0,0,1)\rangle = \langle x,y,z \mid x^4 = y^2 = z^2 = 1,\ [x,y] = [y,z] = 1,\ x^z = x^3y\rangle$ which is group 16/9 of [7], or $Y = \langle x = (1,0,0,1),\ y = (1,0,0,0)(1243)\rangle = \langle x,y \mid y^8 = x^2 = 1,\ y^x = y^5\rangle$, group 16/11 of [7].

Now suppose that $Y/(Y \cap H) \cong Z_2 \times Z_2$ so that Y is generated by $Y \cap A_2$, $a = x(14)(23)$ and $b = y(12)(34)$ for some $x, y \in \mathbf{Z}_2^4$. Replacing the first element a by a conjugate in W if necessary we may assume that x is 0, (1,0,0,0) or (1,1,0,0). Further since $b^2 \in Y \cap A_2$ we must have $y_1 + y_2 = y_3 + y_4$ where $y = (y_1, y_2, y_3, y_4)$ and (by premultiplying b by an element of $Y \cap A_2$ if necessary) we may assume that $y = (y,0,y,0)$ with $y = 0$ or $y = 1$.

Suppose first that $x = 0$; if $y = 0$ then $Y = \langle u = (1,0,0,1),\ a = (14)(23),\ b = (12)(34)\rangle = \langle u,a,w = (1,0,0,1)b\rangle = \langle u,w \mid u^2 = w^4 = 1,\ w^u = w^3\rangle \times \langle a\rangle \cong D_8 \times Z_2$, while if $y = 1$ then $Y = \langle u = (1,0,0,1),\ a = (14)(23),\ b = (1,0,1,0)(12)(34)\rangle = \langle u,\ v = (1,0,0,1)a,\ b\rangle = \langle u,b \mid u^2 = b^4 = 1,\ b^u = b^3\rangle \times \langle v\rangle \cong D_8 \times Z_2$. These two groups are conjugate in W ((0,0,0,1) conjugates the first to the second).

Suppose next that $x = (1,0,0,0)$. If $y = 0$ then $a^b = (0,1,0,0)(14)(23) = (1,1,0,0)a$, so $a^b a^{-1} = (1,1,0,0) \in Y$ which is not the case. Similarly if $y = 1$ then $ab = (1,0,0,0)(14)(23)(1,0,1,0)(12)(34) = (1,1,0,1)(13)(24)$ and $(ab)^2 = (1,0,1,0) \in Y$ which again is not so. Thus there are no examples in this case.

Finally suppose that $x = (1,1,0,0)$. If $y = 0$ then $Y = Y_1 = \langle u = (1,0,0,1),\ a = (1,1,0,0)(14)(23),\ b = (12)(34)\rangle = \langle u,a,b \mid u^2 = a^4 = b^2 = 1, [a,u] = [a,b] = 1,\ b^u = a^2b\rangle$, which is group 16/8 of [7]. If $y = 1$ then $Y = \langle u = (1,0,0,1),\ a = (1,1,0,0)(14)(23),\ b = (1,0,1,0)(12)(34)\rangle = \langle u, v = ua,\ w = ub\rangle = Y_1^{(0,0,0,1)}$, that is Y is conjugate in W to the group Y_1 above. Thus we have proved the following Lemma.

Lemma 4.4. *Suppose that $A = S.Y$, H and U are as in Lemma 3.2. Then $Y \leq W = \mathbf{Z}_2 \ wr\ D_8 = \{u\sigma \mid u \in \mathbf{Z}_2^4,\ \sigma \in D\}$ where $D = \langle(1243),\ (12)(34)\rangle \leq S_4$. Further $Y \cap H = Z(W) = \langle(1,1,1,1)\rangle$ and up to conjugacy in* Aut W *the possible groups Y are the following.*

(a) $\langle(1,0,0,1),(0,1,1,0),(1243)\rangle = \langle x = (1243),\ y = \langle(1,1,1,1)(14)(23),\ z = (1,0,0,1)\rangle = \langle x, y, z \mid x^4 = y^2 = z^2 = 1,\ [x,y] = [y,z] = 1,\ x^z = x^3 y\rangle$ *which is group* 16/9 *of* [7].

(b) $\langle x = (1,0,0,1),\ y = (1,0,0,0)(1243)\rangle = \langle x, y \mid x^2 = y^8 = 1,\ y^x = y^5\rangle$ *which is group* 16/11 *of* [7].

These are the only two groups for which the projection of Y *to* D *is* Z_4*; in the remaining examples this projection is* $Z_2 \times Z_2$.

(c) $\langle(1,0,0,1),(14)(23),(12)(34)\rangle = \langle x = (1,0,0,1),\ y = (14)(23),\ z = (1,0,0,1)(12)(34)\rangle = \langle x, z \mid x^2 = z^4 = 1,\ z^x = z^{-1}\rangle \times \langle y \mid y^2 = 1\rangle \cong D_8 \times Z_2$*, which is group* 16/6 *of* [7].

(d) $\langle x = (1,0,0,1),\ y = (1,1,0,0)(14)(23),\ z = (12)(34)\rangle = \langle x, y, z \mid x^2 = y^4 = z^2 = 1,\ [x,y] = [y,z] = 1,\ z^x = y^2 z\rangle$*, which is group* 16/8 *of* [7].

References

[1] W. Jehne, Kronecker classes of algebraic number fields, *J. Number Theory* **9** (1977), 279-320.

[2] N. Klingen, Zahlkörper mit gleicher Primzerlegung, *Journal reine angew. Math.* **299** (1978), 342-384.

[3] N. Klingen, Private communication.

[4] M.W. Leibeck, C.E. Praeger and J. Saxl, On the O'Nan Scott theorem for finite primitive permutation groups, *J. Austral. Math. Soc.* (A), to appear.

[5] C.E. Praeger, Covering subgroups of groups and Kronecker classes of fields, *J. Algebra*, to appear.

[6] J. Saxl, On a question of W. Jehne, (preprint) 1986.

[7] A.D. Thomas and G.V. Wood, *Group Tables*, Shiva Math. Series 2, Orpington, 1980.

[8] J.H. Walter, The characterization of finite groups with abelian Sylow 2-subgroups, *Ann. of Math.* **89** (1969), 405-514.

Department of Mathematics
University of Western Australia
Nedlands, WA 6009
Australia

Geometric invariants and HNN-extensions

Burkhardt Renz

1. Introduction

For every HNN-extension

$$(*) \qquad H = \langle B, t\,;\, t^{-1}B_1 t = B_2 \rangle$$

over a base group B and with stable letter t one has the ***associated homomorphism*** $\chi : H \twoheadrightarrow \mathbf{Z}$ given by $\chi(t) = 1$ and $\chi(B) = 0$. Every homomorphism χ of a group G onto $\mathbf{Z}$ can, of course, be regarded as the associated homomorphism of some HNN-decomposition of G; but in many circumstances G has, in fact, an HNN-decomposition over a ***finitely generated base group*** with associated homomorphism χ. This is, for instance, the case when G is finitely presented, see [2].

We call the HNN-extension $(*)$ ***ascending*** if the first associated subgroup B_1 coincides with the base group B, so that the kernel N of the associated homomorphism χ is the union of the ascending chain

$$\ldots \subseteq t^{-1}Bt \subseteq B \subseteq tBt^{-1} \subseteq t^2Bt^{-2} \subseteq \ldots$$

Correspondingly $(*)$ is ***descending*** if $B_2 = B$. It is interesting to know which homomorphisms $\chi : G \twoheadrightarrow \mathbf{Z}$ are associated to an ***ascending*** HNN-decomposition over a ***finitely generated*** base group. This question is answered in [1] in terms of the 'geometric invariant' Σ of G. The Bieri-Neumann-Strebel invariant Σ of a finitely generated group G is a certain subset of the 'character sphere' $S(G)$, by which we mean the set of all equivalence classes $[\chi] = \{\lambda\chi \mid 0 < \lambda \in \mathbf{R}\}$ of non-zero homomorphisms $\chi : G \longrightarrow \mathbf{R}_{\text{add}}$ under multiplication by positive real numbers. We should mention that Σ captures not only the information about ascending HNN-decompositions over finitely generated base groups but also characterizes the finitely generated normal subgroups of G with Abelian quotient.

In this paper, I go one step further by investigating the question as to which homomorphisms $\chi : G \twoheadrightarrow \mathbf{Z}$ are associated to an ascending HNN-extension over a ***finitely presented base group***. The answer is given in terms of a new geometric invariant ${}^*\Sigma^2$. This is part of a more general concept in my Thesis [6] where I define a chain of ***higher geometric invariants***

$$S(G) \supseteq {}^*\Sigma^1 \supseteq {}^*\Sigma^2 \supseteq \cdots \supseteq {}^*\Sigma^k \cdots$$

generalizing ${}^*\Sigma^2$ and the Bieri-Neumann-Strebel invariant $\Sigma = -{}^*\Sigma^1$. The higher geometric invariant ${}^*\Sigma^k$ allows to decide as to whether a given normal subgroup N of G with G/N Abelian is of type F_k, i.e. has an Eilenberg-MacLane complex $K(G,1)$ with finite k-skeleton.

The paper is organized as follows. In §2 we extend homomorphisms $\chi : G \longrightarrow \mathbf{R}$ to valuations v_χ on the Cayley complex $C = C(X; R)$ of a presentation $\langle X; R\rangle$ of G. In §3 we define the geometric invariants ${}^*\Sigma^1$ and ${}^*\Sigma^2$. The combinatorial characterization of ${}^*\Sigma^1$ in terms of certain loops in the Cayley graph of G shows that, up to a sign, ${}^*\Sigma^1$ coincides with the Bieri-Neumann-Strebel invariant. Generalizing this description to dimension 2 we get a combinatorial characterization of ${}^*\Sigma^2$ in terms of simple diagrams over $\langle X; R\rangle$. We use these descriptions to study ascending HNN-extensions with finitely generated base group in §4. We give the proof of the result in [1] in our geometric setting. Then we give necessary and sufficient conditions (in terms of ${}^*\Sigma^2$) for finite presentation of the base group B in an ascending HNN-extension $G = \langle B, t\,;\, t^{-1}Bt \leq B\rangle$.

Acknowledgement. I would like to thank Robert Bieri for the many fruitful and stimulating discussions during the process of finding the results of my Thesis, which contains the theorem presented in this paper. I would also like to express my thanks to the Hermann-Willkomm-Stiftung, Frankfurt for its financial support of my journey to Singapore.

2. Characters and valuations on the Cayley complex

2.1 Let G be a finitely generated group, and d the $\mathbf{Z}$-rank of the abelianization G^{ab} of G. A *character* of G is a non-zero homomorphism $\chi : G \longrightarrow \mathbf{R}$ into the additive group of real numbers. Two characters are equivalent if they coincide up to multiplication by a positive real number. $Hom(G,\mathbf{R}) = Hom(G^{ab},\mathbf{R})$ is a d-dimensional real vector space which can be identified with $\mathbf{R}^d$. The equivalence class $[\chi]$ of a character thus is the ray from 0 through χ in $Hom(G,\mathbf{R}) \cong \mathbf{R}^d$. The *character sphere* $S(G)$ of G is defined to be $S(G) = \{[\chi] \mid \chi \in Hom(G,\mathbf{R}) \setminus \{0\}\}$. $S(G)$ is homeomorphic to the unit sphere S^{d-1}.

A character χ with infinite cyclic image is called a *discrete* character. The subset of the rational points of $S(G) \cong S^{d-1}$ consists of the classes of discrete characters and is dense in $S(G)$. For a rational point $[\chi] \in S(G)$ we always find a representative χ with $\chi : G \twoheadrightarrow \mathbf{Z} \subseteq \mathbf{R}$.

A character χ allows us to interpret the ordering of $\mathbf{R}$ in the preimage of χ: attached to each $[\chi] \in S(G)$, we consider the submonoid $G_\chi = \{g \mid \chi(g) \geq 0\}$ of G. G_χ does not depend upon the choice of the representative $\chi \in [\chi]$.

2.2 Let $\langle X; R\rangle$ be a presentation of G where R is a set of cyclically reduced words in the free group $F(X)$ with basis X. We do not assume that X embeds in G, but will not distinguish notationally between *words* in $X^{\pm 1}$, i.e. elements of $F(X)$, and their images in G.

The *Cayley graph* $\Gamma = \Gamma(X)$ and the *Cayley complex* $C = C(X;R)$ of G are defined as follows (see [5], III.4):

The set V of vertices of C is the set G of elements of the group. The set E of edges of C is $G \times X^{\pm 1}$. An edge (g,x), by definition, links the vertex g to gx. [Note that gx here is regarded as an element of G.] The inverse oriented edge is (gx, x^{-1}). We have a labelling function $\varphi : E \longrightarrow X^{\pm 1}$ defined by $\varphi((g,x)) = x$. φ extends multiplicatively to edge paths in C: if $p = e_1 e_2 \dots e_n$ is an edge path then $\varphi(p) = \varphi(e_1)\varphi(e_2)\cdots\varphi(e_n)$ is a word in $F(X)$. $\varphi(p)$ is reduced if and only if p is a reduced path. p is a loop if and only if $\varphi(p)$ is in the normal closure of R in $F(X)$. The set F of faces of $C(X;R)$ is $G \times R^{\pm 1}$. A face (g,r) has as boundary the loop p_r at g with label $\varphi(p_r) = r$. The inverse of (g,r) is (g, r^{-1}). The 1-skeleton of the Cayley complex C is called the Cayley graph $\Gamma(X)$ of G with respect to the generators X.

2.3 Let χ be a character of G and C the Cayley complex of G in the presentation $G = \langle X; R\rangle$. We extend χ to a valuation v_χ on C:

If $g \in V$ is a vertex of C, we put $v_\chi(g) = \chi(g)$. For an edge $e = (g, x)$ we define $v_\chi(e) = min\{v_\chi(g), v_\chi(gx)\}$. If $p = e_1 e_2 \cdots e_n$ is an edge path beginning at g then the χ-track of p is the sequence

$$(v_\chi(g), v_\chi(g\varphi(e_1)), v_\chi(g\varphi(e_1)\varphi(e_2)), \ldots, v_\chi(g\varphi(e_1)\varphi(e_2)\cdots\varphi(e_n)))$$

and $v_\chi(p)$ is defined to be the minimum of the χ-track of p. Accordingly we denote by $v_\chi(w)$ the minimum of the χ-track of a word w in $X^{\pm 1}$. If (g, r) is a face of C then $v_\chi((g, r))$ is the minimum of the χ-track of the boundary loop of (g, r).

The automorphisms of the Cayley complex C, i.e. automorphisms of the combinatorial 2-complex C which preserve labels, are exactly those induced by left multiplication of G ([5], III.4.1.). G is the group of deck transformations of C.

A valuation v_χ on C extending a character χ has the following property:

(*) $\quad v_\chi(gc) = \chi(g) + v_\chi(c)$ for $c \in V$ or $c \in E$ or $c \in F$ and all $g \in G$.

Remark. The notion of a valuation on the combinatorial Cayley complex $C(X; R)$ is the special case of a more general notation of valuations v_χ extending a character χ of G. Recall that a G-complex is a CW-complex C together with an operation of G by homeomorphisms which permute the cells. If furthermore the stabilizer of each cell is trivial then C is a *free G-complex.*

Let C be a free G-complex and $\chi \in Hom(G; \mathbf{R}) \setminus \{0\}$. A continuous function $v_\chi : C \longrightarrow \mathbf{R}$ is called a valuation on C associated with χ if

(1) $v_\chi(gc) = \chi(g) + v_\chi(c)$ for all $c \in C, g \in G$

(2) $v_\chi(C^0) \subseteq \chi(G)$ [C^0 is the 0-skeleton of C.]

(3) Let $\sigma \subseteq C$ be a cell with boundary $\partial\sigma$ then

$$min\, v_\chi(\partial\sigma) \leq v_\chi(c) \leq max\, v_\chi(\partial\sigma)$$

for all $c \in \sigma$.

If C is the geometric realization of the Cayley complex $C(X; R)$ of a group G then C is the universal cover of the 2-dimensional CW-complex which is usually called the geometric realization of the presentation $\langle X; R\rangle$ of G (see e.g. [4], p.44). A combinatorial valuation on $C(X;R)$ yields by piecewise linear extension a valuation on C.

2.4 A *full* subcomplex C' of the Cayley complex $C = C(X;R)$ of a group G is a subcomplex with the following property: If e is an edge of C or f a face of C and all vertices g of e or of f are in C' then e or f is in C'. A full subcomplex C' of C is determined by the set of vertices of C'.

Let v_χ be a valuation on C associated with the character χ. The *valuation subcomplex* C_v of C is defined to be the full subcomplex of C spanned by the submonoid G_χ, i.e. by $\{g \mid v_\chi(g) \geq 0\} \subseteq V$. We put $C_{v,\lambda}(\lambda \in \mathbf{R})$ for the full subcomplex of C generated by $\{g \mid v_\chi(g) \geq -\lambda\}$ and C_{-v} for the full subcomplex of C spanned by the subset $\{g \mid v_\chi(g) \leq 0\}$ of the vertices V of C. If $\Gamma = \Gamma(X)$ is the Cayley graph of G with respect to the generating set X then Γ_v is the subgraph spanned by G_χ. Γ_v contains those edges (g,x) of Γ for which $v_\chi(g) \geq 0$ and $v_\chi(gx) \geq 0$. Note that $C_{v_\chi} = C_{v_{\chi'}}$ and $\Gamma_{v_\chi} = \Gamma_{v_{\chi'}}$ if χ and χ' are equivalent characters.

3. The geometric invariants $^*\Sigma^1$ and $^*\Sigma^2$

3.1 We keep the notation and conventions of section 2. Recall that the edge path group of a combinatorial 2-complex is isomorphic with the fundamental group of its geometric realization.

Definition. Let G be a finitely generated group, X a finite generating set of G, and $[\chi] \in S(G)$. We put

$$[\chi] \in {}^*\Sigma^1 :\Leftrightarrow \text{the valuation subgraph } \Gamma_{v_\chi} \text{ of the Cayley graph } \Gamma(X) \text{ of } G \text{ is connected.}$$

Lemma 1. *Let G be a finitely generated group and $[\chi] \in {}^*\Sigma^1$. Then the valuation subgraph $\Gamma_{v_\chi}(Y)$ of $\Gamma(Y)$ is connected for any finite set Y of generators of G.*

Proof. Let X be a finite set of generators of G such that $\Gamma_v(X)$ is connected, and let Y be another finite set of generators. Each $x_i \in X^{\pm 1}$ is expressible as a word w_i in the generators $Y^{\pm 1}$. We fix such expressions and put $\lambda = min\,\{v_\chi(w_i)\}$. Since $\Gamma_v(X)$ is connected, for each vertex h of $\Gamma_v(Y)$ there is an edge path p in $\Gamma(Y)$ connecting 1 and h such that $v_\chi(p) \geq \lambda$. Furthermore, given two vertices h_1, h_2 of $\Gamma_v(Y)$ with $v_\chi(h_1) \geq \mu$ for $i = 1,2$ and some $\mu \geq 0$, we can find an edge path p' in $\Gamma_v(Y)$ which connects h_1 and h_2 and fulfills $v_\chi(p') \geq \mu + \lambda$. This follows from the fact that G acts on $\Gamma(Y)$ by left multiplication together with property $(*)$ of 2.3. Let g be a vertex of $\Gamma_v(Y)$. Choose $t \in Y^{\pm 1}$ with $\chi(t) > 0$, and $k \in \mathbf{N}$ such that $\chi(t^k) \geq |\lambda|$. Then there is an edge path p_2 connecting t^k and gt^k such that $v_\chi(p_2) \geq 0$. Let p_1 be the edge path on $\Gamma_v(Y)$ corresponding to the word t^k and starting at 1, and p_3 the path with $\varphi(p_3) = t^{-k}$ starting at gt^k. Then $p_1p_2p_3$ is an edge path in $\Gamma_v(Y)$ which connects 1 and g, thus $\Gamma_v(Y)$ is connected.

3.2 We give a combinatorial criterion for $[\chi] \in {}^*\Sigma^1$.

Theorem 1 (Criterion for ${}^*\Sigma^1$). *Let G be a finitely generated group, X a finite set of generators, and $[\chi] \in S(G)$.*

Then $[\chi] \in {}^*\Sigma^1$ if, and only if, there is a $t \in X^{\pm 1}$ with $\chi(t) > 0$ such that for every $x \in X^{\pm 1} \setminus \{t, t^{-1}\}$ the conjugate ${}^{t^{-1}}x \in G$ can be expressed as a word w in $X^{\pm 1}$ with $v_\chi(t^{-1}xt) < v_\chi(w)$.

Proof. Let $\Gamma = \Gamma(X)$ be the Cayley graph of G and Γ_v the valuation subgraph of Γ.

If $[\chi] \in {}^*\Sigma^1$ then Γ_v is connected. Take a $t \in X^{\pm 1}$ with $\chi(t) > 0$. Consider the path $t^{-1}xt$ in Γ beginning at 1. If $\chi(x) > 0$ then the endpoint of $t^{-1}xt$ is Γ_v, thus there is also a path w in Γ_v beginning at 1 and ending at $t^{-1}xt$, and therefore $v_\chi(t^{-1}xt) < v_\chi(w)$. If $\chi(x) < 0$ then there exists an integer $l > 0$ such that the endpoint of the path $t^{-1}xt^l$ beginning at 1 lies in Γ_v. Let w' be a word in $X^{\pm 1}$ with $t^{-1}xt^l = w'$ (in G) and $v_\chi(w') > 0$. Thus $t^{-1}xt = w't^{-(l-1)}$ is a desired expression.

Now we consider a vertex g in Γ_v together with a path p connecting 1 and g in Γ. If there is a vertex h in p with $v_\chi(h) < 0$ then we proceed as follows: Choose h such that $v_\chi(h) = v_\chi(p)$, and consider the part $\bullet\overset{x}{\longrightarrow}\underset{h}{\bullet}\overset{y}{\longrightarrow}\bullet$

of p. If $x \neq t$, t^{-1} and $y \neq t$, t^{-1} we have expressions ${}^{t^{-1}}x = w_x$ and ${}^{t^{-1}}y = w_y$ with $v_X(t^{-1}xt) < v_X(w_x)$ and $v_X(t^{-1}yt) < v_X(w_y)$. Thus we can pass to a path p' by using the paths labelled by w_x and w_y instead of x and y. If $x = t^{-1}$ or $y = t$ we proceed in the same manner for y or x, respectively, and reduce the path p' afterwards. In any case the number of vertices c with $v_X(c) = v_X(p)$ decreases. Since X is a finite set, $\{v_X(w_x) - v_X(t^{-1}xt) \mid x \in X^{\pm 1} \setminus \{t, t^{-1}\}\}$ has a minimum > 0, and so iteration of the procedure yields eventually a path in Γ_v connecting 1 and g.

Comparing Theorem 1 with [1], Proposition 2.1, it is easy to see that ${}^*\Sigma^1 = -\Sigma$. [$-$ is the antipodal map of $S(G)$.] Hence ${}^*\Sigma^1$ is an open subset of $S(G)$. This follows easily from Theorem 1 too. Note that Bieri, Neumann, Strebel consider G as acting by the right on G', whereas we use left action according to the occidental custom to read edge paths in Γ from left to right.

3.3 A *diagram* M over the presentation $\langle X; R\rangle$ of G is a finite planar configuration of vertices, edges and faces fulfilling the following conditions: The oriented edges of M are labelled by the set $X^{\pm 1}$. If the edge e has label x then its inverse is labelled by x^{-1}. The boundary path of each face of M corresponds under the labelling to a cyclic permutation of a defining relation $r \in R$ or its inverse r^{-1}.

A connected and simply connected diagram M with boundary ∂M a reduced loop p based at 1 describes the equivalence in the edge path group of $C(X; R)$ of p to the trivial path (see [5], III.4 and V.1). We call a connected and simply connected diagram with reduced boundary loop a *simple diagram*.

If a diagram M has two faces which are neighboured as shown in the following illustration

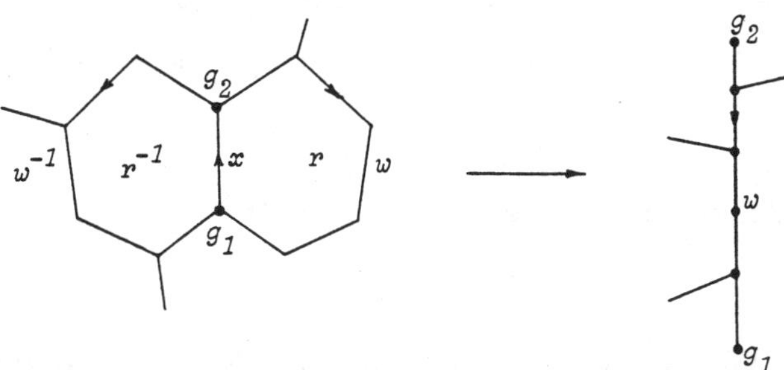

Illustration 1. Lyndon reduction

then we can reduce M by shrinking the interior of the loop labelled ww^{-1}. We refer to this kind of reduction of diagram as *Lyndon reduction.*

Unlike the usual definition of diagrams we do allow trivial faces labelled by $tt^{-1}t^{-1}t$ for a distinguished generator t of G. This deviation simplifies the drawing of diagrams that contain paths coming from conjugation of a word by t.

Each closed path in a simple diagram M is labelled by a relator of G, i.e. a consequence of the defining relations. Thus if we choose a base point of M then every vertex of M can uniquely be labelled by an element of G. For a given valuation v_χ of the Cayley complex $C(X;R)$ of G we get after the choice of a base point in M a *valuated* simple diagram. We denote $v_\chi(M) = min\{v_\chi(g) | g \text{ is vertex in } M\}$.

Obviously, we obtain:

Theorem 2. *Let $G = \langle X;R\rangle$ be a finitely generated group, and C the Cayley complex together with a valuation v_χ associated with the character χ. Then the valuation subcomplex C_v is 1-connected if, and only if, C_v is connected and for each reduced loop p at $1 \in C$ with $v_\chi(p) \geq 0$ there is a simple diagram M with $\partial M = p$ such that $v_\chi(g) \geq 0$ for every vertex g of M.*

3.4. Now we pass to the geometric invariant ${}^*\Sigma^2$.

Definition. Let G be a finitely generated group and $[\chi] \in S(G)$. Then we define

$[\chi] \in {}^*\Sigma^2 :\Leftrightarrow$ there is a finite presentation $\langle X;R\rangle$ of G such that the valuation subcomplex C_{v_χ} of the Cayley complex $C(X;R)$ of G is 1-connected.

From the definitions we see that $S(G) \supseteq {}^*\Sigma^1 \supseteq {}^*\Sigma^2$ for a finitely generated group.

Remark. We should mention that there is a generalization: A character $[\chi] \in S(G)$ is, by definition, in ${}^*\Sigma^k$ $(k \geq 1)$ if there is an Eilenberg-

Maclane complex $K = K(G,1)$ with finite k-skeleton such that the valuation subcomplex C_v of the universal cover C of K is $(k-1)$-connected. [v is a valuation extending χ on the free G-complex C.] [6]

If we change the finite presentation of G, then the valuation subcomplex of the Cayley complex of G will not, in general, remain 1-connected. But a weaker property of the valuation subcomplex is independent of the choice of the finite presentation of G.

Definition. Let C be the Cayley complex of G with respect to the finite presentation $G = \langle X; R\rangle$ and let v_χ be a valuation on C. Suppose C_v is connected. We say that the valuation subcomplex C_v is essentially 1-connected if there is a real number $\lambda \geq 0$ such that the homomorphism $\pi_1(C_v) \longrightarrow \pi_1(C_{v,\lambda})$ induced by the inclusion $C_v \longrightarrow C_{v,\lambda}$ is trivial.

Analogously to Theorem 2, C_v is essentially 1-connected if, and only if, C_v is connected and there exists a $\lambda \geq 0$ such that for every reduced loop based at 1 in C_v we can find a simple diagram M with $\partial M = p$ and $v(g) \geq -\lambda$ for every vertex g of M.

Lemma 2. *Suppose G is a finitely presented group and $[\chi] \in {}^*\Sigma^2$. If $\langle Y; S\rangle$ is a finite presentation of G, then $C_v(Y,S)$ is essentially 1-connected.* [v stands for a valuation on $C(Y,S)$ associated with $[\chi]$.]

Proof. Let $\langle X; R\rangle$ be a finite presentation of G such that $C_v(X;R)$ is 1-connected. We can pass from $\langle X; R\rangle$ to the presentation $\langle Y; S\rangle$ of G by a finite sequence of Tietze transformations. Hence it is sufficient to study the effect of Tietze transformations to the corresponding valuation subcomplexes and to prove that essential 1-connectivity is preserved. Let's fix the following notation:

$$T_1 : \langle X_1; R_1\rangle \longrightarrow \langle X_2; R_2\rangle$$

where $X_2 = X_1$ and $R_2 = R_1 \cup \{r\}$ for a consequence r of R_1.

$$T_2 : \langle X_1; R_1\rangle \longrightarrow \langle X_2; R_2\rangle$$

where $X_2 = X_1 \cup \{y\}$ and $R_2 = R_1 \cup \{r\}$ for a letter $y \notin X_1$ and a relation $r = y^{-1}w$ expressing y as a word w in $X_1^{\pm 1}$.

T_1^{-1} and T_2^{-1} are the transformations in the opposite direction. In both cases, we obviously can view $C(X_1, R_1)$ as a subcomplex of $C(X_2; R_2)$.

(1) If one performs T_1 on $\langle X_1; R_1\rangle$ then $C_v(X_2; R_2)$ is essentially 1-connected, provided that this was the case for $C_v(X_1, R_1)$.

(2) Now we consider T_1^{-1}. Suppose $C_v(X_2, R_2)$ is essentially 1-connected, i.e. for every reduced loop p at 1 in $C_v(X_1, R_1)$ there is a simple diagram M over $\langle X_2; R_2\rangle$ with $\partial M = p$ and $v(g) \geq -\lambda$ for some $\lambda \geq 0$ and for all vertices g of M. Since r is a consequence of R_1 there is a simple diagram M_r over $\langle X_1; R_1\rangle$ with $\partial M_r = r$. Let $\mu = max\{|v_\chi(g) - v_\chi(h)| \mid g, h$ vertices of $M_r\}$. We replace each face of M corresponding to r by M_r and obtain a simple diagram M' with $\partial M' = p$ and $v(g) \geq -\lambda - \mu$ for every g in M'. M' is now a diagram in $C(X_1, R_1)$, hence $C_v(X_1, R_1)$ is essentially 1-connected.

(3) Suppose $C_v(X_1, R_1)$ is essentially 1-connected. Now we perform T_2 on $\langle X_1; R_1\rangle$. Put $\mu = max\{|v_\chi(g) - v_\chi(h)| \mid g, h$ vertices of $r\}$. Each reduced loop based at 1 in $C_v(X_2, R_2)$ is in $C_{v,\mu}(X_2, R_2)$ homotopic to a reduced loop in $C_{v,\mu}(X_1, R_1)$. But $C_{v,\mu}(X_1, R_1)$ is essentially 1-connected, because $C_v(X_1, R_1)$ is so. Therefore $C_v(X_2, R_2)$ is essentially 1-connected.

(4) Suppose $\langle X_1; R_1\rangle$ results from $\langle X_2; R_2\rangle$ by T_2^{-1}, and $C_v(X_2, R_2)$ is essentially 1-connected. For each reduced loop p in $C_v(X_1, R_1)$ we can find a simple diagram M_p over $\langle X_2; R_2\rangle$ with $\partial M_p = p$ and $v(g) \geq -\lambda$ for a $\lambda \geq 0$ and all vertices g in M_p. p has no edge labelled by y or y^{-1}. Since r is the only relation in R_2 involving y, we can remove all occurrences of y (or y^{-1}) in the interior by Lyndon reductions. Thus without loss we can assume that M_p is a diagram over $\langle X_1; R_1\rangle$ and therefore $C_v(X_1, R_1)$ is essentially 1-connected.

Let G be a finitely presented group and suppose that $[\chi] \in {}^*\Sigma^1$. By Theorem 1 there is a finite presentation $\langle X; R\rangle$ of G such that $R \supseteq \{t^{-1}xt = w_x \mid x \in X^{\pm 1} \setminus \{t, t^{-1}\}\}$ where t is a distinguished generator with $\chi(t) > 0$ and $v_\chi(t^{-1}xt) < v_\chi(w_x)$ for all $x \in X^{\pm 1}$, $x \neq t$, t^{-1}. In this situation we can show that $[\chi] \in {}^*\Sigma^2$ implies $\pi_1(C_{v_\chi}(X, R)) = 1$.

Lemma 3. *Let G be a finitely presented group and $[\chi] \in {}^*\Sigma^1$. Suppose the presentation $\langle X; R\rangle$ of G contains the defining relations $t^{-1}xt = w_x$ for all $x \in X^{\pm 1} \setminus \{t, t^{-1}\}$ according to Theorem 1. Then we have:*

If $[\chi] \in {}^*\Sigma^2$ then $C_{v_\chi}(X, R)$ is 1-connected.

Proof. Since $[\chi] \in {}^*\Sigma^2$ $\quad C_{v_\chi}(X,R)$ is essentially 1-connected, i.e. for each loop p based at 1 in $C_v(X,R)$ there is a simple diagram M_p with boundary p such that $v_\chi(g) \geq -\lambda$ for all g in M_p and a certain fixed real number $\lambda \geq 0$. Since $R \supseteq \{t^{-1}xt = w_x | \text{ for all } x \in X^{\pm 1} \setminus \{t, t^{-1}\}\}$ p is freely homotopic in $C_v(X,R)$ to a loop p' based at t^k for $k \in \mathbf{N}$ such that $\chi(t^k) \geq \lambda$. G acts on $C(X,R)$ and the valuation v_χ fulfills $(*)$ of 2.3, thus there is a simple diagram $M_{p'}$ with $\partial M_{p'} = p'$ such that for every vertex g of $M_{p'}$ we have $v_\chi(g) \geq 0$. This implies that $C_v(X,R)$ is 1-connected.

3.5 We assume that G has a presentation $\langle X; R\rangle$ as in Lemma 3. If r is a relation in R, say $r = x_1 x_2 \cdot\cdot x_n$ $(x_i \in X^{\pm 1})$, we write $\hat{r}$ for the word $w_{x_1} w_{x_2} \ldots w_{x_n}$, and say that $\hat{r}$ results from r by conjugation by t. [We put $w_x = t$ or t^{-1} if $x = t$ or t^{-1}.] A connected and simply connected diagram with boundary label $\hat{r}$ is denoted by $M_{\hat{r}}$. We want to choose the base point b_1 of $M_{\hat{r}}$ in dependence of the base point b_0 of r, and accordingly we demand for a valuated diagram with boundary r and $M_{\hat{r}}$ in the interior that $v_\chi(b_1) = v_\chi(b_0) + v_\chi(t)$ See Illustration 2.

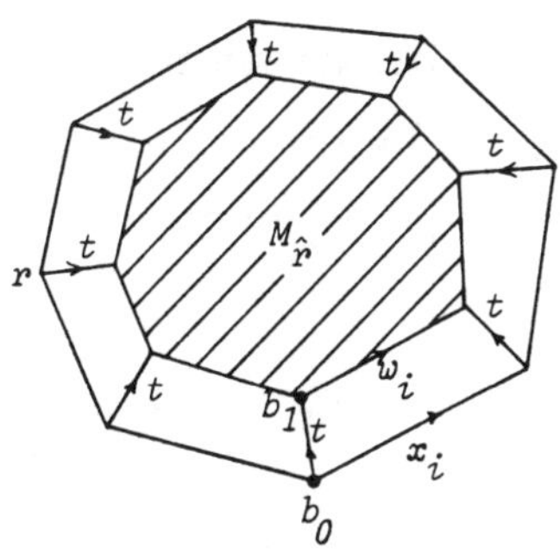

Illustration 2

Theorem 3 (Criterion for ${}^*\Sigma^2$). *Let G be a finitely presented group, and $[\chi] \in {}^*\Sigma^1$. We choose a presentation of G as in Lemma 3. Then $[\chi] \in {}^*\Sigma^2$ if, and only if, for each relation $r \in R^{\pm 1}$ there is a simple diagram $M_{\hat{r}}$ with $\partial M_{\hat{r}} = \hat{r}$ and $v_\chi(r) < v_\chi(M_{\hat{r}})$.*

Proof. If $[\chi] \in {}^*\Sigma^2$ then the valuation subcomplex C_{v_χ} of the Cayley complex C of G with respect to the chosen presentation is, by Lemma 3,

1-connected. Hence for each $\hat{r}$ $(r \in R)$ there is a diagram M based at 1 with $\partial M = \hat{r}$ and $v_\chi(M) \geq v_\chi(\hat{r})$. But we can change the base point and consider M as a diagram M' based at t. Using the notation of Illustration 2, we see that $v_\chi(r) < v_\chi(M')$.

Let p be a reduced loop in the valuation subcomplex C_v based at 1. Since the Cayley complex is 1-connected there is a simple diagram M with $\partial M = p$. If $v_\chi(M) < 0$ we proceed as follows: Let g be a vertex in M with $v_\chi(g) = v_\chi(M)$. For all faces $r_1, r_2, \ldots, r_n$ containing g we have $v_\chi(r_i) = v_\chi(g)$.

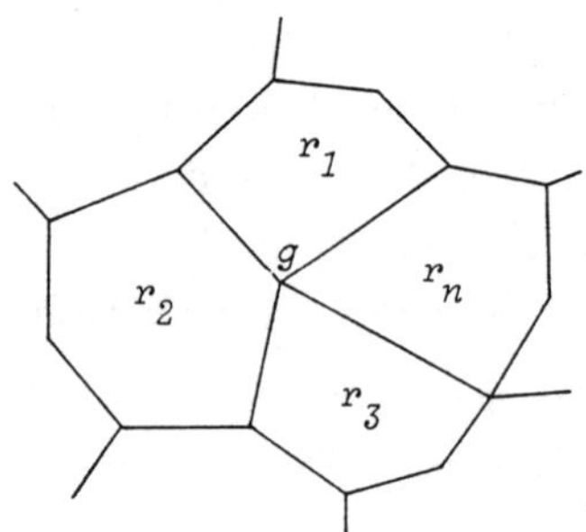

Illustration 3

By replacing each face r_i by the collar coming from conjugation by t together with the diagram $M_{\hat{r}_i}$., we obtain the situation as shown in Illustration 4.

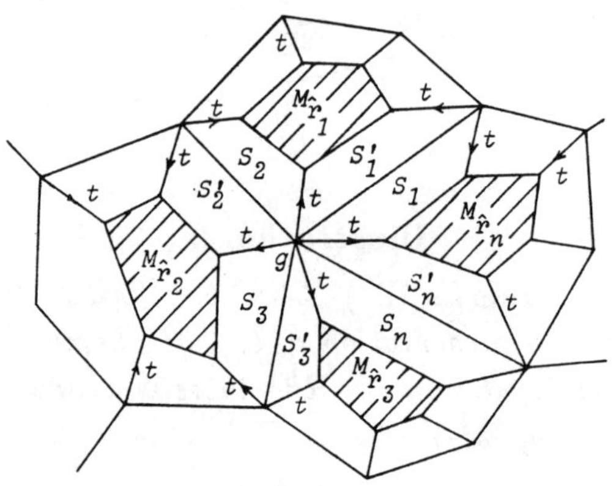

Illustration 4

Obviously s_i and $s_i'(1 \leq i \leq n)$ in Illustration 4 are the same relations, but inverse oriented. We can reduce all the faces s_i by Lyndon reductions and the critical vertex g disappears.

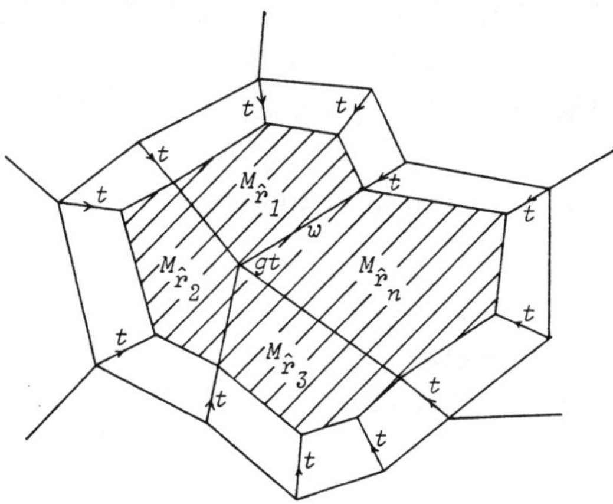

Illustration 5

We get a diagram M_1 which compared with M has only new vertices h with $v_\chi(h) > v_\chi(M)$. By iteration we finally reach a diagram M' with $\partial M' = p$ and $v_\chi(g) \geq 0$ for all vertices $g \in M'$.

As an immediate consequence of Theorem 3 we obtain

Corollary. *Let G be a finitely presented group. Then ${}^*\Sigma^2$ is an open subset of $S(G)$.*

4. Ascending HNN-extensions with finitely presented base group

Suppose $G = \langle B, t\,; t^{-1}Bt = B_2 \leq B \rangle$ is an ascending HNN-extension with finitely generated base group B. For each generator b_i of B, $t^{-1}b_it$ can be written as a word in the generators of B. The associated character χ_t with $\chi_t(t) = 1$ and $B \subset \operatorname{Ker} \chi_t$ fulfills the condition of Theorem 1, thus $[\chi_t] \in {}^*\Sigma^1$.

If, on the other hand, a rational point $[\chi]$ is in ${}^*\Sigma^1$ then for a suitable representative χ there is a finite set X of generators of G such that $\chi(t) = 1$ for a distinguished generator t and $\chi(x) = 0$ for the others. We consider the Cayley graph $\Gamma = \Gamma(X)$ of G and denote by Γ_0 the subgraph spanned by the vertices $g \in V$ with $v_\chi(g) = 0$. Γ_0 has as vertices just the elements of $N = \text{Ker } \chi$ and Γ_0 is a subgraph of the connected valuation subgraph Γ_v of Γ. Hence N is finitely generated over the monoid $\langle t \mid$ generated by t, say $N =^{\langle t|} \langle b_0, b_1, \ldots, b_k\rangle$. Let $B_0 = \langle b_0, b_1, \ldots, b_k\rangle$. The normal closure of B_0 in G is N, and therefore we have $t^{-1}B_0t \subseteq N$. Hence there is a positive integer l such that the group $B = \langle B_0 \cup {}^tB_0 \cup \ldots \cup {}^{t^l}B_0\rangle$ contains $t^{-1}B_0t$. B is closed under conjugation by t^{-1}. Thus G is an ascending HNN-extension with finitely generated base group B and stable letter t.

This proves

Theorem ([1], Proposition 4.3). *Let $\chi : G \twoheadrightarrow \mathbf{Z}$ be a discrete character, $N = \text{Ker } \chi$ and $t \in G$ with $\chi(t) = 1$. The following statements are equivalent:*

(i) $[\chi] \in {}^*\Sigma^1$.

(ii) *N is finitely generated as a $\langle t|$-operator group.*

(iii) *G is an ascending HNN-extension $G = \langle B, t\,;\, t^{-1}Bt = B_2\rangle$ with finitely generated base group $B \subseteq N$.*

(iv) *If G is a descending HNN-extension $G = \langle C, t\,;\, t^{-1}C_1t = C\rangle$ and $C \subseteq N$ then $C = N$.*

Let G be a finitely presented group, and $\chi : G \twoheadrightarrow \mathbf{Z}$ an epimorphism. We characterize those $[\chi] \in {}^*\Sigma^1$ for which the base group B in the ascending HNN-extension of the theorem of Bieri, Neumann, Strebel is finitely presented:

We fix the following notation: If $G = \langle b_1, \ldots, b_n, t\,;\, t^{-1}b_it = u_i\ (1 \leq i \leq n), r_1, \ldots, r_m\rangle$ where the u_i and r_j are words in $\{b_1, \ldots, b_n\}^{\pm 1}$ is an ascending HNN-extension, we write $R = \{r_1, \ldots, r_m\}$, $S = \{t^{-1}b_it = u_i \mid 1 \leq i \leq n\}$ and $X = \{b_1, \ldots, b_n, t\}$.

Theorem 4. *Let G be a finitely presented ascending HNN-extension. $G = \langle B, t\,;\, t^{-1}Bt = B_2\rangle = \langle X\,;\, R \cup S\rangle$ with finitely generated base group $B = \langle b_1, \ldots, b_n\rangle$, and χ the associated homomorphism. Then B is finitely presented if, and only if,*

(1) $[\chi] \in {}^*\Sigma^2$ *and*

(2) *there is a finite set* $R' \supseteq R$ *of words in* $\{b_1, \ldots, b_n\}^{\pm 1}$ *such that* $G = \langle X\,;\, R' \cup S\rangle$ *and the component* D_{-v} *of* 1 *in* $C_{-v}(X, R' \cup S)$ *is* 1*-connected.*

Proof. If B is finitely presented, then there is a finite set $R' \supseteq R$ of words in $\{b_1, \ldots, b_n\}^{\pm 1}$ such that $G = \langle X\,;\, R' \cup S\rangle$ and $\langle b_1, \ldots, b_n\,;\, R'\rangle$ is a finite presentation of B. We use the Cayley complex $C(X\,;\, R' \cup S)$ with respect to this presentation of G. It is easy to see that the associated homomorphism χ fulfills the criterion for $[\chi] \in {}^*\Sigma^2$: Let v be a valuation on C associated with $[\chi]$.

1. If $r \in R'$ then $\hat{r}$ is a word in the generators of B, i.e. a relator of B. Hence there is a diagram $M_{\hat{r}}$ with $v(M_{\hat{r}}) > v(r)$.

2. A defining relation $s \in S$ is of the form $t^{-1} b_i t = u_i$ with $u_i = b_{i_1}^{\varepsilon_1} \ldots b_{i_k}^{\varepsilon_k}$; $(\varepsilon_j = \pm 1)$. We denote the edge path $u_{i_1}^{\varepsilon_1} \ldots u_{i_k}^{\varepsilon_k}$ by $\hat{u}_i$. For all $1 \leq i \leq n$, we have the diagrams M_i as in Illustration 6.

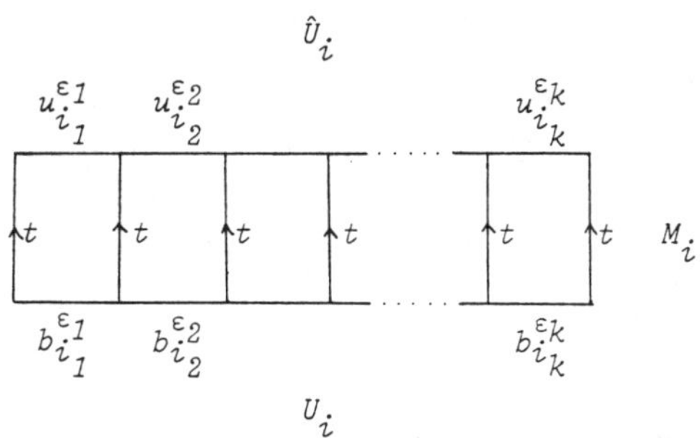

Illustration 6

Now we construct diagrams $M_{\hat{s}}$ for each $s \in S$ such that $v_\chi(s) < v_\chi(M_{\hat{s}})$. See Illustration 7 where $\bar{M}_i$ is the diagram M_i with inverse orientation.

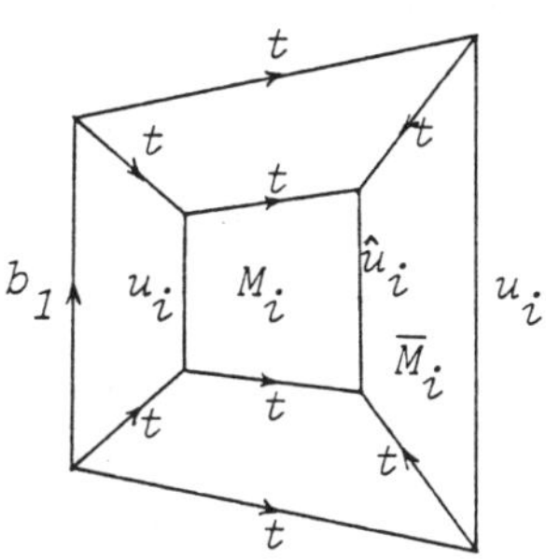

Illustration 7

For the specific choice $u_i = b_2 b_1^{-1} b_3$ e.g., we obtain:

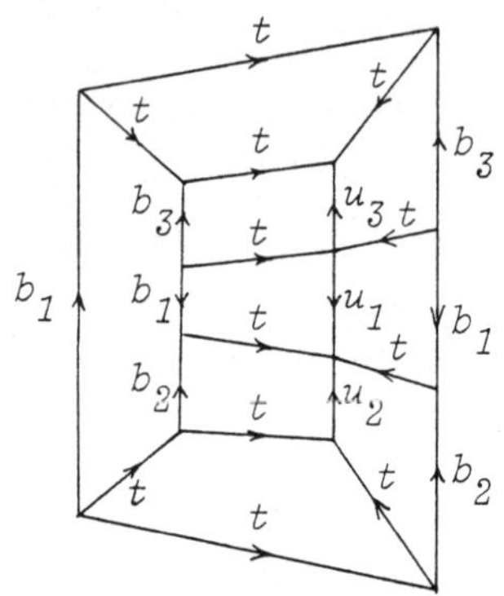

Illustration 8

Let's consider the component D_{-v} of 1 in $C_v(X, R' \cup S)$: We take a loop p at 1 in D_{-v}. If p has edges labelled by t or t^{-1}, we replace a subpath of the form $t^{-1}b_i b_j$ by one labelled $u_i t^{-1} b_j$. The exponent sum of t in the label of p is 0, hence we reach eventually a subword $t^{-1}b_k t$ which can be replaced by u_k. We conclude that there is a path p', not containing t or t^{-1}, homotopic to p (in D_{-v}). Since p' is labelled by a relator of B there is a diagram M with $\partial M = p'$ and $v_\chi(h) = 0$ for all vertices h in M. This means that D_{-v} is 1-connected.

We assume now that (1) and (2) are fulfilled. Note that (1) does not assure automatically that the valuation subcomplex C_v of the Cayley complex $C(X, R' \cup S)$ is 1-connected. But since S is a subset of the set of defining relations, Lemma 3 shows that C_v, in fact, is 1-connected.

The subcomplex C_0 of C spanned by the $g \in V$ with $v_\chi(g) = 0$ contains the elements of B as vertices. Let C_B be the component of 1 in C_0. The 1-skeleton of C_B is the Cayley graph of B with respect to the generators b_1, $b_2, \ldots, b_n$. We show that - after attaching finitely many relations if necessary - C_B is 1-connected. This implies that B is finitely related.

Let p be a reduced loop in C_B. p is in D_{-v}, thus there is a simple diagram M_1 with $\partial M_1 = p$ and $v_\chi(h) \leq 0$ for every vertex h of M_1. Put $a = \max\{v(h) \mid h \in M_{\hat{r}}, r \in R'\}$ for the diagrams $M_{\hat{r}}$ according to Theorem 3. Since C_v is 1-connected we can proceed as in the proof of Theorem 3 to remove vertices h of M_1 with $v_\chi(h) < 0$. We do so until we reach a diagram

M_2 still in D_{-v} with $\partial M_2 = p$ and $v_\chi(M_2) \geq -a$. M_2 is a diagram in the 'strip' of the Cayley complex limited by $-a$ and 0.

For all $j = 1, 2, \ldots, a$ and all $r \in R'$, let $\hat{r}^{(j)}$ be the word in the generators of B resulting from r by conjugation with t^j. Let $R'' = R' \cup \{\hat{r}^{(j)}\}$. We attach the faces determined by $\{\hat{r}^{(j)}\}$ and get the Cayley complex C'' of G which contains C as a subcomplex. Now it is possible to pass from M_2 to a diagram M' with $v_\chi(h) = 0$ for all vertices h in M'. Hence C''_B is 1-connected.

Corollary. *Let G be a finitely presented group. $\chi : G \twoheadrightarrow \mathbf{Z}$ a discrete character, and $N =$ Ker ; χ. Then N is finitely presented if, and only if, both $[\chi]$ and $[-\chi]$ are in ${}^*\Sigma^2$.*

Proof. Choose $t \in G$ with $\chi(t) = 1$.

Suppose N to be finitely presented, say $N = \langle Y\,;\, R\rangle$. Put $X = Y \cup \{t\}$ and $S = \{t^{-1}y_it = u_i \mid y_i \in Y\}$, where the u_i are words in $Y^{\pm 1}$ resulting from y_i by conjugation with t^{-1}. G is the semidirect product $N \rtimes \langle t\rangle$ presented by $G = \langle X\,;\, R \cup S\rangle$, thus $[\chi] \in {}^*\Sigma^2$. On the other hand G is an ascending HNN-extension with base group N and stable letter t^{-1}, i.e. $[-\chi] \in {}^*\Sigma^2$.

Let $[\chi] \in {}^*\Sigma^2$ and $[-\chi] \in {}^*\Sigma^2$. Since $[\chi]$ and $[-\chi]$ are in ${}^*\Sigma^1$ N is finitely generated, say $N = \langle n_1, n_2, \ldots n_l\rangle$, and therefore G has a finite presentation $\langle n_1, n_2, \ldots n_l, t\,;\, R\rangle$ such that R includes all relations expressing ${}^{t^{-1}}n_i$ and ${}^{t}n_i$ for $1 \leq i \leq l$ as words in $\{n_1, n_2, \ldots n_l\}^{\pm 1}$. By Lemma 3 $[\chi] \in {}^*\Sigma^2$ implies that the valuation subcomplex C_v of C with respect to this presentation is 1-connected. By the same argument C_{-v}, which coincides with D_{-v}, is 1-connected. Hence N is finitely presented.

Remark. The Corollary above is a special case of the main theorem in [6]: If G is a group of type F_k and N a normal subgroup of G with G/N Abelian then we have

$$N \quad \text{is of type } F_k \Leftrightarrow {}^*\Sigma^k \supseteq S(G,N) = \{[\chi] \in S(G) \mid \chi(N) = 0\}.$$

5. Examples

5.1 Let G be the metabelian group

$$G = \langle a,\, s\,, t\,;\ {}^{s^{-1}}a = a^2\,,\ {}^{t^{-1}}a = a^3,\ [s,t] = 1\rangle$$

and let χ be the epimorphism $\chi : G \twoheadrightarrow \mathbf{Z}$ defined by $\chi(a) = \chi(s) = 0$, $\chi(t) = 1$. The subgroup $B = \langle a, s\rangle$ is the one-relator group $B = \langle a, s\,;\ {}^{s^{-1}}a = a^2\rangle$. By Theorem 4, $[\chi] \in {}^*\Sigma^2$. We can check this easily by writing down the diagrams which are needed for an application of Theorem 3. See Illustration 9.

Furthermore, D_{-v} is 1-connected in this example: Let p be a reduced loop in D_{-v} based at 1 and $\varphi(p)$ the corresponding word. We observe that the exponent sums of t and s in $\varphi(p)$ are zero, and that the χ-track of each initial segment of $\varphi(p)$ is non-positive. Using the relations $t^{-1}at = a^3$ and $t^{-1}st = s$, p is homotopic in D_{-v} to a loop p' such that $\varphi(p')$ is a word in a, a^{-1}, s, s^{-1}. Since the exponent sum of s in $\varphi(p')$ is zero, we can use the relation $s^{-1}as = a^2$ to produce a loop p'' which is homotopic in D_{-v} to p' and the corresponding word $\varphi(p'')$ involves only the letters a and a^{-1}. But the image of $\varphi(p'')$ in G is 1, i.e. the exponent sum of a in $\varphi(p'')$ is zero. Hence p is homotopic in D_{-v} to the trivial loop.

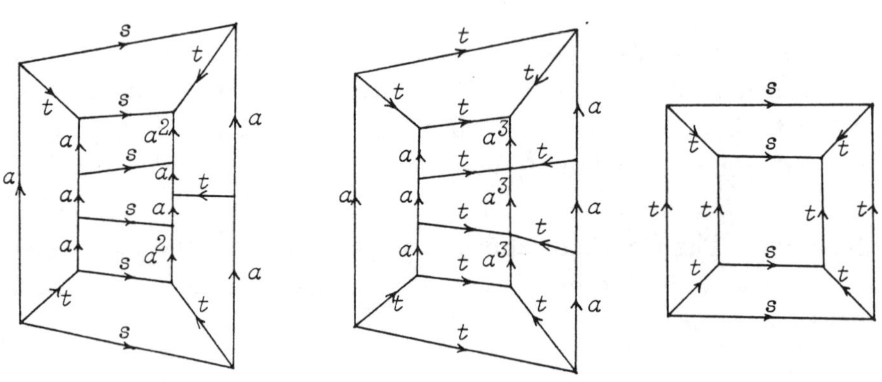

Illustration 9

5.2 Let G be the group in the previous example. Using the criterion for ${}^*\Sigma^2$

and results of Bieri, Strebel [3] about ${}^*\Sigma^1$ of metabelian groups, we calculate in [6] the complement ${}^*\Sigma^{2c}$ of ${}^*\Sigma^2$ for G.

The normal subgroup N of G generated by a and st^{-1} is the kernel of the discrete character χ defined by $\chi(t) = 1$ and $\chi(s) = 1$. We obtain $[\chi] \in {}^*\Sigma^2$ and $[-\chi] \notin {}^*\Sigma^2$. Thus N is not finitely presentable. (See [7] for another argument for this fact.)

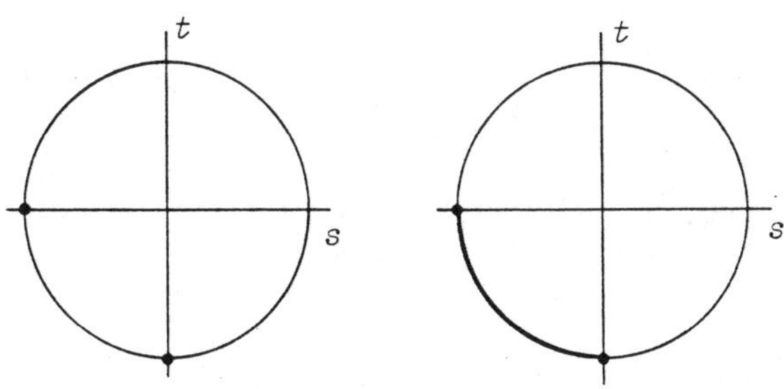

Illustration 10

${}^*\Sigma^{1c}$ and ${}^*\Sigma^{2c}$ for $G = \langle a, s, t\,;\; s^{-1}as = a^2, t^{-1}at = a^3, [s,t] = 1\rangle$.

5.3 In this third example we use Theorem 4 to show that a certain point $[\chi]$ is *not* in ${}^*\Sigma^2$: Let G be the metabelian group of Baumslag and Remeslennikov

$$G = \langle a, s, t\,;\; [a, s^{-1}as] = 1,\, t^{-1}at = as^{-1}as,\, [s,t] = 1\rangle.$$

We consider the character χ with $\chi(t) = 1$ and $\chi(s) = 0$. It is easy to see that $[\chi] \in {}^*\Sigma^1$. Let p be a closed reduced edge path in D_{-v}. We can without loss assume that p has only edges labelled by a, a^{-1} and s, s^{-1}. Since the base group $B = \langle a, s\rangle$ has the presentation $B = \langle a, s\,;\; [a, s^{-j}as^j] = 1,\, j > 0\rangle$ the label of p is a product of conjugates of these relations in the free group with basis $\{a, s\}$. If for all $i \leq n$ a commutes with $s^{-i}as^i$ then

$$\begin{aligned} 1 = t^{-1}[a, s^{-n}as^n]t &= [t^{-1}at, s^{-n}t^{-1}ats^n] = [as^{-1}as, s^{-n}as^{-1}ass^n] \\ &= [a, s^{-(n+1)}as^{(n+1)}]. \qquad \text{(see [7])} \end{aligned}$$

Interpreting these equations geometrically we see that for every $j > 0$ there is a simple diagram M with boundary label $\partial M = [a, s^{-j}as^j]$ such that

$v_\chi(g) \leq 0$ for each vertex g of M. Thus p is in D_{-v} homotopic to the trivial loop. Since B being the wreath product of two infinite cyclic groups is not finitely presented, we obtain $[\chi] \notin {}^*\Sigma^2$.

Recently, Bieri and Strebel proved that ${}^*\Sigma^2$ in this example is, in fact, empty.

References

[1] Bieri, R., Neumann, W.D., Strebel, R., A geometric invariant of discrete groups, *Invent. Math.* **90** (1987), 451-477.

[2] Bieri, R., Strebel, R., Almost finitely presented soluble groups. *Comment. Math. Helv.* **53** (1978), 258-278.

[3] Bieri, R., Strebel, R., Valuations and finitely presented groups. *Proc. London Math. Soc.* (3) **41** (1980), 439-464.

[4] Brown, K.S., *Cohomology of groups*, Grad. Texts Math. 87, New York Heidelberg Berlin 1982.

[5] Lyndon, R.C., Schupp, P.E., *Combinatorial Group Theory*, Ergebnisse der Math. und ihrer Grenzgebiete 89, Berlin Heidelberg New York 1977

[6] Renz, B., *Geometrische Invarianten und Endlichkeitseigenschaften von Gruppen*, Thesis, Frankfurt 1988.

[7] Strebel, R., Finitely presented soluble groups, in: *Group Theory - essays for Philip Hall*, edit. K.W. Gruenberg, J.E. Roseblade, London 1984.

Mathematisches Seminar
Johann Wolfgang Goethe-Universität
Robert-Mayer-Str. 6-8
6000 Frankfurt
Federal Republic of Germany

Groups with permutable subgroup products

A.H. Rhemtulla and *A.R. Weiss*

Abstract

Let G be a group. If there exists an integer $n > 1$ such that for each n-tuple $(H_1, \ldots, H_n)$ of subgroups of G there is a non-identity permutation σ of S_n such that the complexes $H_1 \ldots H_n$ and $H_{\sigma(1)} \cdots H_{\sigma(n)}$ are equal, then G is said to have the property of permutable subgroup products. The main result of this article is to show that if G is a finitely generated solvable group, then G has the above property if and only if it is finite-by-abelian.

Introduction

Call a group G a PSP-group if there exists an integer $n > 1$ such that for each n-tuple $(H_1, \ldots, H_n)$ of subgroups of G, there is a non-identity permutation σ of S_n such that the two complexes $H_1 \ldots H_n$ and $H_{\sigma(1)} \cdots H_{\sigma(n)}$ are equal. Thus PSP-groups are those that have permutable subgroup products. When $n = 2$, we are reduced to the case where every subgroup of G is quasinormal. For an up to date reference to quasinormal subgroups we refer the reader to Chapter 7 of the recent book [7]. It is an easy corollary of our main theorem that if G is torsion-free and every subgroup of G is quasinormal then G is abelian.

A similar notion of permutable products, for elements of G instead of subgroups of G, has been considered by M. Curzio, P. Longobardi and M. Maj in [2]; these three together with D.J.S. Robinson in [3], and by R.D. Blyth in [1]. They introduce the concept of P_n and Q_n groups. An n-tuple $(x_1, \ldots, x_n)$ of elements of G is said to be permutable (rewritable) if there is a non-trivial permutation σ in S_n such that $x_1 \ldots x_n = x_{\sigma(1)} \cdots x_{\sigma(n)}$. G is a P_n-group if every n-tuple of elements of G is permutable. The main

---

The authors thank NSERC Canada for partial support.

result of [3] is that G is a P_n-group for some $n > 1$ if and only if it is finite-by-abelian-by-finite.

Whereas P_n-groups are defined in terms of products of elements, the PSP-groups in this paper deal with products of complexes. So it is not surprising that characterizing PSP-groups in general is presently a hopeless task. Our main result is the following.

Theorem. *Let G be a finitely generated solvable group. Then G is a PSP-group if and only if G is finite-by-abelian.*

Proofs. The reduction from solvable to finite-by-abelian will require three steps; these are considered by Lemma 1 through Lemma 3.

Lemma 1. *The wreath product of a cyclic group of order p with the infinite cyclic group is not a PSP-group.*

Lemma 2. *Let $G = \langle A, t\rangle$ where A is a torsion-free abelian group of finite rank on which $\langle t\rangle$ acts rationally irreducibly. If G is a PSP-group, then $\langle t\rangle$ acts trivially on A.*

Lemma 3. *A torsion-free nilpotent PSP-group is abelian.*

Proof of Theorem. Let G be a finitely generated solvable group and $n > 1$ such that for any n-tuple $(H_1, \ldots, H_n)$ of subgroups of G, there is a permutation $\sigma \in S_n$ such that

$$H_1 \ldots H_n = H_{\sigma(1)} \ldots H_{\sigma(n)}.$$

Since this property is inherited by subgroups and quotient groups, it follows by Lemma 1, that G has no section isomorphic to the wreath product of a cyclic group of order p with the infinite cyclic group. Thus G is "constrained" in the terminology of P.H. Kropholler and by the easy special case of his main theorem in [6], G has finite rank. Arguing by induction on the derived length of G, we may suppose that G has an abelian normal subgroup A such that G/A is finite-by-abelian.

We show that A is finitely generated. Consider A as a G-module. By Theorem 3 of [5], A satisfies the maximal condition for G-submodules. If A is not finitely generated as $\mathbf{Z}$-module, then let L be a maximal submodule such that A/L is not finitely generated as a $\mathbf{Z}$-module. Hence we may assume that A is not finitely generated as a $\mathbf{Z}$-module, but every proper quotient of A is finitely generated. If A is not torsion-free, then for some prime p, the submodule

$$P = \{a \in A,\ \ pa = 0\}$$

is non-trivial. Since A and hence P has finite-rank as abelian groups, $|P| < \infty$ and hence A would be finitely generated, a contradiction. Thus A may be assumed to be torsion-free as $\mathbf{Z}$-module. Now $A = \sum_{i=1}^{r} a_i \mathbf{Z}G$, for suitable $a_1, \ldots, a_r$ in A. If $a_i\mathbf{Z}\langle g\rangle$ is finitely generated as $\mathbf{Z}$-module for every $g \in G$, then A would be finitely generated since G/A is finitely generated finite-by-abelian. Thus we may assume that $B = a\mathbf{Z}\langle g\rangle$ is not finitely generated for some $a \in A$ and some $g \in G$. Pick a suitably so that the rank of B is minimal. Then B is rationally irreducible as $\mathbf{Z}\langle g\rangle$-module. Lemma 2 applies and we get a contradiction.

Now G is finitely generated, $A \triangleleft G$, A is finitely generated abelian and G/A is finite-by-abelian. Let T be the torsion subgroup of A. Then T is finite, and since we want to prove that G is finite-by-abelian, there is no harm in assuming that $T = 1$. We claim that for each $g \in G$, $\langle g\rangle$ acts nilpotently on A. This follows by applying Lemma 2 to each factor in a maximal $\langle g\rangle$-invariant series for A with torsion-free factors. Thus $\langle A, g\rangle$ is nilpotent and by Lemma 3, it is finite-by-abelian. Thus $[A, \langle A, g\rangle] = 1$. Since this holds for all g in G, A lies in the centre of G. Now G/A is finitely generated finite-by-abelian, hence there is a normal abelian subgroup $H \geq A$ of finite index in G such that $[G, H] \leq A$. Since a group is finite by nilpotent if and only if a finite term of its upper central series has finite index ([8], Theorem 4.25), it follows that G is finite by nilpotent. One more application of Lemma 3 shows that G is in fact finite by abelian.

Conversely let G be a finitely generated finite by abelian group. Such a group is an FC-group and hence G is a finite extension of its centre Z ([8], Theorem 4.32). Thus there are only a finite number, say m, of distinct subgroups of G containing Z. If X is any one of these then note that $|X/Z| < |G/Z|$, hence bounded by the index of Z in G. Now Z is of finite rank, say r. Thus if K is any subgroup of Z then Z/K is the direct sum of at most r cyclic groups. Thus there are boundedly many subgroups of Z/K

which are of order not exceeding $|G/Z|$. Let

$$\alpha = \max\{|\mathrm{Hom}(H/K, Z/K)|;\ |H/K| \le |G/Z|\}.$$

Then $\alpha = \alpha(r, |G/Z|)$ is bounded independently of K. Pick n to be any integer greater than $m\alpha$. Let $H_1, \ldots, H_n$ be subgroups of G. We show that there is a non-trivial permutation σ in S_n such that $H_1 \ldots H_n = H_{\sigma(1)} \cdots H_{\sigma(n)}$.

Let $K_i = H_i \cap Z$ and let $K = K_1 \ldots K_n$. Note that $H_1 \ldots H_n = KH_1 \ldots H_n = (KH_1) \ldots (KH_n)$ and likewise, $H_{\sigma(1)} \cdots H_{\sigma(n)} = (KH_{\sigma(1)}) \ldots (KH_{\sigma(n)})$. Thus we may assume that $K \le H_i$ for all $i = 1, \ldots, n$ and hence $H_i \cap Z = K$. It is sufficient to show that $H_1 \ldots H_n \equiv H_{\sigma(1)} \cdots H_{\sigma(n)}$ *modulo* K. Hence we may replace G by $\bar{G} = G/K$; H_i by $\bar{H}_i = H_i/K$ and Z by $\bar{Z} = Z/K$. Observe that $\bar{H}_i \cap \bar{Z} = 1$ and $|\bar{H}_i| \le |G/Z|$. Since there are at most m distinct subgroups of G containing Z, there are at least $\alpha + 1$ subgroups $\bar{H}_{i_0}, \ldots, \bar{H}_{i_\alpha}$ such that $\overline{ZH}_{i_0} = \overline{ZH}_{i_r}$ for all $r = 0, 1, \ldots, \alpha$. Let $\bar{H}$ denote $\bar{H}_{i_0}$ and $\bar{L}_r = \bar{H}_{i_r}$ for $r = 0, 1, \ldots, \alpha$. Then $\bar{Z} \times \bar{H} = \bar{Z} \times \bar{L}_r$; each $h \in \bar{H}$ has a unique expression of the form $h = z_r l_r$ with $z_r \in \bar{Z}$, $l_r \in \bar{L}_r$. The map $\phi_r : \bar{H} \to \bar{Z}$ given by $\phi_r(h) = z_r$ is a homomorphism and $\phi_r(\bar{H})$ is a subgroup of order at most $|G/Z|$. There are at most α distinct such maps ϕ_r. It follows that for at least 2 values u, v of r, the subgroups $\bar{L}_u$ and $\bar{L}_v$ are the same. This implies that $\bar{H}_{i_u} = \bar{H}_{i_v}$ and we may pick the permutation σ of S_n given by $\sigma(i_u) = i_v$, $\sigma(i_v) = i_u$ and $\sigma(i) = i$, $i \ne i_u$, i_v. This completes the proof.

Proof of Lemma 1. Let $G = \langle a_0 \rangle$ wreath $\langle t \rangle$ where $\langle a_0 \rangle$ is cyclic of order p and $\langle t \rangle$ is infinite cyclic. Write a_i for $t^i a_0 t^{-i}$ and use additive notation for the base group $B = Dr_{i \in Z} \langle a_i \rangle$. Let n be any given integer greater than 1. Choose l larger than n and let $H_i = \langle t^{a_{2l+i}} \rangle$, $i = 0, 1, \ldots, n-1$. We will show that $H_0 \ldots H_{n-1} = H_{\sigma(0)} \cdots H_{\sigma(n-1)}$ holds only if σ is the identity permutation of the set $0, 1, \ldots, n-1$. $H_0 \ldots H_{n-1}$ contains the element

$$\begin{aligned}\Pi &= (t^{a_{2l}})(t^{a_{2l+1}}) \ldots (t^{a_{2l+n-1}}) \\ &= (-a(l,0) + a(l,1) - a(l+1,1) + a(l+1,2) \ldots - a(l+k,k) \\ &\qquad + a(l+k,k+1) \ldots - a(l+n-1,n-1) + a(l+n-1,n))t^n\end{aligned} \tag{1.1}$$

where $a(k,j)$ denotes a_{2^k+j} to lessen the difficulty of deciphering the subscripts.

Suppose $\Pi \in H_{\sigma(0)} \dots H_{\sigma(n-1)}$. Then

$$\begin{aligned}\Pi &= \left(t^{a_{2^{l+\sigma(0)}}}\right)^{\beta_0} \dots \left(t^{a_{2^{l+\sigma(n-1)}}}\right)^{\beta_{n-1}} \\ &= \big(-a(l+\sigma(0),0) + a(l+\sigma(0),\beta_0) - a(l+\sigma(1),\beta_0) + a(l+\sigma(1),\beta_0+\beta_1) \\ &\quad \dots - a(l+\sigma(k),\beta_0+\dots+\beta_{k-1}) + a(l+\sigma(k),\beta_0+\dots\beta_k)\dots \\ &\quad \dots - a(l+\sigma(n-1),\beta_0+\dots+\beta_{n-2}) + a(l+\sigma(n-1),\beta_0+\dots \\ &\quad \dots + \beta_{n-1})\big)t^{\beta_0+\dots+\beta_{n-1}}.\end{aligned} \tag{1.2}$$

Compare the two expressions for Π and deduce the following:

$$\beta_0 + \dots + \beta_{n-1} = n.$$

Each term in (1.1) is distinct from the others and their negatives. Thus the same must hold for (1.2).

There is only one term in (1.1) where the suffix is congruent to n *modulo* 2^l, viz. $a(l+n-1,n)$. Hence $a(l+n-1,n) = a(l+\sigma(n-1),n)$. Thus $\sigma(n-1) = n-1$. Similarly $\sigma(0) = 0$. Suppose $\beta_0 \equiv i_0$ *modulo* 2^l, where $0 \le i_0 < 2^l$. Then $2^l + \beta_0 \equiv i_0 \pmod{2^l}$ and $2^{l+\sigma(1)} + \beta_0 \equiv i \pmod{2^l}$. Since $\sigma(1) > 0$, $2^l + \beta_0 = 2^{l+(i_0-1)} + i_0$ and $2^{l+\sigma(1)} + \beta_0 = 2^{l+i_0} + i_0$. Conclude $i_0 - 1 = 0$ and hence $\beta_0 = 1$ and $\sigma(1) = 1$. We can continue this process and inductively show that $\sigma(k) = k$ for all $k = 0, 1, \dots, n-1$. Thus σ is the identity permutation. This completes the proof.

We shall need the corollary to the following technical lemma in the proof of Lemma 2.

Lemma 4. *Suppose α is an algebraic number so that the equation*

$$\alpha^{\lambda_1} + \alpha^{\lambda_2} + \dots + \alpha^{\lambda_m} = \beta$$

has, for some fixed β, solutions in rational integers $\lambda_1, \dots, \lambda_m$ with $\lambda = \max(\lambda_1, \lambda_2, \dots, \lambda_m)$ taking arbitrarily large positive values.

Then for each absolute value $|\ |_v$ *of the field* $\mathbf{Q}(\alpha)$ *we have*

$$|\alpha|_v \le f_v(m) = \begin{cases} m, & v \text{ archimedean} \\ \max\{|r|_v^{-1} : r = 1, 2, \ldots, m\}, & v \text{ non-archimedean.} \end{cases}$$

Proof. Depends on the

Claim. $|\alpha|_v > f = f_v(m)$ implies $|\alpha^{\lambda_1} + \cdots + \alpha^{\lambda_m}|_v \ge |\alpha|_v^{\lambda}/f$.

Proof of Claim. Let r be the number of indices with $\lambda_i = \lambda$ and write

$$\alpha^{\lambda_1} + \cdots + \alpha^{\lambda_m} = r\alpha^{\lambda} + \sum_{\lambda_i < \lambda} \alpha^{\lambda_i}.$$

Now if v is non-archimedean then $|\alpha|_v > f$ implies that $|\alpha^{\lambda_i}|_v < |r\alpha^{\lambda}|_v$ whenever $\lambda_i < \lambda$ so the ultrametric inequality yields

$$|r\alpha^{\lambda} + \sum_{\lambda_i < \lambda} \alpha^{\lambda_i}|_v = |r\alpha^{\lambda}|_v \ge \frac{|\alpha|_v^{\lambda}}{f}$$

as claimed. Similarly if v is archimedean the triangle inequality gives

$$|r\alpha^{\lambda} + \sum_{\lambda_i < \lambda} \alpha^{\lambda_i}|_v \ge r|\alpha|_v^{\lambda} - (m - r)|\alpha|_v^{\lambda - 1}$$

$$\ge |\alpha|_v^{\lambda}(r - \frac{m - r}{m}) \ge \frac{|\alpha|_v^{\lambda}}{m},$$

also as claimed.

The Lemma follows immediately: for if it were false then the Claim would yield $|\alpha|_v^{\lambda}/f \le |\beta|_v$ for arbitrarily large positive λ, a contradiction.

Corollary. *Suppose* α *is a non-zero algebraic number so that* α^k *satisfies the hypothesis of the Lemma for all integers* $k \ne 0$*. Then* α *is a root of unity.*

Proof. The Lemma gives $|\alpha^k|_v \le f_v(m)$ so taking kth roots and letting $k \to +\infty$ gives $|\alpha|_v \le 1$. Since this holds for every absolute value of $\mathbf{Q}(\alpha)$ the desired conclusion follows.

Proof of Lemma 2. We are given that $G = \langle A, t\rangle$ where A is torsion-free abelian of finite rank on which $\langle t\rangle$ acts rationally irreducibly. Assume, if possible, that $[A, t] \neq 1$. Now $V = A \otimes_{\mathbf{Z}} \mathbf{Q}$ is an irreducible $\mathbf{Q}\langle t\rangle$-module, hence, by Schur's Lemma, the centralizer ring $\Gamma = \mathrm{End}_{\mathbf{Q}\langle t\rangle} V$ is a division ring, finite dimensional over $\mathbf{Q}$. The image of $\langle t\rangle$ in $\mathrm{End}_{\mathbf{Q}} V$ clearly lies in and spans Γ so that Γ is an algebraic number field. Moreover, regarded as a Γ-space, V is one dimensional. Thus we may consider A to be an additive subgroup of $\mathbf{Q}(\alpha)$ for some algebraic number α and the action of conjugation by t as multiplication by α. Two cases arise depending on whether α is a root of 1 or not.

If α is a root of 1, say it is a primitive kth root of 1, then pick any $1 \neq b$ in A, let $a = [t, b] = b(1-\alpha)$ and let $H_i = \langle (a^{k^i} t^{-1})\rangle$, $i = 0, \ldots, n-1$.

Let

$$S = H_0 \ldots H_{n-1} \tag{2.1}$$

and suppose

$$S = H_{\sigma(0)} \cdots H_{\sigma(n-1)} \tag{2.2}$$

for some $\sigma \neq 1$ in S_n. We look at elements from $S \cap \langle A, t^k\rangle$. These are of the form $x = (at^{-1})^{\lambda_0}(\alpha^k t^{-1})^{\lambda_1} \ldots (\alpha^{k^{n-1}} t^{-1})^{\lambda_{n-1}}$ where $\lambda_0 + \ldots + \lambda_{n-1}$ is a multiple of k. We will use alternate notation where A is written additively and a^t is expressed as $a\alpha$. Observe that $(a^\mu t^{-1})^k = t^{-k}$ for all $\mu \in Z$. Then

$$\begin{aligned} xt^{\lambda_0+\ldots+\lambda_{n-1}} &= b\big((1-\alpha^{[\lambda_0]}) + k(1-\alpha^{[\lambda_1]})\alpha^{[\lambda_0]} + \ldots \\ &\ldots + k^{n-1}(1-\alpha^{[\lambda_{n-1}]})\alpha^{[\lambda_0+\ldots+\lambda_{n-2}]}\big) \end{aligned} \tag{2.3}$$

where [ ] denotes residue class modulo k. Given that $x \in \langle A, t^k\rangle$, it follows that $[\lambda_0 + \ldots + \lambda_{n-1}] = [0]$. Thus (2.3) yields:

$$\begin{aligned} xt^{(\lambda_0+\ldots+\lambda_{n-1})} =& b\big(1 + (k-1)\alpha^{[\lambda_0]} + (k^2-k)\alpha^{[\lambda_0+\lambda_1]} + \ldots \\ &\ldots + (k^{n-1}-k^{n-2})\alpha^{[\lambda_0+\ldots+\lambda_{n-2}]} - k^{n-1}\big). \end{aligned} \tag{2.4}$$

Notice that for all choices of $\lambda_0, \ldots, \lambda_{n-1}$ other than $([\lambda_0], \ldots, [\lambda_{n-1}]) = ([0], \ldots, [0])$, the real part of $1+(k-1)\alpha^{[\lambda_0]}+\ldots+(k^{n-1}-k^{n-2})\alpha^{[\lambda_0+\ldots+\lambda_{n-2}]}$ $-k^{n-1}$ is strictly negative. On the other hand, if $\sigma \neq 1$ then $\sigma(i) > \sigma(i+1)$ for some i. Pick such an integer i for σ given in (2.2). Then $\big(a^{k^{\sigma(i)}} t^{-1}\big)$ $\big(a^{k^{\sigma(i+1)}} t^{-1}\big)^{-1} \in S \cap \langle A, t^k\rangle$. This element, in the other notation is

$$b\big(k^{\sigma(i)} - k^{\sigma(i+1)}\big)(1-\alpha).$$

The real part of $(k^{\sigma(i)} - k^{\sigma(i+1)})(1-\alpha)$ is non-negative if $\sigma(i) > \sigma(i+1)$. Thus $H_0 \ldots H_{n-1} \neq H_{\sigma(0)} \cdots H_{\sigma(n-1)}$ as required to show.

If α is not a root of unity, then let k be any integer not equal to zero, and let $H_i = \langle (t^k)^{a^i} \rangle$, $i = 0, 1, \ldots, n-1$. Suppose, if possible, that $H_0 \ldots H_{n-1} = H_{\sigma(0)} \cdots H_{\sigma(n-1)}$ for some $\sigma \neq 1$. Then a typical element of $(H_0 \ldots H_{n-1}) \cap A$ is of the form

$$x = \left(t^{a^0}\right)^{k\gamma_0} \ldots \left(t^{a^{n-1}}\right)^{k\gamma_{n-1}}$$

where $\gamma_0 + \ldots + \gamma_{n-1} = 0$. Using additive notation for A this takes the form

$$x = -a[\alpha^{k\gamma_0} + \ldots + \alpha^{k(\gamma_0 \ldots + \gamma_{n-2})} - (n-1)].$$

Denote $\gamma_0 + \ldots + \gamma_i$ by λ_i. Then x takes the form

$$x = -a[\alpha^{k\lambda_0} + \ldots + \alpha^{k\lambda_{n-2}} - (n-1)]. \tag{2.5}$$

Since $\sigma \neq 1$, $\sigma(i) > \sigma(i+1)$ for some i, and the element $(t^{a^{\sigma(i)}})^{k\mu}(t^{a^{\sigma(i+1)}})^{-k\mu}$ lies in $(H_{\sigma(0)} \cdots H_{\sigma(n-1)}) \cap A$. In additive notation this is

$$-a[\sigma(i) - (\sigma(i) - \sigma(i+1))\alpha^{k\mu} - \sigma(i+1)]$$

or

$$-a[\delta - \delta\alpha^{k\mu}] \quad \text{where} \quad \delta = \sigma(i) - \sigma(i+1) > 0.$$

Compare this with the element given by (2.5) and conclude that for any $0 \neq k \in \mathbf{Z}$ and any $\mu \in \mathbf{Z}$, there exist integers $\lambda_0, \ldots, \lambda_{n-2}$ such that

$$\alpha^{k\lambda_0} + \ldots + \alpha^{k\lambda_{n-2}} + \delta\alpha^{k\mu} = (n-1) + \delta.$$

By the corollary to Lemma 4, this can happen only if α is a root of unity. This completes the proof of the lemma.

We could have used Theorem 1 of [9], see also [4], and avoided the use of Lemma 4 and its corollary in the last but crucial part of the proof of Lemma 2. But these results are much deeper and Lemma 4 is elementary.

Proof of Lemma 3. Let G be a torsion-free nilpotent PSP-group and suppose, if possible, that G is not abelian. Then there exist elements

a, b in G such that $[b,a] = c \neq 1$ and c lies in the centre of $\langle a,b\rangle$ which we may call G. We will denote a by $(1,0,0)$; b by $(0,1,0)$ and c by $(0,0,1)$. Then the elements of G are of the form (x,y,z) where x, y, $z \in \mathbf{Z}$ and $(x_1,y_1,z_1) + (x_2,y_2,z_2) = (x_1+x_2, y_1+y_2, z_1+z_2+y_1x_2)$. Note that we are using the additive notation although the group is not commutative. Let $n > 1$ be a given integer. For all i, $j \in 1, \ldots, n$ pick distinct primes $q_{ij} > n$. Let $Q_i = q_{i1} \ldots q_{in}$ and $\beta_i = Q/Q_i$ where $Q = Q_1 \ldots Q_n$. Let $P_j = q_{1j} \ldots q_{nj}$ and $\alpha_j = Q/P_j$.

Let $H_i = \langle(\alpha_i, \beta_i, 0)\rangle$ $i = 1, \ldots n$. Suppose that for some nontrivial permutations σ in S_n,

$$H_1 \ldots H_n = H_{\sigma(1)} \ldots H_{\sigma(n)} \tag{3.1}$$

It is easy to check that the following relations hold in G. For all integers $r, r_1, \ldots, r_n$,

$$r(x,y,0) = (rx, ry, r(r-1)xy/2) \tag{3.2}$$

and

$$\sum_{i=1}^{n} r_i(x_i,y_i,0) = (\sum_{i=1}^{n} r_ix_i, \sum_{i-1}^{n} r_iy_i, \sum_{i=1}^{n} x_iy_ir_i(r_i-1)/2 + \sum_{1\leq i\leq j\leq n} r_ir_jy_ix_j) \tag{3.3}$$

For each element $\sum r_i(\alpha_i, \beta_i, 0)$ in $H_1 \ldots H_n$, there is an element

$$\sum s_{\sigma(i)}(\alpha_{\sigma(i)}, \beta_{\sigma(i)}, 0) \in H_{\sigma(1)} \ldots H_{\sigma(n)}$$

such that

$$\sum_{i=1}^{n} r_i(\alpha_i, \beta_i, 0) = \sum_{i=1}^{n} s_{\sigma(i)}(\alpha_{\sigma(i)}, \beta_{\sigma(i)}, 0). \tag{3.4}$$

Use (3.3) to obtain the equality

$$\left.\begin{aligned}
&\Big(\sum_{i=1}^{n} r_i\alpha_i, \sum_{i=1}^{n} r_i\beta_i, \sum_{i=1}^{n} \alpha_i\beta_i r_i(r_i-1)/2 + \sum_{i<j} r_i r_j \beta_i \alpha_j\Big) \\
&\quad = \Big(\sum_{i=1}^{n} s_{\sigma(i)}\alpha_{\sigma(i)}, \sum_{i=1}^{n} s_{\sigma(i)}\beta_{\sigma(i)}, \sum_{i=1}^{n} \alpha_{\sigma(i)}\beta_{\sigma(i)} s_{\sigma(i)}(s_{\sigma(i)}-1)/2 \\
&\qquad + \sum_{i<j} s_{\sigma(i)} s_{\sigma(j)} \beta_{\sigma(i)} \alpha_{\sigma(j)}\Big) \\
&\quad = \Big(\sum_{i=1}^{n} s_i\alpha_i, \sum_{i=1}^{n} s_i\beta_i, \sum_{i=1}^{n} \alpha_i\beta_i s_i(s_i-1)/2 \\
&\qquad + \sum_{i<j} s_{\sigma(i)} s_{\sigma(j)} \beta_{\sigma(i)} \alpha_{\sigma(j)}\Big).
\end{aligned}\right\} \tag{3.5}$$

Thus, in particular

$$\sum_{i=1}^{n} r_i\alpha_i = \sum_{i=1}^{n} s_i\alpha_i \tag{3.6}$$

and

$$\sum_{i=1}^{n} r_i\beta_i = \sum_{i=1}^{n} s_i\beta_i. \tag{3.7}$$

Let l be the largest integer such that $\sigma(l) \neq l$. Let $t = \sigma^{-1}(l)$. Then $t < l$ and $\sigma(t) > \sigma(l)$. Look at equation (3.5) *modulo* q_{tl}. The left hand side is $(r_l\alpha_l, r_t\beta_t, r_t r_l \beta_t \alpha_l)$ and the other side is $(s_l\alpha_l, s_t\beta_t, s_{\sigma(t)}s_{\sigma(l)}\beta_{\sigma(t)}\alpha_{\sigma(l)})$ which is $(s_l\alpha_l, s_t\beta_t, 0)$ since $\sigma(l) \neq l$.

Hence $r_t r_l \beta_l \alpha_l \equiv 0$ *modulo* q_{tl}. But β_t and α_l are coprime to q_{tl} and we may pick r_t, r_l to be coprime to q_{tl}, resulting in a contradiction. This completes the proof.

References

[1] R.D. Blyth, *Rewriting products of group elements*, Ph.D. Thesis, University of Illinois (1987).

[2] M. Curzio, P. Longobardi and M. Maj, Su di un problema combinatorio in teoria dei gruppi, *Atti Acc. Lincei Rend. fiz.* VIII **74** (1983), 136-142.

[3] M. Curzio, P. Longobardi, M. Maj and D.J.S. Robinson, A permutational property of groups, *Arch. Math.* (Basel) **44** (1985), 385-389.

[4] J.H. Evertse, On sums of S-units and linear recurrences, *Compositio Math.* **53** (1984), 225-244.

[5] P. Hall, Finiteness conditions for soluble groups, *Proc. London Math. Soc.* (3) **4** (1954),419-436.

[6] P.H. Kropholler, On finitely generated soluble groups with no large wreath product sections, *Proc. London Math. Soc.* (3) **49** (1984), 155-169.

[7] J.C. Lennox and S.E. Stonehewer, *Subnormal subgroups of groups*, in "Oxford Mathematical Monographs," Clarendon Press, Oxford, 1987.

[8] D.J.S. Robinson, *Finiteness conditions and generalized soluble groups, Part I*, Springer-Verlag, New York, 1972.

[9] A.J. Van der Poorten, Additive relations in number fields, *Seminaire de theorie des nombres de Paris* (1982-1983), 259-266.

Department of Mathematics
University of Alberta
Edmonton, Alberta
Canada T6G 2G1

Symmetric presentations

E.F. Robertson and *C.M. Campbell*

1. Definitions and examples

Presentations with certain types of symmetry have been considered over many years, see for example [2] and the references quoted therein. Many examples of symmetric presentations are given, for instance by Coxeter in [6], without a formal definition of a symmetric presentation. Most examples arise as special cases of the definitions which follow.

Let G be any finite presentation

$$G = \langle a_1, a_2, \ldots, a_r \mid w_i(a_1, a_2, \ldots, a_r) = 1,\ 1 \leq i \leq s \rangle$$

and let P be any subgroup of the symmetric group S_n where $n \geq r$. Define

$$PG = \langle a_1, a_2, \ldots, a_n \mid w_i(a_{1\theta}, a_{2\theta}, \ldots, a_{r\theta}) = 1,\ 1 \leq i \leq s,\ \theta \in P \rangle.$$

We give three examples of such presentations.

Example 1. Take $\theta = (1\,2\,\ldots\,n) \in S_n$ and let $T = \langle \theta \rangle$. Further let G be a 1-relation, r-generator group presentation where $r \leq n$. Then TG is a cyclically presented group in the sense defined in [8].

Example 2. Let G be the presentation for a free group of rank 2

$$G = \langle a_1, a_2, a_3 \mid a_1 a_2 = a_3 \rangle.$$

Then if T is as in Example 1 we see that TG is the Fibonacci group $\underset{\sim}{F}(2, n)$, see [8]. Similarly $\underset{\sim}{F}(r, n)$ is the group TG_1 where

$$G_1 = \langle a_1, a_2, \ldots, a_{r+1} \mid a_1 a_2 \ldots a_r = a_{r+1} \rangle.$$

Example 3. Take G to be as in Example 2, i.e.

$$G_1 = \langle a_1, a_2, a_3 \mid a_1 a_2 = a_3 \rangle.$$

Let $\theta_1 = (1\,2\,3\,4)$, $\theta_2 = (1\,3\,2\,4)$ be elements of S_4 and put $T_1 = \langle \theta_1 \rangle$, $T_2 = \langle \theta_2 \rangle$. Then $T_1 G \cong C_5$ while $T_2 G \cong A_4$, the alternating group of degree four.

Further let $\phi_1 = (1\,2\,3\,4\,5)$, $\phi_2 = (1\,3\,2\,4\,5)$, $\phi_3 = (1\,2)(3\,4\,5)$ be elements of S_5 and put $P_i = \langle \phi_i \rangle$, $1 \leq i \leq 3$. Then $P_1 G \cong C_{11}$, $P_2 G \cong SL(2,5)$, $P_3 G \cong C_\infty \times C_\infty$. It is also easy to see that $S_3 G \cong C_2 \times C_2$. More information about some of these and other similar examples are given by Coxeter in [6].

Before giving one further example we define what we mean by a symmetric presentation.

Definition. We say that the n-generator group presentation G is a *symmetric presentation* if $S_n G = G$.

Example 4. The presentation

$$G = \langle a, b \mid a^2 b^{-2} = 1,\ abab^{-1} = 1 \rangle$$

is a symmetric presentation for the quaternion group Q_8. For,

$$S_2 G = \langle a, b \mid a^2 b^{-2} = 1,\ b^2 a^{-2} = 1,\ abab^{-1} = 1,\ baba^{-1} = 1 \rangle$$

and using $a^2 = b^2$, $aba = b$ we have

$$bab = b.aba.a^{-1} = bba^{-1} = a,$$

so $S_2 G = \langle a, b \mid a^2 b^{-2} = 1, abab^{-1} = 1 \rangle = G$ as required.

Lemma 1. *If $P_1 \leq P_2 \leq S_n$ and G is any r-generator group presentation where $r \leq n$ then*

$$P_1(P_2 G) = P_2 G = P_2(P_1 G).$$

In particular S_nG is a symmetric presentation.

Proof. The first part of the lemma is straightforward. That S_nG is a symmetric presentation follows on taking $P_1 = P_2 = S_n$.

In this paper we wish to consider the following question. Suppose G is a symmetric 2-generator presentation. What are the groups S_nG for $n \geq 2$? Notice that, as a consequence of Lemma 1, there is no loss of generality in assuming that G is a symmetric presentation since, for any 2-generator presentation G, we have $S_nG = S_n(S_2G)$ for $n \geq 2$ and S_2G is symmetric. Note also that although often $S_{n-1}G$ is a subgroup of S_nG this is not always the case and it is quite possible that a subset of $n-1$ generators of S_nG (and of course any two subsets of the generators with the same cardinality generate isomorphic subgroups) generates a proper homomorphic image of $S_{n-1}G$.

Example 5. If $G = \langle a, b \mid a^m = b^m = [a,b] = 1\rangle$ then $S_nG \cong C_m^n$.

Example 6. If $G = \langle a, b \mid a^2 = b^2 = (ab)^3 = 1\rangle$ then S_nG is infinite for $n \geq 3$. The case $n = 3$ is the group of equilateral triangular tesselations of the plane. See [7, Fig. 4.5r] where this group is given by the notation **p 3 m 1**.

Example 7. If $G = \langle a, b \mid a^3 = b^3 = (ab)^2 = 1\rangle$ then $S_nG \cong A_{n+2}$.

This is proved in [5], see also [7, §6.3].

Generalisations of Example 7 have been studied in [9] by Sidki and these involve orthogonal groups. Other generalisations of Example 7 and of Beetham's symmetric presentation for $PSL(2,p)$ [1] involving simple groups are at present being studied by the authors in collaboration with J. Neubüser and S. Sidki.

After investigating series of groups S_nG using a computer implementation of the Todd-Coxeter coset enumeration algorithm, three questions regarding such series suggest themselves. The results of Section 2 of this paper give a partial answer to the first of these questions.

Question 1. Given any fixed integer $k \geq 2$, is it possible to find a 2-generator symmetric presentation G (which depends on the particular k) such that $S_n G$ is a finite group for $2 \leq n \leq k$ and $S_n G$ is an infinite group for $n > k$?

Question 2. Does there exist a symmetrically presented 2-generator presentation G and some integer $m \geq 2$ such that $S_m G$ is an infinite group but $S_{m+1} G$ is a finite group?

Question 3. Given any integer $k \geq 2$, is it possible to find a 2-generator symmetric presentation G (depending on k) such that k of the generators of $S_{k+1} G$ generate a proper homomorphic image of $S_k G$?

Example 6 provides an affirmative answer to Question 1 for $k = 2$ while Theorem 3 below gives an affirmative answer for $k = 3$. Question 3 has an affirmative answer in the case $k = 2$. For

$$G = \langle a, b \mid a^{13} = b^{13} = a^4 bab^{-3} = b^4 aba^{-3} = 1 \rangle$$

is a presentation for $PSL(3,3)$, see [4], while a computer implementation of the Todd-Coxeter algorithm shows that $S_3 G$ is the trivial group.

2. Some series $S_n G$

First we consider $S_n G$ where G is the presentation of the quaternion group given in Example 4.

Theorem 2. *Let $G = \langle a, b \mid a^2 = b^2, aba = b \rangle$. Then $S_n G$ is a group of order 2^{n+1} being an extension of C_2 by C_2^n.*

Proof. A presentation for $S_n G$ is

$$\langle a_1, a_2, \ldots, a_n \mid a_i^2 = a_j^2, a_i a_j a_i = a_j, \ 1 \leq i, j \leq n, i \neq j \rangle.$$

From the relation $a_i a_j a_i = a_j$, $a_j^{-1} a_i a_j = a_i^{-1}$ and so $a_j^{-1} a_i^2 a_j = a_i^{-2}$ giving $a_i^4 = 1$. Further a_i^2 is central. Let $Z = \langle a_i^2 \rangle$. Then $S_n G/Z$ is isomorphic to C_2^n giving the result.

Before giving our main theorem we define the Lucas sequence of numbers (g_i) by

$$g_0 = 2,\ g_1 = 1,\ g_{i+2} = g_{i+1} + g_i,\ i \geq 0$$

and the Fibonacci sequence of numbers (f_i) by

$$f_0 = 0,\ f_1 = 1,\ f_{i+2} = f_{i+1} + f_i,\ i \geq 0.$$

Theorem 3. *Let $G_m = \langle a, b \mid ab^2 = ba^2,\ a^m = b^m = 1\rangle$. Suppose that m is odd. Then:*
(i) if $(m,3) = 1$, S_nG_m has order mg_m^{n-1};
(ii) if $(m,3) = 3$, S_2G_m has order mg_m, S_3G_m has order $2mg_m^2$ and S_nG_m is infinite for $n \geq 4$.

Proof. $S_nG_m = \langle a_1, a_2, \ldots, a_n \mid a_i^m = 1, a_ia_j^2 = a_ja_i^2, 1 \leq i,j \leq n\rangle$. Define $\theta : S_nG_m \to S_m$ by

$$\theta : a_i \to (1\ 2\ \ldots\ m) \quad 1 \leq i \leq n.$$

Then let $H = \operatorname{Ker}\theta$. Now H is a subgroup of index m in S_nG_m and letting $\tau_1 = 1$ and $\tau_i = \tau_{i-1}a$ for $2 \leq i \leq m$ we obtain a Schreier transversal for H in S_nG_m. Let

$$\alpha_{i,j} = \tau_ia_j\tau_{i+1}^{-1} \quad 1 \leq i \leq m,\ 1 \leq j \leq n$$

be the set of Schreier generators. Then $\alpha_{i,1} = 1$, $1 \leq i \leq m-1$ by definition of τ_i, $1 \leq i \leq m$.

Rewriting $a_1^m = 1$ in terms of Schreier generators of H gives $\alpha_{m,1} = 1$ and rewriting $a_i{}^m = 1$, $2 \leq i \leq n$ gives the relations (P_i)

$$\alpha_{1,i}\alpha_{2,i}\ldots\alpha_{m,i} = 1. \tag{2.1}$$

Rewriting the relations $a_1a_i^2 = a_ia_1^2$, $2 \leq i \leq n$ gives the relations (F_{ji})

$$\alpha_{j,i} = \alpha_{j+1,i}\alpha_{j+2,i},\ 1 \leq j \leq m \tag{2.2}$$

where the subscripts are reduced modulo m.

Rewriting the relations $a_ka_i^2 = a_ia_k^2$, $2 \leq i,\ k \leq n$ gives the relations

$$a_{j,k}\alpha_{j+1,i}\alpha_{j+2,i}\alpha_{j+2,k}^{-1}\alpha_{j+1,k}^{-1}\alpha_{j,i}^{-1} = 1,\ 1 \leq j \leq m. \tag{2.3}$$

Using (F_{ji}) and (F_{jk}) (2.3) becomes

$$\alpha_{j,k}\alpha_{j,i}\alpha_{j,k}^{-1}\alpha_{j,i}^{-1} = 1. \tag{2.4}$$

Next we prove that the subgroups $A_i = \langle \alpha_{j,i} \mid 1 \leq j \leq m \rangle$ are abelian for each i, $2 \leq i \leq n$. First note that from the relations (2.2) it is clear that $A_i = \langle \alpha_{1i}, \alpha_{2i} \rangle$. Hence, to prove that A_i is abelian, we need only show that $[\alpha_{1,i}, \alpha_{2i}] = 1$. From (2.1)

$$\alpha_{1,i}\alpha_{2,i}\ldots\alpha_{m,i} = 1$$

so, using $\alpha_{1,i} = \alpha_{2,i}\alpha_{3,i}, \ \ \alpha_{3,i} = \alpha_{4,i}\alpha_{5,i}, \ldots$ from (2.2) we have

$$\alpha_{1,i}^2\alpha_{3,i}\alpha_{5,i}\ldots\alpha_{m-2,i} = 1.$$

However, again using (2.2),

$$\alpha_{m-2,i} = \alpha_{m-3,i}\alpha_{m-1,i}^{-1}, \quad \alpha_{m-4,i}\alpha_{m-3,i} = \alpha_{m-5,i},$$
$$\alpha_{m-6,i}\alpha_{m-5,i} = \alpha_{m-7,i}, \ldots$$

so, substituting successively, we get

$$\alpha_{1,i}^2\alpha_{2,i}\alpha_{m-1,i}^{-1} = 1. \tag{2.5}$$

However $\alpha_{m-1,i} = \alpha_{m,i}\alpha_{1,i}$ and $\alpha_{m,i} = \alpha_{1,i}\alpha_{2,i}$ so $\alpha_{m-1,i} = \alpha_{1,i}\alpha_{2,i}\alpha_{1,i}$. Substituting in (2.5) gives

$$\alpha_{1,i}^2\alpha_{2,i}\alpha_{1,i}^{-1}\alpha_{2,i}^{-1}\alpha_{1,i}^{-1} = 1$$

that is

$$[\alpha_{1,i}, \alpha_{2,i}] = 1$$

as required.

Next we obtain a relation matrix for the abelian subgroup A_i. From (2.2) $\alpha_{3,i} = \alpha_{1,i}\alpha_{2,i}^{-1}$, $\alpha_{4,i} = \alpha_{1,i}^{-1}\alpha_{2,i}^2$ and, in general, an inductive argument gives $\alpha_{t,i} = \alpha_{1,i}{}^{(-1)^{t+1}f_{t-2}}\alpha_{2,i}{}^{(-1)^t f_{t-1}}$, where f_n denotes the nth

Fibonacci number. Using the fact that m is odd we therefore have the relations

$$\alpha_{1,i}^{-f_{m-3}}\alpha_{2,i}^{f_{m-2}} = \alpha_{m-1,i} = \alpha_{1,i}^2\alpha_{2,i}$$
$$\alpha_{1,i}^{f_{m-2}}\alpha_{2,i}^{-f_{m-1}} = \alpha_{m,i} = \alpha_{1,i}\alpha_{2,i}$$

giving the relation matrix

$$\begin{bmatrix} -f_{m-3}-2 & f_{m-2}-1 \\ f_{m-2}-1 & -f_{m-1}-1 \end{bmatrix}.$$

Subtracting the rows and then the columns we obtain the equivalent relation matrix

$$\begin{bmatrix} f_{m+1}+1 & f_m \\ f_m & f_{m-1}+1 \end{bmatrix}.$$

We consider the cases $(m,3)=1$ and $(m,3)=3$ separately.

Case (i) $(m,3)=1$.

In this case $(f_m, f_{m+1}+1)=1$ (see, for example, Theorem 2.3 of [3]) so A_i is cyclic. In this case the relations (2.4) show that H is abelian.

Since $|A_i| = (f_{m+1}+1)(f_{m-1}+1) - f_m^2 = g_m$, $H = C_{g_m}^{n-1}$ and $G = C_{g_m}^{n-1}\colon C_m$ as required.

Case (ii) $(m,3)=3$.

In this case $(f_m, f_{m+1}+1)=2$ (again see Theorem 2.3 of [3]). Notice that $g_m/2$ is even in this case and A_i is not cyclic. However $\alpha_{1,i}$, $\alpha_{2,i}$ each has order $g_m/2$, $\alpha_{1,i}^2$ is a power of $\alpha_{2,i}$ and $\alpha_{2,i}^2$ is a power of $\alpha_{1,i}$.

When $n=2$, then $H = A_2 \cong C_2 \times C_{g_m/2}$. If $n=3$ note that

$$H = \langle \alpha_{i,2}, \alpha_{i,3} \mid 1 \le i \le m \rangle = \langle \alpha_{1,2}, \alpha_{2,2}, \alpha_{1,3}, \alpha_{2,3} \rangle.$$

Since $[\alpha_{3,2}^{-1}, \alpha_{3,3}] = 1$, $\alpha_{3,2} = \alpha_{1,2}\alpha_{2,2}^{-1}$ and $\alpha_{3,3} = \alpha_{1,3}\alpha_{2,3}^{-1}$ we have

$$\alpha_{1,2}^{-1}\alpha_{2,2}\alpha_{1,3}\alpha_{2,3}^{-1}\alpha_{2,2}^{-1}\alpha_{1,2}\alpha_{2,3}\alpha_{1,3}^{-1} = 1.$$

Therefore, observing that

$$\alpha_{i,j} \quad \text{and} \quad \alpha_{k,l} \quad \text{commute unless both} \quad i \ne k \quad \text{and} \quad j \ne l \tag{2.6}$$

we obtain

$$\alpha_{1,2}^{-1}\alpha_{2,3}^{-1}\alpha_{1,2}\alpha_{2,3}\alpha_{2,2}^{-1}\alpha_{1,3}^{-1}\alpha_{2,2}\alpha_{1,3} = 1.$$

Therefore

$$[\alpha_{1,2}, \alpha_{2,3}][\alpha_{2,2}, \alpha_{1,3}] = 1.$$

But $[\alpha_{1,2}, \alpha_{2,3}]$ commutes with $\alpha_{1,3}$ and $\alpha_{2,2}$, by (2.6), and $[\alpha_{2,2}, \alpha_{1,3}]$ commutes with $\alpha_{1,2}$ and $\alpha_{2,3}$, again by (2.6). Since $[\alpha_{1,2}, \alpha_{2,3}] = [\alpha_{2,2}, \alpha_{1,3}]^{-1}$ and $H = \langle \alpha_{1,2}, \alpha_{2,2}, \alpha_{1,3}, \alpha_{2,3} \rangle$ we see that $N = \langle [\alpha_{1,2}, \alpha_{2,3}] \rangle$ is central in H. Now $H/N \cong (C_2 \times C_{gm/2})^2$ as above. Finally we show that $[\alpha_{1,2}, \alpha_{2,3}]^2 = 1$. For

$$\alpha_{1,2}^{-1}\alpha_{2,3}^{-1}\alpha_{1,2}\alpha_{2,3}\alpha_{1,2}^{-1}\alpha_{2,3}^{-1}\alpha_{1,2}\alpha_{2,3} = \alpha_{1,2}^{-2}\alpha_{2,3}^{-1}\alpha_{1,2}^{2}\alpha_{2,3}.$$

But $\alpha_{1,2}^2$ is a power of $\alpha_{2,2}$ and $[\alpha_{2,3}, \alpha_{2,2}] = 1$ so

$$\alpha_{1,2}^{-2}\alpha_{2,3}^{-1}\alpha_{1,2}^{2}\alpha_{2,3} = 1.$$

The final part of the proof is to show that $S_n G_m$ is infinite if $(m, 3) = 3$ and $n \geq 4$. First note that it is sufficient to prove that $S_n G_3$ is infinite if $n \geq 4$ since this group is a homomorphic image of $S_n G_m$ in this case.

The subgroup of index 3 in $S_n G_3$ has presentation

$$\begin{aligned} [\alpha_{1,i}, \alpha_{1,j}] &= 1 \\ [\alpha_{2,i}, \alpha_{2,j}] &= 1 \\ [\alpha_{3,i}, \alpha_{3,j}] &= 1 \end{aligned} \tag{2.7}$$

$$\alpha_{1,i}\alpha_{2,i}\alpha_{3,i} = 1 \tag{2.8}$$

$$\begin{aligned} \alpha_{1,i} &= \alpha_{2,i}\alpha_{3,i} \\ \alpha_{2,i} &= \alpha_{3,i}\alpha_{1,i} \\ \alpha_{3,i} &= \alpha_{1,i}\alpha_{2,i} \end{aligned} \tag{2.9}$$

where $2 \leq i, j \leq n$. These relations simplify to

$$[\alpha_{i,j}, \alpha_{k,l}] = 1 \quad \text{if} \quad i = k \tag{2.7$'$}$$

$$[\alpha_{i,j}, \alpha_{k,l}] = 1 \quad \text{if} \quad j = l \tag{2.8$'$}$$

$$\begin{aligned}\alpha_{1,i} &= \alpha_{2,i}\alpha_{3,i}\\ \alpha_{2,i} &= \alpha_{3,i}\alpha_{1,i} \qquad 2 \le i \le n.\\ \alpha_{3,i} &= \alpha_{1,i}\alpha_{2,i}\end{aligned} \tag{2.9$'$}$$

Eliminating $\alpha_{3,i}$, $2 \le i \le n$ gives

$$\alpha_{2,i}{}^2 = 1, \ \alpha_{1,i}{}^2 = 1. \tag{2.10}$$

Equation $(2.7)'$ with the $\alpha_{3,i}$ elimination from $(2.9)'$ becomes

$$(\alpha_{1,i}\alpha_{2,i}\alpha_{1,j}\alpha_{2,j})^2 = 1$$

while $(2.8)'$ with the same elimination is redundant.

Now consider the homomorphic image obtained by putting $\alpha_{1,i} = \alpha_{2,i} = 1$ for $i \ge 5$. We are left with the H-subgroup of S_4G_3 and it suffices to prove that S_4G_3 is infinite. Using a Tietze transformation program and a Reidemeister-Schreier program we get as a presentation for the derived group of index 96 the following:

$$\begin{aligned}\langle a,b,c,d,e,f,g \mid a^2 = b^2 = c^2 = d^2 = e^2 = f^2 = g^2 = 1,&\\ (ab)^2 = (ac)^2 = (ad)^2 = (ae)^2 = (bc)^2 = (bd)^2 = (bf)^2&\\ = (ce)^2 = (cf)^2 = (de)^2 = (df)^2 = 1,&\\ (abg)^2 = (acg)^2 = (adg)^2 = (aeg)^2 = (bcf)^2&\\ = (bdf)^2 = (bdg)^2 = (bef)^2 = (ceg)^2 = (abfg)^2 = 1\rangle.&\end{aligned}$$

Put $a = b = e = f = g = 1$ to obtain the homomorphic image

$$\langle c,d \mid c^2 = d^2 = 1\rangle$$

which is the infinite dihedral group and so S_4G_3 is infinite as required.

References

[1] M.J. Beetham, A set of generators and relations for the group $PSL(2,q)$, q odd, *J. London Math. Soc.* **3** (1971), 554-557.

[2] C.M. Campbell, Symmetric presentations and linear groups, *Contemp. Math.* **45** (1985), 33-39.

[3] C.M. Campbell and E.F. Robertson, Deficiency zero groups involving Fibonacci and Lucas numbers, *Proc. Roy. Soc. Edingburgh* **81A** (1978), 273-286.

[4] C.M. Campbell and E.F. Robertson, Some problems in group presentations, *J. Korean Math. Soc.* **19** (1983), 123-128.

[5] H.S.M. Coxeter, Abstract groups of the form $v_i{}^k = v_j{}^3 = (v_i v_j)^2 = 1$, *J. London Math. Soc.* **9** (1934), 213-219.

[6] H.S.M. Coxeter, Symmetrical definitions for the binary polyhedral groups, *Proceedings of Symposia in Pure Mathematics* (American Mathematical Society) **1** (1959), 64-87.

[7] H.S.M. Coxeter and W.O.J. Moser, *Generators and relations for discrete groups*, (Springer, 4th edition, Berlin, 1979).

[8] D.L. Johnson, *Presentations of groups*, London Math Soc Lecture Notes 22 (Cambridge University Press, 1976).

[9] S. Sidki, A generalisation of the alternating groups - a question of finiteness and representation, *J. Algebra* **75** (1982), 324-372.

Mathematical Institute
University of St Andrews
St Andrews KY16 9SS
Scotland

On the periodicity of group operations

Dedicated to Professor Elmar Thoma on the occasion of his 60th birthday

Günter Schlichting

§0. Introduction

Consider a topological space X, a homomorphism $\lambda : \Gamma \to H(X)$ of an abstract group Γ into the group $H(X)$ of all homeomorphisms of X onto itself. We write $\lambda(\gamma)x = \gamma x$ $(\gamma \in \Gamma, x \in X)$ when there is no possibility of confusion. There are three types of periodicity which the action of Γ on X may have or not

$$P_1(\Gamma, X) \;:\Leftrightarrow O_\Gamma(x) = |\Gamma_x| < \infty \quad \text{for all} \quad x \in X$$
$$P_2(\Gamma, X) \;:\Leftrightarrow O_\Gamma \quad \text{is bounded}$$
$$P_3(\Gamma, X) \;:\Leftrightarrow \lambda(\Gamma) \quad \text{is finite}.$$

Our aim is to give, for various choices of Γ and X, conditions on Γ and X such that P_i implies P_{i+1} $(i = 1, 2)$. Given a Γ-invariant Boolean subalgebra $\mathcal{A}$ of $\mathcal{P}(X)$ (all subsets of X) λ induces a homomorphism $\lambda^{\mathcal{A}} : \Gamma \to \operatorname{Aut} \mathcal{A}$. By Stone's duality theorem $\mathcal{A}$ is isomorphic to the algebra of all compact-open subsets of a compact, totally disconnected set $\bar{X}^{\mathcal{A}} = \Delta(\mathcal{A})$ (spectrum of $\mathcal{A}$, cf. [6]). In this way Boolean subalgebras $\mathcal{A}$ of $\mathcal{P}(X)$ describe all the totally disconnected compactifications $X_d \to \bar{X}^{\mathcal{A}}$ of X_d (the set X with its discrete topology). Because of $\operatorname{Aut} \mathcal{A} = H(\bar{X}^{\mathcal{A}})$ and the Γ-invariance of $\mathcal{A}$ there is an induced action of Γ on $\bar{X}^{\mathcal{A}}$.

The obvious examples are $\mathcal{A} = \mathcal{P}(X)$ where $\bar{X}^{\mathcal{A}}$ is the Stone-Cech compactification of X_d and $\mathcal{A} = \{F \subset X \mid F \text{ or } F^c = X - F \text{ is finite}\}$ where $\bar{X}^{\mathcal{A}} = X_d^*$ is the Alexandroff (one point) compactification of X_d.

For Boolean rings $\mathcal{A}$ and subgroups $\Gamma \leq \operatorname{Aut} \mathcal{A}$ it is easy to see (cf. [5]) that $P_2(\Gamma, \mathcal{A})$ implies $P_2(\Gamma, X)$ where $X = \Delta(\mathcal{A})$. We prove a stronger result

Theorem 7. *For Boolean rings $\mathcal{A}$ and subgroups $\Gamma \leq \operatorname{Aut} \mathcal{A}$ property $P_2(\Gamma, \mathcal{A})$ is equivalent to property $P_3(\Gamma, \mathcal{A})$ respectively $P_3(\Gamma, X)$.*

With respect to the topology of pointwise convergence $H(X_d)$ is a topological group, a neighbourhood basis of the identity is given by the subgroups which fix finitely many elements of X. Equivalently, the topology of $H(X_d)$ is the topology of uniform convergence on X_d^*. Therefore

$$\begin{aligned} P_1(\Gamma, X) &\Leftrightarrow \Gamma \quad \text{relatively compact in} \quad H(X_d) \\ &\Leftrightarrow \Gamma \quad \text{equicontinuous in} \quad x_\infty \in X_d^* - X_d. \end{aligned}$$

If there is a group G acting transitively on X and a point $x_0 \in X$ such that $\Gamma = G_{x_0} := \{g \in G | gx_0 = x_0\}$ then $X = G/\Gamma$ and because the closure $\bar{G}$ of G in $H(X_d)$ acts transitively on X as well one obtains $\bar{G} = G(\bar{G})_{x_0}$ and since $(\bar{G})_{x_0}$ is open in $\bar{G}$ evidently $(\bar{G})_{x_0} = \overline{(\bar{G})_{x_0} \cap G} = \bar{\Gamma}$. Hence $\bar{G} = G\bar{\Gamma}$.

Subgroups Γ_1, Γ_2 are called commensurable ($\Gamma_1 \sim \Gamma_2$) if $\Gamma_1 \cap \Gamma_2$ has finite index both in Γ_1 and Γ_2. Denote by $[IN]$ (respectively $[SIN]$) the class of all locally compact topological groups which contain at least one (respectively arbitrarily small) compact neighbourhoods of the identity invariant relative to inner automorphisms. Properties $P_i(\Gamma, G/\Gamma)$ now translate into topological properties of $\lambda(G) \subset H(X_d)$

$P_1(\Gamma, G/\Gamma)$	$\Leftrightarrow$	subgroups in G (respectively $\overline{\lambda(G)}$) fixing only finitely many elements of G/Γ are commensurable
$\overline{\lambda(G)}$ is locally compact	$\Leftrightarrow$	there are finitely many $x_1, \ldots, x_n \in G/\Gamma$ such that property $P_1(G_{x_1 \cdots x_n}, G/\Gamma)$ holds
$P_2(\Gamma, G/\Gamma)$	$\Leftrightarrow$	$\overline{\lambda(G)} \in [IN]$
$P_3(\Gamma, G/\Gamma)$	$\Leftrightarrow$	$\overline{\lambda(G)} \in [SIN]$.

All of this is contained in [7] and implies

Theorem 1. *Let Γ be a subgroup of a group G, then property $P_2(\Gamma, G/\Gamma)$ is equivalent with the existence of a subgroup H of G such that $\Gamma \leq H \leq G$, $[H : \Gamma] < \infty$ and property $P_3(\Gamma, G/H)$ holds. Likewise these properties are equivalent to the existence of a normal subgroup N of G commensurable to Γ. One may obtain N in such a way that it contains a subgroup of G fixing finitely many elements of G/Γ.*

A striking consequence of the above result is

Theorem 2. *Let Γ, A be subgroups of a group G such that $G = \Gamma A$ and $\Gamma \cap A = \{e\}$. Then the induced action of Γ on A satisfies $P_2(\Gamma, A)$ iff there is a finite, Γ-invariant, normal subgroup E of A such that property $P_3(\Gamma, A/E)$ holds for the induced action of Γ on A/E.*

Remark. If A is a normal subgroup of G the induced action of Γ on $G/\Gamma = A$ is given by automorphisms and the above result generalizes a theorem of R. Baer (cf, [2], [7]) where in addition A is assumed to be abelian.

If $X = G/\Gamma$ is a homogeneous space for which topological properties such as compactness or connectedness obtain then $P_1(\Gamma, X)$ implies $P_2(\Gamma, X)$ (cf. Theorem 9).

§1. Properties of the orbit length function which are dependent on the topology of X

Lemma 1. *Let X be a topological Hausdorff space and Γ a subgroup of $H(X)$. Then the orbit length function $x \to O_\Gamma(x) = |\Gamma x|$ is lower semicontinuous on X.*

Proof. Let $x_0 \in X$ and $|\Gamma x_0| > \alpha \in \mathbf{R}$. Then there are $\gamma_1, \dots \gamma_r \in \Gamma$, $r > \alpha$ such that $\gamma_1 x_0, \dots, \gamma_r x_0$ are pairwise distinct. Choose pairwise disjoint neighbourhoods U_ρ of $\gamma_\rho x_0$ $(1 \le \rho \le r)$ then $U := \bigcap_{\rho=1}^r \gamma_\rho^{-1} U_\rho$ is a neighbourhood of x_0 such that $\gamma_\rho U \subset U_\rho$ $(1 \le \rho \le r)$, hence $|\Gamma x| \ge r > \alpha$ for all $x \in U$.

Corollary. *Let X be a Hausdorff Baire space (e.g. locally compact or complete metrizable) and $\Gamma \le H(X)$. Then there is a nonempty open subset $U \subset X$ such that $O_\Gamma | U$ is bounded.*

Proof. This is a well known property of lower semi-continuous functions.

Theorem 3. *Let X be a topological vector space which is a Baire space and Γ a group of invertible continuous linear transformations on X. Then* *property $P_1(\Gamma, X)$ holds iff Γ is finite.*

Proof. O_Γ is bounded on a nonempty open subset U of X, therefore bounded on a neighbourhood $U - f$ $(f \in U)$ of $O \in X$. Since $X = \mathbf{R}(U - f)$, property $P_2(\Gamma, X)$ follows. Since X (as an additive group) has no finite subgroups, $P_3(\Gamma, X)$ is an immediate consequence of Theorem 2.

For a group G and a subgroup $\Gamma \leq G$ the quasinormalizer $Q(\Gamma, G)$ of Γ in G is defined by

$$\begin{aligned} Q(\Gamma, G) &:= \{\tau \in G \mid \Gamma \sim \tau\Gamma\tau^{-1}\} \\ &= \{\tau \in G \mid \Gamma\tau\Gamma/\Gamma \quad \text{and} \quad \Gamma\tau^{-1}\Gamma/\Gamma \quad \text{are finite}\}. \end{aligned}$$

Obviously $Q(\Gamma, G)$ is a subgroup of G which contains Γ, the normalizer of Γ in G and because of $\Gamma\tau\Gamma = [\tau]^\Gamma \Gamma$ all $\tau \in G$ such that $[\tau]^\Gamma := \{\gamma\tau\gamma^{-1} \mid \gamma \in \Gamma\}$ is finite. If $G = Q(\Gamma, G)$ $(\Leftrightarrow P_1(\Gamma, G/\Gamma))$ and $\lambda : G \to H(X)$ defines an operation of B on X then transitivity properties of the action of G on X combined with topological properties of X imply the equivalence of properties $P_1(\Gamma, X)$ and $P_2(\Gamma, X)$. The importance of $Q(\Gamma, G)$ derives from the above and

Lemma 2. $O_\Gamma(\tau x) \leq |\Gamma\tau\Gamma/\Gamma| O_\Gamma(x)$ *for all* $x \in X$, $\tau \in G$.

Proof. If $\Gamma\tau\Gamma = \bigcup_{i=1}^r \tau_i\Gamma$ then $\Gamma\tau x \subset \Gamma\tau\Gamma x \subset \bigcup_{i=1}^r \tau_i\Gamma x$.

Theorem 4. *Let G be a minimal transformation group on a compact space X and Γ a subgroup of $H(X)$ such that $G \subset Q(\Gamma, H(X))$ $(\Leftrightarrow P_1(\Gamma, G/\Gamma))$ then $P_1(\Gamma, X)$ iff $P_2(\Gamma, X)$.*

Proof. Because of Lemma 1 O_Γ is bounded on a nonempty open subset $U \subset X$, then O_Γ is bounded on τU for all $\tau \in G$ (Lemma 2). By minimality $X = G \cdot U$, hence X is covered by a finite number of translates τU, $\tau \in G$, therefore O_Γ is bounded on X.

§2. Group operations on Boolean algebras

Let X be a locally compact space and X^* the Alexandroff (one point) compactification of X. Then the Arens topology of $H(X)$ (cf. [1]) may be described as the topology of uniform convergence on X^*, which in turn is the topology of pointwise convergence of $H(X)$ as a group of linear transformation on the Banach space $C(X^*)$ of all continuous functions on X^*. Since $C(X^*) = C_0(X) + \mathbf{C} \cdot 1$ this topology is also seen to be the topology of pointwise convergence of $H(X)$ as a group of linear transformations on the Banach space $C_0(X)$ of all continuous functions on X vanishing at infinity .

Let $\mathcal{K}(X)$ denote the set of all compact-open subsets of X. Identify $\mathcal{K}(X)$ with the set of all $\{0,1\}$-valued functions in $C_0(X)$ then the linear span of $\mathcal{K}(X)$ is dense in $C_0(X)$ iff X is totally disconnected. Since $H(X)$ operates on $C_0(X)$ by isometries, the topology of pointwise convergence of $H(X)$ on $C_0(X)$ is the same as the topology of pointwise convergence of $H(X)$ on $\mathcal{K}(X)$ if X is totally disconnected. Since $\mathcal{K}(X)$ is a discrete subset of $C_0(X)$ we obtain

Lemma 3. *Let X be locally compact and totally disconnected and let $H(X)$ be given the Arens topology (in which it is a topological group). Then for a subgroup $\Gamma \leq H(X)$ the property $P_1(\Gamma, X)$ is equivalent with the property that Γ is almost periodic on X, respectively relative compact in $H(X)$. A neighbourhood basis of the identity in $H(X)$ is given by the system of subgroups fixing finitely many $U_1, U_2, \ldots, U_n \in \mathcal{K}(X)$.*

Remark. In particular the algebraic isomorphism of $H(X)$ and Aut $(\mathcal{K}(X))$ given by Stone's duality theory is in fact an isomorphism of topological groups.

One may expect that for compact, totally disconnected spaces X properties $P_1(\Gamma, X)$ and $P_1(\Gamma, \mathcal{K}(X))$ imply property $P_3(\Gamma, \mathcal{K}(X))$ (equivalently $P_3(\Gamma, X)$). But this is not true since for an infinite discrete space X_0 and its one point compactification X, properties $P_1(\Gamma, X_0)$, $P_1(\Gamma, X)$ and $P_1(\Gamma, \mathcal{K}(X))$ are all equivalent and do not in general imply P_3. Very illuminating is the following

Example (cf. [5]). Let X be the space of all integers compactified by adding ∞. τ_i, the permutation which interchanges $-i$, i and leaves all other elements fixed, is a homeomorphism of X for all $i \in \mathbf{Z}$. The group Γ generated by τ_i, $i \in \mathbf{Z}$ operates on X with orbits of length ≤ 2, therefore property $P_2(\Gamma, X)$ holds. On a system of generators of $\mathcal{K}(X)$ (the orbital atoms, cf. [5]) all orbits of Γ have length ≤ 2, therefore property $P_1(\Gamma, \mathcal{K}(X))$ holds as well, but $P_3(\Gamma, X)$ fails.

Relative to the action of Γ on X a subset E of X is called a domain of imprimitivity (orbital atom) provided that for all $\gamma \in \Gamma$ either $\gamma E = E$ or $\gamma E \cap E = \emptyset$. Denote by $\mathcal{A}_0$ the set of all domains of imprimitivity $E \subset X$, $E \in \mathcal{A} := \mathcal{K}(X)$. Evidently $\mathcal{A}_0$ is Γ-invariant, moreover

Theorem 5 (cf. [5]). *Let X, Γ, $\mathcal{A}$ and $\mathcal{A}_0$ be as above. Then properties $P_1(\Gamma, \mathcal{A})$ and $P_2(\Gamma, \mathcal{A}_0)$ hold iff Γ has separated orbits on X and property $P_2(\Gamma, X)$ holds. Provided $P_1(\Gamma, X)$ holds Γ has separated orbits iff property $P_1(\Gamma, \mathcal{A})$ holds.*

Proof. All of this is due to [5] under the additional assumption that $\mathcal{A}$ is a Boolean algebra (this means that X is compact). For Boolean rings (X locally compact, totally disconnected) the desired result follows from this because of the following

Remark. Set $\mathcal{A}^* = \{U \subset X | U \in \mathcal{A} \text{ or } U^c \in \mathcal{A}\}$. Then $\mathcal{A}^*$ is a Γ-invariant Boolean subalgebra of $\mathcal{P}(X)$ isomorphic to the algebra $\mathcal{K}(X^*)$ of all compact-open subsets of the one point compactification X^* of X. The isomorphism is explicitly given by $\mathcal{K}(X^*) \ni U^* \to U = U^* \cap X \in \mathcal{A}^*$.

Theorem 6. *Let X, Γ, $\mathcal{A}$ and $\mathcal{A}_0$ be as above and assume property $P_1(\Gamma, X)$ holds. Then property $P_1(\Gamma, \mathcal{A}_0)$ holds and $\mathcal{A}_0$ is a basis of the topology of X iff property $P_1(\Gamma, \mathcal{A})$ holds.*

Proof. Given $U \in \mathcal{A}_0$ and $x \in U$. Then $\Gamma_x \subset \Gamma_U$ and therefore $O_\Gamma(U) \leq O_\Gamma(x) < \infty$. If $\mathcal{A}_0$ is a basis of the topology of X then $\mathcal{A}$ as an additive group ($U \Delta V = UV^c \cup U^c V$) is generated by $\mathcal{A}_0$. Since $O_\Gamma(U \Delta V) \leq O_\Gamma(U) O_\Gamma(V)$ for all U, $V \in \mathcal{A}$, evidently $P_1(\Gamma, \mathcal{A}_0)$ implies $P_1(\Gamma, \mathcal{A})$. Conversely property $P_1(\Gamma, \mathcal{A})$ implies that $\mathcal{A}_0$ is a basis of the topology of X (cf. 2.2. [5]).

Up to now we have described how properties $P_1(\Gamma, X)$, $P_1(\Gamma, \mathcal{A}_0)$, $P_1(\Gamma, \mathcal{A})$, $P_2(\Gamma, X)$ and $P_2(\Gamma, \mathcal{A}_0)$ depend on each other. All of them are consequences of the much stronger property $P_2(\Gamma, \mathcal{A})$. In fact as an application of Theorem 2 we get the following result.

Theorem 7. *Let X be a locally compact, totally disconnected space, Γ a subgroup of the Boolean ring $\mathcal{K}(X)$ of all compact-open subsets of X. Then property $P_2(\Gamma, \mathcal{A})$ holds iff Γ is finite.*

Proof. Because of Theorem 2 there is a finite, Γ-invariant additive subgroup $\mathcal{C}$ of $\mathcal{A}$ such that Γ acts on $\mathcal{A}/\mathcal{C}$ as a finite group. Without loss of generality we may assume that $\mathcal{C}$ consists of all unions of finitely many, pairwise disjoint, nonempty elements $E_1, \dots E_n$ of $\mathcal{A}$, $\gamma E_i = E_i$ for all $\gamma \in \Gamma$, $1 \leq i \leq n$ and $\gamma A \,\Delta\, A \in \mathcal{C}$ for all $\gamma \in \Gamma$, $A \in \mathcal{A}$. Therefore $\gamma A \,\Delta\, A \in \{E_1, \phi\}$ for all $\gamma \in \Gamma$, $A \subset E_1$, $A \in \mathcal{A}$, hence $\gamma A \in \{A, E_1 - A\}$ for all $\gamma \in \Gamma$, $A \subset E_1$, $A \in \mathcal{A}$. Let $E := E_1 \cup \dots \cup E_n$ and let $A \in \mathcal{A}$ be a subset of $X - E = E^c$. Then $\gamma A \subset X - E$ (since $\gamma E = E$), therefore $\gamma A \,\Delta\, A \subset X - E$, but $\gamma A \,\Delta\, A \subset E$ (since $\gamma A \,\Delta\, A \in \mathcal{C}$). Hence $\gamma A = A$ for all $\gamma \in \Gamma$, $A \in \mathcal{A}$, $A \subset X - E$. In particular $\gamma a = a$ for all $\gamma \in \Gamma$, $a \in X - E$. Hence we may suppose without loss of generality $X = E = E_1$ and $\gamma A \in \{A, X - A\}$ for all $\gamma \in \Gamma$, $A \in \mathcal{A}$. Therefore $O_\Gamma(A) \leq 2$ for all $A \in \mathcal{A}$. Assume there is an $A \in \mathcal{A}$ such that $O_\Gamma(A) = 2$, that is there is a $\gamma_0 \in \Gamma$ with $\gamma_0 A = X - A$. Then $\gamma_0 B \subset \gamma_0 A = X - A$ for all $B \in \mathcal{A}$, $\phi \neq B \subset A$. Hence $\gamma_0 B \neq B$ and therefore $\gamma_0 B = X - B$ which implies $B = A$, hence $|A| = 1$. Therefore $\Gamma A = A$ for all $A \in \mathcal{A}$, $|A| \geq 2$. This implies $|\Gamma x| \leq 2$ for all $x \in X$. Assume there are different points x, y, a, $b \in X$ such that $\{x, y\} = \Gamma x$ and $\{a, b\} = \Gamma a$. Then there is $A \in \mathcal{A}$ such that a, $x \in A$ but y, $b \notin A$, and $\Gamma A \neq A$ in contrast to $|A| \geq 2$. Hence there is at most one Γ-orbit of length ≥ 2 in X and Γ has to be finite.

In this context we want to mention the beautiful main-result of [5].

Theorem 8 (K.H. Hofmann - F.B. Wright). *Let $\mathcal{A}$ be a Boolean σ-algebra and Γ a subgroup of* Aut $\mathcal{A}$. *Then $P_1(\Gamma, \mathcal{A})$ holds iff Γ is finite.*

§3. Periodic operations on homogeneous spaces

A classic result of Montgomery (cf. [9]) asserts that $P_1(\Gamma, X)$ implies $P_3(\Gamma, X)$ if X is a connected locally Euclidean space and Γ a cyclic group of homeomorphisms. We are going to derive similar results under the assumption that X is the homogeneous space G/Γ, G being a locally compact group.

Theorem 9. *Let G be a locally compact group, Γ a closed subgroup of G and X the quotient space G/Γ. Then*
(i) $P_1(\Gamma, X) \Leftrightarrow P_2(\Gamma, X)$ *if X is compact or connected,*
(ii) $P_1(\Gamma, X) \Leftrightarrow P_3(\Gamma, X)$ *if G is compact or connected.*

Proof. If X is compact assertion i) is a special case of Theorem 4 and ii) is just Theorem 1 in [8]. We now consider O_Γ as a function on G $(O_\Gamma(g) = O_\Gamma(g\Gamma),\, g \in G)$. Then $\psi(g) := (O_\Gamma(g)O_\Gamma(g^{-1}))^{-\frac{1}{2}}$ is upper semicontinuous because of lemma 1, in particular ψ is measurable. Moreover ψ is positive-definite (cf. [7]). Therefore one has a decomposition $\psi = \varphi + \psi_0$ into positive-definite functions φ, ψ_0 where φ is continuous and ψ_0 locally null almost everywhere (cf. [3], 32.12). Because of $\psi = \mathrm{Re}\psi = \mathrm{Re}\varphi + \mathrm{Re}\psi_0$ one may suppose that φ, ψ_0 are real valued. Since $\varphi = \psi - \psi_0$ is continuous and non-negative locally a.e. one obtains $\varphi \geq 0$ everywhere. Since $\psi_0 = 0$ locally a.e. and upper semicontinuous, $\{\psi_0 < -\epsilon\}$ has to be empty for all $\epsilon > 0$, hence $\psi_0 \geq 0$. Because of $\varphi|\Gamma + \psi_0|\Gamma = \psi|\Gamma = 1$ and because the inequality $|\varphi| \leq \varphi(e)$ (e=identity of G) is valid for positive-definite functions one obtains $\varphi|\Gamma = \varphi(e)$ and $\varphi(\gamma g) = \varphi(g\gamma) = \varphi(g)$ for all $g \in G$, $\gamma \in \Gamma$. Being upper semicontinuous the points of continuity of ψ are dense in G. Because of $\psi(G) \subset \{\frac{1}{\sqrt{n}} | n \in \mathbf{N}\}$ ψ has to be constant and equal to φ in suitable small neighbourhoods of points of continuity. Therefore $\varphi(G) \subset \overline{\psi(G)} \subset \{0, \frac{1}{\sqrt{n}} | n \in \mathbf{N}\}$. If X is connected this implies $\varphi(G) = \varphi(G/\Gamma) = \{\varphi(e)\}$, hence $\psi = \varphi + \psi_0 \geq \varphi = \varphi(e) > 0$. In fact we have proved something slightly more than boundedness of O_Γ, namely

Remark. If $X = G/\Gamma$ is a connected homogeneous space, then property $P_1(\Gamma, X)$ implies $O_\Gamma(g\Gamma)O_\Gamma(g^{-1}\Gamma) = \max$ locally a.e. and the set of points where the max value is obtained is open and dense.

For the proof of (ii) it is enough to suppose that $G = \Gamma G_0$, where G_0 denotes the connected component of the identity. By Theorem 1 and (i) from above property $P_1(\Gamma, X)$ implies the existence of a (closed) normal subgroup N of G which is commensurable to Γ. Evidently Γ is totally disconnected, hence N as well, therefore $N \subset \mathbf{C}(G_0)$ (= centralizer of G_0 in G). Hence $\Gamma \cap N$ is normal in $G_0\Gamma = G$ and therefore contained in the kernel of the action of Γ on G/Γ, this means $P_3(\Gamma, X)$.

§4. Finiteness conditions for groups of automorphisms

We now want to discuss the interrelations between the properties P_i in the special case that $\Gamma \leq \mathrm{Aut}\ A$ is a subgroup of the group Aut A of automorphisms of an arbitrary locally compact group A. If one forms the semidirect product $G = \Gamma ⓢ A$ one may apply the previous results because the action of Γ on A is just the action of Γ on the homogeneous space $X = G/\Gamma$.

Denote by $\hat{A}$ the set of unitary equivalence classes of continuous, unitary, irreducible representations of A. As usual we do not distinguish between the class $\{\pi\} \in \hat{A}$ and its representative π. Subgroups $\Gamma \leq \mathrm{Aut}\ A$ operate on $\hat{A}$ according to $\pi \to \pi o \gamma^{-1}$ $(\pi \in \hat{A}, \gamma \in \Gamma)$.

If A is abelian $\hat{A}$ is precisely the set of all continuous homomorphisms $\pi : A \to S^1 = \{z \in \mathbf{C} \mid |z| = 1\}$. In general there is the following important Γ-invariant subset $\hat{A}_{\mathrm{fin}} := \{\pi \in \hat{A},\ \deg \pi = \mathrm{Tr}\, \pi(e) < \infty\}$ of $\hat{A}$. $N := \cap\{ker\pi \mid \pi \in \hat{A}_{\mathrm{fin}}\}$ is a closed normal subgroup of A, it is called the v. Neumann kernel von A and equals the kernel of the universal group compactification $A \to \bar{A}^B$ (Bohr compactification). A is called maximal almost periodic ($G \in [MAP]$) if $N = \{e\}$ (for a discussion of such groups cf. [4]).

Theorem 10. *Let A be a compact group and Γ a subgroup of* Aut A. *Then Γ is finite iff properties $P_1(\Gamma, A)$ and $P_1(\Gamma, \hat{A})$ hold.*

Proof. The Banach space $\mathbf{C}(A)$ of all continuous (complex-valued) functions on A is by the theorem of Peter-Weyl the translation invariant, norm closed linear hull of $\{\alpha = \mathrm{Tr}\, \pi \mid \pi \in \hat{A}\}$. Therefore $P_1(\Gamma, \hat{A})$ implies that all Γ-orbits of functions in $\mathbf{C}(A)$ are relatively compact, this is equivalent to the fact that Γ is relatively compact in Aut A ($\subset H(A)$). Therefore the semidirect product $G = \bar{\Gamma}Ⓢ A$ is a compact group and property $P_1(\bar{\Gamma}, G/\bar{\Gamma})$ holds. By Theorem 9 Γ is finite.

Theorem 11. *Let A be a locally compact group, $A \in [MAP]$ and $\Gamma \leq$ Aut A. Then Γ is finite iff properties $P_2(\Gamma, A)$ and $P_1(\Gamma, \hat{A}_{\text{fin}})$ hold.*

Proof. Each topological automorphism of A extends uniquely to a topological automorphism of $\bar{A}^B$ the Bohr compactification of A. In this way $\Gamma \leq$ Aut $A \leq$ Aut $\bar{A}^B$. Because of lemma 1 property $P_2(\Gamma, A)$ implies property $P_2(\Gamma, \bar{A}^B)$. Since $\hat{A}_{\text{fin}} = \widehat{\bar{A}^B}$ the assertion follows from Theorem 10.

Theorem 12. *Let A be a locally compact abelian group and Γ a subgroup of* Aut A. *Then Γ is finite iff properties $P_2(\Gamma, A)$ and $P_1(\Gamma, \hat{A})$ hold.*

Proof. This follows from Theorem 11, since $A \in [MAP]$ and $\hat{A} = \hat{A}_{\text{fin}}$.

Literature

[1] R. Arens, Topologies for homomorphism groups, *Amer. J. Math* **68** (1946) pp. 593-610.

[2] R. Baer, Automorphismengruppen von Gruppen mit endlichen Bahnen gleichmäßig beschränkter Mächtigkeit, *J. reine angew. Math.* **262/263** (1973) S. 93-119.

[3] E. Hewitt - K.A. Ross, *Abstract harmonic analysis* Vol. II, New York 1970.

[4] H. Heyer, *Dualität lokalkompakter gruppen.* LNM 150, New York 1970.

[5] K.H. Hofmann - F.B. Wright, Pointwise periodic groups, *Fund. Math.* L II (1963) pp. 103- 122.

[6] J.L. Kelley, *General topology,* New York 1955.

[7] G. Schlichting, Operationen mit periodischen Stabilisatoren, *Arch. Math.* Vol. **34** (1980) S. 97-99.

[8] G. Schlichting, Polynomidentitäten und Permutationsdarstellungen lokal-kompakter Gruppen, *Inv. math.* **55** (1979) S. 97-106.

[9] D. Montgomery, Pointwise periodic homeomorphisms, *Amer. J. Math.* **59** (1937) pp. 118-120.

* I wish to thank DFG who gave me financial support to present these results to the Singapore Group Theory Conference.

Institut für Mathematik
der Technischen Universität
Arcisstraße 21
D-8000 München
Federal Republic of Germany

An application of groups to the topology design of connection machines

S. C. Shee and *H. H. Teh*

1. Introduction

Connection Machines belong to a new breed of super-computers which contain a large number of processors inter-linked to form a homogeneous graph called their topology. W. D. Hillis is the first person to propose a workable model for the implementation of connection machines on a 12-dimensional hypercube. In choosing a topology (graph) Hillis suggests we look for a combination of the following: small diameter, uniformity, extendability, short wires, redundant paths, minimum number of wires, efficient layout, a simple routing algorithm, a fixed degree and adaptability to available topology [3]. It is not easy to construct a graph with all these properties.

The 12-dimensional hypercube now used as the topology of connection machines has the drawback that when the dimension of the hypercube becomes large, its degree and diameter are also large. The performance of a connection machine depends a great deal on the degree and the diameter of its topology. It therefore seems there is a need to find other types of graphs of small degree and small diameters as well as some of the desirable properties as mentioned by Hillis. As the uniformity property is almost a must for the topology of a connection machine, we therefore be mainly concerned with graphs which are homogeneous (vertex-transitive) and have a fixed degree and relatively small diameters.

As every Cayley graph is vertex-transitive and up to now people can exhibit only 4 vertex-transitive graphs which are not Cayley graph, we shall therefore look for our graphs from the collection of Cayley graphs. Cayley graphs are constructed using groups. Cayley graphs constructed using commutative groups in general do not have adequately small diameters. In this paper we shall introduce a class of non-commutative groups which can be used to construct Cayley graphs of degree 4 and relatively small diameters. We also discuss the routing problem of these Cayley graphs.

2. The group $\Phi(p) \times \mathbf{Z}_p$

Take a prime $p \geq 5$. Consider the Euler group $\langle \Phi(p), \cdot \rangle$ where $\Phi(p) = \{1, 2, \ldots, p-1\}$, and the field $\langle \mathbf{Z}_p, +, \cdot \rangle$ of integers modulo p. It is well-known that $\langle \Phi(p), \cdot \rangle$ is a cyclic group. Let h be a generator of $\Phi(p)$. We have the following lemma whose proof can be found in [1].

Lemma 1. $h^{\frac{p-1}{2}} \equiv -1 \pmod{p}$.

We form the Cartesian product

$$G = \Phi(p) \times \mathbf{Z}_p = \{(i, a) \mid i \in \Phi(p) \quad \text{and} \quad a \in \mathbf{Z}_p\}.$$

and define a binary operation "$*$" on G as follows: for every $(i, a), (j, b) \in G$

$$(i, a) * (j, b) = (i \cdot j, a \cdot j + b).$$

It is easily verified that $\langle G, * \rangle$ is a non-commutative group with identity $(1, 0)$ and the inverse of (i, a) is $(i^{-1}, -a \cdot i^{-1})$. We shall henceforth write $(i, a)(j, b)$ instead of $(i, a) * (j, b)$.

We shall use the group $\langle G, * \rangle$ to construct a vertex-transitive graph of degree 4. Let

$$A = \{(h, 0), (h^{-1}, 0), (1, 1), (1, -1)\}$$

Suppose $(i, a) \in G$. Let $i = h^r$. Then $(i, a) = (h^r, a) = (h, 0)^r (1, 1)^a$. Hence A generates G. If $g \in G$ and

$$g = a_1 a_2 \ldots a_m, \quad a_i \in A.$$

we say that the length of g is at most m and we write $\ell(g) \leq m$. If $\ell(g) \leq m$ but $\ell(g) \not\leq m - 1$, we say the minimum length of g is m and write $\ell(g) = m$.

In the following discussion the symbols G and A will always refer to the group $G = \Phi(p) \times \mathbf{Z}_p$ and the above subset respectively. For some reasons to be explained in the next section we are interested in expressing an element of G as a product of elements of A.

Proposition 1. Every element of the group G is expressible as a product of elements of A of length at most $\frac{p+1}{2}$.

Proof. Let $(k,c) \in G$. Then $h^r \equiv k \pmod{p}$ for some positive integer $0 \leq r \leq p-1$. If $c = 0$, then

$$(1,0) = (h,0)(h^{-1},0)$$

and for $0 < r < p-1$,

$$\begin{aligned}(h^r,0) &= (h,0)^r \\ &= (h^{-1},0)^{(p-1)-r},\end{aligned}$$

and one of these products has length at most $\frac{p+1}{2}$. If $c \neq 0$, then by Lemma 1 there exists a $0 \leq t \leq \frac{p-1}{2}$ such that $h^t \equiv c$ or $-c \pmod{p}$. We have $\ell(h^r,h^t), \ell(h^r,-h^t) \geq \min\{r+1, p-r\}$ for $0 \leq r \leq p-1$ and $0 \leq t \leq \frac{p-1}{2}$.

Case (i). $t < r \leq \frac{p-1}{2}$. We find

$$(h^r,1) = (h,0)^r(1,1),$$

and for $t > 0$

$$(h^r,\pm h^t) = (h,0)^{r-t}(1,\pm 1)(h,0)^t$$

whose length is $r+1 \leq \frac{p+1}{2}$.

Case (ii). $r < t \leq \frac{p-1}{2}$. We find

$$\begin{aligned}(h^r,\pm h^t) &= (h^{-1},0)^{t-r}(1,\pm 1)(h,0)^t \\ &= (h,0)^{\frac{p-1}{2}-t+r}(1,\mp 1)(h^{-1},0)^{\frac{p-1}{2}-t}.\end{aligned}$$

The lengths of these products are respectively $2t-r+1$ and $(p+1)-(2t-r+1)$, and the sum of these lengths is $p+1$. It follows that one of the lengths must be at most $\frac{p+1}{2}$.

Case (iii). $\frac{p-1}{2} < r \leq \frac{p-1}{2}+t$. In this case

$$(h^r,\pm h^t) = (h^{-1},0)^{\frac{p-1}{2}+t-r}(1,\mp 1)(h^{-1},0)^{\frac{p-1}{2}-t}$$

whose length is $p-r \leq \frac{p+1}{2}$.

Case (iv). $\frac{p-1}{2} + t < r \leq p - 1$. Then

$$\begin{aligned}(h^r, \pm h^r) &= (h^{-1}, 0)^{(p-1)+t-r}(1, \pm 1)(h, 0)^t \\ &= (h, 0)^{r-(\frac{p-1}{2}+t}(1, \mp 1)(h^{-1}, 0)^{\frac{p-1}{2}-t}.\end{aligned}$$

As the lengths of these products are respectively $p + 2t - r$ and $r - 2t + 1$, one of them must be at most $\frac{p+1}{2}$.

Hence (k, c) can always be expressed as a product of elements of A of length at most $\frac{p+1}{2}$.

Remark 1. The inequality in Proposition 1 is best possible since when the element $(h^{\frac{p-1}{2}}, 1)$ is expressed as a product of elements of A, the minimum length is $\frac{p+1}{2}$ as indicated below

$$\begin{aligned}(h^{\frac{p-1}{2}}, 1) &= (h, 0)^{\frac{p-1}{2}}(1, 1) \\ &= (h^{-1}, 0)^{\frac{p-1}{2}}(1, 1).\end{aligned}$$

For reason to be explained in the next section we are also interested in expressing an element of G as a product of elements of A of minimum length. From the proof of Proposition 1, we obtain the following

Corollary 1.

(i) $\ell(h^r, 0) = r$ or $p - r - 1$, according as $0 < r \leq \frac{p-1}{2}$, or $\frac{p-1}{2} < r < p - 1$.

(ii) $\ell(h^r, h^t) = \ell(h^r, -h^t) = r + 1$ or $p - r$, according as $0 < t \leq r \leq \frac{p-1}{2}$, or $0 \leq t \leq \frac{p-1}{2}$ and $\frac{p-1}{2} < r \leq \frac{p-1}{2} + t$.

3. The Cayley graph $X(p, h)$

A graph X consists of a finite non-empty set $V(X)$ together with a collection $E(X)$ of 2-element subsets of $V(X)$. Elements of $V(X)$ are called vertices (or nodes), and those of $E(X)$ edges (or lines) of the graph X. Two vertices u and v are said to be adjacent if $\{u, v\} \in E(X)$; in this case the edge $\{u, v\}$ is incident with u as well as v. The degree of a vertex u is the number of edges incident with it. If the degree of every vertex of the graph

X is the same, say k, then X is regular of degree k. A sequence of distinct edges of the form

$$\{v_0, v_1\}, \{v_1, v_2\}, \ldots, \{v_{m-1}, v_m\}$$

is called a path connecting v_0 and v_m of length m. We shall denote this path simply by $P(v_0, v_1, \ldots, v_m)$ and we say that v_0 and v_m are connected. If the vertices u and v are connected, the distance between u and v is the length of a shortest path connecting them. If there is a path connecting any two vertices of the graph X, then X is connected. In a connected graph X, the diameter of X is the maximum distance between two vertices of X. The size of a graph X is $|E(X)|$.

Let G be a group with identity element e. Suppose A is a Cayley subset of G, that is, $e \notin A$, whenever $g \in A$ then $g^{-1} \in A$ and A generates G. The Cayley graph $X = X(G, A)$ of G with respect to A has the elements of G as its vertices, with two vertices g and h adjacent if $gh^{-1} \in A$. It is well-known that Cayley graphs are vertex-transitive.

Consider the group $G = \Phi(p) \times \mathbf{Z}_p$ defined in Section 2. Let $A = \{(h,0), (h^{-1},0), (1,1), (1,-1)\}$, where h is a generator of $\Phi(p)$. Then A is a Cayley subset of G. We shall denote the Cayley graph of G with respect to A by $X(p,h)$.

For example, when $p = 5$ and $h = 2$, the graph $X(p,h)$ is shown in Fig. 1(a). The same graph $X(p,h)$ can be drawn more compactly in a torus as shown in Fig. 1(b).

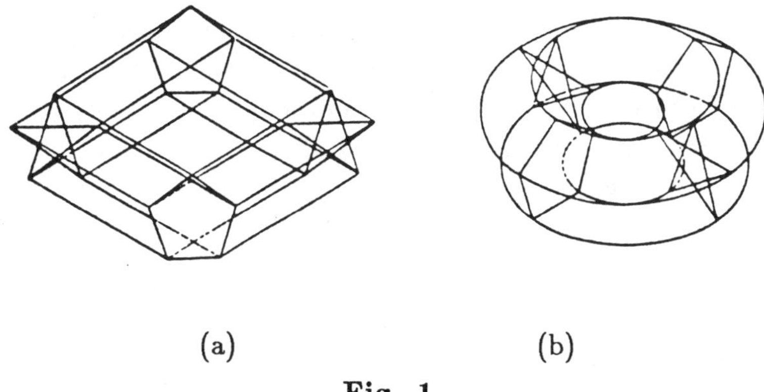

(a) (b)

Fig. 1

Evidently the graph $X(p,h)$ has order $(p-1)p$, degree 4, size $2p(p-1)$, and is connected.

Proposition 2. For each $h^r \in \Phi(p)$, let $X(h^r) = \{(h^r, a) \mid a \in \mathbf{Z}_p\}$. Then the subgraphs of $X(p,h)$ induced by $X(h^r)$ and $X(h^{\frac{p-1}{2}+r})$, $r = 1, 2, \ldots$, are isomorphic.

Proof. It follows from Lemma 1 that the mapping $\theta : X(h^r) \to X(h^{\frac{p-1}{2}+r})$ defined by $(h^r, a)\theta = (h^{\frac{p-1}{2}+r}, a)$, $a \in \mathbf{Z}_p$, is an isomorphism.

Remark 2. In fact the subgraph induced by $X(h^r)$ is a cycle of length p.

Paths between two distinct vertices (i,a) and (j,b) of $X(p,h)$ can be easily determined as follows: Let $(k,c) = (j,b)(i,a)^{-1}$. Since A generates G,

$$(k,c) = (\ell_1, d_1)(\ell_2, d_2) \ldots (\ell_m, d_m)$$

for some $(\ell_r, d_r) \in A$, $r = 1, 2, \ldots, m$. Then

$$P((i,a), (\ell_m, d_m)(i,a), (\ell_{m-1}, d_{m-1})(\ell_m, d_m)(i,a), \\ \ldots (\ell_1, d_1)(\ell_2, d_2) \ldots (\ell_m, d_m)(i,a) = (j,b))$$

is a path between (i,a) and (j,b) of length m. Every expression of (k,c) as a product of elements of A of minimum length m gives rise to a shortest path of length m between (i,a) and (j,b) and conversely. The determination of the shortest paths between two distinct vertices is important because messages from one vertex can reach the other one more quickly through these paths. That is why we investigate expressions of an element of G as a product of elements of A of minimum length. Let (i,a) and (j,b) be two distinct vertices, and let $(k,c) = (j,b)(i,a)^{-1}$. Assume that $h^r \equiv k$ and $h^t \equiv c \pmod{p}$, where $0 \le t \le \frac{p-1}{2}$. In case that $t \le r \le \frac{p-1}{2}$ or $\frac{p-1}{2} < r \le \frac{p-1}{2} + t$, then

$$P((i,a), (h,0)(i,a), \ldots, (h^t, 0)(i,a), (h^t, h^t)(i,a), (h^{t+1}, h^t)(i,a), \\ \ldots (h^r, h^t)(i,a))$$

or

$$P((i,a),(h^{p-2},0)(i,a),\ldots,(h^{\frac{p-1}{2}+t},0)(i,a),(h^{\frac{p-1}{2}+t},-h^{\frac{p-1}{2}+t})(i,a),$$
$$(h^{\frac{p-1}{2}+t-1},-h^{\frac{p-1}{2}+t})(i,a),\ldots,(h^{r},-h^{\frac{p-1}{2}+t})(i,a))$$

is a shortest path between (i,a) and (j,b). A similar shortest path between (i,a) and (j,b) can be found when $h^t \equiv -c \pmod{p}$. In case that $c=0$, then

$$P((i,a),(h,0)(i,a),...,(h^r,0)(i,a))$$

or

$$P((i,a),(h^{p-2},0)(i,a),(h^{p-3},0)(i,a),...,(h^r,0)(i,a))$$

is a shortest path between (i,a) and (j,b).

Table 1

k \ c	0	2	$2^2=4$	$2^3=8$	$2^4=5$	$2^5=10$	$2^6=9$	$2^7=7$	$2^8=3$	$2^9=6$	$2^{10}=1$
1	0	2	4(2)	3	3	1	2	4(2)	3	3	1
2	1	2	3	3	4	2	2	3	3	4	2
$2^2=4$	2	3	3	4(2)	4(2)	3	3	3	4(2)	4(2)	3
$2^3=8$	3	4	4	4	5(4)	4	4	4	4	5(4)	4
$2^4=5$	4	5	5	5	5	5	5	5	5	5	5
$2^5=10$	5	6	6	6	6	6	6	6	6	6	6
$2^6=9$	4	5	5	5	5	5	5	5	5	5	5
$2^7=7$	3	5	4	4	4	4	5	4	4	4	4
$2^8=3$	2	4	4	3	3	3	4	4	3	3	3
$2^9=6$	1	3	3	4(2)	2	2	3	3	4(2)	2	2

As we have not yet found a general algorithm to express an arbitrary (k, c) as a product of elements of A of minimum length, we are unable to provide the shortest path(s) between any two vertices (i, a) and (j, b). But for small p, it is not too difficult to determine the shortest path(s) between any two vertices of $X(p, h)$. Table 1 above shows the distance between the two vertices $(1, 0)$ and (k, c) of $X(p, h)$, or equivalently the minimum length in which (k, c) can be expressed as a product of elements of A, for $p = 11, h = 2$. The number within each bracket indicates the number of possible shortest paths. A distance without a bracketed number beside it means that the corresponding shortest path is unique.

For example, the distance between (1,0) and (7,9) is 5, and that between (1,0) and (4,3) is 4. We also find

$$(7,9) = (2^{-1}, 0)^3 (1, -1)^2$$

and

$$\begin{aligned}(4,3) &= (2,0)(1,1)(2,0)(1,1) \\ &= (1,1)(2,0)^2(1,-1).\end{aligned}$$

Proposition 3. The diameter of the graph $X(p, h)$ is $\frac{p+1}{2}$.

Proof. As the graph $X(p, h)$ is vertex-transitive, it is sufficient to show that the supremum of the distances between any vertices and the vertex $(1,0)$ is $\frac{p+1}{2}$. Equivalently we shall show that any non-identity element of G can be expressed as a product of elements of A of length at most $\frac{p+1}{2}$, and that there are non-identity elements whose expressions as a product of elements of A have a minimum length of $\frac{p+1}{2}$. Hence Proposition 3 is an immediate consequence of Proposition 1 and Remark 1. □

4. Some comparisons

For the purpose of comparison, let us consider the Euler group $\Phi(p)$ and the additive group $\langle \mathbf{Z}_p, + \rangle$ of integers modulo p. The direct product $G^* = \Phi(p) \times \mathbf{Z}_p$ is an abelian group. Assume that h is a generator of $\Phi(p)$ and again let $A = \{(h,0),(h^{-1},0),(1,1),((1,-1)\}$. Then the Cayley graph $X = (V(X), E(X))$ with

$$V(X) = G^*$$

and

$$E(X) = \{\{(i,a),(j,b)(i,a)\} | (i,a) \in G^* \quad \text{and} \quad (j,b) \in A\}$$

will have a diameter equal to $p-1$ which is almost double that of the graph $X(p,h)$. Fig. 2 shows the graph X for $p=5$, $h=2$.

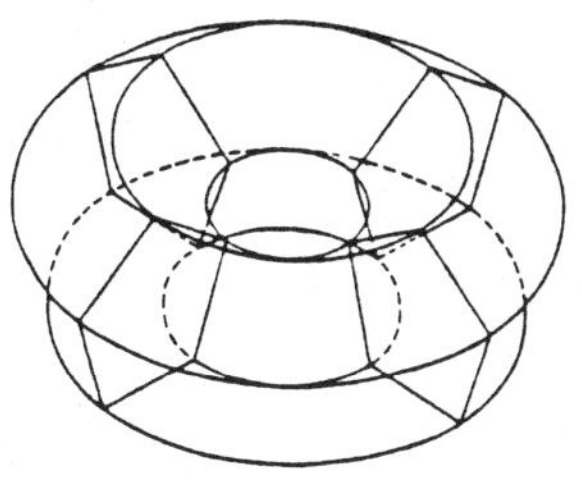

Fig. 2

Let $X_i = (V(X_i), E(X_i)), i = 1,2,...,n$ be n graphs. The product of the graphs $X_i, i = 1,2,...,n$, denoted by $X^* = X_1 \times X_2 \times ... \times X_n$, is a graph with vertex set $V(X^*) = V(X_1) \times V(X_2) \times ... \times V(X_n)$, in which $(x_1, x_2, ..., x_n)$ is adjacent to $(y_1, y_2, ..., y_n)$ if and only if $(x_i, y_i) \in E(X_i)$ for exactly one $i = 1,2,...,n$ and $x_j = y_j$ for all $j \neq i$. It is known that the product of vertex-transitive graphs is again vertex-transitive. The n-dimensional hypercube or n-cube used in Connection Machines is the product of n copies of the graph P_2, a path on 2 vertices. The n-cube has 4096 vertices, degree 12 and diameter 12, while the product $Y = X(5,2) \times X(5,2) \times X(5,2)$ of 3 copies of $X(5,2)$ has more vertices, namely 8000, the same degree 12 and a smaller diameter 9.

Remark 3. The method for the construction of the Cayley graph $X(p,h)$ can be extended to the case that a subgroup of $\Phi(p)$ is used, or p is not a prime. The extension often leads to Cayley graphs of relatively small diameters. For example, take $p = 9$. Then $H = \{1,4,7\}$ is a subgroup of the Euler group $\Phi(9)$, and $H \times \mathbf{Z}_9$ is a non-commutative subgroup of the group $\langle \Phi(9) \times \mathbf{Z}_9, * \rangle$. Take the subset $A = \{(4,0),(7,0),(1,1),(1,-1)\}$. Then the Cayley graph Y^* of the group $H \times \mathbf{Z}_9$ with respect to A has 27 vertices, degree 4 and diameter 3. We believe that among all Cayley graphs of degree 4 and diameter 3, the above Y^* has the greatest number of vertices. Figure 3 shows the above Cayley graph Y^*.

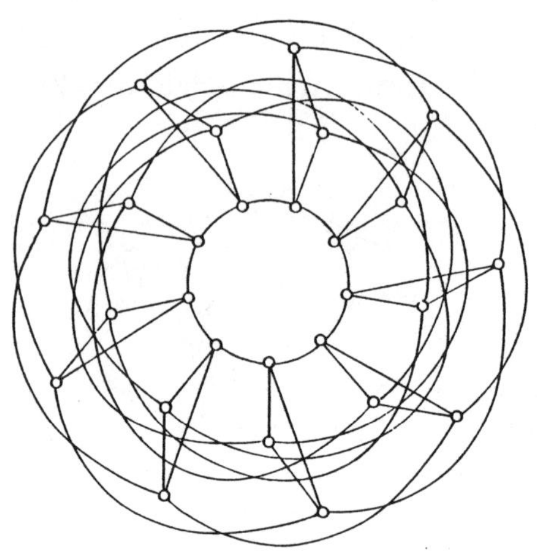

Fig. 3

References

[1] E.D. Bolker, *Elementary Number Theory* (W.A. Benjamin Inc. New York 1970).

[2] J.A. Bondy and U.S.R. Murty, *Graph Theory with Applications* (MacMillan, London 1976).

[3] W.D. Hillis, *The Connection Machine* (The MIT Press, Cambridge, Massachusetts, London, England 1986).

[4] Y. Saad & M.H. Schultz, *Topological Properties of Hypercubes* (Research Report YALEO/DCS/RR-389 (1985)).

[5] D.B. Skillicorn, Graph Problems in Network Design, *Ars Combinatoria* **21-A** (1986), 169-188.

[6] H.H. Teh and S.C. Shee, *Algebraic Theory of Graphs* (Lee Kong Chian Institute of Mathematics and Computer Science, Nangany University, Singapore 1976).

[7] A.Y. Wu, Embedding of Tree Networks into Hypercubes, *J. of Parallel and Distributed Computing*, **2** (1985), 238-249.

Department of Mathematics
National University of Singapore
Kent Ridge
Singapore 0511

A new characterization of the sporadic simple groups

Shi Wujie

There are many different characterizations to the attractive sporadic groups (see the bibliography of [1]). In this paper I continue the former discussions ([5 – 9]) and characterize uniformly the 26 sporadic simple groups using only the set $\pi_e(G)$ of the orders of elements in group G and $|G|$. We prove the following theorem.

Theorem. *Let G be a group and H a sporadic simple group. Then the necessary and sufficient condition that $G \cong H$ is:*

(1) $\pi_e(G) = \pi_e(H)$, *and*

(2) $|G| = |H|$.

The notation is taken from [1, 2]; in addition, we let $\pi(G)$ denote the set of prime factors of $|G|$, $|\pi(G)|$ the number of different prime factors of $|G|$, G_p the Sylow p-subgroup of G and $\mathbf{Z}_m$ the cyclic group of order m.

Lemma 1. *Let G be a finite group and H one of the following groups: M_{11}, M_{22}, M_{23}, M_{24}, J_1, HS or Co_2. Then $G \cong H$ if and only if $\pi_e(G) = \pi_e(H)$.*

Proof. See [5 – 8].

Lemma 2. *Let G be a group and H one of the following groups: M_{12}, J_2, J_3 or J_4. Then $G \cong H$ if and only if* (1) $\pi_e(G) = \pi_e(H)$, *and* (2) $|G| = |H|$.

Proof. See [6, 9].

Lemma 3. *Let G be a finite group and H a sporadic simple group. If*

$\pi_e(G) = \pi_e(H)$, then G is nonsolvable. Moreover G is not Frobenius and G has a normal series which contains a noncyclic simple factor group.

Proof. By Lemma 1 we can assume H is one of those groups listed in Table I.

1. The case $H \neq J_2$. If G is solvable we consider the $\{p,q,r\}$-Hall subgroup K of G, where p,q,r are the largest three different prime factors of $|G|$. Since $\pi_e(G) = \pi_e(H)$, $H \neq J_2$, K does not contain any element of order pq, pr or qr from Table I. So K is a solvable group all of whose elements are of prime power order. By [3] Theorem 1 we have $|\pi(K)| \leq 2$ and get a contradiction.

2. The case $H = J_2$. Let G be solvable and N the minimal normal subgroup of G. Then N is an elementary abelian group. If N is a 7-group, then G_3 is a group of order 3 as $9, 21 \notin \pi_e(G)$ and G_3 acts fixed-point-free on N. Similarly we derive that G_5 is a group of order 5 and G_2 is a cyclic group of order 8 or a generalized quaternion group of order 16. Considering the $\{2,3,5\}$-Hall subgroup C of G, $|C| = 2^3.3.5$ and $\pi_e(C) = \{1-6, 8, 10, 12, 15\}$ when G_2 is cyclic. Since G_2 is cyclic, C has a normal cyclic 2-complement T of order 15. It is contrary to $T = C_C(T)$ and

$$G_2 \cong C/T = N_C(T)/C_C(T) \cong \text{ a subgroup of } (\mathbf{Z}_2 \times \mathbf{Z}_4).$$

When G_2 is generalized quaternion, $|C| = 2^4.3.5$. If $O(C) = 1$, then C has a central element of order 2. It is contrary to $30 \notin \pi_e(G)$. If $O(C) \neq 1$, we can assume C has a normal subgroup L of order 3 or 5. Put $|L| = 3$. We have $|C_C(L)| = 2^3.3.5$ as $C/C_C(L) \cong$ a subgroup of $\mathbf{Z}_2$. Consider the factor group $S = C_C(L)/L$. $|S| = 2^3.5$. S has a normal subgroup of order 5 as $\pi_e(S) \subseteq \pi_e(C)$. Thus $C_C(L)$ has a normal cyclic subgroup of order 15. We get a contradiction similarly. This is the same for $|L| = 5$, so N is not a 7-group.

If $O_{7'}(G) \neq 1$, then $G_7 \cong \mathbf{Z}_7$ and $O_{7'}(G)$ is nilpotent ([2] Theorem 10.2.1) by Thompson's theorem. Considering $G/O_{7'}(G)$ we infer that $O_{7'}(G)$ contains elements of order 4 as $\mathrm{Aut}(\mathbf{Z}_7) \cong \mathbf{Z}_6$ and $8 \in \pi_e(G)$. It also follows that $5 \in \pi(O_{7'}(G))$ as $5 \in \pi_e(G)$. So G has elements of order 20, a contradiction. Therefore G is nonsolvable.

Table I

H	$\|H\|$	$\pi_e(H)$
M_{12}	$2^6.3^3.5.11$	1-6,8,10,11
J_2	$2^7.3^3.5^2.7$	1-8,10,12,15
J_3	$2^7.3^5.5.17.19$	1-6,8-10,12,15,17,19
J_4	$2^{21}.3^5.5.7.11^3.23.29.31.$ 37.43	1-8,10-12,14-16,20-24,28-31,33, 35,37,40,42-44,66
McL	$2^7.3^6.5^3.7.11$	1-12,14,15,30
Suz	$2^{13}.3^7.5^2.7.11.13$	1-15,18,20,21,24
Ru	$2^{14}.3^3.5^3.7.13.29$	1-8,10,12-16,20,24,26,29
He	$2^{10}.3^3.5^2.7^3.17$	1-8,10,12,14,15,17,21,28
Ly	$2^8.3^7.5^6.7.11.31.37.67$	1-12,14,15,18,20-22,24,25,28,30, 31,33,37,40,42,67
$O'N$	$2^9.3^4.5.7^3.11.19.31$	1-8,10-12,14-16,19,20,28,31
Co_1	$2^{21}.3^9.5^4.7^2.11.13.23$	1-16,18,20-24,26,28,30,33,35,36, 39,40,42,60
Co_3	$2^{10}.3^7.5^3.7.11.23$	1-12,14,15,18,20-24,30
Fi_{22}	$2^{17}.3^9.5^2.7.11.13$	1-16,18,20-22,24,30
Fi_{23}	$2^{18}.3^{13}.5^2.7.11.13.17.23$	1-18,20-24,26-28,30,35,36,39,42,60
Fi'_{24}	$2^{21}.3^{16}.5^2.7^3.11.13.17.23.$ 29	1-18,20-24,26-30,33,35,36,39,42, 45,60
Th	$2^{15}.3^{10}.5^3.7^2.13.19.31$	1-10,12-15,18-21,24,27,28,30,31, 36,39
HN	$2^{14}.3^6.5^6.7.11.19$	1-12,14,15,19-22,25,30,35,40
B	$2^{41}.3^{13}.5^6.7^2.11.13.17.19.$ 23.31.47	1-28,30-36,38-40,42,44,46-48,52, 55,56,60,66,70
M	$2^{46}.3^{20}.5^9.7^6.11^2.13^3.17.$ 19.23.29.31.41.47.59.71	1-36,38-42,44-48,50-52,54-57,59, 60,62,66,68-71,78,84,87,88,92-95, 104,105,110,119

If G is Frobenius, then the complement F is nonsolvable. Thus F has a normal subgroup F_0 with $|F : F_0| = 1$ or 2 such that $F_0 = SL(2,5) \times Z$, where every Sylow subgroup of Z is cyclic and $\pi(Z) \cap \{2,3,5\} = \emptyset$ ([4] Theorem 18.6). If $H \neq M_{12}$, then H contains elements of order 15 from Table I. Thus $15 \in \pi_e(F)$ as $\pi_e(G) = \pi_e(H)$. This is contrary to the fact that $SL(2,5)$ does not contain any element of order 15. If $H = M_{12}$ we have $Z = 1$ and the kernel is a 11-group. Since $22 \notin \pi_e(M_{12})$ the Sylow 2-subgroup F_2 of F is generalized quaternion. Moreover we can infer that $|F| = 2^4.3.5$ as $8 \in \pi_e(M_{12})$ and $16 \notin \pi_e(M_{12})$. Considering the centralizer of F_3 in F we derive that F has elements of order 12 by Sylow's theorem, a contradiction. Hence G is not a Frobenius group.

Thus [14] Theorem A and its corollary implies that G can only be one of the following types: (1) noncyclic simple; (2) an extension of a π_1-group by a simple group, where π_1 is the prime graph component containing 2 (see [14]); (3) simple by π_1-solvable; or (4) π_1 by simple by π_1. Whatever G is, G has a normal series

$$G \geq G_1 \geq N \geq 1,$$

where $\pi(G/G_1) \subseteq \pi_1$, $\pi(N) \subseteq \pi_1$ and $\bar{G}_1 = G_1/N$ is a noncyclic simple group. The lemma is proved.

Remark. We cannot weaken the conclusion of Lemma 3 into "H be a noncyclic simple group". For example, $\pi_e(L_3(3)) = \{1-4, 6, 8, 13\}$, and it is not difficult to make a solvable group G such that $\pi_e(G) = \pi_e(L_3(3))$.

Lemma 4. *Let G be a finite group in which every nonidentity element has prime order. Then $G \cong A_5$ or G has a normal Sylow p-subgroup and $|G| = p^m$ or qp^n, where $n = kb$, b is the exponent of p* (mod q).

Proof. See [10].

Now we discuss the noncyclic simple factor group $\bar{G}_1$ in Lemma 3 using the classification of the finite simple groups and the results of [14]. For this we first list the components of some sporadic groups (see [14]) in Table II.

Table II

H	π_1	π_2	π_3	π_4
Co_1	2,3,5,7,11,13	23		
Co_3	2,3,5,7,11	23		
Fi_{23}	2,3,5,7,11,13	17	23	
Fi'_{24}	2,3,5,7,11,13	17	23	29
McL	2,3,5,7	11		
He	2,3,5,7	17		
Ru	2,3,5,7,13	29		
Suz	2,3,5,7	11	13	
$O'N$	2,3,5,7	11	19	31
Fi_{22}	2,3,5,7,11	13		
HN	2,3,5,7,11	19		
Ly	2,3,5,7,11	31	37	67
Th	2,3,5,7,13	19	31	
B	2,3,5,7,11,13,17,19,23	31	47	
M	2,3,5,7,11,13,17,19,23,31,47	41	59	71

Lemma 5. *Let G be a group and H a sporadic group. If $\pi_e(G) = \pi_e(H)$ and $|G| = |H|$, then the simple factor group $\bar{G}_1$ in* Lemma 3 *is not a simple group of Lie type.*

Proof. By Lemma 1, 2 we need only prove that H is one of those groups listed in Table II.

1. The case $H = Co_1$, Co_3, Fi_{23} or Fi'_{24}. If H is one of the above groups, then $23 \in \pi(\bar{G}_1)$ by Table II. First we prove that $\bar{G}_1$ is not a simple group of Lie type in characteristic 2 by calculating its order.

(a) $\bar{G}_1$ is not of type $A_n(2^m)$, where

$$|A_n(2^m)| = 2^{mn(n+1)/2}.\Pi_{i=1}^{n}((2^m)^{i+1} - 1)/d,$$

where $d = (n+1, 2^m - 1)$. If $\bar{G}_1 \cong A_n(2^m)$, then $23 \in \pi(A_n(2^m))$. Since the exponent of 2 (mod 23) is 11, $2^{11} - 1 = 23.89$, it follows that

$89 \in \pi(\bar{G}_1)$ or $89|d$. If $89^e|d$ and $89^{e+1} \nmid d$, $e \geq 1$, then $n \geq 88^e$, we have $89^{8e}|\Pi_{i=1}^n((2^m)^{i+1}-1)$. It also follows that $89 \in \pi(\bar{G}_1)$, a contradiction. Similarly we can prove that $\bar{G}_1$ is not of type $B_n(q)$, $n>1$, $C_n(q)$, $n>2$, $D_n(q)$, $n>3$, $G_2(q)$, $F_4(q)$, $E_6(q)$, $E_7(q)$ or $E_8(q)$, where $q=2^m$.

(b) Since the congruence equations $2^k+1 \equiv 0 \pmod{23}$ and $2^{8k}+2^{4k}+1 \equiv 0 \pmod{23}$ all have no solution, so $\bar{G}_1 \not\cong {}^2A_n(q)$, $n>1$, ${}^2B_2(q)$, ${}^2D_n(q)$, $n>3$, ${}^3D_4(q)$, ${}^2F_4(q)$ or ${}^2E_6(q)$, $q=2^m$. It is obvious that $\bar{G}_1 \not\cong {}^2F_4(2)'$.

Because the exponent of 3 (mod 23) is 11, $3^{11}-1 = 2.23.3851$, $3851 \notin \pi(\bar{G}_1)$, the congruence equations $3^k+1 \equiv 0 \pmod{23}$ and $3^{8k}+3^{4k}+1 \equiv 0 \pmod{23}$ all have no solution we see that $\bar{G}_1$ is not a simple group of Lie type in characteristic 3. Similarly by $5^{11}+1 = 2.3.23.67.5281$, $7^{11}+1 = 2^3.23.10746341$, 10746341 is a prime, $5^{12} \nmid |H|$ and $7^{12} \nmid |H|$ we infer that $\bar{G}_1$ is not a simple group of Lie type in characteristic 5 or 7.

Now as $r^2 \nmid |H|$, $r = 11,13,17,23$ or 29, we see that $\bar{G}_1$ can only be $A_1(r)$ when $\bar{G}_1$ is of Lie type in characteristic r. But $23 \nmid |A_1(r)|$, $r = 11, 13, 17$ or 29, so $\bar{G}_1$ can only be $A_1(23)$. If $\bar{G}_1 \cong A_1(23)$, then as $23 \notin \pi(N)$ and G does not contain any element of order $r.23$, $r = 2,3,5,7,11,13$, we derive that G_{23} acts fixed-point-free on N and N is nilpotent. Since $7 \nmid |\mathrm{Aut}(A_1(23)|$ and $7 \notin \pi(C_{\bar{G}}(\bar{G}_1))$, $\bar{G}/C_{\bar{G}}(\bar{G}_1) \cong$ a subgroup of $\mathrm{Aut}(\bar{G}_1)$, where $\bar{G} = G/N$, we have $7 \in \pi(N)$. Thus we can assume that N is a 7-group. Now $G_{23}N$ is a solvable group in which every nonidentity element has prime order as $7^2, 7.23, 23^2 \notin \pi_e(G_{23}N) \subseteq \pi_e(H)$. From Lemma 4 we have $|G_{23}N| = 23.7^n, n = 22k$. It is contrary to $7^{22} \nmid |H|$. Therefore $\bar{G}_1$ is not a simple group of Lie type in characteristic r, $r = 11, 13, 17, 23, 29$.

Next we prove the results of this lemma using immediately the list of orders of simples groups of [1].

2. The case $H = McL$, He, Ru, Suz or $O'N$ (order $< 10^{12}$). If $\bar{G}_1$ is of Lie type, then as $|\bar{G}_1| \mid |G|$, $|G| = |H|$ we have $|\bar{G}_1| < 10^{12}$. Moreover, $11 \in \pi(\bar{G}_1)$ for $H = McL$, $17 \in \pi(\bar{G}_1)$ for $H = He$, $29 \in \pi(\bar{G}_1)$ for $H = Ru$, $11,13 \in \pi(\bar{G}_1)$ for $H = Suz$ and $11,19,31 \in \pi(\bar{G}_1)$ for $H = O'N$ by Table II. From the list of order of simple groups in [1] we compare the orders and their powers of prime factors, and find $\bar{G}_1$ can only be $A_1(q)$ for $H = McL$, Ru, Suz or $O'N$; $C_2(4)$ or $A_1(q)$ for $H = He$. By further calculation $\bar{G}_1$ can only be $A_1(11)$ for $H = McL$, $A_1(29)$ for $H = Ru$; $C_2(4)$, $A_1(17)$ or $A_1(16)$ for $H = He$; and $\bar{G}_1$ does not exist for $H = Suz$ or $O'N$. If $\bar{G}_1 = A_1(11)$ for $H = McL$ then as $11 \notin \pi(N)$ and G does not contain any element of

order $r.11$, $r = 2, 3, 5$ or 7, so N is nilpotent. Moreover we can assume N is a 7-group as $7 \nmid |\mathrm{Aut}(A_1(11))|$ and $7.11 \notin \pi_e(G)$. Considering $G_{11}N$ we have $|G_{11}N| = 11.7^n, n = 10k$, a contradiction. We can also prove the same conclusion for $H =$ He or Ru. Hence $\bar{G}_1$ is not a simple group of Lie type in this case.

3. $H = Fi_{22}$ or HN (order $< 10^{16}$). If $\bar{G}_1$ is of Lie type, since $13 \in \pi(\bar{G}_1)$ so $\bar{G}_1$ can only be $A_1(13)(2^2.3.7.13)$, $A_1(25)(2^3.3.5^2.13)$, $A_1(27)(2^2.3^3.7.13)$, $A_2(3)(2^4.3^3.13)$, $A_2(9)(2^7.3^6.5.7.13)$, ${}^2A_2(4)(2^6.3.5^2.13)$, ${}^2B_2(8)(2^6.5.7.13)$,$G_2(3)(2^6.3^6.7.13)$,$A_3(3)(2^7.3^6.5.13)$,${}^2F_4(2)'(2^{11}.3^3.5^2.13)$, ${}^3D_4(2)(2^{12}.3^4.7^2.13)$, $G_2(4)(2^{12}.3^3.5^2.7.13)$, $C_3(3)(2^9.3^9.5.7.13)$ or $B_3(3)(2^9.3^9.5.7.13)$ for $H = Fi_{22}$; and since $19 \in \pi(\bar{G}_1)$, $\bar{G}_1$ can only be $A_1(19)(2^2.3^2.5.19)$, ${}^2A_2(8)(2^9.3^4.7.19)$ for $H = HN$ by comparing their orders ([1]) and a calculation. For any above-cited group we have $11 \notin \pi(\bar{G}_1)$ and $11 \notin \pi(\mathrm{Out}(\bar{G}_1))$, so we can assume N is a 11-group. Considering $G_{13}N$ and $G_{19}N$ we get a contradiction as above.

4. The case $H = Ly$ or Th (order $< 10^{20}$). Since $67 \in \pi(\bar{G}_1)$, $\bar{G}_1$ can only be $A_1(67)$ for $H = Ly$ by the comparison of orders and calculation. But $37 \notin \pi(A_1(67))$, thus $\bar{G}_1$ is not of Lie type in this case. Similarly since 19,31 $\in \pi(\bar{G}_1)$ and $19^2 \nmid |\bar{G}_1|$, $31^2 \nmid |\bar{G}_1|$ we can derive that $\bar{G}_1$ is not of Lie type for $H = Th$.

5. The case $H = B$ of M (order $> 10^{25}$). In this case we cannot use the list of orders of simple groups of [1]. However we can use the conclusion of 1. of this lemma. For this we first prove $23 \in \pi(\bar{G}_1)$.

From Lemma 3 G has a normal series $G \geq G_1 \geq N \geq 1$, where $\pi(G/G_1) \subseteq \pi_1$, $\pi(N) \subseteq \pi_1$. If $23 \in \pi(N)$ then we can assume N is a 23-group as $\pi_e(G) = \pi_e(H)$ and 47 ($H = B$), 71 ($H = M$) $\notin \pi(N)$. Considering $G_{47}N$ and $G_{71}N$ we derive a contradiction. If $23 \notin \pi(\bar{G}_1)$, then $23 \in \pi(G/G_1)$. As $\bar{G}/C_{\bar{G}}(\bar{G}_1) \cong$ a subgroup of $\mathrm{Aut}(\bar{G}_1)$, where $\bar{G} = G/N$, and $23 \notin \pi(C_{\bar{G}}(\bar{G}_1))$ ($|\pi(\bar{G}_1)| \geq 3$, $\pi_e(G) = \pi_e(H)$), we have $23 \in \pi(\mathrm{Out}(\bar{G}_1))$. Since $\bar{G}_1$ is of Lie type, $|\mathrm{Out}(\bar{G}_1)|$ is given by [1; xvi Table 5], so $\bar{G}_1$ can only be $A_n(q)$, ${}^2A_n(q)$ $(23|n+1)$ or any group of Lie type with $q = p^{23}$. But the orders of these groups are all contrary to $|G| = |H|$ and $31 \in \pi_e(\bar{G}_1)$ ($H = B$) or $71 \in \pi_e(\bar{G}_1)$ ($H = M$), hence $23 \in \pi(\bar{G}_1)$. Thus $\bar{G}_1$ is not of Lie type in characteristic r, $r = 2$, 3, 5 or 7, by the proof of 1. of this lemma.

For $H = B$, as $r^2 \nmid |B|$, $r = 11$, 13, 17, 19, 23, 31 or 47, we see that

$\bar{G}_1$ can only be $A_1(r)$. But $47 \nmid |A_1(r)|$, $r = 11, 13, 17, 19, 23$ or 31; and $31 \nmid |A_1(47)|$, so $\bar{G}_1$ is not of Lie type. For $H = M$, as $11^3 \nmid |M|$, $13^4 \nmid |M|$ and $r^2 \nmid |M|$, $r = 17, 19, 23, 29, 31, 41, 47, 59$ or 71, $\bar{G}_1$ can only be $A_1(r)$, $r = 11, 13, 17, 19, 23, 29, 31, 41, 47, 59$ or 71, $A_1(11^2)$, $A_1(13^2)$, $A_1(13^3)$, $A_2(13)$ or ${}^2A_2(13)$. But 71 does not divide exactly the orders of these groups except $A_1(71)$ and $59 \nmid |A_1(71)|$, so $\bar{G}_1$ is not of Lie type. This lemma is proved.

Proof of Theorem. First we prove $\bar{G}_1$ is not an alternating group. Comparing H with the order of alternating groups we have $\bar{G}_1$ can only be A_{13} for $H = Suz$ or Fi_{22}. But $35 \in \pi_e(A_{13})$ and $35 \notin \pi_e(H)$, $H = Suz$ or Fi_{22}. Thus $\bar{G}_1$ can only be a sporadic group. For $H = McL$, since $|G_1| \mid |H|$ and $11 \in \pi_e(\bar{G}_1)$, $\bar{G}_1$ can only be M_{11}, M_{12}, M_{22}, J_2 or McL. If $\bar{G}_1 \ncong McL$, then as $5^3 \nmid |\mathrm{Aut}(\bar{G}_1)|$ we can assume that N is a 5-group as above. Considering $G_{11}N$ we get a contradiction by Lemma 4. Hence $\bar{G}_1 \cong McL$. It follows that $G \cong McL$ as $|G| = |H|$. Similarly $\bar{G}_1$ can only be He, Ru, Suz, $O'N$, Ly, Th, Fi_{23}, Fi_{24}, B or M for $H = He$, Ru, Suz, $O'N$, Ly, Th, Fi_{23}, Fi_{24}, B or M respectively. So the theorem holds for these groups.

For $H = Co_3$, $\bar{G}_1$ can only be M_{23}, M_{24} or Co_3. Considering $G_{23}N$, N is a 5-group and $\bar{G}_1$ can only be Co_3. For $H = Fi_{22}$, $\bar{G}_1$ can only be Suz or Fi_{22}. Considering $G_{13}N$, N is a 3-group and $\bar{G}_1$ can only be Fi_{22}. For $H = HN$, $\bar{G}_1$ can only be J_1 or HN. Considering $G_{19}N$, N is a 5-group and $\bar{G}_1$ can only be HN. For $H = Co_1$, $\bar{G}_1$ can only be M_{23}, M_{24}, Co_3, Co_2 or Co_1. Considering $G_{23}N$, N is a 7-group and $\bar{G}_1$ can only be Co_1. Thus $G \cong H$ for all cases. The theorem is proved.

In Lemma 1 we have characterized M_{11}, M_{22}, M_{23}, M_{24}, J_1, HS and Co_2 using only the condition of $\pi_e(G)$. How many sporadic simple groups can be characterized using only the set $\pi_e(G)$? This is a unsolved problem.

Since we have proved the same conclusion for the groups of alternating type and some Lie type ([11,12]), so we can put forward the following conjecture.

Conjecture. Let G be a group and H a finite simple group. Then $G \cong H$ if and only if (a) $\pi_e(G) = \pi_e(H)$, and (b) $|G| = |H|$.

A problem related to this conjecture is the following open problem put forward by J. G. Thomson ([13]). For each finite group G and each integer $d \geq 1$, let $G(d) = \{x \in G | x^d = 1\}$.

Definition. G_1 and G_2 are of the same order type iff $|G_1(d)| = |G_2(d)|$, $d = 1, 2, \ldots$.

Open Problem. Suppose G_1, G_2 are finite groups of the same order type. Suppose also that G_1 is solvable. Is it true that G_2 is also necessarily solvable?

References

[1] J.H. Conway, R.T. Curtis, S.T. Norton, R.A. Parker, and R.A. Wilson, *Atlas of finite groups*, Oxford Univ. Press, London, 1985.

[2] D. Gorenstein, *Finite groups* , Harper and Row, London/New York, 1968.

[3] G. Higman, Finite groups in which every element has prime power order, *J. London Math. Soc.*, **32** (1957), 335-342.

[4] D.S. Passman, *Permutation groups*, W.A. Benjamin, New York, 1968.

[5] Shi Wujie, A characteristic property of J_1 and $PSL_2(2^n)$, *Adv. Math.* (Chinese), **16** (1987), 397-401.

[6] Shi Wujie, A characteristic property of the Mathieu groups, to be published in *Chin. Ann. of Math.* (Chinese).

[7] Shi Wujie, A characterization of the Higman-Sims simple group, to be published in *Houston J. of Math.*.

[8] Shi Wujie, A characterization of the Conway simple group Co_2, to be published in *J. of Math.* (Chinese).

[9] Shi Wujie, Simple groups of order $2^a.3^b.5^c.7^d$ and Janko's simple groups, *J. Southwest-China Teachers Univ.* (Ser. B) (Chinese), **4** (1987), 1-8.

[10] Shi Wujie, and Yang Wenze, A new characterization of A_5 and the finite group in which every nonidentity element has prime order, *J.*

Southwest-China Teachers College (Ser. B) (Chinese), **1** (1984), 36-40.

[11] Shi Wujie and Bi Jianxing, A new characterization of the alternating groups, to appear.

[12] Shi Wujie, A new characterization of some simple groups of Lie type, to appear in *Contemporary Maths.*.

[13] J.G. Thomson, A letter to Shi Wujie.

[14] J. S. Williams, Prime graph components of finite groups, *J. Algebra*, **69** (1981), 487-513.

Department of Mathematics
Southwest-China Teachers University
Chongqing, CHINA

Equicentralizer subgroups of sporadic simple groups

Wang Efang

A subgroup H of a group G is called an equicentralizer subgroup (EC-subgroup) of G if the centralizers of its non-identity elements are all equal. We discuss the equicentralizer subgroups of sporadic simple groups. Using the orders of the centralizers of elements and the structures of the maximal subgroups, we prove that the equicentralizer subgroups of a sporadic simple group should be elementary abelian p-groups and determine all the equicentralizer subgroups of the Mathieu groups and some other sporadic simple groups.

In order to prove our theorems we use the results of [3] and the following propositions about the properties of equicentralizer subgroups of finite groups.

Proposition 1. *Let a be an element of prime order p of a finite group G. Assume that $q\,(\neq p)$ is a prime divisor of the order of $\mathbf{C}(a)$, the centralizer of a. If G has an equicentralizer subgroup of order p^m containing a, then there is an element b of order q in G, such that p^m divides the order of $\mathbf{C}(b)$.*

Let p, q be two distinct prime numbers, a an element of order p in G. If q divides the order of $\mathbf{C}(a)$ and p^2 does not divide the order of the centralizer of any element of order q in G, then by Proposition 1, G cannot have an EC-subgroup of order p^2 containing a. In this case, we say a satisfies $N(p,q;a)$. If any element a of order p in G satisfies $N(p,q;a)$, then we say G satisfies $N(p,q)$. In this situation, G cannot have any EC-subgroup of order p^2.

Proposition 2. *If H is an equicentralizer subgroup of order $p^m\,(m \geq 1)$ of G, p prime, then either H is a Sylow subgroup of G or p^{m+1} divides the order of the centralizer of nonidentity elements of H.*

Let a be an element of prime order p in G. If p^3 divides the order of G, but p^3 does not divide the order of $\mathrm{C}(a)$, then by Proposition 2, G cannot have an EC-subgroup of order p^2 containing a. In this case, we say a satisfies $S(p;a)$. If any element a of order p in G satisfies $S(p;a)$, then we say G satisfies $S(p)$. In this situation, G does not contain any EC-subgroup of order p^2.

Proposition 3. *Let a be an element of prime order p in G. If $\mathrm{C}(a)$ is a maximal subgroup of G and the index $|\mathrm{C}(a)|$ is a composite number, then G cannot have an equicentralizer subgroup of order p^2 which is generated by a and one of its conjugate elements.*

Proposition 4. *Let G be a finite simple group and a an element of prime order p. If the centralizer of a in G is a maximal subgroup of G, then G cannot have an equicentralizer subgroup of order p^2 which is generated by a and a conjugate element of a.*

Now we are going to prove our main results. A detailed discussion about the structure of sporadic simple groups can be found in [1].

Theorem 1. (I) *All the nontrivial equicentralizer subgorups of sporadic simple groups are elementary abelian p-groups, which are generated by conjugate elements.*

(II) *For all non-cyclic equicentralizer p-subgroups of sporadic simple groups, $p = 2$ or 3.*

Proof. (I) The equicentralizer subgroups are abelian, so they are direct products of cyclic p-subgroups and their cyclic direct facotrs are also EC-subgroups of G. It follows from the character tables of sporadic simple groups that for any sporadic group G, the following holds :

(1) For two distinct prime divisors p and q of $|G|$, $G \neq J_1$, the centralizers of any element of order p and any element of order q have distinct orders. In J_1, the classes of order 3 and 5 contain elements with the same centralizer order 30, but the classes of order 15 have centralizer order 15.

(2) For those primes p with $p^2 \mid |G|$, the centralizers of any element of order p and any element of order p^2 (if it exists) have distinct orders.

(3) For any prime p with $p^2 \mid |G|$, $G \neq J_2$, He or HN, the centralizer orders of non-conjugate elements or order p in G are different from each other. In J_2, there are non-conjugate elements of order 5 whose centralizers have the same order. But J_2 satisfies $N(5,2)$. In He, there are non-conjugate elements of order 7 whose centralizers have the same order. But any element a of order 7 satisfies $N(7,2;a)$ or $N(7,3;a)$. In HN, there are two different conjugate classes of order 5 whose elements have centralizers of order 15000. But every element a in these two classes satisfies $N(5,3;a)$. The orders of the centralizers of other elements of order 5 are not equal to 15000 and are different from each other.

Hence the equicentralizer subgroups of sporadic simple groups must be elementary abelian p-groups generated by conjugate elements.

(II) Any subgroup of an EC-subgroup of G in an EC-subgroup of G and any conjugate subgroup of an EC-subgroup is also an EC-subgroup of G. Hence we need only to prove that for any sporadic simple group G and any element a of prime order $p\,(\geq 5)$ in G, G cannot contain an EC-subgroup $H \ni a$ of order p^2. In the following, we discuss different cases according to different prime divisors of the orders of sporadic groups. We list all primes p satisfying $p^2 \mid |G|$, $p \geq 5$, and prove that G cannot have an EC-subgroup with order p^2 which is generated by conjugate elements.

(1) $G = M_{11},\ M_{12},\ M_{22},\ M_{23},\ M_{24},\ J_1,\ J_3$.

For any prime $p \geq 5$, we have $p^2 \nmid |G|$.

(2) $G = J_2,\ Suz,\ HS,\ McL,\ Co_2,\ Co_3,\ Fi_{22},\ Fi_{23},\ HN,\ Ru,\ Ly$; $p = 5$.

(i) J_2, Suz, Co_2, Co_3, Fi_{22} and Ru satisfy $N(5,2)$.

(ii) Fi_{23} satisfies $N(5,7)$.

(iii) HS has three conjugate classes of elements of order 5. Denote their representative elements by a_5, b_5, c_5 respectively. We see from the character table of HS that they satisfy $N(5,2;a_5)$, $N(5,2;b_5)$ and $S(5;c_5)$ respectively. McL has two classes of conjugate elements of order 5, their representative elements a_5 and b_5 satisfy $N(5,2;a_5)$ and $S(5;c_5)$ respectively.

(iv) HN has five conjugate classes of elements of order 5. Denote their representative elements by a_5, b_5, c_5, d_5 and e_5. The orders of their centralizers are :

$$|\mathbf{C}(a_5)| = 630000,$$
$$|\mathbf{C}(b_5)| = 500000,$$
$$|\mathbf{C}(c_5)| = |\mathbf{C}(d_5)| = 15000,$$
$$|\mathbf{C}(e_5)| = 2500.$$

The elements a_5 , c_5 and d_5 satisfy $N(5,3;a_5)$, $N(5,3;c_5)$ and $N(5,3;d_5)$ respectively. Hence HS cannot have an EC-subgroup of order 5^2 containing a_5, c_5, or d_5.

If HN has an EC-subgroup H generated by b_5 and one of its conjugate elements, then nonidentity elements of H are all conjugates. Hence $|\mathbf{N}_G(H) : \mathbf{C}_G(H)| = |\mathbf{N}_G(H) : \mathbf{C}(b_5)| = 5^2 - 1 = 24$ and $|\mathbf{N}_G(H)| = 24 \times 500000$. But HN has no subgroup of this order. So HN cannot have such an EC-subgroup. If HN contains an EC-subgroup H generated by e_5 and one of its conjugate elements, then $|N_G(H)| = 24 \times 2500 = 60000$. $N_G(H)$ is contained in the maximal subgroup K of HN, where $K \cong (D_{10} \times U_3(5))^{\cdot}2$, and H is also an EC-subgroup of K. Since any element x of order 5 in $U_3(5)$ satisfies $N(5,2;x)$ or $S(5;x)$, $U_3(5)$ and thereby HN does not contain an EC-subgroup of order 5^2 containing e_5.

(v) Ly has two conjugate classes of elements of order 5. The orders of their centralisers are

$$|\mathbf{C}(a_5)| = 2250000,$$
$$|\mathbf{C}(b_5)| = 3750.$$

Moreover $\mathbf{C}(a_5) < \mathbf{N}(\langle a_5\rangle) \cong 5_+^{1+4} : 4S_6$ and $|\mathbf{N}(\langle a_5\rangle) : \mathbf{C}(a_5)| = 4$. In S_6, an element of order 5 cannot commute with any element of order 3, but the center of 5_+^{1+4} is of order 5. Hence $\mathbf{N}(\langle a_5\rangle)$, thereby Ly cannot contain an EC-subgroup of order 5^2 containing a_5. There are three possible maximal subgroups K of Ly containing $\mathbf{C}(b_5)$:

(a) $K \cong G_2(5)$.

In this case, $\mathbf{C}(b_5)$ is contained in a maximal subgroup K_1 of K. According to the order of $\mathbf{C}(b_5)$, we have either $K_1 \cong 5_+^{1+4} : GL_2(5)$ or $K_1 \cong 5^{2+3} : GL_2(5)$.

(a.1) $K_1 \cong 5_+^{1+4} : GL_2(5)$.

In $GL_2(5)$, an element of order 5 cannot commute with any element of order 3, so b_5 is contained in 5_+^{1+4}. But any two non-conjugate elements of order 5 have different centralizers. Hence there is no EC-subgroup of order 5^2 containing b_5.

(a.2) $K_1 \cong 5^{2+3} : GL_2(5)$.

K is the normalizer of the elementary abelian subgroup of order 5^2 generated by two conjugate elements of order 5, and these two elements are not conjugate to b_5. Hence there is no EC-subgroup of order 5^2 containing b_5.

(b) $K \cong 5^3 : L_3(5)$.

K is the normalizer of an elementary abelian subgroup $\langle x, y, z\rangle$ of order 5^3, where x, y, z are not conjugate with b_5. An element of order 5 in $L_3(5)$ cannot commute with any element of order 3, so $\mathbf{C}(b_5)$ cannot be contained in $5^3 : L_3(5)$.

(c) $K \cong 5_+^{1+4} : 4S_6$.

There are elements of order 3 in $\mathbf{C}(b_5)$. Just as mentioned above, there is no EC-subgroup of order 5^2 containing b_5.

(3) $G = O'N$; $p = 7$.

Any element a of order 7 in G satisfies $N(7,2;a)$ or $S(7;a)$.

(4) $G = He,\ Fi'_{24},\ Th,\ Co_1,\ B$; $p = 5, 7$.

(i) He satisfies $N(5,2)$. Any element a of order 7 in He satisfies $N(7,2;a)$ or $N(7,3;a)$.

(ii) Fi'_{24} satisfies $N(5,7)$ and $N(7,2)$.

(iii) Th satisfies $N(5,2)$ and $N(7,2)$.

(iv) $G = Co_1$.

G satisfies $N(7,2)$, hence G has no EC-subgroups of order 7^2. There are three conjugate classes of elements of order 5. Denote their representative elements by a_5, b_5 and c_5. The orders of their centralizers are

$$|\mathbf{C}(a_5)| = 3024000,$$
$$|\mathbf{C}(b_5)| = 36000,$$
$$|\mathbf{C}(c_5)| = 15000.$$

The element a_5 satifies $N(5,7;a_5)$, hence there is no EC-subgroup of order 5^2 containing a_5; and $\mathbf{C}(b_5) \leq \mathbf{N}(\langle b_5\rangle) \cong (D_{10} \times (A_5 \times A_5).2).2$. Every element of D_{10} commutes with any element of $A_5 \times A_5$, and the center of A_5 is $\{e\}$. Hence $\mathbf{N}(\langle b_5\rangle)$ cannot have an EC-subgroup of order 5^2, and therefore G cannot have an EC-subgroup of order 5^2 containing b_5. We have $\mathbf{C}(c_5) \leq \mathbf{N}(\langle c_5\rangle) \cong 5_+^{1+2} : GL_2(5)$, $|\mathbf{N}(\langle c_5\rangle) : \mathbf{C}(c_5)| = 4$. $\mathbf{C}(c_5)$ contains a Sylow 5-subgroup of $\mathbf{N}(\langle c_5\rangle)$. Every element of order 5 in $GL_2(5)$ does not commute with any element of order 3. Hence $\mathbf{N}(\langle c_5\rangle)$ and thereby G cannot have an EC-subgroup of order 5^2 containing c_5.

(v) $G = B$.

G satisfies $N(7,5)$, so G does not contain an EC-subgroup of order 7^2. G has two conjugate classes of elements of order 5, the orders of their centralizers are

$$|\mathbf{C}(a_5)| = 44352000,$$
$$|\mathbf{C}(b_5)| = 6000000.$$

The element a_5 satifies $N(5,7;a_5)$. For b_5, we have $\mathbf{C}(b_5) < \mathbf{N}(b_5) \cong 5_+^{1+4} : 2^{1+4}.A_5.4$, $|\mathbf{N}(\langle b_5\rangle) : \mathbf{C}(b_5)| = 4$, as in (iv). Hence G does not contain an EC-subgroup of order 5^2.

(5) $G = J_4$; $p = 11$.

J_4 satisfies $N(11,2)$.

(6) $G = M$; $p = 5,7,11,13$.

G satisfies $N(7,5)$, $N(11,5)$ and $N(13,5)$. G has two conjugate classes of elements of order 5, the orders of their centralizers are

$$|\mathbf{C}(a_5)| = 136515456 \cdot 10^7,$$
$$|\mathbf{C}(b_5)| = 945 \cdot 10^8.$$

We have $\mathbf{C}(a_5) < \mathbf{N}(\langle a_5 \rangle) = (D_{10} \times HN)^{\cdot}2$. The orders of the centralizers of elements of order 5 in HN are less than $\frac{|\mathbf{C}(a_5)|}{20}$, and $5^2 \nmid |D_{10}|$, thus G does not have an EC-subgroup of order 5^2 containing a_5. For element b_5, $\mathbf{C}(b_5) < \mathbf{N}(\langle b_5 \rangle) = K \cong 5_+^{1+6} : 4J_2{}^{\cdot}2$, $|K : \mathbf{C}(b_5))| = 4$. In J_2, for any element x of order 5, $|J_2 : \mathbf{C}_J(x)| > 4$. The group 5_+^{1+6} does not have an EC-subgroup of order 5^2. Hence K and thereby G does not have an EC-subgroup of order 5^2 containing b_5.

Theorem 2. *Any nontrivial equicentralizer subgroup of the Mathieu groups is a cyclic group of prime order.*

Proof.

(1) M_{11} with $|M_{11}| = 2^4 \cdot 3^2 \cdot 5 \cdot 11$.

M_{11} satisfies $N(2,3)$ and $N(3,2)$.

(2) M_{12} with $|M_{12}| = 2^6 \cdot 3^2 \cdot 5 \cdot 11$.

M_{12} satisfies $N(3,2)$. M_{12} has two conjugate classes of elements of order 2, the orders of their centralizers are

$$|\mathbf{C}(a_2)| = 2^5 \cdot 3 \cdot 5,$$
$$|\mathbf{C}(b_2)| = 2^6 \cdot 3 = 192.$$

The element a_2 satisfies $N(2,5;a_2)$. All the subgroups of order 192 of M_{12} are maximal subgroups of M_{12}, hence by Proposition 4, M_{12} cannot have an EC-subgroup of order 2^2 which is generated by b_2 and one of its conjugate elements.

(3) $G = M_{22}$ with $|M_{22}| = 2^7 \cdot 3^2 \cdot 5 \cdot 7 \cdot 11$.

G has only one conjugate class of elements of order 2 : $|\mathbf{C}(a_2)| = 2^7 \cdot 3$ and only one conjugate class of elements of order 3 : $|\mathbf{C}(a_3)| = 2^2 \cdot 3^2$. Hence G satisifes $N(3,2)$ and hence contains no EC-subgroups of order 3^2. The centralizer $\mathbf{C}(a_2)$ is contained in a maximal subgroup K_1 of order 1920 or in a maximal subgroup K_2 of order 5760. If $\mathbf{C}(a_2) < K_1 = 2^4 : S_5$, then $|K_1 : \mathbf{C}(a_2)| = 5$. As K_1 does not contain a normal subgroup with index 5, $\mathbf{N}_G(\mathbf{C}(a_2)) = \mathbf{C}(a_2)$. If $\mathbf{C}(a_2) < K_2 \cong 2^4 : A_6$, then $|K_2 : \mathbf{C}(a_2)| = 15$. Since the subgroups of order 24 of A_6 are maximal and A_6 is simple, so

$\mathbf{N}_G(\mathbf{C}(a_2)) = \mathbf{C}(a_2)$. Hence the number of conjugate elements of a_2 is equal to the number of conjugate subgroups of $\mathbf{C}(a_2)$. Different elements of order 2 in G have different centralizers. Thus G does not have an EC-subgroup of order 2^2.

(4) M_{23} with $|M_{23}| = 2^7 \cdot 3^2 \cdot 5 \cdot 7 \cdot 11 \cdot 23$.
M_{23} satisfies $N(2,7)$ and $N(3,5)$.

(5) $G = M_{24}$ with $|M_{24}| = 2^{10} \cdot 3^3 \cdot 5 \cdot 7 \cdot 11 \cdot 23$.

G satisfies $N(3,2)$. G has two conjugate classes of elements of order 2, the orders of their centralizers are

$$|\mathbf{C}(a_2)| = 2^{10} \cdot 3 \cdot 7,$$
$$|\mathbf{C}(b_2)| = 2^9 \cdot 3 \cdot 5.$$

The element a_2 satisfies $N(2,7;a_2)$. There are two possible maximal subgroups containing $\mathbf{C}(b_2)$: the maximal subgroup K_1 of order 322560 or the maximal subgroup K_2 of order 138240.

(i) $\mathbf{C}(b_2) < K_1 \cong 2^4 : A_8$, $|K_1 : \mathbf{C}(b_2)| = 42$.

The orders of the centralizers of elements of order 2 in A_8 are $2^5 \cdot 3$ or $2^6 \cdot 3$. Since $2^5 \cdot 3 \cdot 2^4$ and $2^6 \cdot 3 \cdot 2^4$ are less than the order of $\mathbf{C}(b_2)$, so $b_2 \in 2^4$, $\mathbf{C}(b_2) = 2^4 : B$, $B < A_8$, $|A_8 : B| = 42$. But A_8 has no subgroups of index 42. Hence $\mathbf{C}(b_2)$ cannot be contained in K_1.

(ii) $\mathbf{C}(b_2) < K_2 \cong 2^6 : 3^{\cdot} S_6$, $|K_2 : \mathbf{C}(b_2)| = 18$.

An element of order 2 in $3^{\cdot} S_6$ cannot commute with any element of order 5, so $b_2 \in 2^6$, $\mathbf{C}(b_2) \cong 2^6 : B$, $B < 3^{\cdot} S_6$ and $|3^{\cdot} S_6 : B| = 18$. Since S_6 has no subgroups of index 18 and the subgroups of S_6 with index 6 are isomorphic to S_5, we can assume that $\mathbf{C}(b_2) \cong 2^6 : S_5$. If G has an EC-subgroup $\langle b_2, c_2\rangle$ of order 2^2, then $\mathbf{C}(c_2) = \mathbf{C}(b_2) < K_2$, and $c_2 \in 2^6$. Therefore $K_2 \geq \mathbf{N}_G(\mathbf{C}(b_2)) > \mathbf{C}(b_2)$ and $\mathbf{N}_G(\mathbf{C}(b_2)) \cong 2^6 : 3 . S_5$. The group $3 . S_5$ has a subgroup $\langle x\rangle$ of order 3. Since $2^6 : S_5 \lhd 2^6 : 3 . S_5$, x commutes with every element of S_5. Take an element y of order 6 in S_5. We have $\mathbf{C}(y) \geq \langle y, x, b_2, c_2\rangle$, so $|\mathbf{C}(y)| = 72$. But the order of the centralizer of any element of order 6 is equal to 24. Hence M_{24} cannot have an EC-subgroup of order 2^2 containing b^2.

Since any prime order cyclic subgroup of G is an EC-subgroup of G, all the EC-subgroups of the five Mathieu groups are determined. Using analogous method, we can determine all the EC-subgroups of some other sporadic simple groups.

Theorem 3. *All the equicentralizer subgroups of sporadic simple groups M, HS, HN, He, Co_3, and J_1 are cyclic groups of prime order.*

However, it is not true that any EC-subgroup of every sporadic simple group has to be a cyclic group of prime order. We have the following results.

Theorem 4. *Apart from cyclic subgroups of prime order, the only non-cyclic equicentralizer subgroups of the sporadic group J_2 are elementary abelian groups of order 4, in which all nonidentity elements are conjugates. Moreover, all equicentralizer subgroups of order 4 of J_2 are conjugates and with centralizers isomorphic to $2^2 \times A_5$. For the sporadic group Ru, we have the same results, but the centralizers of the equicentralizer subgroups of order 4 are isomorphic to $2^2 \times Sz(8)$.*

Proof. (1) J_2 satisfies $N(3,2)$. So J_2 has no EC-subgroups of order 3^2. J_2 has two conjugate classes of elements of order 2, the orders of their centralizers are

$$|\mathbf{C}(a_2)| = 1920,$$
$$|\mathbf{C}(b_2)| = 240 = 2^4 \cdot 3 \cdot 5.$$

The subgroups of order 1920 are maximal subgroups of J_2, so J_2 has no EC-subgroups of order 2^2 containing a_2. According to the order of $\mathbf{C}(b_2)$, there are three possible maximal subgroups K containing $\mathbf{C}(b_2)$:

(i) $K \cong 3 \cdot PGL_2(9)$, $|K : \mathbf{C}(b_2)| = 9$.

Since $PGL_2(9)$ has no subgroups of index 3 or 9, this case is impossible.

(ii) $K \cong 2_+^{1+4} : S_5$

Any element of order 2 in S_5 does not commute with an element of order 5, so the order of the centralizer of any noncentral element is equal to 2^4. But S_5 has no subgroups of order 15. This case is also impossible.

(iii) We must have $K \cong A_4 \times A_5$.

Then $\mathbf{C}(b_2) \cong 2^2 \times A_5$. Take a conjugate element b_2' of b_2 in A_4. The subgroup $\langle b_2, b_2' \rangle$ is an EC-subgroup of order 4 of K and also of J_2. Those elements of order 2, for which the orders of their centralizers are equal, are all conjugates, and moreover A_4 has only one subgroup of order 4. So the EC-subgroups of order 4 are all conjugates. Of course, the elements of order 2 in these EC-subgroups are conjugates.

(2) Ru satisfies $N(3,5)$.

Ru has two conjugate classes of elements of order 2, the orders of their centralizers are

$$|\mathbf{C}(a_2)| = 245760,$$
$$|\mathbf{C}(b_2)| = 116480.$$

$\mathbf{C}(a_2) = \mathbf{N}(\langle a_2 \rangle)$ is maximal in Ru, so Ru does not have an EC-subgroup of order 2^2 containing a_2.

$\mathbf{C}(b_2)$ is contained in a maximal subgroup $K \cong (2^2 \times Sz(8)) : 3$, and $|K : \mathbf{C}(b_2)| = 3$. Since $Sz(8)$ does not have a subgroup of index 3 and its center is $\{e\}$, we have $\mathbf{C}(b_2) \cong 2^2 \times Sz(8)$, $b_2 \in 2^2 < K$. Take another element $b_2' \in 2^2$ conjugate to b_2, then $\langle b_2, b_2' \rangle$ is an EC-subgroup with order 4 of K also of Ru. The elements of order 2, whose centralizers have the same order 116480, are all conjugates, and $\langle b_2, b_2' \rangle$ is the only subgroup of order 4 in K. Thus the EC-subgroups of order 4 of Ru are all conjugates, and the elements of order 2 in any EC-subgroup are all conjugates.

Remark. There are sporadic simple groups which contain equicentralizer subgroups of order 9. Their structure would be discussed in a forthcoming paper.

Acknowledgement. The author would like to thank the referee for his useful comments and suggestions.

References

[1] J. H. Conway et al, *Atlas of finite groups*, Clarendon Press, Oxford (1985).

[2] D. Gorenstein, *Finite simple groups : an introduction to their classification*, Plenum Press, New York and London (1982).

[3] Wang Efang, Equicentralizer subgroups of symmetric and alternating groups, *Scientia Sinica (Ser. A)*, XXVII **9** (1984), 915-923.

Peking University
Beijing
People's Republic of China

Some transfer theorems for finite groups

Tomoyuki Yoshida

The purpose of this paper is to state some transfer theorems and related topics. The details will be found in the following papers: [YO88a] Character-theoretic transfer (II), J. Algebra, to appear, [YO87] Fisher's inequality for block designs with finite group action, J. Fac. Sci. Univ. Tokyo, to appear.

1. Dedekind sums and a transfer theorem

Throughout this section, G denotes a finite group, G' the commutator subgroup of G, and p a prime.

(1.1) *Dedekind sums.* For integers $n > 0$ and r, the Dedekind sum is defined by

$$s(r,n) := \sum_{j=1}^{n-1} \left(\left(\frac{j}{n}\right)\right)\left(\left(\frac{rj}{n}\right)\right),$$

where $$((x)) := \begin{cases} x - [x] - \frac{1}{2} & \text{if } x \in \mathbf{R} - \mathbf{Z} \\ 0 & \text{if } x \in \mathbf{Z}. \end{cases}$$

Refer to [Ap76], [GR72] for Dedekind sums. By an easy calculation, we have the following:

(a) $s(r+n,n) = s(r,n)$.
(b) $s(-r,n) = -s(r,n)$.
(c) $s(1,n) = (n-1)(n-2)/12n$.
(d) $s(1,2) = 0, \quad s(\pm 1, 3) = \pm 1/18$.

Dedekind sums appeared first in the work of Dedekind about the transformation formula for the Dedekind eta function

$$\eta(z) := q^{1/24} \prod_{n=1}^{\infty} (1 - q^n), \quad q := e^{2\pi i z}.$$

In fact,

$$\eta\Big(\frac{az+b}{cz+d}\Big) = exp\{\pi i\Big(\frac{a+d}{12c} - s(d,c)\Big)\}\Big(\frac{cz+d}{i}\Big)^{1/2}\eta(z).$$

The fact that $\eta(z)^{24}$ is an automorphic function shows that $24ns(r,n)$ is an integer, but in general $ns(r,n)$ is not an integer. This delicate difference causes the existence of some special kind of transfer theorems for the primes $p = 2, 3$. To tell the truth, there is the following better result:

(e) $2(3,n)ns(r,n)$ is an integer.

(1.2) Let A be an abelian group of order a and λ a linear character of A. Define a function on A as follows:

$$\Delta[\lambda] := \Delta_A[\lambda] := \sum_{j=1}^{a-1}\Big(\frac{j}{a} - \frac{1}{2}\Big)\lambda^j.$$

Then $\overline{\Delta[\lambda]} = \Delta[\bar{\lambda}] = -\Delta[\lambda]$ and $\Delta[\lambda]_{|B} = \Delta_B[\lambda_{|B}]$ for $B \le A$.

Lemma. *Let A be a cyclic subgroup of G' of order a and let λ be a linear character of A of order a. Then the following hold:*

(a) $\Delta[\lambda]^G + \frac{1}{2}(\rho_G - 1_A^G)$ *is a virtual character of G, where ρ_G is the regular character and 1_A is a trivial character.*

(b) *If q is an integer prime to a, then*

$$\langle \Delta[\lambda]^G,\ \Delta[\lambda^q]^G\rangle + \frac{1}{4}(|G| - 2|G:A| + |A\backslash G/A|)$$

is an integer.

(1.3) We use Mackey decomposition and Frobenius reciprocity in order to calculate the above inner product. For each $x \in G$, we write $A(x) := A^x \cap A$, where $A^x := x^{-1}Ax$. Then we obtain

$$\begin{aligned}\langle \Delta[\lambda]^G,\ \Delta[\lambda^q]^G\rangle &= \sum_{x\in A\backslash G/A} \langle \Delta[\lambda^x]_{|A(x)}{}^A,\ \Delta[\lambda^q]\rangle \\ &= \sum_{x\in A\backslash G/A} \langle \Delta[\lambda^x{}_{|A(x)}],\ \Delta[\lambda^q{}_{|A(x)}]\rangle.\end{aligned}$$

By the following lemma, we see that each term can be presented by Dedekind sums.

Lemma. *Let B be a cyclic group of order b, μ a linear character of B of order b, and q an integer prime to b. Then*

$$\langle \Delta_B[\mu],\ \Delta_B[\mu^q]\rangle = s(q,b).$$

(1.4) It is convenient to introduce some new notation. Let C be a cyclic normal p-subgroup of a finite group N with $p^l := |C| > 1$. Then there exist integers $r(g)$ such that

$$g^{-1}xg = x^{r(g)} \quad \text{for } x \in C,\ g \in N.$$

Let q be an integer prime to p, and put

$$\sigma(C,N,q) := \frac{1}{|N|}\sum_{g\in N}\Big(\frac{p-1}{4} + ps(qr(g),p^l) - s(qr(g),p^{l-1})\Big).$$

(1.5) Theorem A. *Let A be a cyclic p-subgroup of G of order $p^m > 1$ and let q be an integer prime to p. For each i $(0 \le i \le m)$, put*

$A_i := \Omega_i(A)$, *the unique subgroup of* A *of order* p^i,
$n(i) := |N_G(A_i)|$,
$\sigma_i(q) := \sigma(A_i, N_G(A_i), q)$.

Then the following hold:

(i) $n(i) \equiv n(i-1) \bmod p^{2m-i+1}$, $1 \le i \le m$.
(ii) *If A is contained in G', then*

$$\sum(q) := \frac{1}{4}(1-p^{-m})^2|G| + p^{-2m}\sum_{i=1}^{m} p^{i-1}\sigma_i(q)n(i) \in \mathbf{Z}.$$

(1.6) Example. We give here only the simplest example. Further examples are found in another paper. Assume that A is a subgroup of G' of order p and that $N_G(A) = C_G(A)$ $(=: N)$. Then

$$\sigma_1(q) = \sigma(A, N, q) := \frac{p-1}{4} + ps(q,p).$$

Thus

$$\Sigma(q) = \tfrac{1}{4}(1-p^{-1})^2|G| + p^{-2}(\tfrac{p-1}{4} + ps(q,p))|N| \in \mathbf{Z}.$$

$$\Sigma(1) = \tfrac{1}{4}(1-p^{-1})^2|G| + p^{-2} \cdot \tfrac{p^2-1}{12} \cdot |N| \in \mathbf{Z}.$$

Case $p = 2$. $\sigma(1) = \frac{1}{16}(|G| + |N|) \in \mathbf{Z}$.

Thus $16 \mid |N|$ or $|G|_2 = |N|_2 \leq 8$.

Case $p = 3$. $\Sigma(\pm 1) = \frac{1}{9}|G| + \frac{1}{9}(\frac{1}{2} \pm \frac{1}{6})|N| \in \mathbf{Z}$.

Thus $27 \mid |N|$.

Case $p \geq 5$. The integrality of $\Sigma(1)$ implies only a trivial result $p^2 \mid |N|$. Conversely by a property of Dedekind sums we can prove that this condition implies the integrality of $\Sigma(q)$ for each q.

(1.7) Problems. (1) What can we say about various kinds of generalized Dedekind sums?

(2) Apply the above theorem to the study of finite groups of low 2- or 3-rank.

2. Fisher's inequality for Ext

It is well-known that if P is a Sylow p-subgroup of a finite group G, then $(G/G')_{(p)}$ (= a Sylow p-subgroup of the abelianized group G/G') is isomorphic to a direct summand of P/P'. This follows easily from the fact that the composition

$$G/G' \xrightarrow{V} P/P' \xrightarrow{nat} G/G',$$

where V is the transfer map, equals to the $|G : P|$-th power map. A similar argument is found in the proof of Fisher's inequality for block designs. The purpose of this and the next section is to give Fisher's inequalities for block designs with finite group action which include the above result.

(2.1) *Block designs.* Let (X, B) be a block design (possibly with repeated blocks) with parameters (v, b, r, k, λ) and with order $n := r - \lambda$, where $v(= |X|)$ is the number of points, $b(= |B|)$ is the number of blocks, r is the number of blocks incident to any point, k is the number of points incident to a block, and λ is the number of blocks incident to two distinct points. The incidence matrix A of (X, B) is a matrix of size $X \times B$ of which (x, β)-entry is 1 if $x \in \beta$ and $= 0$ otherwise. Then $AA^t = nI + \lambda J$, where J is the all-one matrix. Assume that a finite group G acts on this design (X, B).

(2.2) *Incidence maps.* Let R be a commutative ring. We denote by RX and RB the free R-modules generated by X and B, so that RX and RB are RG-modules . We use incidence maps on free R-modules RX and RB instead of incidence matrices.

$\alpha : RB \longrightarrow RX;\ \ \beta \mapsto \sum_{x \in \beta} x,$

$\alpha' : RX \longrightarrow RX;\ \ x \mapsto \sum_{x \in \beta} \beta,$

Furthermore, there are several (co-)augmentation maps:

$\varepsilon_X : RX \longrightarrow R;\ \ x \mapsto 1,$

$\varepsilon'_X : R \longrightarrow RX;\ \ 1 \mapsto \sum_{x \in X} x,$ etc.

These maps satisfy the following incidence equations:

$\alpha\alpha' : RX \longrightarrow RX;\ \ x \mapsto nx + \lambda \sum_{y \in X} y.$

$\alpha\varepsilon'_B = r\varepsilon'_B,\ \ \varepsilon_B\alpha' = r\varepsilon_B,$

$\alpha'\varepsilon'_X = k\varepsilon'_X,\ \ \varepsilon_X\alpha = k\varepsilon_X.$

Note that $det(\alpha\alpha') = rkn^{v-1}$ and $(r^2 - b\lambda)k = nr^2$. Thus if rn is invertible in R, then α' is a split monomorphism of RX to RB, and so

$$RX \mid RB \quad \text{as } RG\text{-modeles} \quad ,$$

that is, RX is isomorphism to a direct summand of RB as RG-modules. Of course, this fact implies original Fisher's inequality $v \leq b$ and the well-known theorem of Parker-Hughes-Dembowski which states that $|X/G| \leq |B/G|$.

(2.3) Now, let N an RG-modele. For each $m \geq 0$, put

$$E^m(M) := Ext^m_{RG}(M, N).$$

Then incidence maps and (co-)augmentation maps induces the following maps:

$$E^m(R) \underset{\varepsilon'^*_X}{\overset{\varepsilon^*_X}{\rightleftarrows}} E^m(RX) \underset{\alpha'^*}{\overset{\alpha^*}{\rightleftarrows}} E^m(RB) \underset{\varepsilon^*_B}{\overset{\varepsilon'^*_B}{\rightleftarrows}} E^m(R).$$

Furthermore, for any abelian group A and an integer m, we put

$${}_mA := \{a \in A|\ ma = 0\},\ A/(m) := A/mA.$$

Theorem B. *Assume that n is invertible in R.*

(1) *The following sequence is exact:*

$$0 \longrightarrow {}_{rk}E^m(R) \longrightarrow E^m(RX) \xrightarrow{\alpha^*} E^m(RB) \longrightarrow E^m(R)/(rk) \longrightarrow 0.$$

(2) *If $r \cdot id$ is an isomorphism on $E^m(R)$ ($\cong H^m(G, N)$), then*

$$E^m(RX) \mid E^m(RB).$$

(3) *$Ker\ \varepsilon'^*_X | Ker\ \varepsilon'^*_B$, $Cok\ \varepsilon^*_X | Cok\ \varepsilon^*_B$.*

(4) *When the design is symmetric (that is, $v = b$), we can replace "$|$" in (2), (3) by RG-isomorphism "$\cong$".*

(2.4) In general, there is an R-isomorphism

$$E^m(RX) := Ext^m_{RG}(RX, N) \cong \bigoplus_{x \in X/G} H^m(G_x, N),$$

where x runs over a complete set of representatives of G-orbits in X, and G_x is the stabilizer at x. Thus the above theorem can be represented by cohomology groups of finite groups. Then the maps induced by (co-) augmentation maps are written by (co-)restriction maps.

It is well-known that $H^1(G, \mathbf{C}^*)$ is isomorphic to the dual group of the abelian group G/G', where G' is the commutator group of G. Applying the theorem to this special case, we have the following transfer theorem. (For a prime p and a finite abelian group A, we denote a unique Sylow p subgroup of A by $A_{(p)}$).

(2.5) Theorem. *Assume that G acts transitively on B (and also on X). Let $x \in X$, $\beta \in B$, $H := G_x$, $K := G_\beta$. Let p be a prime which does not divide n.*

(1) *If p is prime to r or $|G/G'|$, then*

$$(H/H')_{(p)} | (K/K')_{(p)}.$$

(2) *Let P, Q be Sylow p-subgroups of H, K, respectively. Then*

$$(P \cap G')/(P \cap H') | (Q \cap G')/(Q \cap K').$$

(3) *When $v = b$, the RG-isomorphisms instead of direct summands in* (1), (2) *hold.*

Applying this theorem to the design with $|X| = 1$, $B = G/P$, where P a Sylow p-subgroup of G, we have that $(G/G')_{(p)}$ is isomorphic to a direct summand of P/P', as is stated before.

3. Fisher's inequality for poly-Hecke functors

In the previous section, we considered the permutation RG-modules RX and RB simply as RG-modules. In this section we introduce an R-additive and self-dual category in which objects are finite G-sets and X is a direct summand of B for a block design (X, B) on which G acts. Of course, the category of permutation RG-modeles (Hecke category) also possesses this property. But the category we want to construct here is larger than the Hecke category. To do this, we must begin by thinking the concept of matrices in "equivariant theory" over again.

Throughout this section, G denotes a finite group, R a commutative ring and p a prime. The disjoint union of two finite G-sets, A, B is denoted by $A + B$.

(3.1) *Mackey category.* We define an R-additive category $Mc(G, R \otimes \Omega)$ by the following data:

(1) Objects are finite G-sets.

(2) The hom-set $Mc(G, R\otimes\Omega)(Y,X)$ of Y to X is an R-module given by the following generators and relations:

(a) Generators: any element of the generator set has the form

$$[X \xleftarrow{f} A \xrightarrow{f'} Y] \quad (\text{simply} \quad [f,f'] \quad \text{or} \quad [f,A,f'])$$

where A is a finite G-set and f, f' are G-maps.

(b) Relations: For two generators $[X \xleftarrow{f} A \xrightarrow{f'} Y]$, $[X \xleftarrow{g} B \xrightarrow{g'} Y]$,

(b.1) if there exists a G-isomorphism $h : A \longrightarrow B$ such that $f = g\circ h$ and $f' = g'\circ h$, then $[f,f'] = [g,g']$.

(b.2) $[X\leftarrow A\rightarrow Y] + [X\leftarrow B\rightarrow Y] = [X\leftarrow A+B\rightarrow Y]$.

(3) The composition is defined by the fibre product:

$$[X\leftarrow A\rightarrow Y]\circ[Y\leftarrow B\rightarrow Z] = [X\leftarrow A\times_Y B\rightarrow Y].$$

Remark. If X is a regular G-set (resp. a trivial G-set), then the endomorphism ring $Mc(G, R\otimes\Omega)(X,X)$ is isomorphic to the group ring RG (resp. the Burnside ring $\Omega(G)$). When $G=1$, a morphism of a set X to a set Y is regarded as a matrix of size $X\times Y$ and the compositions then are products of matrices. The Mackey category is familiar to topologists as the category of stable G-maps.

(3.2) *Hecke category.* We define the Hecke category $Hec(G,R)$ as follows:

(1) Objects are finite G-sets.

(2) Each morphism of Y to X is a matrix $(a_{xy})_{x\in X, y\in Y}$ of size $X\times Y$ that entries are elements of R such that $a_{xg,yg} = a_{x,y}$ for $x\in X$, $y\in Y$, $g\in G$.

(3) The compositions are the multiplications as matrices.

Clearly this category is equivalent to the category of permutation RG-modules.

(3.3) Lemma. *For any subgroup H of G, there is an R-additive functor Φ_H of $Mc(G, R\otimes\Omega)$ to $Hec(\mathrm{N}_G(H), R)$ such that $\Phi_H(X) = X^H$ (the set* *of H-fixed points) and*

$$\Phi_H([X \xleftarrow{f} A \xrightarrow{f'} Y]) = (|f^{-1}(x) \cap f'^{-1}(y) \cap A^H|)_{x \in X^H, y \in Y^H}.$$

(3.4) *Mackey category* $Mc_p = Mc(G, e^p_{G,1}(R \otimes \Omega))$. Suppose $R/J(R)$ is of characteristic p. Then the category Mc_p is defined to be the quotient category of $Mc(G, R \otimes \Omega)$ by the intersection of the kernels of the functors Φ_P, where P's are p-subgroups of G, that is, this category is defined by the conditions of (3.1) together with another relation

$$(b.3) \qquad [X \xleftarrow{f} A \xrightarrow{f'} Y] = [X \xleftarrow{g} B \xrightarrow{g'} Y] \quad \text{provided}$$

$$|f^{-1}(x) \cap f'^{-1}(y) \cap A^P| = |g^{-1}(x) \cap g'^{-1}(y) \cap A^P|$$

for all p-subgroups P of G and $x \in X^P$, $y \in Y^P$.

(3.5) The functor Φ_1 induces an additive functor

$$\begin{aligned} \Psi : Mc_p &\longrightarrow Hec(G, R) \\ : X \mapsto X, \ \ &[f, f'] \mapsto (|f^{-1}(x) \cap f'^{-1}(y)|)_{x,y}. \end{aligned}$$

Proposition. *The functor Ψ reflects split mono, split epi and iso, that is, for a morphism α in Mc_p, if $\Psi(\alpha)$ is split mono (resp. split epi, iso), so is α.*

The proof is hard. Essentially, we need to show that the (categorical) radical of Mc_p contains the kernel of the functor Ψ.

(3.6) R-additive functors from $Mc(G, R \otimes \Omega)$, $Hec(G, R)$, Mc_p to the category of R-modules are called *Mackey functors, Hecke functors, poly-Hecke functors*, respectively.

A Hecke functor is a poly-Hecke functor, and a (poly-) Hecke functor is a Mackey functor. If M is a Mackey functor, then each component $M(X)$ is a module over the Burnside ring $\Omega(G)$ by the composition

$$\pi_* \circ \pi^* : M(X) \longrightarrow M(X \times A) \longrightarrow M(X)$$

for any $[A] \in \Omega(G)$. Let $e^p_{G,1}$ be the idempotent of the Burnside ring $R \otimes \Omega(G)$ corresponding to the trivial subgroup. See [Y083]. When $pR \subseteq J(R)$, the functor $X \mapsto e^p_{G,1}M(X)$ is poly-Hecke functor and conversely any poly-Hecke functor M satisfies the property $e^p_{G,1}M(X) = M(X)$.

Mackey functors M are bijectively corresponding to Green's G-functors a by $M(G/H) = a(H)$. Furthermore Hecke functors are corresponding to cohomological G-functors. A Mackey functor is poly-Hecke functor if and only if the corresponding G-functor a satisfies the property that for any subgroup H of G and a Sylow p-subgroup P of H, the "restriction map" $res^H_P : a(H) \longrightarrow a(P)$ is a monomorphism.

Examples. (1) $X \mapsto Ext^m_{RG}(RX, N)$ is a Hecke functor. The corresponding G-functor is $H(\leq G) \mapsto H^m(H, N)$.

(2) Let $R(H)$ denote the character ring of a finite group H. Then $H \mapsto e^p_{G,1}(R \otimes R(H))$ is a G-functor corresponding to a poly-Hecke functor and $e^p_{G,1}(R \otimes R(H))$ consists of all elements of $R \otimes R(H)$ which vanishes on non p-elements of G.

(3) Let $\Omega(H)$ be the Burnside ring of a finite group H, that is, the Grothendieck ring of the category of finite H-sets with respect to disjoint sums and cartesian products. Then $H \mapsto \Omega(H)$ is a G-functor.

(3.7) A pair of functors (Mc^*, Mc_*) is defined by

$$Mc^* : X \mapsto X, \quad (f : X \longrightarrow Y) \mapsto [f, 1_Y],$$

$$Mc_* : X \mapsto X, \quad (f : X \longrightarrow Y) \mapsto [1_X, f].$$

Thus the definition of Mackey functors is equivariant to ordinary one by Dress. Applying these functors to the Amitsur complex $\{d^n_i : X^n \longrightarrow X^{n-1}\}$, we have (co-)chain complexes in Mc_p (or $Mc(G, R \otimes \Omega)$):

$$Am(X) : 0 \longrightarrow 1 \longrightarrow X \longrightarrow X^2 \longrightarrow \dots$$

$$Am(X)^{op} : 0 \longleftarrow 1 \longleftarrow X \longleftarrow X^2 \longleftarrow \dots$$

As a corollary of the above proposition, we have the following useful result:

Corollary. *Assume that p is prime to $|X|$. Then $Am(X)$ and $Am(X)^{op}$ have contracting homotopies in Mc_p. In particular, for any poly-Hecke functor M, the chain complexes $M(Am(X))$ and $M(Am(X)^{op})$ are exact sequences.*

(3.8) Assume that $pR \subseteq J(R)$. As before, let (X, B) be a block design with parameters v, b, r, k, λ, n on which G acts. Let M be a poly-Hecke functor. For any G-set Y, we put

$$Y^* := M([Y, 1_Y]) : M(1) \longrightarrow M(Y),$$

$$Y_* := M([1_y, Y]) : M(Y) \longrightarrow M(1).$$

Theorem C. *Assume that $(n, p) = 1$. Then the following hold:*

(1)(a) *If $(p, r) = 1$, then $M(X) \mid M(B)$.*

(b) *If $(p, v) = 1$, then $Ker\ X_* \mid M(B), Cok\ X^* \mid M(B)$.*

(c) *If $(p, b) = 1$, then $Ker\ X_* \mid Ker\ B_*, Cok\ X^* \mid Cok\ B^*$.*

(2) *Assume that $v = b$.*

(a) *If $(p, r) = 1$, then $M(X) \cong M(B)$.*

(b) *If $(p, v) = 1$, then $Ker\ X_* \cong Ker\ B_*$, $Cok\ X^* \cong Cok\ B^*$.*

Proof. We give here only the proof of (1a). Consider the set of flags $F := \{(x, \beta) \mid x \in \beta \in B\} \subseteq X \times B$. There are projections of F to X and B, and so we have "incidence morphism" and its transpose in Mc:

$$\alpha = [X \longleftarrow F \longrightarrow B] : B \longrightarrow X,$$

$$\alpha' = [B \longleftarrow F \longrightarrow X] : X \longrightarrow B.$$

Then the images of α, α' by the functor $\Psi : Mc \longrightarrow Hec(G, R)$ are the incidence matrices A, A^t. Thus

$$det\ \Psi(\alpha \circ \alpha') = det(AA^t) = rkv^{n-1}.$$

By the assumption, we have that $det\ \Psi(\alpha \circ \alpha')$ is invertible, and so $\Psi(\alpha \circ \alpha')$ is an automorphism. Furthermore by proposition in (3.5), $\alpha \circ \alpha'$ is an automorphism in Mc. Thus

$$M(\alpha \circ \alpha') = M(\alpha) \circ M(\alpha') : M(X) \longrightarrow M(B) \longrightarrow M(X)$$

is also an automorphism, and so $M(\alpha)$ is a split monomorphism of $M(X)$ to $M(B)$.

(3.9) Applying Theorem C to the Burnside ring functors, we have the following:

Corollary. *Let P be a normal p-subgroup of the finite group G. Assume that $(p,n) = 1$.*
(1)(a) *If p is prime to r or b, then $|X^P/G| \leq |B^P/G|$.*
(b) *If $p \mid r$, then $|X^P/G| - 1 \leq |B^P/G|$.*
(2) *If $v = b$, then $|X^P/G| = |B^P/G|$.*

Remark. An elementary proof of this corollary is given by T. Atsumi of Kagoshima University.

(3.10) Problem. Study the equivariant version of results (e.g. Schutzenberger's theorem and Bruck-Ryser-Chowla's theorem for symmetric designs, many well-known results about (strongly) regular graphs) in combinatorics without group action. (There is a version of "equivariant" Macwilliams identity for a linear code and its dual code. See [Y088b].)

References

[Ap76] Apostol, T.M., *Modular functions and Dirichlet Series in Number Theory*, Springer, 1976.

[De68] Dembowski, P., *Finite Geometries*, Springer, 1968.

[La83] Lander, E.S., *Symmetric Designs: An algebraic Approach,* London Math. Soc. LNS 74, 1983.

[RG72] Rademacher, H. and Grosswald, E., *Dedekind sums*, Carus Math. Monograph, 16, Math. Assoc. of America, 1972.

[Y083] Yoshida, T., Idempotents of Burnside rings and Dress induction theorem, *J. Algebra* **80** (1983), 90-105.

[Y087] Yoshida, T., Fisher's inequality for block designs with finite group action, *J. Fac. Tokyo Univ.*, to appear.

[Y088a] Yoshida, T., Character-theoretic transfer (II), *J. algebra*, to appear.

[Y088b] Yoshida, T., MacWilliams identity for linear codes with group action, to appear.

Department of Mathematics
Hokkaido University
Sapporo 060
Japan

Appendices

List of participants

S.I. ADIAN, Steklov Mathematical Institute, 117966 Moscow GSP-1, U.S.S.R.

Bernhard AMBERG, F.B. Mathematik, Universität Mainz, 6500 Mainz, Federal Republic of Germany

R.P. AGARWAL, Department of Mathematics, National University of Singapore, Kent Ridge, Singapore 0511, Republic of Singapore

S. BACHMUTH, Department of Mathematics, University of California, Santa Barbara CA 93106, U.S.A.

Bernd BAUMANN, Mathematisches Institut, Universität Giessen, Arndtstrasse 2, 6300 Giessen, Federal Republic of Germany

Katalin A. BENCSÁTH, Department of Mathematics & Computer Science, Manhattan College, Riverdale (Bronx), New York 10471, U.S.A.

Ronald D. BERCOV, Department of Mathematics, University of Alberta, Edmonton, Alberta T6G 2G1, Canada

A.J. BERRICK, Department of Mathematics, National University of Singapore, Kent Ridge, Singapore 0511, Republic of Singapore

F. Rudolf BEYL, Department of Mathematical Science, Portland State University, Portland OR 97207, U.S.A.

Ashwani K. BHANDARI, Department of Mathematics, Panjab University, Chandigarh 160014, India

Thomas BIER, Department of Mathematics, National University of Singapore, Kent Ridge, Singapore 0511, Republic of Singapore

C.J.B. BROOKES, Department of Mathematics, Corpus Christi College, Cambridge, CB2 1RH, United Kingdom

Michel BROUÉ, Service de Math, Ecole Normale Superieure, 45 Rue d'Ulm, 75005 Paris Cedex 05, France

Robert A. BRYCE, Department of Mathematics, Australian National University, Faculty of Science, GPO Box 4, Canberra ACT 2601, Australia

Robert G. BURNS, Department of Mathematics, York University, North York Toronto, Ontario M3J 1P3, Canada

C.M. CAMPBELL, Mathematical Institute, University of St Andrews, North Haugh, St Andrews, Fife KY16 9SS, Scotland

John CANNON, Department of Pure Mathematics, University of Sydney, NSW 2006, Australia

Florita CAPAO-AN, Ipil Residence Hall, University of Philippines, Diliman, Quezon City 3004, Philippines

Charles CASSIDY, Department de Mathematiques, Statistique et Actuariat, Universite Laval Cite Universitaire, Quebec, Canada G1K 7P4

CHAN Gin Hor, Department of Mathematics, National University of Singapore, Kent Ridge, Singapore 0511, Republic of Singapore

CHAN Kai Meng, Department of Mathematics, National University of Singapore, Kent Ridge, Singapore 0511, Republic of Singapore

CHAN Yiu Man, Department of Mathematics, National University of Singapore, Kent Ridge, Singapore 0511, Republic of Singapore

Dhamrong CHANTHORN, Department of Mathematics, Chiangmai University, Chiangmai 50002, Thailand

CHEN Chuan Chong, Department of Mathematics, National University of Singapore, Kent Ridge, Singapore 0511, Republic of Singapore

CHENG Kai Nah, Department of Mathematics, National University of Singapore, Kent Ridge, Singapore 0511, Republic of Singapore

CHEW Tuan Seng, Department of Mathematics, National University of Singapore, Kent Ridge, Singapore 0511, Republic of Singapore

CHONG Chi Tat, Department of Mathematics, National University of Singapore, Kent Ridge, Singapore 0511, Republic of Singapore

CHOW Yoong Ming, Department of Mathematics, National University of Singapore, Kent Ridge, Singapore 0511, Republic of Singapore

Stephen T.L. CHOY, Department of Mathematics, National University of Singapore, Kent Ridge, Singapore 0511, Republic of Singapore

CHUA Seng Kee, Department of Mathematics, Rutgers University, Piscataway, NJ 08854, U.S.A.

CHUA Seng Kiat, Blk 63 Commonwealth Drive #02-265, Singapore 0314, Republic of Singapore

Leo P. COMERFORD, Jr., Department of Mathematics, University of Wisconsin-Parkside Kenosha, WI 53141-2000, U.S.A.

Martyn R. DIXON, Department of Mathematics, University of Alabama, P.O. Box 1416, Tuscoloosa, Alabama 35486, U.S.A.

FAN Yun, Department of Mathematics, Wuhan University, Wuhan, People's Republic of China

Valfria FEDRI, Universita Degli Studi, Istituto Matematico, Ulisse Dini, Viale Morgagni 67/A, 1 50134 Firenze, Italy

Walter FEIT, Department of Mathematics, Yale University, Box 2155, Yale Station, New Haven CT 06520, U.S.A.

René P. FELIX, Mathematics Department, University of Philippines, College of Science, Diliman, Quezon City 3004, Philippines

Daniel E. FLATH, Department of Mathematics, National University of Singapore, Kent Ridge, Singapore 0511, Republic of Singapore

Peter FÖRSTER, Department of Mathematics, Australian National University, Institute of Advanced Studies G.P.O. Box 4, Canberra ACT 2601, Australia

Anthony M. GAGLIONE, Department of Mathematics, United States Naval Academy, Annapolis, Maryland 21402-5000, U.S.A.

Anna L. GILOTTI, Universita Degli Studi, Istituto Matematico, Ulisse Dini, Viale Morgagni 67/A, 1 50134 Firenze, Italy

Stephen P. GLASBY, Department of Pure Mathematics, University of Sydney, Carslaw Building, NSW 2006, Australia

J.P.C. GREENLEES, Department of Mathematics, National University of Singapore, Kent Ridge, Singapore 0511, Republic of Singapore

J.R.J. GROVES, Department of Mathematics, University of Melbourne, Parkville, Victoria 3052, Australia

Karl W. GRUENBERG, School of Mathematical Sciences, Queen Mary College, University of London, Mile End Road, London E1 4NS, United Kingdom

C.K. GUPTA, Department of Mathematics & Astronomy, University of Manitoba, Winnipeg, Manitoba R3T 2N2, Canada

N.D. GUPTA, Department of Mathematics & Astronomy, University of Manitoba, Winnipeg, Manitoba R3T 2N2, Canada

HANG Kim Hoo, Blk 408, Jurong West St 42 #02-689, Singapore 2264, Republic of Singapore

Koichiro HARADA, Department of Mathematics, Ohio State University, Columbus, OH 43210, U.S.A.

Brian HARTLEY, Department of Mathematics, University of Manchester, Manchester M13 9PL, United Kingdom

Graham HIGMAN, Mathematical Institute, 24-29 St Giles, Oxford OX1 3LB, United Kingdom

C.S. HOO, Department of Mathematics, University of Alberta, Edmonton, Alberta T6G 2G1, Canada

HU Kaiyuan, Department of Mathematics, National University of Singapore, Kent Ridge, Singapore 0511, Republic of Singapore

Paul IGODT, Mathematics Department, Katholieke Universiteit, Leuven, Campus Kortrijk, B-8500 Kortrijk, Belgium

Bambang IRAWAN, Jln Kemuning 6/160, Blok 5 Helvetia, Medan, Indonesia

IRYANTO, Jln Enggang No. 15A, Medan, North Sumatera, Indonesia

Noboru ITO, Department of Mathematics, Konan University, Okamoto, Kobe 658, Japan

Tatsuro ITO, Department of Mathematics, Ohio State University, Columbus, Ohio 43210, U.S.A.

Judith P. JESUDASON, Department of Mathematics, National University of Singapore, Kent Ridge, Singapore 0511, Republic of Singapore

Dave JOHNSON, Department of Mathematics, University of Nottingham, University Park, Nottingham NG7 2RD, United Kingdom

Naoki KAWAMOTO, Department of Mathematics, Hiroshima University, Faculty of Science, Hiroshima 730, Japan

Otto H. KEGEL, Mathematishes Institut, Albert-Ludwigs-Universität, Albertstrasse 23b, D-7800 Freiburg i. Br., Federal Republic of Germany

Ann Chi KIM, Department of Mathematics, Pusan National University, Pusan 607, Republic of Korea

KOAY Hang Leen, Department of Mathematics, University of Malaya, 59100 Kuala Lumpur, Malaysia

KOH Khee Meng, Department of Mathematics, National University of Singapore, Kent Ridge, Singapore 0511, Republic of Singapore

Aleksej Ivanović KOSTRIKIN, Mathematics Department, Moskovskii University, Mehmat, Moscow 119899, GSP-1, U.S.S.R.

LAM Siu Por, Department of Mathematics, Chinese University of Hong Kong, New Territories, Hong Kong

LANG Mong-Lung, Department of Mathematics, Ohio State University, Columbus, Ohio 43210, U.S.A.

LEONG Yu Kiang, Department of Mathematics, National University of Singapore, Kent Ridge, Singapore 0511, Republic of Singapore

Fred P.F. LEUNG, Department of Mathematics, National University of Singapore, Kent Ridge, Singapore 0511, Republic of Singapore

LIU Zhishan, Department of Mathematics, Huhehaote College of Education, Huhehaote, Inner Mongolia, People's Republic of China

Emilio LLUIS, Instituto de Matemáticas, Universidad Nacional Autónoma de México, Ciudad Universitaria, 04510 México P.F., Mexico

Patrizia LONGOBARDI, Department of Mathematics, University of Napoli, Via Giuseppe Orsi, 15A 80128 Naples, Italy

LUM Kai Hing, Blk 18, Telok Blangah Crescent, #10-164, Singapore 0409, Republic of Singapore

Olga N. MACEDONSKA, 40-282 Katowice UI, Sikorskiego 10, M.17, Poland

Toru MAEDA, Department of Mathematics, Kansai University, Suita Osaka 564, Japan

Sri Harini MAHMUDI, Jurusan Matematika, F.M.I.P.A. UI, Jl Salemba - 4, Jakarta, Indonesia

Mercede MAJ, PTTA Arenella 7/H, 80127 - Napoli, Italy

Osamu MARUO, Department of Mathematics Education, Faculty of Education, Hiroshima University, Hiroshima, Japan

Abu Osman bin MD TAP, Department of Mathematics, Universiti Kebangsaan Malaysia, Ctr of Quantitative Studies, 43600 UKM Bangi, Selangor, Malaysia

C.F. MILLER, Department of Mathematics, University of Melbourne, Parkville, Victoria 3052, Australia

H.Y. MOCHIZUKI, Department of Mathematics, University of California, Santa Barbara CA 93106, U.S.A.

Jamshid MOORI, Department of Mathematics, University of Bophuthatswana, Postbag x 2046, Mafi Keng, Republic of Bophuthatswana

Qaiser MUSHTAQ, Department of Mathematics, Quaid-I-Azam-University, Islamabad, Pakistan

B.H. NEUMANN, Division of Mathematics & Statistics, CSIRO, P.O. Box 1965, Canberra ACT 2601, Australia

M.F. NEWMAN, Department of Mathematics, Australian National University, Institute of Advanced Studies, GPO Box 4, Canberra ACT 2601, Australia

Markku NIEMENMAA, Department of Mathematics, University of Oulu, Linnanmaa, 90570 Oulu, Finland

Luz NOCHEFRANCA, Ipil Residence Hall, University of Philippines, Diliman, Quezon City 3004, Philippines

Elizabeth ORMEROD, Department of Mathematics, Australian National University, Faculty of Science GPO Box 4, Canberra ACT 2601, Australia

Haludin PANJAITAN, Jl. Pelajar Timur GG., Mestika Medan, Sumatera Utara, Indonesia

I.B.S. PASSI, Department of Mathematics, Panjab University, Chandigarh 160014, India

PENG Tsu Ann, Department of Mathematics, National University of Singapore, Kent Ridge, Singapore 0511, Republic of Singapore

Roger K.S. POH, Department of Mathematics, National University of Singapore, Kent Ridge, Singapore 0511, Republic of Singapore

Cheryl E. PRAEGER, Department of Mathematics, University of Western Australia, Nedlands, WA 6009, Australia

Narong PUNNIM, Department of Mathematics, Srinakharinwirot University, Faculty of Science, Soi Sukhumvit 23, Bangkok 10110, Thailand

QUEK Tong Seng, Department of Mathematics, National University of Singapore, Kent Ridge, Singapore 0511, Republic of Singapore

Burkhardt RENZ, F.B. Mathematik, Johann Wolfgang Goethe Universität, Robert Mayer Strasse 6 - 10, 6 Frankfurt am Main 11, Federal Republic of Germany

Akbar H. RHEMTULLA, Department of Mathematics, University of Alberta, Edmonton, T6G 2G1, Canada

Luis RIBES, Department of Mathematics, Carleton University, Ottawa, ONT K1S 5B6, Canada

Edmund F. ROBERTSON, Mathematical Institute, University of St Andrews, North Haugh, St Andrews, Fife KY16 9SS, Scotland

Derek J.S. ROBINSON, Department of Mathematics, University of Illinois, 1409 W. Green Street, Urbana IL 61801, U.S.A.

Frank ROEHL, Department of Mathematics, University of Alabama, Tuscaloosa, Alabama 35487-1416, U.S.A.

Klaus W. ROGGENKAMP, Mathematisches Institute B/3, Universität Stuttgart, P.O. Box 801140, D-7000 Stuttgart 80, Federal Republic of Germany

G. ROSENBERGER, Fachbereich Mathematik, Universität Dortmund, Postfach 500500, D-4600 Dortmund 50, Federal Republic of Germany

Leanne J. RYLANDS, Department of Pure Mathematics, University of Sydney, NSW 2006, Australia

Clive SAUNDERS, Department of Mathematics, University of Sydney, Sydney 2006, New South Wales, Australia

Günther SCHLICHTING, Institut für Mathematik, Der Technischen Universität, Postfach 202420, D-8000 Munich, Federal Republic of Germany

Carlo M. SCOPPOLA, Dipartmento di Matematica, Universita di Trento, 38050 Povo (Trento), Italy

Luigi SERENA, Universite Degli Studi, Istituto Matematico, Ulisse Dini, Viale Morgagni 67/A, 1 50134 Firenze, Italy

Jean-Pierre SERRE, College de France, 11 Place Marcelin Berthelot, 75231 Paris Cedex 05, France

Fahana SHAHEEN, Department of Mathematics, Quaid-e-Azam University, Islamabad, Pakistan

SHEE Sze Chin, Department of Mathematics, National University of Singapore, Kent Ridge, Singapore 0511, Republic of Singapore

SHI Wujie, Mathematics Department, Southwest-China Teachers University, Chongqing Sichuan, People's Republic of China

G.C. SMITH, Department of Mathematics, University of Bath, Bath BA2 7AY, United Kingdom

J.B. SRIVASTAVA, Department of Mathematics, IIT Delhi, New Delhi 110016, India

V.J. STOKES, 40B Denton Road, E Twickenham, Middx, United Kingdom

Michio SUZUKI, Department of Mathematics, University of Illinois, 1409 W. Green Street, Urbana, IL 61801, U.S.A.

TAN Eng Chye, 297-H, Mansfield Street, New Haven CT 06511, U.S.A.

TAN Sie Keng, Department of Mathematics, National University of Singapore, Kent Ridge, Singapore 0511, Republic of Singapore

F.C.Y. TANG, Department of Mathematics, University of Waterloo, Waterloo, Ontario N2L 3F1, Canada

D.E. TAYLOR, Department of Pure Mathematics, University of Sydney, NSW 2006, Australia

John G. THOMPSON, Department of Pure Mathematics & Mathematical Statistics, University of Cambridge, 16 Mill Lane, Cambridge CB2 1SB, United Kingdom

TON Daorong, University of Science & Technology of China, Hefei, Anhui 230029, People's Republic of China

Ronald F. TURNER-SMITH, Department of Mathematics, Hong Kong Polytechnic, Hung Hom, Hong Kong

G.E. WALL, Department of Pure Mathematics, University of Sydney, NSW 2006, Australia

WAN Fook Sun, Department of Mathematics, National University of Singapore, Kent Ridge, Singapore 0511, Republic of Singapore

WAN Zhexian, Institute of Systems Science, Academia Sinica, Beijing, People's Republic of China

WANG Efang, Department of Mathematics, Peking University, Beijing, People's Republic of China

P.J. WEBB, Department of Mathematics, The University, Manchester M13 9PL, United Kingdom

M.J. WICKS, Department of Mathematics, National University of Singapore, Kent Ridge, Singapore 0511, Republic of Singapore

A.L. WILLIAMS, Department of Mathematics, Universiti Brunei Darussalam, Bandar Seri Begawan, Negara Brunei Darussalam

J.S. WILSON, Department of Mathematics, Christ's College, Cambridge, CB2 3BU, United Kingdom

S.J. WILSON, Department of Mathematics, National University of Singapore, Kent Ridge, Singapore 0511, Republic of Singapore

WONG Peng Choon, Mathematics Department, University of Malaya, 59100 Kuala Lumpur, Malaysia

S.K. WONG, Department of Mathematics, Ohio State University, Colombus OH 43210, U.S.A.

YAP Hian Poh, Department of Mathematics, National University of Singapore, Kent Ridge, Singapore 0511, Republic of Singapore

Tomoyuki YOSHIDA, Department of Mathematics, Hokkaido University, Faculty of Science, Sapporo 060, Japan

List of lectures

Workshop lectures

O.H. KEGEL
Sylow theory in locally finite groups

D.J.S. ROBINSON
Homological methods in infinite group theory

J.-P. SERRE
(1) *Galois groups over* **Q**
(2) *A short introduction to Tits buildings*
(3) *Trees, and groups acting on trees*

Invited lectures

S.I. ADIAN
Lower and upper bounds for nilpotency classes of Lie algebras with Engel conditions

M. BROUÉ
Some questions on the group algebras of finite groups

J.J. CANNON
The computation of conjugacy classes and character tables for a finite group

W. FEIT
$6A_6$ *and* $6A_7$ *are Galois groups over certain number fields*

K.W. GRUENBERG
Relation modules and related matters

B. HARTLEY
Group actions on lower central factors of free groups

G. HIGMAN
Countably free groups

N. ITO
Automorphism groups of DRADs

O.H. KEGEL
Some groups of homeomorphisms of R

A.I. KOSTRIKIN
Invariant lattices in Lie algebras and their automorphism groups

B.H. NEUMANN
Finite groups with few defining relations

D.J.S. ROBINSON
Rewriting products of group elements

J.-P. SERRE
Kazhdan's property, and groups acting on trees

M. SUZUKI
Elementary proof of the simplicity of sporadic groups

J.G. THOMPSON
(1) *Fricke, free groups and* SL_2
(2) *Hecke operators on non-congruence subgroups*
(3) *Stable Hurwitz numbers*

Contributed papers

(An asterisk after a name indicates the speaker at the conference.)

B. AMBERG
On soluble products of groups

S. BACHMUTH
Can automorphism groups be infinitely generated in more than a finite number of dimensions?

B. BAUMANN
Modules for linear groups acting on p-groups

A.J. BERRICK
Acyclic group theory

F.R. BEYL
Efficient presentations for certain groups $SL(2, \mathbf{Z}/2^k\mathbf{Z})$

A.K. BHANDARI
On group theoretic properties of unit groups of group rings

T. BIER* & K.N. CHENG
Totient ideals of group rings of cyclic groups

C.J.B. BROOKES
Automorphisms of infinite soluble groups

R.A. BRYCE
Fitting class constructions

C.M. CAMPBELL
On groups related to Fibonacci groups

H. CÁRDENAS & E. LLUIS*
On the Chern classes of representation of the symmetric groups

L.P. COMERFORD, Jr.* & C.C. EDMUNDS
Solutions of equations in free groups

Y. FAN
Block covers and module covers of finite groups

D.E. FLATH
Decomposing tensor products

B. FINE, F. LEVIN & G. ROSENBERGER*
One-relator products of cyclics

A.M. GAGLIONE
A theorem in the commutator calculus

S.P. GLASBY
The composition length and derived length of a finite soluble group

J.P.C. GREENLEES
Equivariant cohomology

C.K. GUPTA
On the conjugacy problem for free centre-by-metabelian groups

N.D. GUPTA
On groups which grow sub-exponentially

W. HERFORT & L. RIBES*
Solvable subgroups of free products of profinite groups

P. IGODT
Torsionfree nilpotent groups and polynomial 2-cocycles: the 3-step case

D.L. JOHNSON
The tensor square of the Baumslag-Hilton group

A.C. KIM
Laws in a variety of groupoids

M.L. LANG
On Conway group .0 and its theta functions

H.Y. MOCHIZUKI
Automorphism groups of solvable groups

J. MOORI
Action tables for $\bar{F}_{22}$

Q. MUSHTAQ & F. SHAHEEN*
On groups $\triangle(2,3,k)$

M.F. NEWMAN
Tabulation of groups of 2-power order

I.B.S. PASSI
The rational group algebra

C.E. PRAEGER
Covering subgroups and a problem on Kronecker classes of algebraic number fields

H. REN, Z. WAN* & X. WU
Automorphisms and isomorphisms of $PSL_2(K)$ *over skew fields*

B. RENZ
Geometric invariants and finiteness conditions of groups

A.H. RHEMTULLA
Groups with permutable subgroup products

E.F. ROBERTSON
Symmetric presentations and some simple groups

F. ROEHL
On the isomorphism problem for modular group rings

K.W. ROGGENKANP
Some new progress on the isomorphism problem for integral group rings

C. SAUNDERS
The reflection character of a finite Chevalley group

G.H. SCHLICHTING
On the periodicity of group operations

S.C. SHEE* & H.H. TEH
An application of groups to the topology design of connection machines

W. SHI
A new characterization of the finite simple groups

J.B. SRIVASTAVA
Group rings of infinite groups with trivial locally finite radical

F.C.Y. TANG
Residual finiteness of polygonal products

D. TON
Maximal subgroups in a finite classical group

G.E. WALL
Multilinear Lie relators for varieties of groups

E. WANG
Equicentralizer subgroups of sporadic simple groups

T. YOSHIDA
Some transfer theorems for finite groups

de Gruyter Studies in Mathematics

H.-O. Georgii
Gibbs Measures and Phase Transitions
1988. XIV, 525 pages. Cloth ISBN 3 11 010455 5 (Vol. 9)

M. Hervé
Analyticity in Infinite Dimensional Spaces
1989. VIII, 206 pages. Cloth ISBN 3 11 010995 6 (Vol. 10)

W. Fenchel
Elementary Geometry in Hyperbolic Space
1989. Approx. 240 pages. Cloth ISBN 3 11 011734 7 (Vol. 11)

H. Amann
Ordinary Differential Equations
An Introduction to Nonlinear Analysis
1989. Approx. 450 pages. Cloth ISBN 3 11 011515 8 (Vol. 12)

Previously published:

W. Klingenberg
Riemannian Geometry
1982. X, 396 pages. Cloth
ISBN 3 11 008673 5 (Vol. 1)

M. Métivier
Semimartingales
A Course on Stochastic Processes
1982. XII, 287 pages. Cloth
ISBN 3 11 008674 3 (Vol. 2)

L. Kaup / B. Kaup
Holomorphic Functions of Several Variables
An Introduction to the Fundamental Theory
1983. XVI, 350 pages. Cloth
ISBN 3 11 004150 2 (Vol. 3)

C. Constantinescu
Spaces of Measures
1984. 444 pages. Cloth
ISBN 3 11 008784 7 (Vol. 4)

G. Burde / H. Zieschang
Knots
1984. XII, 400 pages. Cloth
ISBN 3 11 008675 1 (Vol. 5)

U. Krengel
Ergodic Theorems
1985. VIII, 357 pages. Cloth
ISBN 3 11 008478 3 (Vol. 6)

H. Strasser
Mathematical Theory of Statistics
Statistical Experiments and Asymptotic Decision Theory
1985. XII, 492 pages. Cloth
ISBN 3 11 010258 7 (Vol. 7)

T. tom Dieck
Transformation Groups
1987. X, 312 pages. Cloth
ISBN 3 11 009745 1 (Vol. 8)

de Gruyter · Berlin · New York

Walter de Gruyter & Co., Genthiner Str. 13, D-1000 Berlin 30, FRG, Phone: (0 30) 2 60 05-0 · Telex 1 84 027
Walter de Gruyter, Inc., 200 Saw Mill River Road, Hawthorne, N.Y. 10532, USA, Phone (914) 747-0110 · Telex 6 46 677

Volume 1 · 1989

Forum Mathematicum

An international journal devoted to pure and applied mathematics as well as mathematical physics

Editorial Board:

M. Brin (College Park, USA) · F. R. Cohen (Lexington, USA)
V. Enss (Berlin, FRG) · R. Fintushel (East Lansing, USA)
M. Fliess (Gif-sur-Yvette, France) · M. Fukushima (Osaka, Japan)
G. Gallavotti (Rome, Italy) · R. Göbel (Essen, FRG)
K. H. Hofmann (Darmstadt, FRG) · J. Lindenstrauss (Jerusalem, Israel)
D. H. Phong (New York, USA) · D. Ramakrishnan (Ithaca, USA)
A. Ranicki (Edinburgh, GB) · P.-A. Raviart (Palaiseau, France)
D. S. Scott (Pittsburgh, USA) · D. Segal (Oxford, GB)
B. Shiffman (Baltimore, USA) · F. Skof (Torino, Italy)
K. Strambach (Erlangen, FRG) · G. Talenti (Florence, Italy)
H. Triebel (Jena, GDR) · R. B. Warfield, Jr. (Seattle, USA)

Forum Mathematicum is devoted entirely to the publication of original research articles in all fields of pure and applied mathematics, including mathematical physics. Expository surveys, research announcements, book reviews, etc. will not be published.

Manuscripts should be submitted to any of the editors or to:

Forum Mathematicum

Mathematisches Institut der Universität
Bismarckstraße 1 1/2
D-8520 Erlangen, Federal Republic of Germany

de Gruyter · Berlin · New York

Walter de Gruyter & Co., Genthiner Str. 13, D-1000 Berlin 30, FRG, Phone (0 30) 2 60 05-0 · Telex 1 84 027
Walter de Gruyter, Inc., 200 Saw Mill River Road, Hawthorne, N.Y. 10532, USA, Phone (914) 747-0110 · Telex 6 46 677